XXXII IUPAP Conference on Computational Physics

Journal of Physics: Conference Series Volume 2207

Online
2-5 August 2021

ISBN: 978-1-7138-5170-7
ISSN: 1742-6588

Printed from e-media with permission by:

Curran Associates, Inc.
57 Morehouse Lane
Red Hook, NY 12571

Some format issues inherent in the e-media version may also appear in this print version.

Institute of Physics
Dirac House, Temple Back
Bristol BS1 6BE UK

Phone: 44 1 17 929 7481
Fax: 44 1 17 920 0979

techtracking@iop.org

Additional copies of this publication are available from:

Curran Associates, Inc.
57 Morehouse Lane
Red Hook, NY 12571 USA
Phone: 845-758-0400
Fax: 845-758-2633
Email: curran@proceedings.com
Web: www.proceedings.com

TABLE OF CONTENTS

Preface

Peer Review Statement

STATISTICAL PHYSICS

Comparison of Benchmark HP Model Proteins on Face-Centered Cubic and Simple Cubic Lattices Using Replica-Exchange Wang-Landau Sampling 1
Matthew S. Wilson, David P. Landau

A First Look at Structural Properties of Long HP Model Sequences 7
Alfred C. K. Farris, David P. Landau

Surface Critical Behavior of the Three-Dimensional O(3) Model 13
F Parisen Toldin

Corrections to Scaling in Geometrical Clusters of the 2D Ising Model 19
Michail Akritidis, Nikolaos G Fytas, Martin Weigel

Study of Diluted Kinetic Ising Model Under Sinusoidal External Field 25
Sourav Chattopadhyay, S. B. Santra

Molecular Dynamics Study of a Nano-Scale B-type Stirling Engine 31
K Kitaya, M Isobe

Quantifying Uncertainty with Stochastic Collocation in the Kinematic Magentohydrodynamic Framework 37
E. Rajbhandari, N. L. Gibson, C. R. Woodside

Multicanonical Simulations of the 2D Spin-1 Baxter-Wu Model in a Crystal Field 43
Nikolaos G Fytas, Alexandros Vasilopoulos, Erol Vatansever, Anastasios Malakis, Martin Weigel

Efficient Algorithms for Computing Ground States of the 2D Random-Field Ising Model 49
Argyro Mainou, Nikolaos G Fytas, Martin Weigel

Sharpening the Universality of Network Resilience Patterns Using Motifs 55
Huixin Zhang, Sean P. Cornelius

Diffusional Characteristics of a Newtonian Event-Chain Monte Carlo in Hard Disk Systems 61
H Banno, D Mugita, M Isobe

Resampling Schemes in Population Annealing — Numerical Results 67
Denis Gessert, Martin Weigel, Wolfhard Janke

An Agent-Based Modeling and Simulation for the First Stage of Honeycomb Construction 73
Takayuki Narumi, Takumi Akiyama, Maya Kageyama, Kenta Uemichi, Hisao Honda, Koichi Osaki

Pursuit and Evasion: from Singles to Groups 79
Sota Yoshihara, Toru Ohira

Phase Diagram of the Dodecahedron Spin Model on Square Lattice with Diluted Bonds 83
S. Yunita, R. Rahmat, T. Surungan

SOFT MATTER AND BIOPHYSICS

Characterizing Flexibility and Mobility in the Natural Mutations of the SARS-CoV-2 Spikes 89
James Panayis, Navodya S. Römer, Dom Bellini, A. Katrine Wallis, Rudolf A. Römer

Nonequilibrium Dynamics of a Fluid Vesicle: Turing Patterns and Traveling Waves 95
Hiroshi Noguchi, Naoki Tamemoto

A Simple Method to Accelerate Configurational Sampling for a Generalized Hybrid Monte Carlo Method 99
D Suzuki, T Hori, S Miura

Optical Kerr Effect and Structural Tetrahedrality of Supercooled Water at Ambient Pressure 105
Ping-Han Tang, Yu-Wei Kuo, Ten-Ming Wu

Effect of Coulomb Interaction Between Secondary Electrons on Plasma Formation Due to Heavy-Ion Irradiation 111
Kengo Moribayashi

Theoretical Study of Dissociation Process of Plastocyanins by PaCS-MD Simulation 115
D Fitrasari, M S Arwansyah, K Kawaguchi, A Purqon, Suprijadi, H Nagao

Theoretical Study of Complex Aspirin and Hydroxypropyl-B-cyclodextrin in Solvent Phase 120
H Jayyinunnisya, L S Solihat, K Sugimori, K Kawaguchi, H Nagao

Collective Pattern Formation in a Binary Mixture of Self-Propelled Particles 126
Sagarika Adhikary, S. B. Santra

Simulation Study of the Function of Domain Swapping in the HSP90 Chaperone Cycle 131
L Matsukura, N Miyashita

Interaction Between PET Tracer and the Specific Residues Around the Gate of the Open Form of Monoamine Oxidase B (MAO-B) 135
M Ottawa, N Miyashita

Enhancing Efficient Computation of Long-Wavelength Relaxation Dynamics in a 2D Liquid Involving Millions of Particles 139
Hayato Shiba

Role of Temperature and Alignment Activity on Kinetics of Coil-Globule Transition of a Flexible Polymer 145
Subhajit Paul, Suman Majumder, Wolfhard Janke

QUANTUM MANY-BODY SYSTEMS, QUANTUM COMPUTING

Numerical Simulations of Spectroscopic Properties in Two-Dimensional Mott Insulator.............................. 151
Takami Tohyama, Kazuya Shinjo, Shigetoshi Sota, Kenji Tsutsui

Higher Order Auxiliary Field Quantum Monte Carlo Methods 157
Florian Goth

QMC Study of the Chiral Heisenberg Gross-Neveu Universality Class 163
Yuichi Otsuka, Kazuhiro Seki, Sandro Sorella, Seiji Yunoki

Thermal Pure Quantum Matrix Product States: A Simple Numerical Protocol for Finite Temperature 168
Atsushi Iwaki, Chisa Hotta

Quantum Monte Carlo Simulations of Highly Frustrated Magnets in a Cluster Basis: The Two-Dimensional Shastry-Sutherland Model 174
Andreas Honecker, Lukas Weber, Philippe Corboz, Frédéric Mila, Stefan Wessel

Compressed Sensing Based on QUBO Formulation 180
Kazue Kudo

Quantum Game Theory Approach for Data Network Routing: A Solution for the Congestion Problem 186
Silva Agustin, G. Zabaleta Omar, M. Arizmendi Constancio

Four-Body Variational Calculation of a Hydrogen-like Atom Involving an Excited Muonic Molecule 192
Takuma Yamashita, Motoaki Niiyama, Kazuhiro Yasuda, Yasushi Kino

MATERIALS SCIENCE

Density-Matrix Renormalization-group Study for a Two-dimensional Lattice-gas on the Ih-ice Prism Surface 198
Noriko Akutsu

Unified Time Evolution Approach for the Electronic Structure Calculation 204
Q Wang, Y Zempo

Numerical Analysis of Electronic State of CNT/BNNT Heterojunction 210
Mitsuyoshi Tomiya, Shoichi Sakamoto, Toshihiro Sagara, Masaki Yamamoto, Masahiro Ishii

Optical Properties of OLED Materials by TDDFT 216
N Akino, Y Zempo

Phenomenological Simulations of Intermediate States in a Type I Superconducting Wire 222
Yohitaka Mizuhara, Masaru Kato

Dynamics of Photoinduced Entanglement Generation Between Remote Electron-Phonon Systems 228
Kunio Ishida, Hiroaki Matsueda

The Phonon Quasiparticle Approach for Anharmonic Properties of Solids 234
Zhen Zhang, Dong-Bo Zhang, Tao Sun, Renata M. Wentzcovitch

The Unstable Analysis of Soliton Compression in Silicon Photonic Crystal 246
Sarun Phibanchon

DFT-NEGF Study of Electron Transfer Property in Graphene Nanoribbon Double Barrier System 252
S Sakamoto, M Tomiya, R IIda

Excitation and Transport of Bound Magnon Clusters in Frustrated Ferromagnetic Chain 258
Hiroaki Onishi

FLUID DYNAMICS AND ASTROPHYSICS

Plasma Propagation Via Radiation Transfer in Millimeter-Wave Discharge Under Subcritical Condition 264
S Suzuki, C Kato, M Takahashi, N Ohnishi

Coupling Simulation on Two-Dimensional Axisymmetric Beaming Propulsion System 270
Masayuki Takahashi

3D Hydrodynamics Simulations of Massive Stars with the PROMPI Code 276
F Rizzuti, R Hirschi, C Georgy, W D Arnett, C Meakin, M Mocák

The Formation of Dark-Matter-deficient Galaxies Through Galaxy Collisions 282
Koki Otaki, Masao Mori

Predicting the Expansion of Supernova Shells for High-Resolution Galaxy Simulations Using Deep Learning 288
K. Hirashima, K. Moriwaki, M. S. Fujii, Y. Hirai, T. Saitoh, J. Makino

ISTEDDAS: A New Direct N-Body Code to Study Merging Compact-object Binaries 294
Mattia Mencagli, Natalia Nazarova, Mario Spera

LATTICE FIELD THEORY

Efficient Computations of Continuous Action Densities of States for Lattice Models 306
Biagio Lucini, Olmo Francesconi, Markus Holzmann, David Lancaster, Antonio Rago

General Purpose Lattice QCD Code Set Bridge++ 2.0 for High Performance Computing 312
Yutaro Akahoshi, Sinya Aoki, Tatsumi Aoyama, Issaku Kanamori, Kazuyuki Kanaya, Hideo Matsufuru, Yusuke Namekawa, Hidekatsu Nemura, Yusuke Taniguchi

Tempered Lefschetz Thimble Method as a Solution to the Numerical Sign Problem 318
Masafumi Fukuma, Nobuyuki Matsumoto

Lattice QCD at Nonzero Temperature and Density 324
G. Aarts, C. Allton, S. Hands, B. Jäger, S. Kim, M. P. Lombardo, A. A. Nikolaev, S. M. Ryan, J.-I. Skullerud

MACHINE LEARNING AND ALGORITHMS

Quantum Field Theories, Markov Random Fields and Machine Learning 333
Dimitrios Bachtis, Gert Aarts, Biagio Lucini

Machine Learning the 2D Percolation Model 339
Djénabou Bayo, Andreas Honecker, Rudolf A. Römer

Machine Learning the Square-Lattice Ising Model 345
Burak Çivitcioglu, Rudolf A. Römer, Andreas Honecker

Author Index

Journal of Physics: Conference Series **2207** (2022) 011001 doi:10.1088/1742-6596/2207/1/011001

XXXII IUPAP Conference on Computational Physics

Damien Foster[1], Nikolaos G. Fytas[1], Charo del Genio[1], Ran Holtzman[1], Abhishek Kumar[1], Susanne Horn[1], Alban Potherat[1], Martin Weigel[2], and Taras Yavors'kii[1]

[1] Centre for Fluid and Complex Systems, Coventry University, Coventry CV1 5FB, United Kingdom

[2] Institut für Physik, Technische Universität Chemnitz, 09107 Chemnitz, Germany

E-mail: `martin.weigel@physik.tu-chemnitz.de`

The 32nd IUPAP Conference in Computational Physics was held online August 2–5, 2021 with organizational support from Coventry University. The Conference in Computational Physics series of meetings is held annually under the auspices of the International Union of Pure and Applied Physics (IUPAP) at international locations alternating typically between Europe, the Americas, and the Asian and Pacific regions. It covers all areas of the computational sciences with a relation to physical applications, making it the largest and leading recurrent international meeting in computational physics. Typical topics covered include statistical mechanics and complex systems, soft matter and biophysics, materials and nano-science, fluid dynamics, quantum many-body physics, quantum computing, lattice field theory, astrophysics, gravitation and cosmology, novel hardware and software, computational physics education, machine learning and algorithms as well as geophysics and porous media. The format combines plenary overview lectures with a broad range of topical parallel sessions featuring invited impact talks as well as oral contributions and poster sessions, complemented by discussion sessions.

The organisation of the 2020 meeting was given to the present group of organisers for a meeting to be held in physical presence in Coventry. It was planned for August 2020, but with the advent of the Coronavius pandemic it was decided in April 2020 to postpone the meeting to 2021 with the hope of holding it on site. We were quite confident back then that the meeting could be held face-to-face in 2021, but the virus proved to be more persistent than we had envisaged, such that CCP2021 was held entirely online. While this disappointed some, it also offered a number of opportunities to explore new forms of scientific and social interaction. We used a virtual conference centre constructed in Gather Town that included spaces for poster sessions and exhibition space for sponsors as well as many opportunities for formal and informal discussions among participants. All oral presentations were delivered on Zoom, with the vast majority of talks delivered live and a handful of pre-recorded videos. To support discussions revolving around the presentations, we also ran a discussion space on Slack with channels for each session and a common discussion forum. All oral presentations were recorded and made available to registered participants on the conference website at `https://ccp2021.complexity-coventry.org`.

The conference featured 425 registrations, with 296 fully completed, coming from a wide range of countries, with the most represented being Japan, the United Kingdom, the United States of America, Germany, India, France, Russia and China. The program consisted of 15 plenary speakers, listed below, as well as 33 invited talks, 100 contributed talks and 72 poster contributions. The scientific sessions were complemented by cultural presentations as well as the award session for the IUPAP Young Scientist Prizes in Computational Physics which for

2020 and 2021 were awarded to Zhijun Wang and Prineha Narang, respectively. In total, these contributions led to a very exciting program, and the technical setup allowed for a rather interactive meeting. Although the actual experience might not quite have matched up to that of a traditional face-to-face conference, it was felt by many that this meeting format also has its upsides as it is more inclusive, more ecological, more diverse, and more family friendly that the traditional one.

All authors of accepted contributions to the conference were invited to submit a full paper to the proceedings. These papers are collected in the present volume, covering all major areas represented at the conference and thus providing an excellent representation of where computational physics research stands today.

We would like to thank all participants of the conference, all contributors to the scientific program and all authors of papers in this proceedings volume. Finally, we would like to acknowledge support for the conference by the IUPAP, the European Physical Society (EPS), the Institute of Physics (IOP), Coventry University, Technische Universität Chemnitz, the European Physical Journal (EPJ), Computation (MDPI), Entropy (MDPI), and MapleSoft.

Local organising Committee

- Damien Foster
- Nikolaos G. Fytas
- Charo del Genio
- Ran Holtzman
- Abhishek Kumar
- Susanne Horn
- Alban Potherat
- Martin Weigel (chair)
- Taras Yavors'kii

International advisory board

Joan Adler, Technion, Haifa
Nithaya Chetty, Wits University, Johannesburg
Mei-Yin Chou, Academia Sinica, Taipei
Bismarck Vaz da Costa, Universidade Federal de Minas Gerais, Belo Horizonte
Yuan Ping Feng, National University of Singapore
Suklyun Hong, Sejong University, Seoul
Andrew Horsfield, Imperial College, London
Trevis Humble, Oak Ridge National Laboratory
Barry Klein, University of California Davis
Georg Kresse, University of Vienna
David Landau, University of Georgia, Athens
Hai-Qing Lin, Computational Science Research Center, Beijing
Richard Liska, Czech Technical University, Prague
Priya Mahadevan, Bose Centre, Kolkata
Regina Maphanga, Council for Scientific and Industrial Research, Pretoria
Sitangshu Bikas Santra, Indian Institute of Technology, Guwahati
Lev Shchur, Landau Institute of Theoretical Physics, Chernogolovka
Laurette Tuckerman, École Supérieure de Physique et de Chimie Industrielles, Paris
Roser Valentí, University of Frankfurt

Daniel Vizman, West University of Timișoara
Rodolphe Vuilleumier, École Normale Supérieure, Paris
Xiaoqun Wang, Shanghai Jiao Tong University
Renata Wentzcovitch, Columbia University, New York
Junyi Zhu, Chinese University of Hong Kong

Program Committee

Statistical mechanics and complex systems
Wolfhard Janke, University of Leipzig (chair)
Youjin Deng, University of Science and Technology of China, Hefei
Jonathan Machta, University of Massachussetts, Amherst
Lev Shchur, Landau Institute of Theoretical Physics, Chernogolovka

Soft matter, biophysics
Julia Yeomans, University of Oxford (chair)
Anna Balazs, University of Pittsburgh
Changbong Hyeon, Korea Institute for Advanced Study, Seoul

Materials and nano-science
Luca Ghiringhelli, Fritz Haber Institute, Berlin (chair)
Francesca Baletto, King's College London
Silvana Botti, University of Jena
Bryan Goldsmith, University of Michigan, Ann Arbor
James Kermode, University of Warwick, Coventry
Sergey Levchenko, Skolkovo Innovation Center, Moscow

Fluid dynamics
Greg Sheard, Monash University, Melbourne (chair)
Bruno Carmo, University of São Paulo
Wisam Al Saadi, Australian College of Kuwait

Quantum many-body physics
Anders Sandvik, Boston University (chair)
Sylvain Capponi, University of Toulouse
Kedar Damle, Tata Institute of Fundamental Research, Mumbai
Chisa Hotta, University of Tokyo
Shiwei Zhang, Flatiron Institute, New York

Quantum computing
Stephen Jordan, Microsoft Corporation (chair)
David Gosset, University of Waterloo
Artur Izmaylov, University of Toronto
Bei Zeng, University of Guelph

Lattice field theory
Constantina Alexandrou, Cyprus Institute, Nikosia (chair)
Gert Aarts, Swansea University
Karl Jansen, Deutsches Elektronen-Synchrotron, Zeuthen
Aida X. El-Khadra, University of Illinois, Urbana-Champaign

Astrophysics, gravitation, cosmology
Raphael Hirschi, University of Keele (chair)

Novel hardware and software
Massimo Bernaschi, National Research Council, Rome (chair)
Valeri Halyo, Princeton University
Victor Martin-Mayor, Complutense University of Madrid

Computational physics education
Joan Adler, Technion, Haifa (chair)
Amy Graves, Swarthmore College

Machine learning and algorithms
Lenka Zdeborova, École Polytechnique Fédérale de Lausanne (chair)
Chiara Cammarota, King's College London
Alexander Hartmann, University of Oldenburg
Ehsan Katami, San José State University
Maria Schuld, University of KwaZulu-Natal, Durban
Pan Zhang, Chinese Academic of Sciences, Beijing

Geophysics and porous media
Steve Tobias, University of Leeds (chair)
Emmanuel Dormy, Centre National de la Recherche Scientifique, Paris
Geoff Vallis, University of Exeter

Plenary Speakers

Dwight Barkley, University of Warwick, Coventry
Ludovic Berthier, University of Montpellier
Sergio Boixo, Google Research
Gábor Csányi, University of Cambridge
Christine Davies, University of Glasgow
Claudia Draxl, Humboldt University Berlin
Steven Gottlieb, Indiana University, Bloomington
Prineha Narang, Harvard University, Cambridge
Annalisa Pillepich, Max Planck Institute for Astronomy, Heidelberg
Frank Pollmann, Technical University of Munich
Andrew Saxe, University of Oxford
Olga Shishkina, Max Planck Institute for Dynamics and Self-Organization, Göttingen
David Thirumalai, University of Texas, Austin
Zhijun Wang, Chinese Academy of Sciences, Beijing
Andy Woods, University of Cambridge

Invited Speakers

Gert Aarts, Swansea University
Ronojoy Adhikari, University of Cambridge
Tim Atherton, Tufts University, Medford
Titus Beu, Babeș-Bolyai University, Cluj

Juan Carrasquilla, Vector Institute, Toronto
Philippe Corboz, University of Amsterdam
Sean Couch, Michigan State University, East Lansing
Stefano Curtarolo, Duke University, Durham
Thomas Guillet, University of Exeter
Vera Gülpers, University of Edinburgh
Itay Hen, University of Southern California, Los Angeles
Silke Henkes, University of Bristol
Susanne Horn, Coventry University
Estelle Inack, Perimeter Institute, Waterloo
Jaehyeok Jin, University of Chicago
Naoki Kawashima, University of Tokyo
Boris Kozinsky, Harvard University, Cambridge
Justin Leontini, Swinburne University of Technology, Melbourne
Valerio Lucarini, University of Reading
Cristian Micheletti, Scuola Internazionale Superiore di Studi Avanzati, Trieste
Mario Motta, International Business Machines Corporation, Almaden
Kostas Orginos, College of William & Mary, Williamsburg
Ludivine Oruba, Sorbonne Université, Paris
Laura Ratcliff, Imperial College, London
Martin Roetteler, Microsoft Corporation
Josh Romero, Nvidia, Portland
Stefan Schnabel, University of Leipzig
Hui Shao, Beijing Normal University
Jemma Shipton, University of Exeter
Mario Spera, Scuola Internazionale Superiore di Studi Avanzati, Trieste
Miles Stoudenmire, Flatiron Institute, New York
Lei Wang, Chinese Academic of Sciences, Beijing
Christopher White, University of Maryland, College Park

Peer Review Statement

All papers published in this volume have been reviewed through processes administered by the Editors. Reviews were conducted by expert referees to the professional and scientific standards expected of a proceedings journal published by IOP Publishing.

- **Type of peer review:** Single Anonymous
- **Conference submission management system:** Morressier
- **Number of submissions received:** 65
- **Number of submissions sent for review:** 64
- **Number of submissions accepted:** 58
- **Acceptance Rate (Submissions Accepted / Submissions Received × 100):** 89.2
- **Average number of reviews per paper:** 1
- **Total number of reviewers involved:** 21
- **Contact person for queries:**
  Name: Martin Weigel
  Email: martin.weigel@physik.tu-chemnitz.de
  Affiliation: Technische Universität Chemnitz

Comparison of Benchmark HP Model Proteins on Face-Centered Cubic and Simple Cubic Lattices using Replica-Exchange Wang-Landau Sampling

Matthew S. Wilson[1] **and David P. Landau**[1]

[1] Center for Simulational Physics, The University of Georgia, Athens, GA 30602, USA

E-mail: msw.wilson@uga.edu

Abstract. The HP model represents protein sequences by coarse-graining amino acid types as either hydrophobic (H) or polar (P). This model has been studied extensively, where the simple cubic (sc) lattice is a commonly used for simplicity and efficiency, despite a limited ability to geometrically represent protein structures. One straightforward improvement is to instead use the face-centered cubic (fcc) lattice, which offers a more realistic geometry while retaining simplicity. Using the replica-exchange Wang-Landau algorithm, the ground state energies and full density of states are predicted for a set of biologically-motivated HP sequences on the fcc lattice. A direct thermodynamic comparison is made between the folding on fcc and sc lattices, and structural properties are calculated to further examine the folding on the fcc lattice.

1. Introduction

Protein folding is a challenging topic to study computationally due to the large timescales and length scales involved in macromolecular structural transitions. Such challenges often motivate the use of coarse-grained models simulated with Monte Carlo methodology to extract essential physics, rather than looking at dynamics or atomic-level details. However, the enormous number of available states in addition to entropic barriers from polymeric backbone constraints make finding the ground state for even simple models like lattice proteins extremely difficult.

The hydrophobic-polar (HP) model [1, 2] has been widely employed for the study of protein folding, where lattice geometry, model Hamiltonian, and amino acid classifications are chosen to simplify the problem. Square and simple cubic (sc) lattices have a long history of use in the computational biophysics community, although higher-coordination lattices can be used. Statistical analyses [3, 4] of experimentally observed protein structures show that the fcc lattice (actually, the cubo-octahedral lattice, which is a subset of fcc) gives a good approximation for the packing of amino acids. The HP model has also been used on the fcc lattice to test algorithms with protein structure prediction [5–7].

Here, we use the replica-exchange Wang-Landau(REWL) algorithm [8] to determine the ground states and full density of states for biologically-motivated HP sequences on the fcc lattice and make a comparison with results [9, 10] for folding thermodynamics on the sc lattice.

2. Simulation model

2.1. HP lattice protein

The HP model [1, 2] provides a minimal representation of a protein system by classifying the 20 possible amino acid types into two groups depending on their side chain: either hydrophobic (H) or polar (P). Amino acids are reduced to point-like residues that lie on the vertices of some discrete lattice. The protein backbone is a self-avoiding walk with ℓ adjacent, rigidly bonded (peptide bonded) residues. Primary structure, a protein's identifying amino acid sequence, is incorporated as a mapped sequence of H's and P's, and is the main input of a folding simulation.

Folding takes place in an aqueous solvent which is implicitly modeled by including an interaction between H residues in the protein model. Shown in Eq. 1, the HP model Hamiltonian is defined by the number (n_{HH}) of adjacent, non-bonded H residues on the lattice (HH contacts), and is scaled by some energetic coupling (ε_{HH}).

$$\mathcal{H} = -n_{HH}\varepsilon_{HH} \tag{1}$$

The formation of globular, low-energy states is entirely determined by the self-avoidance, HP sequence, and the underlying lattice. The congregation of H residues, or a hydrophobic core (H-core), maximizes the number of HH contacts given the constraint of the specified HP sequence.

2.2. Face-centered cublic lattice

The HP model can be implemented on any rigid lattice, but we focus on the fcc lattice and its comparison with the sc lattice. Unlike the sc lattice, the fcc lattice has no parity problem, where residues must be separated by an odd number in the length of the HP sequence in order to interact as non-bonded nearest neighbors. The 12 fcc lattice directions are $\pm\frac{1}{\sqrt{2}}(\hat{x}+\hat{y})$, $\pm\frac{1}{\sqrt{2}}(\hat{x}-\hat{y})$, $\pm\frac{1}{\sqrt{2}}(\hat{x}+\hat{z})$, $\pm\frac{1}{\sqrt{2}}(\hat{x}-\hat{z})$, $\pm\frac{1}{\sqrt{2}}(\hat{y}+\hat{z})$, and $\pm\frac{1}{\sqrt{2}}(\hat{y}-\hat{z})$. On the fcc lattice, the values for interior angles between any two adjacent backbone bonds are $\pm 60°, \pm 90°, \pm 120°$, and $180°$. The coordination number $\mathbb{c}_n$ is used to normalize temperature scales when comparing results between lattices, where $\mathbb{c}_n = 6$ for the sc lattice and $\mathbb{c}_n = 12$ for the fcc lattice.

2.3. HP mappings of protein sequences

The HP sequences in our fcc lattice simulations are mapped from real proteins and protein fragments [11, 12], and they are chosen to make a direct comparison with established sc lattice results from the folding study of Wüst and Landau [10]. The investigated model proteins are: HP58 (BPTI), HP103 (Cytochrome C), HP124 (Ribonuclease A), and HP136 (1-136 Staphylococcal Nuclease). HP sequences are mapped from primary structures of real proteins, although the HP model does not necessarily capture all interactions present in the real proteins.

3. Methodology

3.1. Wang-Landau and replica-exchange Wang-Landau algorithms

The Wang-Landau (WL) algorithm [13, 14] is a flat-histogram method that iteratively calculates an estimate $\hat{g}(E)$ for the density of states. A random walk is performed in energy space, where the probability to transition between states $\boldsymbol{X}_A$ and $\boldsymbol{X}_B$ with energies E_A and E_B is given by:

$$P^{WL}(\boldsymbol{X}_A \to \boldsymbol{X}_B) = \min\left(1, \frac{\hat{g}(E_A)}{\hat{g}(E_B)}\right) . \tag{2}$$

This transition rule is advantageous because it does not depend on temperature, and that $\hat{g}(E)$ changes at each step to drive the sampler towards less-visited regions of energy space.

We utilize a parallel version of the WL sampling, the REWL algorithm [8], which is extremely useful for complex and large systems, where convergence may be unfeasible for serial WL [15].

Multiple WL instances, or replicas, are employed within bounded regions of energy space, or 'windows'. Each replica has its own $\hat{g}(E)$ and is restricted to one window, where MC moves that put it out of bounds are rejected. Windows must overlap (typically $\geq 75\%$), as replicas in neighboring windows periodically attempt replica exchanges, where configurations are swapped.

$$P_{i,j}^{exch.}(\boldsymbol{X}_A \leftrightarrow \boldsymbol{X}_B) = \min\left(1,\ \frac{\hat{g}_i(E_A)\,\hat{g}_j(E_B)}{\hat{g}_i(E_B)\,\hat{g}_j(E_A)}\right). \tag{3}$$

An exchange between replicas i and j that swaps states $\boldsymbol{X}_A$ and $\boldsymbol{X}_B$ is accepted with the probability in Eq. 3, and only depends on each $\hat{g}(E)$. Replica exchanges connect the state space across windows and also provide enhanced mixing within each window. At the end of a REWL, independently normalized $\hat{g}(E)$ for the windows are successively merged where the absolute difference in the inverse microcanonical temperature is minimal.

3.2. Monte Carlo trial move set

We use an unbiased implementation of the following reversible and ergodic MC trial moves: pull moves [16], pivot moves [6, 17], bond-rebridging moves [18], and diagonal moves (kink-flip, crankshaft moves). For the fcc lattice, pull moves need not displace two residues, and proposed bonds in internal bond-rebridging moves need not be parallel, as with the sc lattice. Moves are randomly selected with ratios tailored separately for each REWL window.

3.3. Calculation of average quantities

After a REWL simulation, the ensemble average of any quantity $\mathcal{O}$ is calculated as

$$\langle \mathcal{O}(T) \rangle = \frac{\sum_E \overline{\mathcal{O}}(E)\ g(E) e^{-E/(k_B T)}}{\sum_E g(E) e^{-E/(k_B T)}}, \tag{4}$$

where $\langle\rangle$ denotes the ensemble average and $\overline{\mathcal{O}}(E)$ is the weighted average for the quantity at each energy. This reweighting gives the average at all temperatures from a single REWL simulation.

Eq. 4 is use to determine the fluctuation in the average energy, which is related to the heat capacity as shown in the left-most equality in in Eq. 5.

$$C_V = \frac{\langle E^2 \rangle - \langle E \rangle^2}{k_B T^2} = -\varepsilon_{HH} \frac{d\,\langle n_{HH} \rangle}{dT} \tag{5}$$

This is proportional to the derivative of $\langle n_{HH} \rangle$, as shown in the right-most equality of Eq. 5.

In addition, we perform a large multicanonical [19] production run using the $\hat{g}(E)$ as weights to collect samples for the calculation of average structural properties. Shown in Eq. 6, the gyration tensor is equivalent to the moment of inertia where all residues have equal mass.

$$S_{\alpha\beta} = \frac{1}{\ell} \sum_{i=1}^{\ell} (r_\alpha^{(i)} - r_\alpha^{cm})(r_\beta^{(i)} - r_\beta^{cm}) \tag{6}$$

The center of mass $\vec{r}^{\,cm}$ is the average of the positions $\vec{r}^{\,(i)}$ between all ℓ residues. Diagonalizing $\boldsymbol{S} = diag(\lambda_1, \lambda_2, \lambda_3)$ gives access to the principle moments, $\lambda_1 \geq \lambda_2 \geq \lambda_3$. From these, various shape descriptors can be calculated [20], where one such quantity is the relative shape anisotropy,

$$\kappa^2 = 1 - 3\left(\frac{\lambda_1\lambda_2 + \lambda_2\lambda_3 + \lambda_1\lambda_3}{(\lambda_1 + \lambda_2 + \lambda_3)^2}\right), \tag{7}$$

which has a value of 1.0 for a linear configuration and a value of 0.0 for a spherical configuration.

Another simple structural observable that is used to study structural transitions in the backbone configuration is the end-to-end distance, $R_{EE} = |\vec{r}_1 - \vec{r}_\ell|$.

4. Thermodynamic results for model protein folding

4.1. Predicted ground states

REWL simulations are performed for the set of HP sequences, with 8 windows for sequence HP58, and 16 windows for all other sequences. Averages and error estimates are calculated from 5-20 independent REWL simulations, each performing 20 iterations that sample until all histogram entries deviate $\leq 20\%$ from the average value. The range of energy states are not known at the start of simulations, and are added adaptively [10] when found.

Predicted ground state energies and structures are shown in the top-left corner of each plot in Fig. 1. All subsequent images show P residues with orange and H residues with white. For all sequences, the minimal energies from REWL agree with those reported by the HPStruct tool from the CPSP web server [21–24]. We find that the HP model ground states have a high structural degeneracy on the fcc lattice (more than 10^6 observed for sequences with $\ell > 100$).

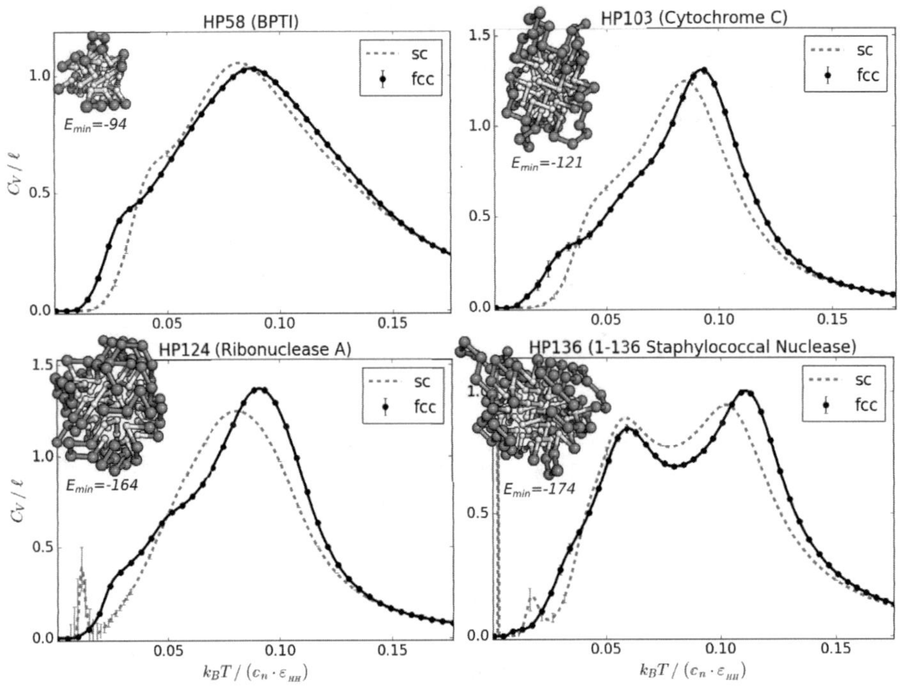

Figure 1. Thermodynamics of HP sequences on the sc (red dashed) and fcc (black) lattices. Temperatures are normalized by $\mathbb{c}_n$. Ground state energies and structures are shown in the top-left of each plot. Error bars are smaller than the marker size if not visible.

4.2. Folding thermodynamics for the fcc lattice

Using the existing results for the HP sequences' folding thermodynamics on the sc lattice [10], we make a direct comparison with our fcc lattice simulations. Fig. 1 shows C_V/ℓ for the sc lattice plotted with red, dashed curves and the fcc lattice with black curves. The C_V/ℓ are used to interpret the thermal rate of H-core formation through the relation in Eq. 5. Temperature axes are normalized by each lattice's coordination number $\mathbb{c}_n$ to align results, and the shorthand $T^* = k_B T/(\mathbb{c}_n \varepsilon_{HH})$ is used in subsequent discussions.

In each plot of Fig. 1, there is a distinct maximum in C_V/ℓ between $0.08 < T^* < 0.125$. These peaks are indicative of a hydrophobic collapse 'transition', where conformations change from an extended, random coil to a state with one or more disordered, globular regions. A large increase in $\langle n_{HH} \rangle$ in this transition signals the initial formation of a disordered H-core. Both the fcc and sc lattices show comparable collapse temperatures when shifted by $\mathbb{c}_n$, with the fcc lattice exhibiting a slightly greater transition temperature and magnitude of C_V/ℓ.

At temperatures below the hydrophobic collapse, $T^* \leq 0.075$, additional structural rearrangements are indicated by peaks or 'shoulders' in C_V/ℓ. In these cases, disordered, globular configurations undergo one or more folding events ($T^* \approx 0.05$ for HP136 on fcc) or small rearrangements ($T^* \leq 0.05$ for HP124 on fcc) to obtain a ground state with the optimal H core. For HP124 and HP136 on the sc lattice, the peaks below $T^* \leq 0.025$ likely result from an undersampled $\hat{g}(E)$ at the ground state, and should be regarded as spurious.

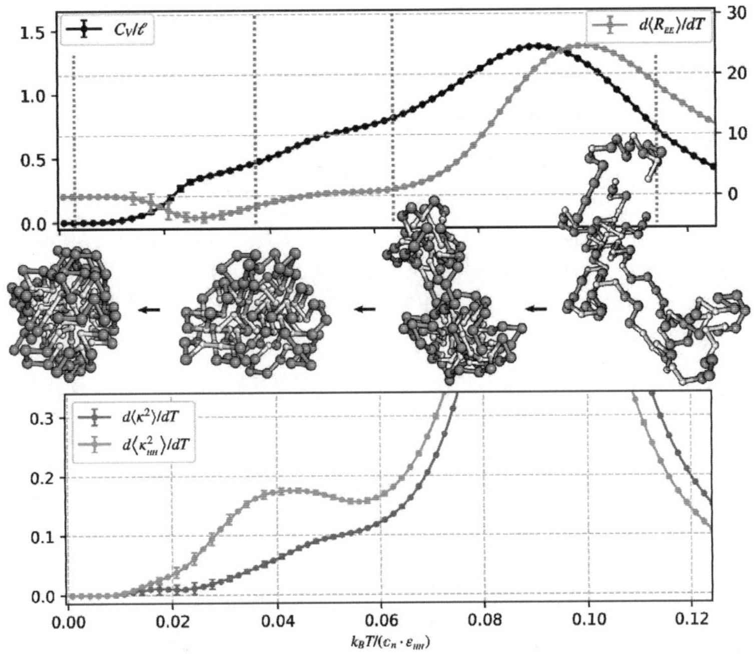

Figure 2. Structural analysis for HP124 of the fcc lattice. The top plot shows C_V/ℓ (black) and $d\langle R_{EE} \rangle/dT$ (purple), with vertical lines marking each configuration's temperature. Derivatives of the average relative shape anisotropy for all residues (blue), and for just H residues (orange) are shown in the bottom plot. Error bars are smaller than the marker size if not visible.

4.3. Structural properties

To better understand the signals in C_V/ℓ, average structural properties are calculated for the fcc lattice, where Fig. 2 shows such analysis for HP124. The top plot shows C_V/ℓ again for reference, with dashed, vertical lines indicating temperatures before and after each suspected structural rearrangement, for which corresponding configurations are shown. In the bottom panel, thermal derivatives of the average relative shape anisotropies are shown for the entire structure ($\langle \kappa^2 \rangle$, blue), and for just the H residues ($\langle \kappa^2_{HH} \rangle$, orange).

The collapse transition between $0.08 < T^* < 0.10$ is clearly evident by the rapid decrease of both $\langle \kappa^2 \rangle$ and $\langle \kappa^2_{HH} \rangle$. Configuration snapshots show a significant structural change near $T^* \approx 0.05$, where two clusters of HH contacts consolidate to a single disordered H-core. There is a noticeable decrease in both anisotropy values beginning at this temperature, but a decrease in $\langle \kappa^2_{HH} \rangle$ appears to extend to lower temperatures. From the anisotropy data, behavior is less clear at the lowest-temperature signal in C_V/ℓ, where the statistical error limits what can be interpreted in the anisotropy values near $T^* \approx 0.025$. However, the structures show a small rearrangement of the dense H-core, and $\langle \kappa^2_{HH} \rangle$ continues to decrease at this temperature.

The value of $\langle R_{EE} \rangle$ is at its global minimum where $T^* \approx 0.05$ and 0.025, matching the temperature in which a single H-core is formed. Below this temperature, the value of $\langle R_{EE} \rangle$ is increasing, where its temperature of maximum increase corresponds to $T^* \approx 0.025$. This behavior could indicate that there is a small structural rearrangement upon the surface of the folded state during the acquisition of the ground state H-core on the fcc lattice.

5. Conclusions

This work describes a method for simulating HP model proteins on the fcc lattice, where the folding thermodynamics are directly compared with results for the sc lattice. Using the REWL algorithm with a set of unbiased MC trial moves, the ground state energies and density of states are determined for a set of biologically-motivated HP sequences.

We observe very similar folding behaviors on the fcc and sc lattices, with slightly higher and lower T^* associated with the hydrophobic collapse and ground state transitions, respectively, for the fcc lattice. For HP124 on the fcc lattice, shoulders in C_V/ℓ below $T^* \leq 0.05$ suggest that additional structural rearrangements are involved either in the hydrophobic collapse, or the eventual acquisition of the native state. This is further supported by an analysis of the shape anisotropies extracted from the gyration tensor along with the end-to-end distance.

Acknowledgments

We thank T. Wüst for sharing the sc lattice data and A. Farris for helpful suggestions. This study was supported in part by resources and technical expertise from the Georgia Advanced Computing Resource Center, a partnership between the University of Georgia's Office of the Vice President for Research and Office of the Vice President for Information Technology.

References

[1] K. A. Dill. *Biochemistry*, 24:1501, 1985.
[2] K. F. Lau and K. A. Dill. *Macromolecules*, 22:3986, 1989.
[3] G. Raghunathan and R. L. Jernigan. *Protein Science*, 6(10):2072, 1997.
[4] M. Mann, R. Saunders, C. Smith, R. Backofen, and C. M. Deane. *Adv Bioinformatics*, 2012:148045, 2012.
[5] J. Liu, B. Song, Y. Yao, Y. Xue, W. Liu, and Z. Liu. *Phys. Rev. E*, 90:042715, 2014.
[6] J.-J. Tsay and S.-C. Su. *Proteome Sci.*, 11:S19, 2013.
[7] I. Dotu, M. Cebrian, P. Van Hentenryck, and P. Clote. *IEEE/ACM Trans. Comput. Bio. Bioinform.*, 8(6):1620, 2011.
[8] T. Vogel, Y. W. Li, T. Wüst, and D. P. Landau. *Phys. Rev. E*, 90:023302, 2014.
[9] T. Wüst and D. P. Landau. *Phys. Rev. Lett.*, 102:178101, 2009.
[10] T. Wüst and D. P. Landau. *J. Chem. Phys.*, 137:064903, 2012.
[11] K. A. Dill, K. M. Fiebig, and H. S. Chan. *Proc. Natl. Acad. Sci. USA*, 90:1942, 1993.
[12] E. E. Lattman, K. M. Fiebig, and K. A. Dill. *Biochemistry*, 33:6158, 1994.
[13] F. Wang and D. P. Landau. *Phys. Rev. E*, 64:056101, 2001.
[14] F. Wang and D. P. Landau. *Phys. Rev. Lett.*, 86:2050, 2001.
[15] Y. W. Li, T. Vogel, T. Wüst, and D. P. Landau. *J. Phys. Conf. Ser.*, 510:012012, 2014.
[16] H. J. Böckenhauer, A. Ullah, L. Kapsokalivas, and K. Steinhöfel. In *WABI*, volume 5251, page 369, 2008.
[17] N. Madras and A. D. Sokal. *J. Stat. Phys.*, 50:109, 1988.
[18] J. M. Deutsch. *J. Chem. Phys.*, 106:8849, 1997.
[19] B. A. Berg and T. Neuhaus. *Phys. Lett. B*, 267:249, 1991.
[20] H. Arkın and W. Janke. *J. Chem. Phys.*, 138(5):054904, 2013.
[21] M. Mann, S. Will, and R. Backofen. *BMC Bioinformatics*, 9:230, 2008.
[22] M. Mann, C. Smith, M. Rabbath, M. Edwards, S. Will, and R. Backofen. *Bioinformatics*, 25:676, 2009.
[23] M. Mann, R. Backofen, and S. Will. In *Proc. 5th WCB*, volume WCB09, 2009.
[24] R. Backofen and S. Will. *Constraints*, 11:5, 2006.

Journal of Physics: Conference Series **2207** (2022) 012002 doi:10.1088/1742-6596/2207/1/012002

A First Look at Structural Properties of Long HP Model Sequences

Alfred C.K. Farris

Department of Physics and Astronomy, Oxford College of Emory University, Oxford, GA 30054, U. S. A.

E-mail: alfred.farris@emory.edu

David P. Landau

Center for Simulational Physics, The University of Georgia, Athens, GA 30602, U. S. A.

Abstract. The longest sequence in the literature of the HP model of protein folding is studied on a simple cubic lattice using replica-exchange Wang-Landau sampling. We find a lower ground state energy than found in previous studies, and, for the first time, study the structural and thermal behavior of this sequence during the folding process.

1. Introduction

Understanding protein folding persists as a Grand Challenge in modern science, with many computational methods being applied to study this problem [1]. Such calculations are extremely demanding because of the many complicated interactions between the many atoms which make up proteins. In an attempt to simplify the problem but retain the essential physics, Ken Dill introduced a coarse-grained model in which all amino acids were classified as either hydrophobic (H) or polar (P) [2, 3], and the only interaction was an attractive coupling between hydrophobic monomers.

Despite the apparent simplicity of the HP model, even determining the ground state of a relatively short sequence is an NP-complete problem [4]. As a result, the HP model has been of interest to researchers across many disciplines, including biochemistry, physics, statistics, and computer science [5, 6, 7, 8, 9].

With modern computers, it is now possible to study HP model sequences of up to $N \approx 130$ monomers with relative success [10, 11]. Recently, Thachuk et *al.* looked at 10 long sequences (between 209 to 248 amino acids) from the Protein Data Bank and studied them in the context of the HP model framework on a simple cubic lattice [12]. Their goal was to compare the ground state energies using two different algorithms – pruned-enriched Rosenbluth method (PERM) [13] and replica exchange Monte Carlo (REMC), also known as parallel tempering [14, 15, 16]. Recent work has shown that replica-exchange Wang-Landau sampling (REWL) [17, 18] is able to not only find lower

Journal of Physics: Conference Series **2207** (2022) 012002 doi:10.1088/1742-6596/2207/1/012002

A First Look at Structural Properties of Long HP Model Sequences

Table 1. HP Sequence 1BPB mapping. Subscripts indicate repeating monomer type.

Sequence	Length	HP Mapping
1BPB	248	$HP_8HPH_2P_2HPH_4PH_2P_2H_2P_2H_2P_2HP_2HP_6HP_5H_3P_2HP_2HP_3H_2P_3$ $H_2PHP_2H_3P_2HP_2HPHP_2H_2PH_3PHP_2H_2P_3HPHPH_3P_2HPHP_7HPH_2$ $P_2H_2P_2HP_2HPH_2P_3HP_2HP_3H_5PH_2P_8HP_3HPHPH_3P_5H_4PHPHP_2$ $H_2P_3HPHPH_2P_2H_2PHP_4HPH_4PH_4PH_4P_5H_2P_2HPHP_4HP_5$

ground state energies for these two sequences than either PERM or REMC, but also to extract specific heat curves for these long sequences [19].

The goal of the current study was to revisit the longest sequence (248 monomers) and study the folding process in more detail using REWL and Multicanonical sampling. To this end, we wished to compute the temperature dependence of various structural properties to understand the major structural transitions which occur during the folding process.

2. Models and Algorithms

2.1. The HP Model

The HP lattice protein model simplifies the 20 amino acids into just two types of (united-atom) monomers, hydrophobic (H) and polar (P), based on their hydrophobicity. The monomers are connect by rigid, unbreakable bonds and the monomer sequence if often chosen to mimic a real protein [2, 3]. There is an energetic coupling ϵ_{HH} between non-bonded, nearest neighbor hydrophobic monomers which is introduced to incorporate the hydrophobic "driving force" into the model. The Hamiltonian is given by

$$\mathcal{H} = -\epsilon_{\text{HH}} n_{\text{HH}}, \tag{1}$$

where n_{HH} is the number of non-bonded, nearest neighbor H-H contacts. Note that although the simplicity of the model makes studying the basics of protein folding computationally tractable, because there is only one energetic interaction, the energy levels (including the ground state) are often highly degenerate. One feature the HP model shares with real proteins is the highly complex nature of the free energy landscape, due to the combination of the entropic entanglements and the rough internal energy landscape [20]. Thus, simulations of even this simple model are very difficult due to the possibility of trapping in free energy minima.

In this study, we examine HP sequence 1BPB (shown in Table 1) on a simple cubic lattice in an attempt to understand, for the first time, the folding process of long HP model sequences. In addition, we continue to look for lower ground state energies than previously reported [12, 19].

2.2. Replica Exchange Wang-Landau Sampling

Replica exchange Wang-Landau sampling (REWL) [17, 18] is a parallel version of simple Wang-Landau sampling [21, 22]. In REWL, random walkers sample their own subset of the energy range, and periodically exchange configurations with walkers whose windows overlap in energy space. At the end of the REWL simulation, we obtain a precise estimate of the density of states $g(E)$ and are thus able to compute thermodynamic properties such as the heat capacity

$$C_V(T) = \frac{\langle E^2 \rangle - \langle E \rangle^2}{k_B T^2} \tag{2}$$

and specific heat C_V/N.

In this work, 17 energy windows were used per simulation, with either 1 or 5 walkers per window, depending on the simulation. The Mersenne Twister random number generator [23] was used, and trial moves consisted of pivot moves [24], bond-rebridging moves [25], and pull moves [26]. The Wang-Landau simulations used a flatness criterion of $p = 0.8$ and a minimum modification factor of $\ln f_{min} = 10^{-8}$.

2.3. Replica Exchange Multicanonical Sampling

We use multicanonical (MUCA) sampling [27, 28] to obtain ensemble averages of structural properties. In (replica exchange) Multicanoncial sampling (REMUCA), we follow the same procedure as in (replica exchange) Wang-Landau sampling, but the instantaneous estimate for the density of states no longer changes, and the simulation ends after a specified number of Monte Carlo steps (10^8 in this study). During the REMUCA simulation, we accumulate the average $\bar{Q}_{MUCA}(E)$ of structural property of interest Q. The canonical ensemble average of structural property Q at temperature T can then be computed via

$$\langle Q \rangle(T) = \frac{\sum_E \bar{Q}_{MUCA}(E)\, g(E) e^{-E/k_B T}}{Z}, \tag{3}$$

where k_B is the Boltzmann constant, and Z is the Canonical partition function.
Two structural properties of interest in this study are the radius of gyration and the end-to-end distance, given by

$$r_g = \left(\frac{1}{N} \sum_{i=1}^{N} (\vec{r}_i - \vec{r}_{CM})^2 \right)^{1/2} \tag{4}$$

and

$$r_{ee} = |\vec{r}_N - \vec{r}_1|, \tag{5}$$

respectively, where r_i is the position of monomer i and r_{CM} is the location of the center of mass.

A First Look at Structural Properties of Long HP Model Sequences

3. Results and Discussion

3.1. Ground State Energy

Using REWL, REMC [12], and PERM [12], ground state energies of (in units of ϵ_{HH}) -182, -176, and -163, were obtained, respectively. In addition to the ground state energy found using REWL being lower than from PERM or REMC, here we report an even lower energy than in our previous work [19]. However, it was very difficult to sample the lowest energy states found during the REWL simulations, so it is possible there are even lower ground state energies to be discovered.

3.2. The Folding Process

From the density of states found using REWL, we computed the specific heat as a function of temperature for sequence 1BPB, shown in the top panel of Figure 1. Dash-dot lines correspond to locations of specific heat signals. Using REMUCA, we then computed average values of the radius of gyration and end-to-end distance as functions of temperature (and their thermal derivatives); these results are shown in the middle and bottom panels of Figure 1, respectively.

Due to the difficulty sampling low energy states, even using REWL, the low-temperature specific heat is not completely reliable. Using our histograms from the REMUCA simulations to recalculate the specific heat, we have verified that results from REWL and REMUCA agree to within statistical error bars, but unfortunately these error bars are quite large in both cases. As a result, low-temperature behavior is rather unreliable in both cases. Better data are needed before a clear conclusion can be drawn about the behavior below $T \approx 0.3$.

For the remainder of the temperature range the error bars are quite small, and conclusions can be drawn about the folding behavior. At higher temperature ($T \approx 0.68$) there is a peak in the specific heat, as well as in the thermal derivatives of both the radius of gyration and end-to-end distance. These peaks mark the coil-globule transition. A second peak in the specific heat occurs at $T \approx 0.41$, again with corresponding signals in the thermal derivatives of the radius of gyration and end-to-end distance. In this case, the signal in the thermal derivative of the radius of gyration is not as pronounced. This second transition is not simply a further collapse of the globule, but primarily a reorganization of the monomers which also changes the end-to-end distance. Although there appears to be a third signal in the end-to-end distance at even lower temperature, we are reluctant to conclude this is real due to the aforementioned sampling difficulties. Thus, better data are needed in order to study the very low temperature behavior.

Shorter HP sequences also show a two-step folding process [11], but the visual appearance of the specific heat of sequence 1BPB (and a previously studied sequence of 209 monomers [19]) is distinct. In these long sequences, the two specific heat peaks are of equal magnitude and are separated by only a shallow minimum. Further investigation is necessary to see if this is a common theme among "long" HP sequences.

A First Look at Structural Properties of Long HP Model Sequences

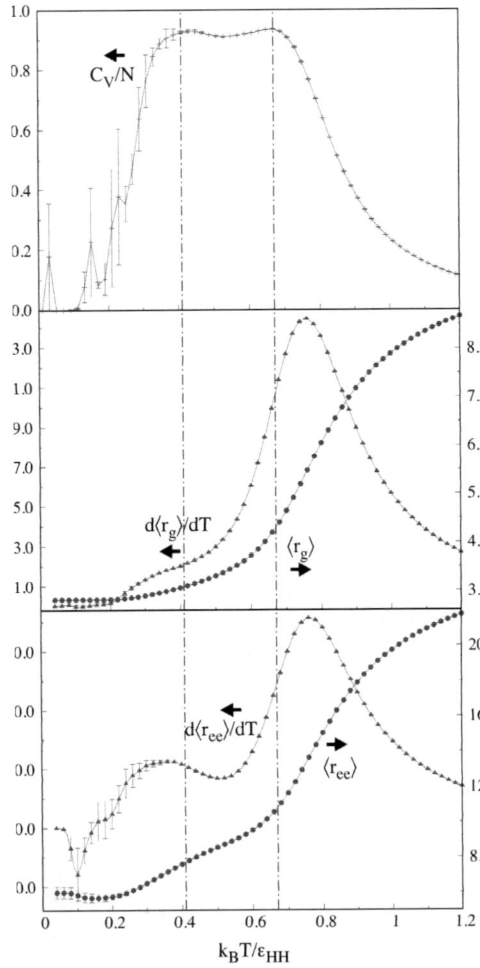

Figure 1. Thermodynamic and structural properties for HP model sequence 1BPB. Dash-dot lines correspond to locations of specific heat signals. Error bars smaller than the symbols are not shown.

4. Conclusions

Using REWL simulations, we have found a new lower ground state energy for sequence 1BPB and extracted the specific heat. Using REMUCA simulations, we have extracted the temperature dependence of the radius of gyration and end-to-end distance, and studied the folding process of the longest HP model in the literature. Although the folding process is a two-step process as in shorter sequences, there is equal contribution

A First Look at Structural Properties of Long HP Model Sequences

to the specific heat from the coil-globule transition and the lower temperature transition to the native state. The lower temperature transition in this long sequence primarily involves a reorganization of the monomers which also changes the end-to-end distance.

Acknowledgements

This work was supported by resources from the Georgia Advanced Computing Resource Center, a partnership between the University of Georgia's Office of the Vice President for Research and the Office of the Vice President for Information Technology.

References

[1] Dill K A and MacCallum J L 2012 *Science* **338** 1042–6
[2] Dill K A 1985 *Biochemistry* **24** 1501–9
[3] Lau K F and Dill K A 1989 *Macromolecules* **22** 3986–3997
[4] Berger B and Leighton T 1998 *Journal of Computational Biology* **5** 27–40
[5] Lattman E E, Fiebig K M and Dill K A 1994 *Biochemistry* **33** 6158–6166
[6] Unger R and Moult J 1993 *Journal of Molecular Biology* **231** 75
[7] Yue K, Fiebig K M, Thomas P D, Chan H S, Shakhnovich E I and Dill K A 1995 *Proceedings of the National Academy of Sciences* **92** 325–329
[8] Bachmann M and Janke W 2004 *The Journal of Chemical Physics* **120** 6779–91
[9] Shi G, Vogel T, Wüst T, Li Y W and Landau D P 2014 *Physical Review E* **90** 033307
[10] Wüst T and Landau D P 2009 *Physical Review Letters* **102**(17) 178101
[11] Wüst T and Landau D P 2012 *The Journal of Chemical Physics* **137** 064903
[12] Thachuk C, Shmygelska A and Hoos H H 2007 *BMC Bioinformatics* **8** 342 ISSN 1471-2105
[13] Grassberger P 1997 *Physical Review E* **56** 3682–3693
[14] Swendsen R H and Wang J S 1986 *Physical Review Letters* **57**(21) 2607–2609
[15] Geyer C J 1991 URL `https://hdl.handle.net/11299/58440`
[16] Hukushima K and Nemoto K 1996 *Journal of the Physical Society of Japan* **65** 1604–1608
[17] Vogel T, Li Y W, Wüst T and Landau D P 2013 *Physical Review Letters* **110** 210603
[18] Vogel T, Li Y W, Wüst T and Landau D P 2014 *Physical Review E* **90** 023302
[19] Farris A C K and Landau D P 2021 *Physica A* **569** 125778 ISSN 0378-4371
[20] Farris A C K, Shi G, Wüst T and Landau D P 2018 *The Journal of Chemical Physics* **149** 125101
[21] Wang F and Landau D P 2001 *Physical Review Letters* **86** 2050–2053
[22] Wang F and Landau D P 2001 *Physical Review E* **64** 056101
[23] Matsumoto M and Nishimura T 1998 *ACM Transactions on Modeling and Computer Simulations* **8** 3–30
[24] Sokal A D 1995 *Monte Carlo and Molecular Dynamics Simulations in Polymer Science* (Oxford University Press) edited by K. Binder
[25] Deutsch J M 1997 *The Journal of Chemical Physics* **106** 8849 ISSN 00219606
[26] Lesh N, Mitzenmacher M and Whitesides S 2003 *RECOMB* 188–195
[27] Berg B A and Neuhaus T 1991 *Physics Letters B* **267** 249 – 253 ISSN 0370-2693
[28] Berg B A and Neuhaus T 1992 *Physical Review Letters* **68**(1) 9–12

Journal of Physics: Conference Series **2207** (2022) 012003 doi:10.1088/1742-6596/2207/1/012003

Surface critical behavior of the three-dimensional $O(3)$ model

F Parisen Toldin

Institut für Theoretische Physik und Astrophysik, Universität Würzburg, Am Hubland, D-97074 Würzburg, Germany

E-mail: francesco.parisentoldin@physik.uni-wuerzburg.de

Abstract. We report results of high-precision Monte Carlo simulations of a three-dimensional lattice model in the $O(3)$ universality class, in the presence of a surface. By a finite-size scaling analysis we have proven the existence of a special surface transition, computed the associated critical exponents, and shown the presence of an extraordinary phase with logarithmically decaying correlations.

1. Introduction

A system in the vicinity of a critical point exhibits a variety of interesting features, such as power-law singularities and scaling behavior in many observables. One of the most fascinating aspects is the emergence of universality: critical exponents associated to the aforementioned singularities, and other quantities, are independent of the local details of interactions. They are rather determined by the global features of the system, such as the symmetry group, the pattern of symmetry-breaking, dimensionality and range of interactions. The theory of Renormalization Group (RG) provides a framework for understanding and predicting the emergence of universality through a suitably defined flow of Hamiltonians, the fixed points of which control the critical behavior and define the so-called Universality Classes (UCs) [1].

While the singular behavior associated to the onset of a phase transition occurs, in principle, in the thermodynamic limit only, real physical systems naturally have boundaries. Their presence is the source of rich phase diagrams and critical behavior, that has been the target of many experimental [2] and theoretical [3–5] investigations. According to RG theory, a given *bulk* fixed point controlling the critical behavior in the thermodynamic limit potentially splits into several fixed points associated to the critical behavior on the boundary [1,4], thereby defining surface or, more generally, boundary UCs. This implies that critical exponents and other universal quantities on the boundary differ from bulk ones. Furthermore, for a given system at a critical point the boundary may exhibit diverse critical behavior, depending on the strength of boundary interactions. Surface UCs also determine the critical Casimir force [6–11].

The simplest setup where this physics is realized is the case of a semi-infinite $d-$dimensional system bounded by a $(d-1)-$dimensional surface. In this context, due to its physical relevance, the critical behavior of the classical 3D $O(N)$ model represents one of the most significant UC [12]. In Fig. 1 we sketch the bulk-surface phase diagram, as a function of the bulk and surface coupling constants. For $N = 1, 2$ a surface transition line, in the presence of a disordered bulk, separates a disordered surface from an ordered one, for $N = 1$, or from a surface possessing quasi-long range order (QLRO), for $N = 2$. For a critical bulk, and as a function of the surface coupling, we distinguish an ordinary and extraordinary transition lines. Surface, ordinary and extraordinary lines meet at a multicritical point, the so-called

Journal of Physics: Conference Series **2207** (2022) 012003 doi:10.1088/1742-6596/2207/1/012003

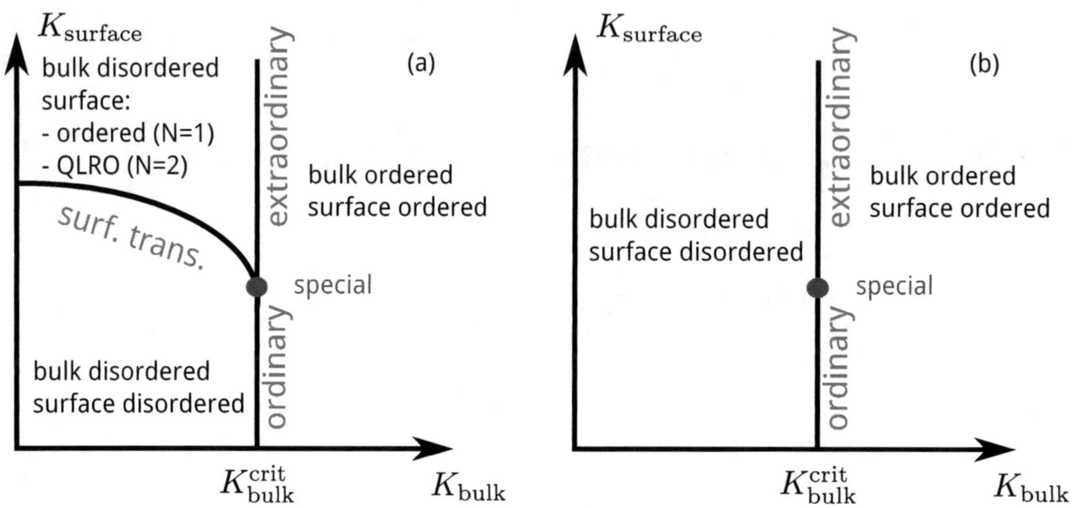

Figure 1. Bulk-surface phase diagram of the three-dimensional O(N) model, bounded by a 2D surface, for $N = 1, 2$ (a) and $N = 3$ (b).

special UC [3, 4]. For $N = 3$, no surface transition exists [12], hence the phase diagram topology does not necessarily mandate a special point. While early Monte Carlo (MC) studies supported the absence of a special point [13], a later MC investigation reported a possible Berezinskii-Kosterlitz-Thouless-like surface transition [14]. More recently, the problem has received a renewed attention in the context of conformal field-theory approaches [15–27], classical [28, 29] and quantum critical behavior [30–38]. In particular, quantum MC investigations have focused on dimerized spin-$1/2$ [30–33] and spin-1 [34] systems in $d = 2$, which exhibit a second-order quantum phase transition in the classical 3D $O(3)$ UC. In the presence of an edge, these models have been shown to display ordinary, as well as nonordinary boundary exponents, depending on the geometrical setup. A recent field-theoretical study has predicted the existence of a so-called "extraordinary-log" phase at the surface of a critical 3D $O(N)$ model. This phase exists for $N < N_c$, with $N_c > 2$ [1], and is characterized by surface correlations decaying as a power of a logarithm. The associated exponent is universal, and it is determined by some amplitudes of the normal UC [25]. This is realized by applying a symmetry-breaking field on the boundary [3,4,39,40].

Motivated by these advancements, in Ref. [28] we have investigated the classical surface $O(3)$ UC by means of MC simulations of an improved lattice model, where leading scaling corrections are suppressed. A finite-size scaling (FSS) analysis has shown the existence of a special transition and of an extraordinary phase consistent with the extraordinary-log scenario of Ref. [25]. In the following we summarize the results of Ref. [28].

2. Model

We have simulated the ϕ^4 model on a three-dimensional lattice of size L in all directions, applying periodic boundary conditions (BCs) along two directions, and open BCs on the remaining one. The reduced Hamiltonian $\mathcal{H}$, such that the Gibbs weight is $\exp(-\mathcal{H})$, is

$$\mathcal{H} = -\beta \sum_{\langle \vec{\imath}\, \vec{\jmath} \rangle} \vec{\phi}_{\vec{\imath}} \cdot \vec{\phi}_{\vec{\jmath}} - \beta_{s,\downarrow} \sum_{\langle \vec{\imath}\, \vec{\jmath} \rangle_{s\downarrow}} \vec{\phi}_{\vec{\imath}} \cdot \vec{\phi}_{\vec{\jmath}} - \beta_{s,\uparrow} \sum_{\langle \vec{\imath}\, \vec{\jmath} \rangle_{s\uparrow}} \vec{\phi}_{\vec{\imath}} \cdot \vec{\phi}_{\vec{\jmath}} + \sum_{\vec{\imath}} [\vec{\phi}_{\vec{\imath}}^{\,2} + \lambda(\vec{\phi}_{\vec{\imath}}^{\,2} - 1)^2], \tag{1}$$

where $\vec{\phi}_{\vec{x}}$ is a three-components real field on the lattice site $\vec{x} = (x_1, x_2, x_3)$, indicated as a 3D vector. In Eq. (1) the first sum extends over nearest-neighbor pairs of sites where at least one belongs to the

[1] We remark that N_c does not need to be an integer.

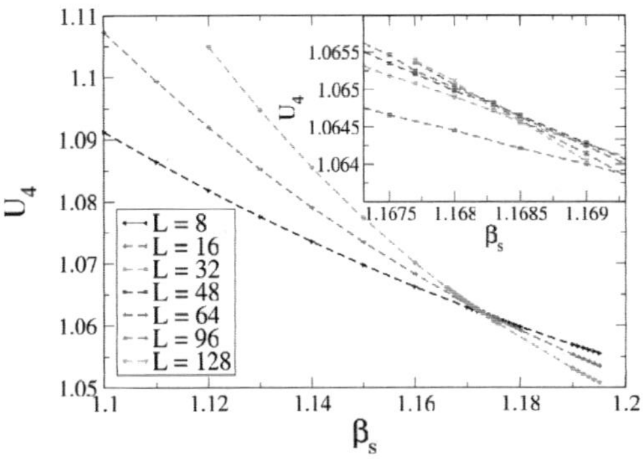

Figure 2. Surface Binder ratio U_4 at the special transition, as a function of β_s; Inset: MC data close to the special transition. Data are taken from Ref. [28].

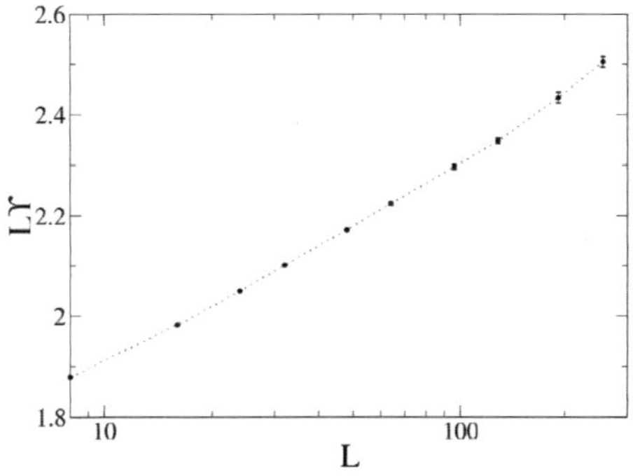

Figure 3. Helicity modulus at $\beta_s = 1.5$, in the extraordinary phase, in a semilogarithmic scale. From Ref. [28].

inner bulk, the second and third sum extend to lattice sites on the lower and upper surface, and the last sum is over all lattice sites. The coupling constants β and λ determine the bulk critical behavior. In the (β, λ) plane the bulk displays a line of continuous phase transitions in the $O(3)$ UC [12, 41], and in the limit $\lambda \to \infty$ the model reduces to the standard $O(3)$ hard spin model. At $\lambda = 5.17(11)$ the model is *improved* [42], i.e., leading bulk scaling corrections $\propto L^{-\omega}$, $\omega = 0.759(2)$, are suppressed. Next-to-leading scaling corrections due to a nonrotationally invariant irrelevant operator have an exponent $\omega_{\rm nr} \approx 2$ [42], hence they decay very fast. Improved lattice models are particularly useful in high-precision numerical studies of critical phenomena [12], in particular for boundary critical phenomena [28, 43–52] because they allow a better control of scaling corrections. In our MC simulations we have fixed $\lambda = 5.2$ and $\beta = 0.68798521$, for which the model is critical [42]. The coupling constants $\beta_{s,\downarrow}$, $\beta_{s,\uparrow}$ in Eq. (1) control the strength of boundary interactions. To study the surface critical behavior we have set $\beta_{s,\downarrow} = \beta_{s,\uparrow} = \beta_s$ and examined various surface observables as a function of β_s. MC simulations have been performed by combining Metropolis, overrelaxation, and Wolff single-cluster updates [28, 53].

3. Results

3.1. Special transition

A standard method to locate the onset of a continuous phase transition consists in a FSS analysis of RG-invariant observables [12, 54]. To study the special UC we have proceeded in two steps. First, we have studied an RG-invariant quantity, determining its critical-point value. Subsequently, we have employed this value in a FSS analysis of other observables, to compute the critical exponents at the special transition.

According to RG, close to a surface phase transition at $\beta_s = \beta_{s,c}$, an RG-invariant observable R behaves as

$$R = f((\beta_s - \beta_{s,c})L^{y_{\rm sp}}), \tag{2}$$

where $y_{\rm sp}$ is the scaling dimension of the relevant scaling field associated with the transition and we have for the moment neglected scaling corrections. We have analyzed the surface Binder ratio U_4, defined as

$$U_4 = \frac{\langle (\vec{M}_s^2)^2 \rangle}{\langle \vec{M}_s^2 \rangle^2}, \qquad \vec{M}_s \equiv \sum_{\vec{x} \in \text{surface}} \vec{\phi}_{\vec{x}}. \tag{3}$$

In Fig. 2 we show U_4 as a function of β_s. A scan over a wide range in β_s reveals a crossing, indicative of a surface transition. Close to the putative transition, we have sampled U_4 for lattice sizes up to

$L = 128$; the data are shown in the inset of Fig. 2. We observe a rather slow increase of the slope of U_4 with L, such that a precision of $\approx 10^{-5}$ is needed in order to resolve the crossing. A fit of U_4 to a suitable Taylor expansion of Eq. (2), including also scaling corrections, allowed us to estimate the critical-point value $U_4^* \equiv U_4(\beta_{s,c}) = 1.0652(4)$. We have used this value to analyze MC data, supplemented by additional simulations at $L = 192$, using FSS analysis at fixed RG-invariant [55–57]. In this method, one fixes a chosen RG-invariant R (here, $R = U_4$), thereby trading the statistical fluctuations of R with fluctuations of a coupling constant driving the transition (here, β_s). A discussion of the method can be found in Ref. [57]. To estimate the exponent $y_{\rm sp}$, we have computed derivatives with respect to β_s of RG-invariants $R = U_4$ and $R = Z_a/Z_p$, the ratio of the partition functions with antiperiodic and periodic BCs on a direction parallel to the surfaces; this ratio can be conveniently sampled with the boundary-flip algorithm [58, 59]. At fixed U_4, $dR/d\beta_s$ behaves as $dR/d\beta_s \propto L^{y_{\rm sp}}$. Suitable fits of $dU_4/d\beta_s$ and $d(Z_a/Z_p)/d\beta_s$ at fixed U_4, and as a function of L, including also scaling corrections, have delivered the estimate

$$y_{\rm sp} = 0.36(1), \qquad \nu_{\rm sp} \equiv 1/y_{\rm sp} = 2.78(8). \tag{4}$$

Next, we have sampled the surface susceptibility χ_s. At fixed U_4 its leading scaling behavior is $\chi_s \propto L^{2-\eta_\parallel}$. Fits of χ_s resulted in the estimate

$$\eta_\parallel = -0.473(2). \tag{5}$$

FSS analysis at fixed U_4 has also allowed us to estimate the critical surface couplint at the onset of the special transition $\beta_{s,c} = 1.1678(2)$.

3.2. Extraordinary phase

The existence of a special transition implies that for $\beta_s > \beta_{s,c}$ the surface displays an extraordinary phase. To study it, we have simulated the model at $\beta_s = 1.5$, for lattice sizes $8 \leq L \leq 384$. We have computed the helicity modulus Υ which measures the response of the model to a torsion in the lateral BCs. To compute it, one replaces the nearest-neighbor interaction along one lateral boundary in the Hamiltonian (1) as

$$\vec{\phi}_{\vec{x}} \cdot \vec{\phi}_{\vec{x}+\hat{e}_l} \rightarrow \vec{\phi}_{\vec{x}} R_{\alpha,\beta}(\theta) \vec{\phi}_{\vec{x}+\hat{e}_l}, \tag{6}$$

with $\hat{e}_l$ the unit vector along one of the lateral directions, where periodic BCs are applied. In Eq. (6) $R_{\alpha,\beta}(\theta)$ is a rotation matrix that rotates the α and β components of $\vec{\phi}$ by an angle θ. For the present geometry, the helicity modulus is then defined as [60]

$$\Upsilon \equiv \frac{1}{L} \frac{\partial^2 F(\theta)}{\partial \theta^2}\Big|_{\theta=0}, \tag{7}$$

where F is the total free energy. In Fig. 3 we show the product ΥL, which exhibits a remarkable logarithmic growth, neither compatible with a standard critical phase, where $\Upsilon L \sim$ const, nor with an ordered phase, where $\Upsilon \sim$ const. A logarithmic violation of FSS is also found in the ratio ξ/L of the finite-size second-moment correlation length ξ [2] over the size L. Moreover, the two-point function on the surface exhibits a rather slow, but visible, decay. These findings are indicative of the extraordinary-log scenario put forward in Ref. [25]. In such a phase, the surface correlations decay as $C(x \rightarrow \infty) \propto \ln(x)^{-(N-1)/(2\pi\alpha)}$, where $N = 3$ for the present case, and α is a universal RG parameter, determined by some amplitudes at the normal UC. Furthermore, in the extraordinary-log phase a logarithmic violation of FSS is predicted, such that $\Upsilon L \simeq 2\alpha \ln(L)$ and $(\xi/L)^2 \sim (\alpha/2) \ln L$ [62]. To further check this scenario, we have performed fits of various observables. Fits of the surface two-point function provided an estimate $\alpha = 0.15(2)$. The quoted error bar has been estimated by comparing various fit results and should be taken with some caution, because in fitting the data we

[2] See Appendix A of [61] for a discussion on the definition of a finite-size correlation length.

did not consider subleading corrections; these are potentially important, as found, e.g., in other critical models with marginal operators [63]. Fits of ξ/L to the expected logarithmic growth delivered $\alpha \approx 0.14$, consistent with the estimate coming from the correlations, although we observed some drift in the fitted value of α as a function of the minimum lattice size L used in the fits. Fits of ΥL gave less stable results, with $\alpha \gtrsim 0.11$. All in all, despite the intrinsic difficulty in estimating a logarithmic exponent, we found a rough quantitative consistence of the scaling behavior with the scenario of an extraordinary-log phase.

4. Summary

In Ref. [28] we have elucidated the boundary critical behavior of the three-dimensional $O(3)$ UC, in the presence of a 2D surface. A FSS analysis of high-precision MC data allowed us to conclude that there is a special transition on the surface, in the presence of a critical bulk, and to compute the associated critical exponents. The exponent $\eta_{\|}$ that we found is remarkably close to the nonordinary $\eta_{\|}$ exponent found in quantum MC studies of dimerized spin models [30–34, 36, 37]. This suggests that, for the geometrical settings where such a nonordinary exponent is found, these models are "accidentally" close to the special transition. Unlike the ordinary and extraordinary surface phases, the special UC has a relevant non symmetry-breaking surface scaling field with dimension $y_{\rm sp}$ [see Eq. (2)]. Therefore, one generically needs a fine tuning of boundary interactions in order to realize the special UC. Nevertheless, we notice that the value of $y_{\rm sp}$ [Eq. (4)] is unusually small: this implies a slow crossover from the special fixed point when surface interactions are tuned away from the special transition. Therefore, it may be possible to observe critical exponents similar to those of the special UC even if a model is not very close to the special transition. This can provide an explanation to the observed nonordinary edge exponent $\eta_{\|}$ found in the aforementioned quantum spin models, without the need of a fine-tuning. To further substantiate this hypothesis, it would be desirable to study in more detail the quantum-to-classical mapping [64] of these spin models.

In the extraordinary phase, our MC data display slowly-decaying correlations and a remarkable logarithmic violation of FSS, which is indicative of the extraordinary-log phase scenario put forward in Ref. [25]. Recently, this scenario has been put on a firmer ground in Ref. [65], where we have computed the universal amplitudes of the normal UC that determine the onset and the value of the logarithmic exponent of the extraordinary-log phase. We found a good agreement with MC simulations of the extraordinary phase presented in Refs. [28], [29], thus providing a nontrivial check of the connection between the normal and the extraordinary-log phases outlined in Ref. [25]. A concurrent conformal bootstrap study [27] found results in agreement with Ref. [65].

Acknowledgments

FPT is funded by the Deutsche Forschungsgemeinschaft (DFG, German Research Foundation)–Project No. 414456783. The author gratefully acknowledges the Gauss Centre for Supercomputing e.V. for funding this project by providing computing time through the John von Neumann Institute for Computing (NIC) on the GCS Supercomputer JUWELS at Jülich Supercomputing Centre (JSC) [66].

References

[1] Cardy J 1996 *Scaling and Renormalization in Statistical Physics* (Cambridge: Cambridge University Press)
[2] Dosch H 2006 *Critical Phenomena at Surfaces and Interfaces: Evanescent X-Ray and Neutron Scattering* (Berlin: Springer Berlin Heidelberg)
[3] Binder K 1983 Critical behavior at surfaces *Phase Transitions and Critical Phenomena* vol 8 ed Domb C and Lebowitz J L (London: Academic Press) p 1
[4] Diehl H W 1986 Field-theoretical approach to critical behaviour at surfaces *Phase Transitions and Critical Phenomena* vol 10 ed Domb C and Lebowitz J L (London: Academic Press) p 75
[5] Pleimling M 2004 *J. Phys. A: Math. Gen.* **37** R79
[6] Fisher M E and de Gennes P G 1978 *C. R. Acad. Sci. Paris Ser. B* **287** 207
[7] Krech M 1994 *The Casimir Effect in Critical Systems* (London: World Scientific)
[8] Krech M 1999 *J. Phys.: Condens. Matter* **11** R391

[9] Gambassi A 2009 *J. Phys.: Conf. Ser.* **161** 012037
[10] Gambassi A and Dietrich S 2011 *Soft Matter* **7** 1247
[11] Maciołek A and Dietrich S 2018 *Rev. Mod. Phys.* **90** 045001
[12] Pelissetto A and Vicari E 2002 *Phys. Rep.* **368** 549–727
[13] Krech M 2000 *Phys. Rev. B* **62** 6360–6371
[14] Deng Y, Blöte H W J and Nightingale M P 2005 *Phys. Rev. E* **72** 016128
[15] McAvity D M and Osborn H 1995 *Nucl. Phys. B* **455** 522–576
[16] Liendo P, Rastelli L and van Rees B C 2013 *J. High Energy Phys.* JHEP07(2013)113
[17] Gliozzi F, Liendo P, Meineri M and Rago A 2015 *J. High Energy Phys.* JHEP05(2015)036
[18] Billò M, Gonçalves V, Lauria E and Meineri M 2016 *J. High Energy Phys.* JHEP04(2016)091
[19] Liendo P and Meneghelli C 2017 *J. High Energy Phys.* JHEP01(2017)122
[20] Lauria E, Meineri M and Trevisani E 2018 *J. High Energy Phys.* JHEP11(2018)148
[21] Mazáč D, Rastelli L and Zhou X 2019 *J. High Energy Phys.* JHEP12(2019)004
[22] Kaviraj A and Paulos M F 2020 *J. High Energy Phys.* JHEP04(2020)135
[23] Dey P, Hansen T and Shpot M 2020 *J. High Energy Phys.* JHEP12(2020)051
[24] Behan C, Di Pietro L, Lauria E and van Rees B C 2020 *J. High Energy Phys.* JHEP12(2020)182
[25] Metlitski M A 2020 *Preprint* arXiv:2009.05119
[26] Gimenez-Grau A, Liendo P and van Vliet P 2021 *J. High Energy Phys.* JHEP04(2021)167
[27] Padayasi J, Krishnan A, Metlitski M A, Gruzberg I A and Meineri M 2021 *Preprint* arXiv:2111.03071
[28] Parisen Toldin F 2021 *Phys. Rev. Lett.* **126** 135701
[29] Hu M, Deng Y and Lv J P 2021 *Phys. Rev. Lett.* **127** 120603
[30] Suzuki T and Sato M 2012 *Phys. Rev. B* **86** 224411
[31] Zhang L and Wang F 2017 *Phys. Rev. Lett.* **118** 087201
[32] Ding C, Zhang L and Guo W 2018 *Phys. Rev. Lett.* **120** 235701
[33] Weber L, Parisen Toldin F and Wessel S 2018 *Phys. Rev. B* **98** 140403(R)
[34] Weber L and Wessel S 2019 *Phys. Rev. B* **100** 054437
[35] Jian C M, Xu Y, Wu X C and Xu C 2021 *SciPost Phys.* **10** 033
[36] Zhu W, Ding C, Zhang L and Guo W 2021 *Phys. Rev. B* **103** 024412
[37] Weber L and Wessel S 2021 *Phys. Rev. B* **103** L020406
[38] Ding C, Zhu W, Guo W and Zhang L *Preprint* arXiv:2110.04762
[39] Bray A J and Moore M A 1977 *J. Phys. A: Math. Gen.* **10** 1927–1962
[40] Burkhardt T W and Cardy J L 1987 *J. Phys. A: Math. Gen.* **20** L233–L238
[41] Campostrini M, Hasenbusch M, Pelissetto A, Rossi P and Vicari E 2002 *Phys. Rev. B* **65** 144520
[42] Hasenbusch M 2020 *Phys. Rev. B* **102** 024406
[43] Hasenbusch M 2009 *J. Stat. Mech.* (2009)P07031
[44] Hasenbusch M 2010 *Phys. Rev. B* **82** 104425
[45] Parisen Toldin F and Dietrich S 2010 *J. Stat. Mech.* (2010)P11003
[46] Hasenbusch M 2011 *Phys. Rev. B* **83** 134425
[47] Hasenbusch M 2011 *Phys. Rev. B* **84** 134405
[48] Hasenbusch M 2012 *Phys. Rev. B* **85** 174421
[49] Parisen Toldin F, Tröndle M and Dietrich S 2013 *Phys. Rev. E* **88** 052110
[50] Parisen Toldin F 2015 *Phys. Rev. E* **91** 032105
[51] Parisen Toldin F, Tröndle M and Dietrich S 2015 *J. Phys.: Condens. Matter* **27** 214010
[52] Parisen Toldin F, Assaad F F and Wessel S 2017 *Phys. Rev. B* **95** 014401
[53] Wolff U 1989 *Phys. Rev. Lett.* **62** 361–364
[54] Privman V 1990 Finite-Size Scaling Theory *Finite Size Scaling and Numerical Simulation of Statistical Systems* ed Privman V (Singapore: World Scientific) p 1
[55] Hasenbusch M 1999 *J. Phys. A: Math. Gen.* **32** 4851–4865
[56] Hasenbusch M, Parisen Toldin F, Pelissetto A and Vicari E 2007 *J. Stat. Mech.* (2007)P02016
[57] Parisen Toldin F 2011 *Phys. Rev. E* **84** 025703(R)
[58] Hasenbusch M 1993 *Physica A* **197** 423–435
[59] Campostrini M, Hasenbusch M, Pelissetto A, Rossi P and Vicari E 2001 *Phys. Rev. B* **63** 214503
[60] Fisher M E, Barber M N and Jasnow D 1973 *Phys. Rev. A* **8** 1111–1124
[61] Parisen Toldin F, Hohenadler M, Assaad F F and Herbut I F 2015 *Phys. Rev. B* **91** 165108
[62] Metlitski M *private communication.*
[63] Hasenbusch M, Parisen Toldin F, Pelissetto A and Vicari E 2008 *Phys. Rev. E* **78** 011110
[64] Sachdev S 2011 *Quantum Phase Transitions* (Cambridge: Cambridge University Press)
[65] Parisen Toldin F and Metlitski M A 2021 *Preprint* arXiv:2111.03613
[66] Jülich Supercomputing Centre 2019 *Journal of large-scale research facilities* **5** A135

Corrections to scaling in geometrical clusters of the 2D Ising model

Michail Akritidis[1], Nikolaos G Fytas[1], and Martin Weigel[2 1]

[1] Centre for Fluid and Complex Systems, Coventry University, Coventry, CV1 5FB, United Kingdom

[2] Institut für Physik, Technische Universität Chemnitz, 09107 Chemnitz, Germany

E-mail: akritidm@uni.coventry.ac.uk

Abstract.
We study the scaling of the average cluster size and percolation strength of geometrical clusters for the two-dimensional Ising model. By means of Monte Carlo simulations and a finite-size scaling analysis we discuss the appearance of corrections to scaling for different definitions of cluster sets. We find that including all percolating clusters, or excluding only clusters that percolate in one but not the other direction, leads to smaller corrections to scaling for the average cluster size as compared to the other definitions considered. The percolation strength is less sensitive to the definitions used.

1. Introduction

Percolation theory [1] is closely linked to critical phenomena in spin models, as it provides geometrical insights regarding the nature of the transition. The percolation problem is defined through clusters of occupied sites or bonds on a lattice, and one studies the geometrical transition at the percolation threshold, characterized by an incipient spanning cluster that becomes infinite in the thermodynamic limit. The q-state Potts model can be mapped onto a percolation process, by constructing clusters of neighbouring spins with the same orientation. Unfortunately, these *geometrical* clusters do not in general percolate at the critical temperature T_c of the actual phase transition, and they hence do not properly describe the thermal correlations of the system (see [2, 3] for a review). Using a different cluster definition where bonds of neighbouring like spins are deleted with a certain temperature-dependent probability, Fortuin and Kasteleyn (FK) [4] showed that the model can be mapped onto a modified percolation process, where clusters percolate at T_c and, more importantly, they encode the critical behaviour of the system, resulting in the same set of critical exponents. The FK clusters are also the core idea behind the powerful Monte Carlo algorithms of Swendsen and Wang [5] and Wolff [6], where entire FK clusters are flipped instead of individual spins as in the standard Metropolis algorithm [7]. The main advantage of these non-local cluster updates over the local spin flips is the reduction of autocorrelation times in the vicinity of the critical point, significantly improving the simulation efficiency.

Although geometrical clusters do not capture the critical behaviour, in the two-dimensional (2D) Potts model they at least do percolate at T_c. It turns out that for $0 \leq q \leq 4$ the critical behaviour of geometrical clusters for the pure model is related to the tricritical behaviour of a diluted Potts model [8, 9]. Stella and Vanderzande [10, 11] showed that the geometrical clusters of the $q = 2$ Potts model, i.e., the Ising model, follow the tricritical behaviour of the $q = 1$ site diluted Potts model. This aspect was numerically verified in references [8, 9].

In the present study we investigate numerically the critical behaviour of the geometrical clusters of the square-lattice Ising model, whose Hamiltonian reads

$$\mathcal{H} = -J \sum_{\langle ij \rangle} S_i S_j, \tag{1}$$

where the spins take on the values ± 1, $J > 0$ is the ferromagnetic exchange interaction, and the sum extends over all nearest neighbours of the lattice. Specifically, we estimate the exponents of the average cluster size and percolation strength for different sets of clusters, while monitoring the existence of corrections to scaling in each case.

2. Observables

As in random percolation [1], the quantities of interest are the percolation strength P_∞, which gives the fraction of sites belonging to the infinite cluster in the thermodynamic limit, and the average cluster size

$$S = \frac{\sum_{s'} s^2 \, n_s}{\sum_{s'} s \, n_s}, \tag{2}$$

where s denotes the size (number of spins) of the cluster and n_s the number of clusters with size s. The notation s' indicates that the sums are restricted to different sub-sets of clusters, which for the present study are:

(i) All clusters are included: C.
(ii) Exclude the largest cluster in each measurement: $C \setminus \max C$.
(iii) Exclude all percolating clusters: $C \setminus P$.
(iv) Exclude all clusters percolating in horizontal and in vertical direction: $C \setminus P_{\text{x and y}}$.
(v) Exclude all clusters percolating in one specific direction, e.g., horizontal: $C \setminus P_{\text{x}}$.
(vi) Exclude all clusters percolating in one but not the other direction, e.g., horizontal and not vertical: $C \setminus P_{\text{x and } \overline{\text{y}}}$.

For finite size systems P_∞ is usually estimated by the fraction of sites belonging to the largest cluster. Following the idea of cluster sets for S, we can introduce similar sets for P_∞. In the present study we consider, in each configuration, the fraction of sites that belong to the:

(i) Largest cluster: $\max C$.
(ii) Largest percolating cluster: $\max P$.
(iii) Largest cluster that percolates in horizontal and in vertical direction: $\max P_{\text{x and y}}$.
(iv) Largest cluster that percolates in one specific direction, e.g., horizontal: $\max P_{\text{x}}$.
(v) Largest cluster that percolates in one but not the other direction, e.g., horizontal and not vertical: $\max P_{\text{x and } \overline{\text{y}}}$.

In the vicinity of the critical point these observables exhibit a scaling behaviour of the form [1]

$$P_\infty (L, T) = L^{-\beta/\nu} f_{P_\infty} \left[(T - T_\text{c}) \, L^{1/\nu} \right], \tag{3}$$

$$S (L, T) = L^{\gamma/\nu} f_S \left[(T - T_\text{c}) \, L^{1/\nu} \right], \tag{4}$$

where ν is the critical exponent of the correlation length, γ the critical exponent that captures the divergence of the average cluster size, and β the critical exponent which describes how the strength of the percolating cluster goes to zero as T_c is approached from below. f_{P_∞} and f_S are universal finite-size scaling functions.

3. Results

We simulated the 2D Ising model with periodic boundary conditions using the Swendsen-Wang algorithm [5]. We considered systems of linear size $L = 16, 32, 64, 128, 256, 512, 600, 1000, 1200, 1600$ and 2000, at the exact critical temperature $T_c = 2/\ln\left(1+\sqrt{2}\right)$. For all L the total number of simulation steps was $1.1 \times \tau_{\text{int, E}} \times 10^5$ sweeps, where $\tau_{\text{int, E}}$ is the integrated autocorrelation time of the energy [12], and $\tau_{\text{int, E}} \times 10^4$ sweeps were discarded during equilibration. After every $\tau_{\text{int, E}}$ sweeps a measurement was taken, leading to up to 10^5 measurements per run. The estimates of $\tau_{\text{int, E}}$, rounded up to the next largest integer, in ascending order of the system size are: $\tau_{\text{int, E}} = 4, 5, 5, 6, 7, 9, 9, 10, 11, 11, 12$ sweeps. Statistical errors were estimated by means of jackknife blocking [13]. We considered a cluster to percolate in one direction if it wraps around this direction and is connected back to itself. To identify the wrapping clusters, we employed the method of Machta *et al.* [14].

At T_c, according to equations (3) and (4), $f_P(0) = \text{const.}$ and $f_S(0) = \text{const.}$, allowing the estimation of γ/ν and β/ν as a function of L. For all sets of clusters, fits were performed using the least-squares Levenberg-Marquardt algorithm [15]. In order to obtain the influence of corrections to scaling in the observables we performed fits including $L_{\text{min}} \leq L \leq L_{\text{max}}$, by systematically increasing the lower cut-off L_{min}, while the upper cut-off is kept fixed $L_{\text{max}} = 2000$.

In figure 1(a) the average cluster size as a function of L is plotted for the different definitions introduced in section 2. Visually, the data of all definitions seem to follow straight lines, on a log-log plot, indicating merely small corrections to scaling. Estimates of γ/ν for different fit intervals and the respective quality-of-fit parameter Q are reported in table 1. For C and $C \setminus P_{\text{x and }\bar{\text{y}}}$, γ/ν converges relatively quickly to its asymptotic value, $\gamma/\nu = 91/48 \approx 1.896$ [10] as L_{min} increases. Specifically, the discrepancy from the asymptotic value is less than 3 standard deviations (3σ) for $L_{\text{min}} \geq 64$ and $L_{\text{min}} \geq 256$ respectively, so that corrections to scaling are not substantial. This is nicely illustrated in figure 1(b), where the exponent γ/ν is plotted as a function of $1/L_{\text{min}}$. For $C \setminus P_{\text{x and y}}$, $C \setminus P_{\text{x}}$, $C \setminus \max C$, and $C \setminus P$, on the other hand, the convergence to the asymptotic value is found to be very slow. Even for $L_{\text{min}} = 1200$, the deviation from the asymptotic values is around 6σ for all definitions, indicating that corrections to scaling are important. This is illustrated in figure 1(b), where the estimates of the exponent seem to converge to a value significantly below the predicted one. The fact that C and $C \setminus P_{\text{x and }\bar{\text{y}}}$ give similar results is to be expected. Percolating clusters in one but not the other direction are very rare, thus excluding them from the sums of equation (2) will not affect the average cluster size significantly. This can also be seen from figure 1(a), where the data points for the two definitions, at the same L, are equal within error bars. Note that previous estimates of the exponent using the C definition are reported in reference [16] and these for the $C \setminus P_{\text{x}}$ definition in reference [17].

For the percolation strength, data for all sets seem to follow a straight line on a log-log plot as a function of L, indicating small corrections to scaling, as is shown in figure 2(a). This observation is supported by table 2, where the estimates of β/ν and Q are reported. Setting aside $\max P_{\text{x and }\bar{\text{y}}}$, for all sets the deviations from the asymptotic value $\beta/\nu = 5/96 \approx 0.052$ [10] is less than 3σ for all $L_{\text{min}} \geq 64$, indicating rather small corrections to scaling. Notably, for $\max P_{\text{x and y}}$ deviations are even smaller than 1.7σ for all L_{min}. According to table 2 the estimates of β/ν for $\max P_{\text{x and }\bar{\text{y}}}$ are less than 3σ from the asymptotic value (except $L_{\text{min}} = 16$), but the statistical errors are large, thus not allowing a reasonable estimate of the exponent. The explanation for this behaviour of $\max P_{\text{x and }\bar{\text{y}}}$ is in line with the $C \setminus P_{\text{x and }\bar{\text{y}}}$ for the average cluster size, but going in the opposite direction. As already stated above, clusters percolating in one but not the other direction are rare, leading to poor statistics and larger statistical errors for the percolation strength; see figure 2(b), where the exponent β/ν is plotted as a function of $1/L_{\text{min}}$.

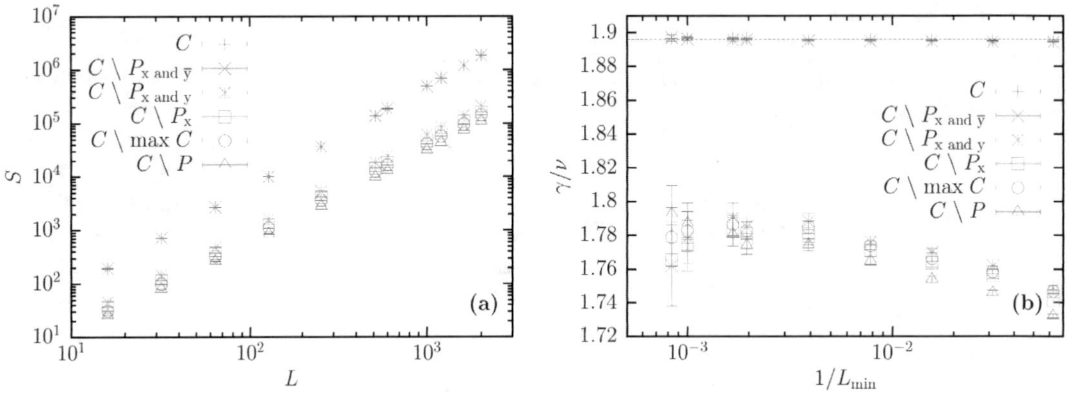

Figure 1. (a) Log-log plot of the average cluster size as a function of L for the different definitions. (b) Exponent ratio γ/ν as a function of $1/L_{\text{min}}$. The dashed horizontal line marks the asymptotic value $91/48 \approx 1.896$ [10].

Table 1. Estimates of γ/ν, quality-of-fit parameter Q, and deviations from the exact value, $\gamma/\nu = 91/48 \approx 1.896$, in multiples of the estimated statistical errors $\Delta\,(\sigma)$, for the cluster sets of the average cluster size as introduced in section 2.

	C			$C \setminus P_{\text{x and }\overline{\text{y}}}$			$C \setminus P_{\text{x and y}}$		
L_{min}	γ/ν	Q	$\Delta\,(\sigma)$	γ/ν	Q	$\Delta\,(\sigma)$	γ/ν	Q	$\Delta\,(\sigma)$
16	1.89490(11)	0.001	8.28	1.89419(11)	0.001	15.03	1.7488(14)	0.000	101.59
32	1.89510(14)	0.010	5.36	1.89440(13)	0.008	10.76	1.7618(19)	0.000	71.51
64	1.89535(17)	0.044	2.83	1.89465(17)	0.042	7.11	1.770(2)	0.002	53.39
128	1.8957(2)	0.123	0.75	1.8950(2)	0.130	3.94	1.776(3)	0.021	36.14
256	1.8957(3)	0.078	0.52	1.8950(3)	0.079	2.78	1.788(5)	0.658	22.21
512	1.8964(4)	0.723	1.30	1.8958(4)	0.747	0.10	1.785(7)	0.575	15.78
600	1.8964(6)	0.613	1.03	1.8956(6)	0.668	0.46	1.791(9)	0.611	11.97
1000	1.8964(10)	0.568	0.60	1.8962(9)	0.623	0.35	1.779(15)	0.639	7.71
1200	1.8964(14)	0.300	0.40	1.8971(14)	0.628	0.86	1.76(2)	0.778	5.59
	$C \setminus P_{\text{x}}$			$C \setminus \max C$			$C \setminus P$		
L_{min}	γ/ν	Q	$\Delta\,(\sigma)$	γ/ν	Q	$\Delta\,(\sigma)$	γ/ν	Q	$\Delta\,(\sigma)$
16	1.7413(16)	0.000	99.00	1.7469(11)	0.000	140.19	1.7321(11)	0.000	148.24
32	1.756(2)	0.000	70.11	1.7584(14)	0.000	97.21	1.7459(15)	0.000	102.89
64	1.764(2)	0.016	54.90	1.7663(18)	0.000	70.59	1.7540(18)	0.000	76.73
128	1.771(3)	0.307	37.30	1.774(2)	0.003	48.41	1.765(3)	0.008	52.35
256	1.779(5)	0.875	23.45	1.785(4)	0.868	30.87	1.774(4)	0.514	33.64
512	1.778(7)	0.794	16.59	1.783(5)	0.806	21.64	1.774(5)	0.374	23.01
600	1.783(9)	0.847	12.52	1.787(7)	0.860	16.57	1.781(7)	0.544	16.65
1000	1.774(15)	0.873	8.11	1.782(11)	0.754	9.87	1.787(12)	0.431	9.04
1200	1.77(2)	0.969	5.85	1.779(17)	0.484	6.76	1.793(17)	0.228	6.15

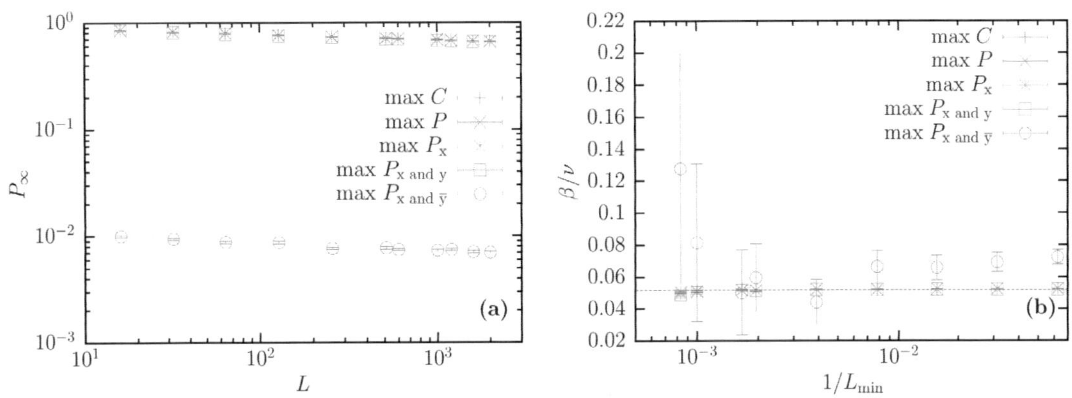

Figure 2. (a) Log-log plot of the strength as a function of L for the different definitions. (b) Exponent ratio β/ν as a function of $1/L_\mathrm{min}$. The dashed horizontal line marks the asymptotic value $5/96 \approx 0.052$ [10].

Table 2. Estimates of β/ν, quality-of-fit parameter Q, and deviations from the exact value, $\beta/\nu = 5/96 \approx 0.052$, in multiples of the estimated statistical errors $\Delta\,(\sigma)$, for the cluster sets of the percolation strength as introduced in section 2.

	max C			max P			max P_x		
L_min	β/ν	Q	$\Delta\,(\sigma)$	β/ν	Q	$\Delta\,(\sigma)$	β/ν	Q	$\Delta\,(\sigma)$
16	0.05271(8)	0.005	7.94	0.05269(8)	0.008	7.65	0.05245(11)	0.230	3.30
32	0.05256(10)	0.030	4.94	0.05255(10)	0.036	4.76	0.05243(14)	0.166	2.56
64	0.05241(12)	0.081	2.66	0.05240(12)	0.092	2.55	0.05236(16)	0.136	1.68
128	0.05222(17)	0.141	0.81	0.05222(17)	0.136	0.83	0.0521(2)	0.216	0.06
256	0.0523(2)	0.109	1.14	0.0523(2)	0.103	1.13	0.0522(3)	0.156	0.43
512	0.0517(3)	0.744	1.06	0.0517(3)	0.718	1.07	0.0515(5)	0.597	1.32
600	0.0519(4)	0.670	0.44	0.0519(4)	0.653	0.42	0.0517(6)	0.494	0.70
1000	0.0514(7)	0.671	0.96	0.0514(7)	0.678	0.99	0.0507(9)	0.681	1.45
1200	0.0509(11)	0.586	1.17	0.0508(11)	0.600	1.19	0.0498(14)	0.955	1.61
	max $P_{\mathrm{x\ and\ y}}$			max $P_{\mathrm{x\ and\ \bar{y}}}$					
L_min	β/ν	Q	$\Delta\,(\sigma)$	β/ν	Q	$\Delta\,(\sigma)$	-	-	-
16	0.05224(14)	0.232	1.17	0.072(5)	0.432	4.38			
32	0.05227(17)	0.169	1.08	0.069(6)	0.402	2.87			
64	0.0522(2)	0.119	0.66	0.066(8)	0.347	1.78			
128	0.0520(3)	0.135	0.42	0.066(10)	0.251	1.43			
256	0.0523(4)	0.153	0.60	0.044(14)	0.730	0.56			
512	0.0514(6)	0.508	1.14	0.06(2)	0.763	0.35			
600	0.0517(7)	0.435	0.48	0.05(3)	0.676	0.07			
1000	0.0505(12)	0.569	1.30	0.08(5)	0.617	0.59			
1200	0.0490(19)	0.881	1.64	0.13(7)	0.651	1.04			

4. Summary

Using several classes of sub-sets of clusters we determined the critical exponents for the average cluster size and percolation strength for geometrical clusters in the 2D Ising model. Our data were generated by means of Monte Carlo simulations for systems of linear size up to $L = 2000$. Estimates of the exponents for all cluster sets using finite-size scaling were reported, while the appearance of corrections to scaling was monitored by performing fits on intervals $L_{\min} \leq L \leq L_{\max}$ with increasing $L_{\min}$. Overall, where corrections to scaling are not very strong, our results are in agreement with exact values within error bars and the numerical accuracy allowed us to draw conclusions for the presence of scaling corrections.

The average cluster size is sensitive to the exclusion of certain classes of clusters. Excluding percolating clusters (except clusters percolating in one but not the other direction) or the largest cluster, which is commonly done in numerical studies of random percolation [1], leads to strong corrections to scaling. On the other hand, when all clusters are included, minimal corrections to scaling appear. This is also valid when percolating clusters in one but not the other direction are excluded, but mainly because these clusters appear less often.

The percolation strength is less sensitive to the definitions used. Except from the case of max $P_{\mathrm{x\ and\ \bar{y}}}$, the rest provide reasonable estimates of the involved exponent with rather small corrections. The low probability of clusters that percolate in one but not the other direction, leads to poor statistics resulting in estimates of very low accuracy compared to the other definitions been used.

As the average cluster size of the full set C does not show any maximum as a function of temperature, estimates of pseudo-critical temperatures need to be extracted from other quantities in cases where the transition temperature is not known exactly. A more systematic analysis, allowing one to quantify the corrections to scaling would be an interesting task for future work. The inclusion of additional sets of clusters is also feasible, and could presumably lead to optimal sets of definitions for the observables under consideration, with minimal corrections to scaling, thus allowing accurate estimates of the critical exponents.

Acknowledgments

We acknowledge the provision of computing time on the parallel computer cluster *Zeus* of Coventry University.

References

[1] Stauffer D and Aharony A 1994 *Introduction to percolation theory* 2nd ed (London: Taylor & Francis)
[2] Coniglio A and Fierro A 2009 *Encyclopedia of Complexity and Systems Science* ed Meyers R A (New York: Springer) p 1596
[3] Saberi A A 2015 *Phys. Rep.* **578** 1
[4] Fortuin C M and Kasteleyn P W 1972 *Physica* **57** 536
[5] Swendsen R H and Wang J S 1987 *Phys. Rev. Lett.* **58** 86
[6] Wolff U 1989 *Phys. Rev. Lett.* **62** 361
[7] Metropolis N, Rosenbluth A W, Rosenbluth M N, Teller A H and Teller E 1953 *J. Chem. Phys.* **21** 1087
[8] Janke W and Schakel A M J 2004 *Nucl. Phys. B* **700** 385
[9] Janke W and Weigel M 2006 *Phys. Lett. B* **639** 373
[10] Stella A L and Vanderzande C 1989 *Phys. Rev. Lett.* **62** 1067
[11] Vanderzande C and Stella A L 1989 *J. Phys. A* **22** L445
[12] Weigel M and Janke W 2010 *Phys. Rev. E* **81** 066701
[13] Efron B and Tibshirani R J 1993 *An Introduction to the Bootstrap* (New York: Chapman and Hall/CRC)
[14] Machta J, Choi Y S, Lucke A, Schweizer T and Chayes L M 1996 *Phys. Rev. E* **54** 1332
[15] Press W H, Teukolsky S A, Vetterling W T and Flannery B P 1996 *Numerical Recipes in FORTRAN 77: The Art of Scientific Computing* vol 1 (Cambridge: Cambridge University Press) p 678 2nd ed
[16] Janke W and Schakel A M J 2005 *Phys. Rev. E* **71** 036703
[17] Fortunato S 2002 *Phys. Rev. B* **66** 054107

Journal of Physics: Conference Series **2207** (2022) 012005 doi:10.1088/1742-6596/2207/1/012005

Study of diluted kinetic Ising model under sinusoidal external field

Sourav Chattopadhyay
Department of Physics, ICFAI University,
Tripura, West Tripura - 799210, India

S. B. Santra
Department of Physics, Indian Institute of Technology Guwahati,
Guwahati - 781039, Assam, India

E-mail: santra@iitg.ac.in

Abstract. The ferromagnet to paramagnet phase transition temperature depends on the dilution concentration in the site diluted Ising ferromagnet. Though this model is a bistable system, few studies reported dynamic phase transition (DPT) in diluted Ising ferromagnet. We study dilution-dependent DPT in diluted Ising ferromagnet via Monte Carlo simulation under a time-varying external field tuning the system's temperature on several system sizes. The nature of the transition is characterized by employing the finite-size scaling study.

1. Introduction

Most natural magnetic materials have several kinds of impurities or disorders. The disorder adds extra complexity to their physical behavior. A physical disorder changes the regular lattice morphology of pure material and the interactions among the magnetic ions. The most common type of physical disorder corresponds to a situation where a few magnetic ions or spins are physically missing at random or replaced by non-magnetic elements in a regular magnetic material [1]. In the site diluted Ising model, impurities have replaced several magnetic spins with a non-magnetic entity [2, 3, 4, 5]. The main aim of the studies so far on the site diluted Ising model [6, 7] was to know the effect of disorder on the critical behavior of the system and its universality class. There exists the Harris criterion [8] which says if the specific heat critical exponent α of a regular system is positive, the values of the critical exponents change with the introduction of weak disorder in the system, and consequently, disorder changes the universality class. On the other hand, there will be no change in the values of the critical exponents with the increase in disorder in the system if $\alpha < 0$ for the corresponding pure system. The site diluted 2d equilibrium Ising model is a case of interest as it is a marginal case, as $\alpha = 0$ for the pure 2d Ising model. There exist two theories. The first one is the strong universality hypothesis [5] which claims that the critical exponents remain unchanged with the disorder. However, scaling has multiplicative logarithmic corrections. The second is the weak universality hypothesis [9, 10, 11, 12], which favors the dilution dependent critical exponents.

Estimates of $T_c(q)$, the phase transition temperature in diluted Ising model are reported for several values of q [3, 10]. Since we are only interested for weak disorder below we quote the

Journal of Physics: Conference Series **2207** (2022) 012005 doi:10.1088/1742-6596/2207/1/012005

values of $T_c = 1.9095(5)$, and $1.5046(10)$ for $q = 0.1$ and 0.2 respectively.

However, hysteresis and dynamic phase transition in site diluted Ising ferromagnets are rarely reported [13, 14, 15]. It will be interesting to know how the underlying diluted lattice morphologies affect nonequilibrium dynamic phase transition (DPT) dynamics. This article will discuss extensive Monte Carlo study results to describe nonequilibrium phase transitions on diluted Ising ferromagnets.

1.1. Model and Simulation

A site on a two dimensional diluted kinetic Ising model (DKIM) is either occupied by an Ising spin $S_i = \pm 1$ or remains empty. A disordered site is a site without a spin; all the regular sites have an Ising spin. The lattice sites are randomly diluted with dilution concentration q using a uniformly distributed random number $r \in [0, 1]$. If $r \leq q$, the site is not assigned any spin and marked as a disordered site. Otherwise, a spin is assigned to the site and marked as a regular site. Each spin on the square lattice can interact with its nearest neighbor spins, if available, with [13, 14, 15] periodic boundary conditions via ferromagnetic interaction strength $J(> 0)$. The spin system is kept under an external time dependent sinusoidal magnetic field $h(t)$ given by $h(t) = h_0 \sin(2\pi t/P)$ where h_0 is the amplitude and P is the time period of the field. The DKIM Hamiltonian in the presence of external field $h(t)$ can be written as,

$$\mathcal{H} = -J \sum_{\langle ij \rangle} c_i c_j S_i S_j - h(t) \sum_i c_i S_i \tag{1}$$

where h_0 is taken in the unit of J, the coefficient $c_i = 0$ denotes a disordered site, and $c_i = 1$ denotes a regular site.

The system is evolved using Glauber Monte Carlo (MC) single spin flip dynamics [16] to equilibrate the system at temperature T. The T is also measured in the unit of J. The Glauber spin flip acceptance ratio is given by $W(S_i \to -S_i) = \frac{1}{1+e^{\Delta E/k_B T}}$, where $\Delta E = 2S_i\Sigma + 2S_i h(t)$ is the change in energy due to spin flip and Σ is the sum of the nearest neighbor spins of the spin S_i. The value of J and Boltzmann constant k_B are taken to be 1 throughout the simulations.

The Monte Carlo (MC) study has been performed on two dimensional square lattices of size $L = 128, 256$ and 512 at different disorder $q = 0.1$ and 0.2 and regular lattice. A MC step is calling and updating $(1-q)L^2$ spins on a diluted square lattice of size L and dilution q, taken as one unit of time. Two initial states are considered throughout all the simulations: all spins up $S_i = +1$ state, and all spins down $S_i = -1$ state. The external magnetic field $h(t)$ is turned on to the system at a constant temperature T. The system is evolved $\mathcal{O}(10^5)$ MC steps to achieve the stable dynamic states in the presence of the applied field before collecting data. Then the variations of different relevant quantities are observed over the next 10^6 MC steps on the stable cycles. The dynamic quantities are averaged over the 1024 stable cycles (chosen from all over the time series for each sample) and 8 random initial disordered configurations to achieve improved averagings, leading to an ensemble of size 8192.

We performed an independent estimate of $T_c(q)$, and found $T_c(q) = 1.90 \pm 0.02$, and 1.52 ± 0.02 for $q = 0.1$ and 0.2 respectively. Our estimates are very close to the reported values within error bars [3, 10].

2. Thermally driven DPT at fixed external field h_0 and Period P

Diluted Ising ferromagnet is a bistable system below the respective ferromagnetic to paramagnetic phase transition temperature $T_c(q)$, q is the dilution concentration. Such a system is expected to show hysteresis effects and nonequilibrium phase transition under external time varying field. The DPT occurs when half period of the external field $P_{1/2}$ becomes comparable with the mean metastable lifetime $\langle \tau \rangle$ of the system [17, 18, 19], $P_{1/2} \sim \mathcal{O}(\langle \tau \rangle)$. The $\langle \tau \rangle$ also

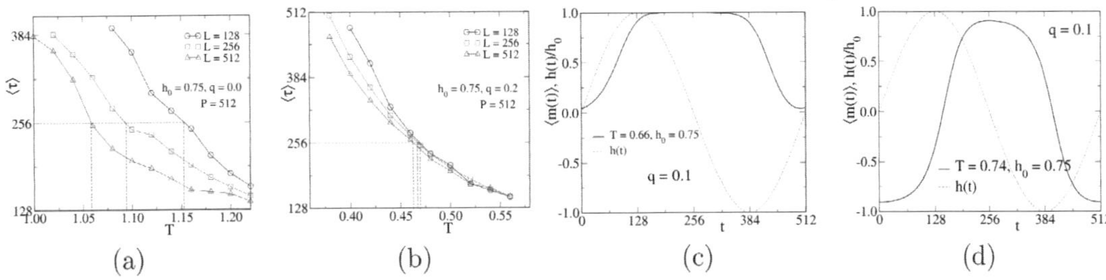

Figure 1. Variation of $\langle\tau\rangle$ against T different system sizes for (a) regular lattice $q = 0.0$ and (b) disorder $q = 0.20$. Instantaneous magnetization $\langle m(t)\rangle$ with MC step t over a complete cycle of the magnetic field $h(t)$ for $P = 512$ at temperatures (c) $T = 0.66$ (DOP) and (d) $T = 0.74$ (DDP). The dashed line in red represents the external sinusoidal field. The external field $h_0 = 0.75$ and system size $L = 512$ for all the figures.

depends on h_0 and L apart from the T. By tuning T keeping P fixed, here we are indeed changing $\langle\tau\rangle$. In this work, we are going to carry out a system size L dependent MC study on DPT in diluted Ising ferromagnet varying the temperature T far below the respective $T_c(q)$, keeping the period $P = 512$ and the field amplitude $h_0 = 0.75$ fixed. The simulations are repeated for all lattices for $L = 128, 256$ and 512.

2.1. Metastable lifetime and critical point

The system reaches a constant magnetization $m(t)$ at a constant external field h_0 at a given temperature T. Then, we reverse the field to $-h_0$ and count the numbers of MC steps it takes to reach zero magnetization that gives the metastable lifetime. The mean metastable lifetime $\langle\tau\rangle$ is determined by taking an average over several such mean lifetimes. The root mean square (rms) value of the field amplitude $h_{\rm rms} = h_0/\sqrt{2}$ is taken as constant field as we have used sinusoidal field of amplitude h_0 for the current study of DPT [20]. The variation of $\langle\tau\rangle$ against temperature T is shown for $q = 0.1$ in Fig.1(b) on $L = 512$. The $\langle\tau\rangle$ corresponds to the half-period $P_{1/2}$ at the dynamic phase transition point [17]. Thus, we can estimate an approximate critical temperature $\theta_c(q)$ of DPT corresponding to a $P_{1/2}$ or $\langle\tau\rangle$ value from the $\langle\tau\rangle$ versus T plot. The $\langle\tau\rangle = 256$ corresponds to temperatures 0.7 for disorder $q = 0.1$ which is the approximate value of $\theta_c(q)$ for $h_0 = 0.75$ and $P = 512$. For a fixed period, $\langle\tau\rangle$ is greater than $P_{1/2}$ for $T < \theta_c(q)$ and the system will be in either of the metastable states dynamic ordered phase (DOP). On the other hand, $\langle\tau\rangle$ is less than $P_{1/2}$ for $T > \theta_c(q)$ and the system continuously moves from one metastable state to the other dynamic disordered phase (DDP).

2.2. Instantaneous magnetization

The ensemble average instantaneous magnetization for a given cycle is defined as $\langle m(t)\rangle = \langle \frac{1}{N_s}\sum_{i=1}^{N_s} S_i(t)\rangle$ where $S_i(t)$ represents the value of the spin at the ith lattice site at time t for a fixed disorder q, $N_s = \sum_{i=1}^{L^2} c_i$ is total number of spins. The variation of $\langle m(t)\rangle$ at lower and higher temperatures are shown against MC time step t over a full period in Fig.1(c), and (d) respectively for dilutions $q = 0.1$ for $h_0 = 0.75$ and the period $P = 512$. The plots are showing the systems that were evolved starting from initial the all spins up ($S_i = +1$) configurations. Firstly, the magnetization lags behind the magnetic field at all temperatures. This leads to hysteresis loss in the system [21]. The magnetization remains at $+1$ (or -1 if started from the negative metastable phase) for most of the time at low temperatures. It varies continuously from a positive to a negative value and vice versa at higher temperatures.

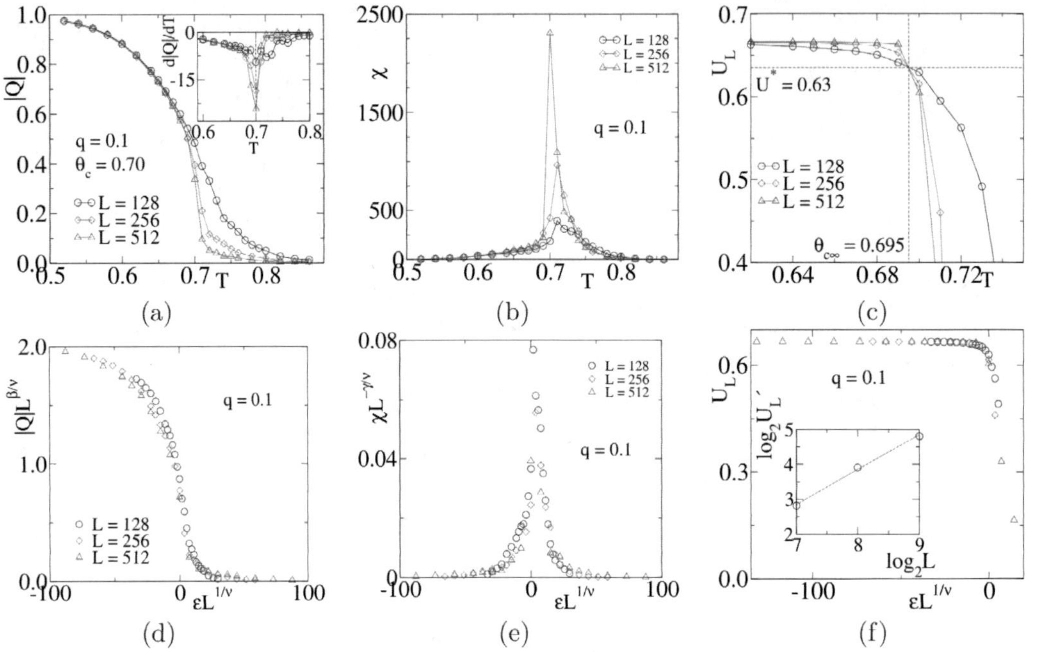

Figure 2. Plot of (a) dynamic order parameters $|Q|$, (b) susceptibility χ, and (c) Binder cumulant U_L with temperature T for $h_0 = 0.75$ and $P = 512$ on system sizes for $q = 0.1$. $d|Q|/dT$ is plotted with T in the insets of (a). The finite size scaled parameters, are plotted against the scaled variable $\varepsilon L^{1/\nu}$ in (d), (e) and (f).

2.3. Dynamic phase transition

The diluted Ising ferromagnet is a bistable system below ferro to para equilibrium phase transition temperature $T_c(q)$. We can define the period average magnetization Q as $Q = \frac{1}{P}\int_0^P \langle m(t)\rangle dt$ where P is the time period of the external field. The system remains in one of the metastable states in DOP, whereas the system moves from one metastable state to the other in a complete cycle in DDP. The system moves from an asymmetric dynamic phase (or DOP) to a symmetric dynamic phase (or DDP) as the temperature T of the system (below the critical temperature $T_c(q)$) is increased from below $\theta_c(q)$ to above $\theta_c(q)$. The DPT occurs around the critical temperature.

The period average magnetization Q is considered to be the dynamic order parameter of DPT. The Q can be either positive or negative in DOP for the two possible metastable states. It is custom to consider the absolute order parameter $|Q|$ to define finite size scaling (FSS) theory. The finite size scaling (FSS) form of $|Q|$ is assumed to be $|Q| = L^{-\beta/\nu}\widetilde{Q}[\varepsilon L^{1/\nu}]$ where $\widetilde{Q}$ is a scaling function. The order parameter exponents is β and the correlation length exponents is ν. The susceptibility χ of the system can be estimated from the fluctuation of dynamic order parameter $|Q|$ as $\chi = L^2(\langle|Q|^2\rangle - \langle|Q|\rangle^2)$. The FSS form of χ is then given by $\chi = L^{\gamma/\nu}\widetilde{\chi}[\varepsilon L^{1/\nu}]$ where $\widetilde{\chi}$ is a scaling function. γ is the susceptibility exponent. The fourth order Binder cumulant is defined as $U_L = 1 - \frac{\langle Q^4\rangle}{3\langle Q^2\rangle^2}$ and its FSS form is given by $U_L = \widetilde{U}[\varepsilon L^{1/\nu}]$ where $\widetilde{U}$ is a universal scaling function. We have at the $T = \theta_c$, $U'_L = \frac{\widetilde{U'}[0]}{\theta_c}L^{1/\nu}$ where $\widetilde{U'}$ is a universal scaling function, and $U'_L = \frac{dU_L}{dT}$.

The $|Q|$, χ, and U_L is plotted with temperature T in the top row of Fig. 2 for disorder $q = 0.1$ and Fig. 3 for disorder $q = 0.2$ on square lattice of sizes $L = 128(\bigcirc)$, $256(\diamond)$ and $512(\triangle)$. The

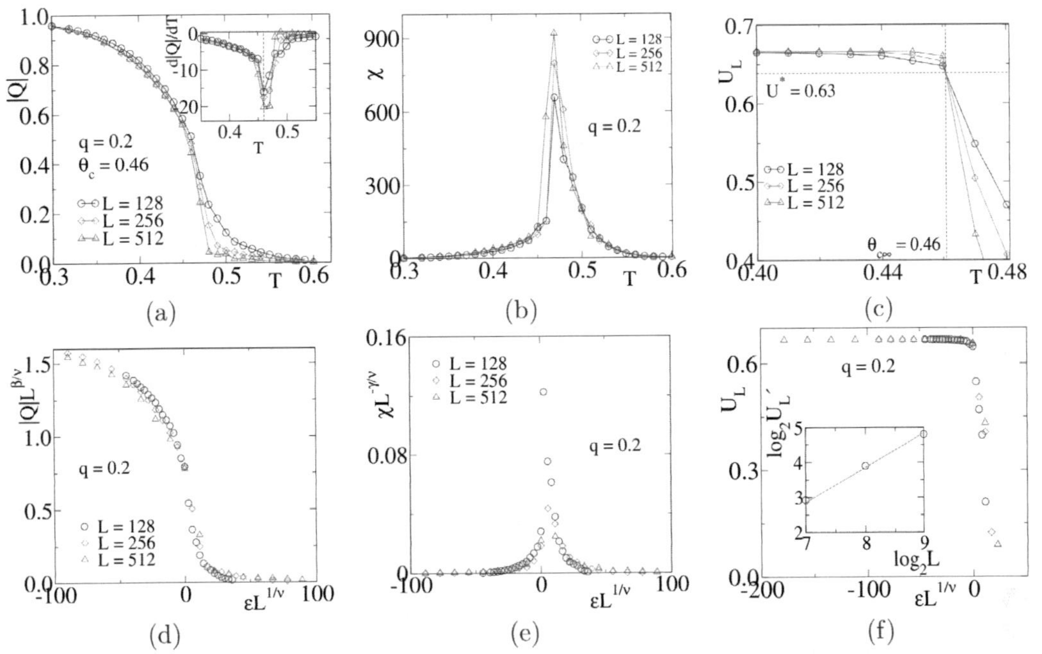

Figure 3. Plot of (a) dynamic order parameters $|Q|$, (b) susceptibility χ, and (c) Binder cumulant U_L with temperature T for $h_0 = 0.75$ and $P = 512$ on system sizes for $q = 0.2$. $d|Q|/dT$ is plotted with T in the insets of (a). The finite size scaled parameters, are plotted against the scaled variable $\varepsilon L^{1/\nu}$ in (d), (e) and (f).

$|Q|$s have strong finite system size dependence and change almost similarly for different q values except they have different $\theta_c(q)$. The value of $\theta_c(q)$ is estimated from the temperature where the change of Q with T *i.e.*, $d|Q|/OT$ is maximum. The $d|Q|/dT$ is plotted with T in the inset and $\theta_c(q)$s are estimated from the trough. The fall of $|Q|$ appears to become more sharper as the dilution increases. The appearance of a peak around the phase transition point is a general observation in the variation of susceptibility with temperature. The peak is the manifestation of the fact that the system goes through huge fluctuation near the phase transition. The largest χ value corresponds to the largest system size in each case as the larger lattice consists of more spins and contributes to maximum fluctuation. The other observation is that the maximum of χ decreases for the largest lattice ($L = 512$) as the dilution q increases. The χ/L^2 for different system sizes are not the same. It indicates the transition is not of the first order or discontinuous type. In each case, U_L corresponds to different system sizes L cross each other at a temperature. The temperature $\theta_{c\infty}$ at which such crossing occurs indicates the phase transition temperature of the system in the thermodynamic limit.

Next, we verify the FSS forms of $|Q|$, χ, and U_L by plotting the respective scaled parameters against the scaled variable $\varepsilon L^{1/\nu}$ in Fig. 2 and Fig. 3 (d), (e), and (f). We have estimated the equilibrium phase transition critical exponents on 2d diluted Ising model and found the exponent values are slightly different from the regular lattice, $\beta/\nu = 0.118 \pm 0.004$, and $\gamma/\nu = 1.76 \pm 0.02$ for $q = 0.1$ and $\beta/\nu = 0.102 \pm 0.006$, and $\gamma/\nu = 1.78 \pm 0.02$ for $q = 0.1$ in accordance with the weak universality hypothesis [10, 22, 23, 24]. The scaling exponents satisfy the relation $2(\beta/\nu) + \gamma/\nu = 2$ approximately. A reasonably good data collapse is obtained for dilution $q = 0.1$, and 0.2 using the respective equilibrium critical exponents on 2d diluted Ising system. The value of $\nu = 1$ is taken for both the cases. The U'_L at θ_c is plotted with different L in double logarithmic scale for regular lattice in the inset of Fig. (f) in symbols. The solid red line shown corresponds to $\nu = 1$ line.

q	$\theta_c(q)$	$\theta_{c\infty}(q)$	U^*
0.0	1.06 ± 0.01	1.06 ± 0.01	0.61
0.1	0.70 ± 0.01	0.70 ± 0.01	0.63
0.2	0.46 ± 0.01	0.46 ± 0.01	0.63

Here we found the universal values U^* slightly increased with the disorder. Heuer reports a similar observation in the case of the diluted equilibrium Ising model [10].

3. Conclusion

The nonequilibrium DPT is studied varying the temperature on the 2d site diluted Ising ferromagnet in a large scale Monte Carlo study on several system sizes $L = 128, 256$, and 512. The DPT occurs when metastable lifetime $\tau(h_0, T)$ becomes comparable to half period of the external field. We found a critical temperature $\theta_c(q)$ separating the DOP and DDP. The DPT results show that the $\theta_c(q)$ value decreases with increasing q. The $m(t)$ follows $h(t)$ with a certain phase lag. The DPT on different diluted Ising ferromagnets was found to be of continuous type. The order parameter $|Q|$ and susceptibility χ obey the finite size scaling with q-dependent exponents. Binder cumulant U_L plot shows the crossing point of different L at the respective $\theta_c(q)$ for different q.

References

[1] Kaul S 1985 *J. Magn. Magn. Mater* **53** 5 – 53
[2] Ballesteros H G, Fernández L A, Martín-Mayor V, Muñoz Sudupe A, Parisi G and Ruiz-Lorenzo J J 1998 *Phys. Rev. B* **58**(5) 2740–2747
[3] Martins P H L and Plascak J A 2007 *Phys. Rev. E* **76**(1) 012102
[4] Kim J K and Patrascioiu A 1994 *Phys. Rev. Lett.* **72**(17) 2785–2788
[5] Kenna R and Ruiz-Lorenzo J J 2008 *Phys. Rev. E* **78**(3) 031134
[6] Mazzeo G and Kühn R 1999 *Phys. Rev. E* **60**(4) 3823–3836
[7] Prudnikov V V and Markov O N 1995 *Eur. Phys. Lett.* **29** 245–250
[8] Harris A B 1974 *J. Phys. C: Solid State Physics* **7** 1671–1692
[9] Ballesteros H G, Fernández L A, Martín-Mayor V, Sudupe A M, Parisi G and Ruiz-Lorenzo J J 1997 *J. Phys. A: Math. Gen.* **30** 8379–8383
[10] Heuer H O 1992 *Phys. Rev. B* **45**(10) 5691–5694
[11] Heuer H O 1993 *J. Phys. A: Math Gen* **26** L333–L339
[12] Heuer H O 1990 *Phys. Rev. B* **42**(10) 6476–6484
[13] Ümit Akıncı, Yüksel Y, Vatansever E and Polat H 2012 *Physica A Stat. Mech. Appl* **391** 5810 – 5817
[14] Vatansever E, Ümit Ak nc, Yüksel Y and Polat H 2013 *J. Magn. Magn. Mater* **329** 14 – 23
[15] Vatansever E and Fytas N G 2018 *Phys. Rev. E* **97**(6) 062146
[16] Glauber R J 1963 *J. Math. Phys.* **4** 294–307
[17] Sides S W, Rikvold P A and Novotny M A 1998 *Phys. Rev. E* **57**(6) 6512–6533
[18] Vatansever E and Fytas N G 2018 *Phys. Rev. E* **97**(1) 012122
[19] Vasilopoulos A, Vatansever Z D, Vatansever E and Fytas N G 2021 *Phys. Rev. E* **104**(2) 024108
[20] Sides S W, Rikvold P A and Novotny M A 1999 *Phys. Rev. E* **59**(3) 2710–2729
[21] Chakrabarti B K and Acharyya M 1999 *Rev. Mod. Phys.* **71**(3) 847–859
[22] Fytas N G and Theodorakis P E 2010 *Phys. Rev. E* **82**(6) 062101
[23] Marro J, Labarta A and Tejada J 1986 *Phys. Rev. B* **34**(1) 347–349
[24] Prudnikov V V, Prudnikov P V, Vakilov A N and Krinitsyn A S 2007 *J. Exp. Theor. Phys.* **105** 371–378

Molecular dynamics study of a nano-scale β-type Stirling engine

K Kitaya and M Isobe

Graduate School of Engineering, Nagoya Institute of Technology, 466-8555, Japan

E-mail: E-mail: cmm12027@ict.nitech.ac.jp, isobe@nitech.ac.jp

Abstract. A Stirling engine is based on a thermodynamic cycle; a temperature difference causes flywheel rotation. Because the arbitrary (stochastic) heat resources contact some of the systems that generate the work (i.e., the energy), Stirling engines remain important in the present era of Sustainable Development Goals because of their broad applicability. This study focuses on a nano-scale "β-type" Stirling engine, specifically concerning a model for numerical simulation. We perform molecular dynamics simulation of the two-dimensional model (using hundreds of particles) and calculate the thermal efficiency. We figured out the minimal necessary conditions for stable rotation, as well as the lower limits of particle numbers and temperature differences for autonomous thermodynamic cycles.

1. Introduction

Heat engines have had considerable impacts. Both external and internal combustion engines drove the industrial revolution and the ubiquity of automobiles.Today, it is imperative to consider heat engines from the "ecological" perspective (i.e., the effective use of limited resources). Although the Stirling engine (SE) was invented almost two centuries ago, it continues to attract attention[1]. First, an SE exploits the temperature difference between the edges of a cylinder; any source of heat can be employed.This distinguishes SEs from gasoline and diesel engines. Second, an ideal Stirling cycle (operating under quasi-static conditions) affords the highest possible thermal efficiency, which is identical to Carnot efficiency. Third, an SE features external combustion; the driving energy (heat) is provided outside the cylinder. This optimizes efficiency.Furthermore, air pollutants are not emitted.Recently, Toyabe and Izumida experimentally investigated the non-equilibrium dynamics and thermodynamics of a low-temperature-differential Stirling engine [2].The driving thermodynamic force and underlying dynamics of the Stirling cycle could be described using the simple dynamic model established by Izumida [3]. The equations of this simple dynamic model have been extensively used in the field of bifurcation; there are only two degrees of freedom. Earlier, the nano-scale (α-type) SEs of particle systems were investigated via molecular dynamics (MD) simulations [4]. The effects of changes in the particle numbers and temperature differences on stable flywheel rotation and thermal efficiency were studied. However, the β-type SE is much more compact than the α-type SE because, in the β-type SE, both the displacer and power pistons lie on the same axis.

Our purposes here are to establish a simple numerical model of an β-type SE using a finite particle system, while defining the minimal necessary conditions for stable rotation and the lower limits of both the temperature difference and the particle number in an infinite system.

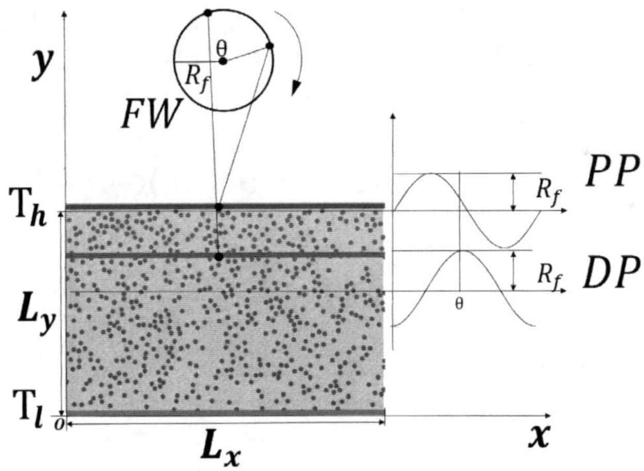

Figure 1. A schematic of a nano-scale (β-type) SE in two dimensions. Two rods are attached to a clockwise-rotating flywheel (FW) with a phase angle difference of $\pi/2$; they are connected to the power piston (PP) and displacer piston (DP), respectively. The elastic particles collide. They also collide with the thermal walls at the bottom and top of the cylinder. The wall temperatures are T_l and T_h, respectively. The regenerator is modeled as part of the DP function.

We use a simplified model of a displacer piston with regenerators; we introduce semi-transparent, asymmetric heat reservoirs. Such a regenerator is expected to enhance thermal efficiency. In Sec. 2, our model and numerical settings are described. Sec. 3 describes the simulations exploring rotational stability and thermal efficiency; it presents the phase diagram. Finally, concluding remarks appear in Sec. 4.

2. Model and simulation methods

To investigate the thermodynamic cycles of a β-type SE at the molecular level, we used the model shown in Fig. 1. Numerical simulation proceeded in two dimensions (2D) (for convenience). The working substance is confined to the cylinder (the rectangular box of area $A = \langle A(t)\rangle_t = L_x \times L_y(\langle\rangle_[wheret$ is the time average]). The systems with the aspect ratio $\alpha = L_y/L_x$ compose of N monodisperse elastic disks of diameter σ. Particles interact solely via repulsive forces within the cutoff distance $r < r_c(= 2^{\frac{1}{6}}\sigma)$ of the Weeks-Chandler-Andersen (WCA) potential [5]: $U(r) = 4\varepsilon[(\sigma/r)^{12} - (\sigma/r)^6] + \varepsilon$. The three basic particle units are the mass m, the diameter σ, and the energy ε in the WCA potential; these contribute to the potential energies of particles in contact. Note that condensation into liquid is avoided; there is no between-particle attractive force. The flywheel (FW) of radius $R_f = 10.0$ and moment of inertia $I_f = M_f R_f^2/2$ is driven by work fluids (composed of particles) moved by the PP and DP. The phase angle difference between the PP and DP is held at $\pi/2$ (see Fig. 1). The masses of the PP and DP are fixed at $M_p = 2 \times 10^4/\alpha$. Movement clockwise from the origin θ is regarded as positive FW rotation; the origin is the point of lowest DP location. Two thermal walls at temperatures T_h (= 1.0: fixed) and T_l are placed at the top (PP) and bottom of the cylinder; for both walls, the Boltzmann constant k_B is unity [6]. In general, when working fluid passes through the regenerators, cold fluid from one side of the cylinder (separated by the displacer from the hot fluid) efficiently gathers heat from the hot fluid on the other side of the cylinder. This can be modeled as a part of the DP function by introducing a probability of particle transparency through the DP. Particles pass through the DP with a probability $p = 0.5$; they gain or lose velocity if their respective temperatures are higher or lower than the hot top or bottom cylinder regions separated by the DP. The effect is thus regenerative. The total heat input to the cylinder is simply estimated by measuring the numbers of collisions with the thermal walls during simulation; the accumulated input heat is Q_{in}. The work W is $\gamma R_f \omega \Delta\theta$, where $\gamma R_f \omega$ is the torque load, $\gamma = 100.0$ is the coefficient of the torque load, and ω and $\Delta\theta$ are the angular velocity and the accumulated displacement of the phase angle, respectively. In the stationary state, the thermal efficiency e is $e = W/Q_{\text{in}}$. We performed the time-step-driven standard molecular dynamics (MD) simulation integrated in the velocity Verlet scheme for N particle systems [7]. The equation of motion for FW rotation is:

$$I_f \frac{d\omega}{dt} = F_R - R_f \gamma \omega \tag{1}$$

where F_R is the constraint force induced by the connecting rods. The number density $\rho(= N/A)$ is set to 0.2 to avoid crystallization. The ratio of the potential to total energy is approximately $0.02 \sim 0.05$, which is sufficiently small to ensure that the particles behave as an ideal gas. No external force (e.g., gravity) and no dissipation between particles are considered. Note that the periodic boundary conditions lie in the horizontal direction.

3. Results

3.1. Stationary angular FW velocity

Figure 2 shows the evolution of angular velocity averaged over a cycle $\tilde{\omega}$ in terms of the number of flywheel rotations N_{cycle}; each simulation at different temperatures $\Delta T = T_l - T_h$ ($T_h = 1.0; fixed$) commences at the same initial angular velocity $\omega_0 = 0.01$ and is performed in up to 4×10^{10} steps with a time width 0.003. We confirmed the stationary behavior of $\tilde{\omega}$ (there was no drift). Furthermore, the behavior was independent of the ω_0 behavior during long runs, but it increased as ΔT increased. It is reasonable that the large rotational kinetic energy (i.e., angular velocity) of a flywheel requires a large ΔT.

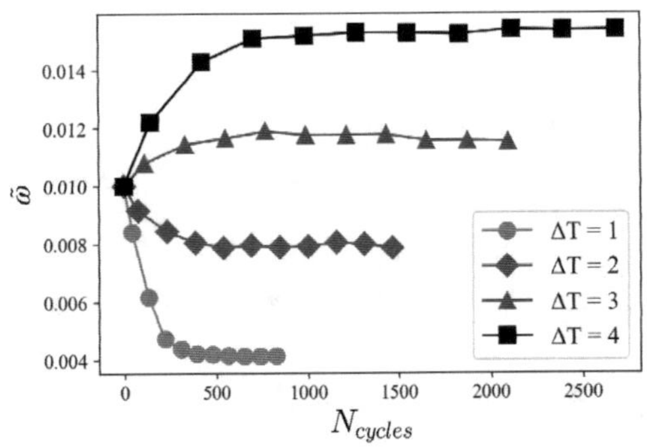

Figure 2. The evolution of the averaged angular velocity $\tilde{\omega}$ by the temperature difference ΔT when T_l changes, as a function of the number of flywheel cycles N_{cycle}. Each simulation commenced at the same initial ω_0 and attained the original stationary values of $\tilde{\omega}$.

3.2. Minimal necessary conditions for stable rotation

To derive the minimal necessary conditions for stable FW rotation, we considered the following two situations.

(A) To establish the minimal necessary conditions at smaller $\Delta T = \Delta T_c$, T_l was changed systematically in $N = 600$ systems.

(B) Based on the results of (A), we derived the minimal necessary conditions at a smaller particle number N_c as the aspect ratio α changed but ρ and ΔT_c were fixed.

The two critical values at $(\Delta T_c, N_c)$ constitute the minimal necessary conditions of our nano-scale SE thermodynamic cycle.

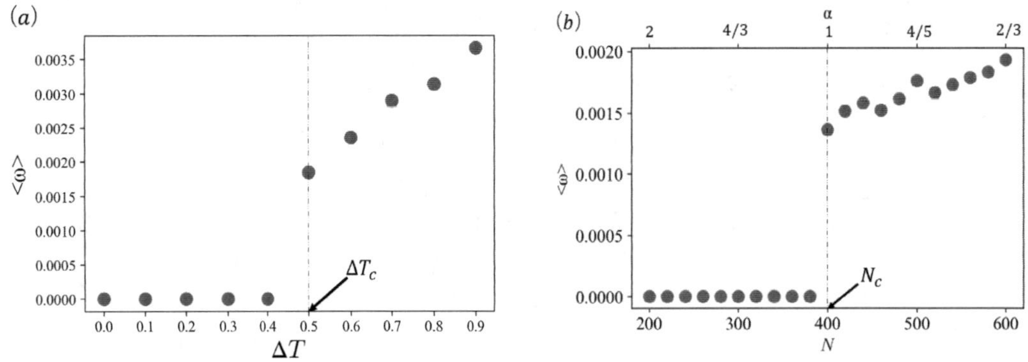

Figure 3. The time-averaged angular velocity $\langle\omega\rangle$ in the steady state as functions of (a) the temperature difference ΔT for $N = 600$, and (b) the particle number N (or aspect ratio α) for $(\rho, \Delta T) = (0.2, 0.5)$.

The lower boundary of ΔT for stable FW rotation is $\Delta T_c \simeq 0.5$ in Fig. 3(a). Above ΔT_c, the stable FW angular velocity increases in a nearly linear manner in the range of $\Delta T = 0.5 \sim 0.9$. Next, using a fixed $(\rho, \Delta T_c) = (0.2, 0.5)$, long-time simulations were performed by changing the aspect ratio α. We show the α dependence on stable FW rotation in Fig. 3(b), in which the lower limit is approximately $N \sim 400$. These are the minimal necessary conditions for stable rotation (i.e., thermodynamic cycle) in terms of both the temperature difference ΔT and the particle number N.

3.3. Deviation of angular velocity from the average

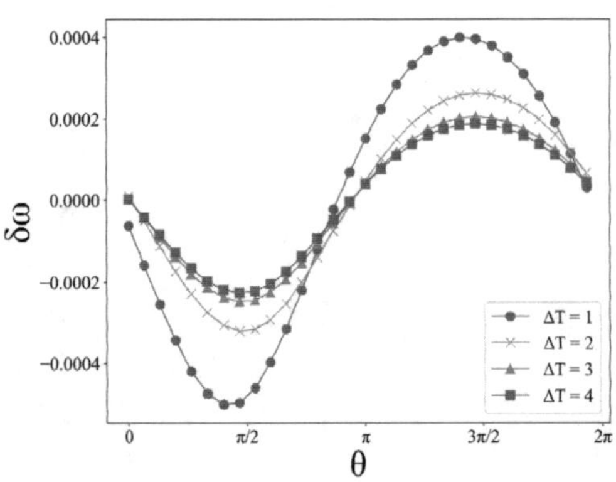

Figure 4. Deviations of the angular velocity from the average $\delta\omega$ are plotted as a function of θ. (the phase angle of the FW)

As shown in Ref. [4], the angular velocity ω fluctuates around the average (in a sinusoidal curve) over a single cycle in a finite system. However, in a nano-scale (finite particle number) system, it would be expected to depend on the FW phase angle θ [4]. To explore the dependence on different ΔT, $\delta\omega(\theta) = \omega(\theta) - \langle\omega\rangle$ were calculated. Figure 4 shows that $\delta\omega$ attains its lowest and highest values at approximately $\pi/2$ and $3\pi/2$, respectively. These respective phase angles correspond to the highest and lowest locations of the PP during a sinusoidal cycle. The amplitude decreases as ΔT increases.

3.4. Thermal efficiency

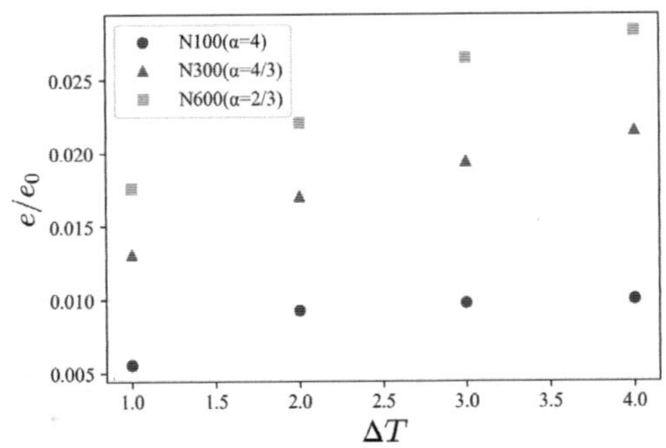

Figure 5. The normalized thermal efficiencies e/e_0 of our model at different system sizes ($N = 100$, 300 and 600) as functions of the temperature difference ΔT.

Fig. 5 shows the thermal efficiencies of our model with different system sizes $N = 100$, 300 and 600 (i.e., the aspect ratio α changes from 4 to 2/3) as functions of the temperature difference ΔT. The vertical axis of e/e_0 is the relative thermal efficiency, where e_0 is the ideal thermal efficiency estimated by $e_0 = 1 - T_h/T_l$. Note that the density dependence of thermal efficiency for a α-type SE was explored in Ref. [4]. In that calculation, the number density ρ was fixed but the width of the cylinder (i.e., the aspect ratio α) changed. Thermal efficiency was higher in larger systems and increased in a generally linear manner with the temperature difference ΔT.

3.5. Pressure diagram

Finally, we calculated the pressure as a function of the area inside the cylinder during the FW cycle. To compare systems with different parameters, we introduced a normalized area $\tilde{A} = (A - A_0)/A_1$, where A is the bulk area inside the cylinder, while A_0 and A_1 are the areas at the lowest and highest PP locations, respectively. The pressure at each time is calculated using the simple equation ($P^* = Nk_BT/A$), assuming that the working fluid is an ideal gas and that the cycles are governed by a quasi-static process. In Fig. 6, the typical pressure curves of $\tilde{A}$ averaged over 200 cycles are shown: (a) $N = 200$ and 600 with $\Delta T = 3$ and (b) $\Delta T = 1$ and 3 in $N = 600$. We confirmed the closed-circuit curves, where their area corresponds to the work.

4. Concluding remarks

We established a simple numerical model of a β-type SE with a DP operating in a finite particle system. We figured out the minimal necessary conditions for stable rotation, as well as the lower limits of both the difference in temperature ΔT and the particle number N. We confirmed that FW rotation stabilizes at a stationary angular velocity independent of the initial angular velocity, which becomes larger as the temperature difference ΔT increases (Sec. 3.1). Notably, the SE cycle operates even in nano-scale, finite particle systems. We also derived the minimal necessary conditions for stable rotation (in Sec. 3.2). A comparison with the MD simulation of an α-type SE [4] revealed that our model exhibited the same trends in terms of angular velocity deviation

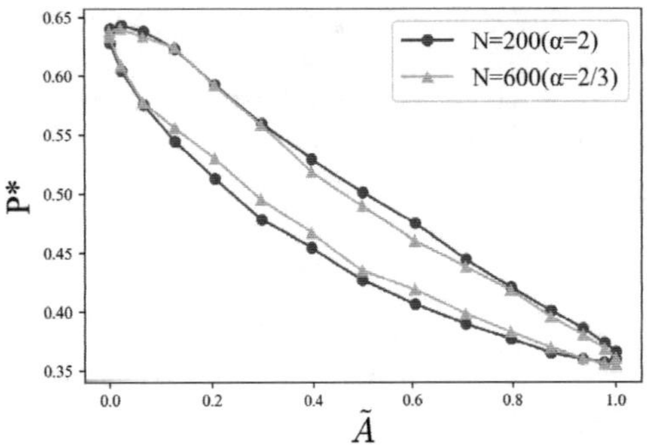

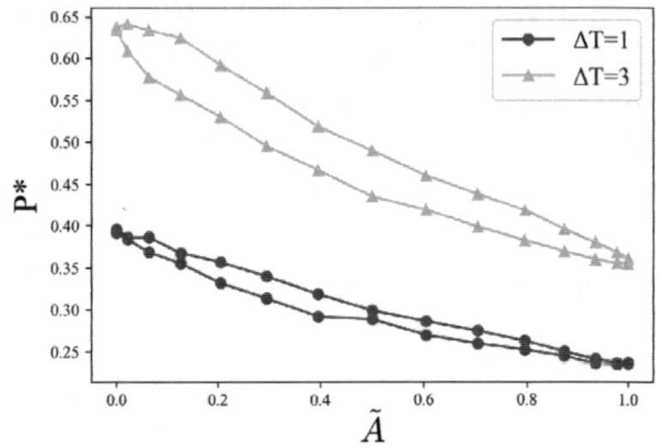

Figure 6. Pressure diagrams as functions of the normalized area of the cylinder for (a) $N = 200$ and 600 with $\Delta T = 3$ and (b) $\Delta T = 1$ and 3 for $N = 600$. The area within the closed circuit generates the work.

from the average (Sec. 3.3). We also explored thermal efficiency in terms of ΔT and N (Sec. 3.4). Finally, the work generated by SE cycles in finite time was illustrated in a pressure diagram (Sec. 3.5). The primary differences between α- and β- type SEs are the number of cylinders and the roles played by the pistons (e.g., two PPs in the α-type, but one PP and one DP in the β-type). In general, the thermodynamic cycles have been established in infinite time at the limit of the continuum. It is imperative to use numerical simulation to reveal the microscopic mechanisms of the stationary thermodynamic cycles and the required conditions. This would greatly aid the theoretical exploration of finite particle systems operating in finite time, as well as the development of microdevices driven by thermal reservoirs.

Acknowledgments

We are grateful to Dr. Y. Izumida for the helpful discussion. M.I. was supported by JSPS KAKENHI Grant Number No. 20K03785. Part of the computations was performed using the facilities of the Supercomputer Center, ISSP, Univ. of Tokyo.

References

[1] Senft J R 2010 *An Introduction to Stirling Engines* (Wisconsin: 8th edition, Moriya Press)
[2] Toyabe S and Izumida Y 2020 *Phys. Rev. Research* **2** 033146
[3] Izumida Y 2018 *Europhys Lett.* **121** 50004
[4] Rapaport D C 2009 *Phys. Rev.* E **79** 047702
[5] Weeks J D, Chandler D and Andersen H C 1971 *J. Chem. Phys.* **54** 5237-47
[6] Tenenbaum A and Ciccotti G and Gallico R 1982 *Phys. Rev. A* **25** 2778
[7] Rapaport D C 2004 *The Art of Molecular Dynamics Simulation* (Cambridge: 2nd ed. Cambridge University Press)

Quantifying uncertainty with stochastic collocation in the kinematic magentohydrodynamic framework

E. Rajbhandari[1,2], N.L. Gibson[1] and C. R. Woodside[2]

[1] Department of Mathematics, Oregon State University

[2] National Energy Technology Laboratory, Albany, OR

E-mail: rajbhane@oregonstate.edu

Abstract. We discuss an efficient numerical method for the uncertain kinematic magnetohydrodynamic system. We include aleatoric uncertainty in the parameters, and then describe a stochastic collocation method to handle this randomness. Numerical demonstrations of this method are discussed. We find that the shape of the parameter distributions affect not only the mean and variance, but also the shape of the solution distributions.

1. Introduction

Magnetohydrodynamics (MHD) is the study of an electrically-conductive medium flowing through a magnetic field [1]. It is a multi-physics problem, governing the behavior of fluid flow, electric fields and currents, magnetic fields, and their interactions. MHD has applications in many different areas of study, such as astrophysics [2], medicine [3] and power generation [1].

These different applications operate under separate conditions, and thus varying assumptions are made on the system. We focus on the power-generation capabilities of MHD, which involves applying a large magnetic field to an electrically conductive plasma, artificially creating an electric field [1]. Following related work to simplify computational complexity [4], we prescribe the fluid-flow, and focus on the electromagnetic behavior of the MHD system. This model is called the kinematic MHD equations, further described in Section 2 and [5].

We are ultimately interested in investigating the feasibility of real-time optimization of an MHD generator. The parameters of the system, including fluid velocity, conductivity, electron mobility, and ion mobility, need to be estimated via observations. This introduces aleatoric uncertainty into the system, as the recovered parameters are then described by probability distributions. Therefore, the optimal design of the generator must consider uncertainty, as each random parameter is governed by some estimated distribution. Related work has investigated the well-posedness of the forward problem under uncertainty [4]. We propose here a non-intrusive approach to quantifying the uncertainty in the forward problem using a numerical model, implemented in COMSOL [13], that has already been validated [5]. This will be incorporated into an optimization under uncertainty problem in future work.

There are many different approaches for dealing with uncertainty in a system in general. Although simplistic for implementation, other methods, such as Monte Carlo, come with a high computational burden [6]. The method we utilize lessens the computational burden both through the choice of points in which we sample the domain, and the way in which we choose to average the solutions. In the following work, we implement an approximation method called *stochastic collocation* (SC) [7]. Utilizing a sparse grid with SC can result in a diminishing of the 'curse of dimensionality' that often affects uncertainty quantification, while maintaining a high degree of accuracy [8], which will allow the utilization of the model in the inverse problem under uncertainty.

We begin by stating the well-posedness of the system under uncertainty. Following, we will go into detail regarding the SC method and the random grid. Finally, we end with a numerical demonstration of the robustness of the method for a variety of different situations, including several complex joint probability distributions of the random variables within the system. These demonstrations will also show that the expected solutions and deterministic solutions vary significantly enough to warrant an approach to optimal design under uncertainty, and that the shape of the solutions' distributions depend on the parameters in the input distributions.

2. Model

For the purposes of this paper, we will not focus on any specific generator model or component, and work in general with the MHD equations. Let the spatial domain be given by $D \subset \mathbb{R}^3$, open with compact closure, and denote the boundary as ∂D. We denote the random space $(\Omega, \mathcal{H}, p)$, where Ω is the set of outcomes, $\mathcal{H}$ is a given sigma algebra of events, and p is some continuous probability measure. As noted in the introduction, we work on the kinematic MHD system, with prescribed fluid-flow, $\mathbf{u}$. We assume that the induced magnetic field is negligible compared to the applied magnetic field [9]. Under the generator model, we then prescribe conductivity, σ, applied magnetic field, $\mathbf{B}$, electron-mobility, μ_e, and ion-mobility, μ_i. We use the standard definition of the l^2 norm of a vector in $\mathbb{R}^3$, as well as the hall parameter and ion-slip parameter, e.g. for $\mathbf{x} \in D,\ \omega \in \Omega$

$$\beta_e(\mathbf{x}, \omega) = \mu_e(\mathbf{x}, \omega)||\mathbf{B}(\mathbf{x})||_{l^2}, \text{ and } \beta_i(\mathbf{x}, \omega) = \mu_e(\mathbf{x}, \omega)\mu_i(\mathbf{x}, \omega)||\mathbf{B}(\mathbf{x})||_{l^2}^2.$$

Finally, we use the conductivity tensor, as defined in [5], given by

$$\underline{\sigma}(\mathbf{x}, \omega) = \sigma(\mathbf{x}, \omega)\left(\mathcal{I} - \frac{\beta_e(\mathbf{x}, \omega)}{||\mathbf{B}||_{l^2}}[\mathbf{B}]_\times - \frac{\beta_i(\mathbf{x}, \omega)}{||\mathbf{B}||_{l^2}^2}[\mathbf{B}]_\times^2\right)^{-1},$$

where $\mathcal{I}$ denotes the identity matrix in $\mathbb{R}^{3\times3}$ and $[\mathbf{B}]_\times$ is the matrix form of the cross-product. Invertibility of this matrix is guaranteed by the physical restriction $\mu_e, \mu_i > 0$. We now turn to defining the random solution spaces. First, define the random function space

$$L_{2,p}(\Omega) := \left\{f : \Omega \to \mathbb{R} \,\middle|\, \int_\Omega f^2(\omega)p(\omega)\, d\omega < \infty\right\}, \text{ with norm } ||f||_{L_{2,p}(\Omega)} := \left(\int_\Omega f^2(\omega)p(\omega)\, d\omega\right)^{1/2}.$$

Next, define the deterministic solution spaces for the induced current density, $\mathbf{J}_i$, and the electric potential, $\mathcal{V}$, respectively

$$V(D) := \left\{\mathbf{f} \in (L^2(D))^3 : \mathbf{f}{\cdot}\mathbf{n} = -(\underline{\sigma}(\mathbf{u}{\times}\mathbf{B})){\cdot}\mathbf{n} \text{ on } \partial D\right\}, \quad W(D) := W_0^{1,2}(D) = \left\{f \in H^1(D) : T(f) = 0\right\},$$

where $\mathbf{n}$ is a vector normal to the boundary and $T(f)$ is the trace of f on D. Finally we define the *random solution space* for $\mathbf{J}_i$ as the tensor product between the deterministic and random function spaces, or

$$\overline{V} := V(D) \times L_{2,p}(\Omega) = \left\{f : D \times \Omega \to \mathbb{R}^3 \,\middle|\, f(\cdot, y) \in \left(L_{2,p}(\Omega)\right)^3,\ f(\mathbf{x}, \cdot) \in V\right\}.$$

Similarly, define $\overline{W} := W(D) \times L_{2,p}(\Omega)$ for $\mathcal{V}$. We define the norm on $\overline{V}$ as the averaging norm, i.e.

$$||F||_{\overline{V}} = \left(\mathbb{E}\left[||F||_V^2\right]\right)^{1/2}$$

and define the norm on $\overline{W}$ to be the averaging norm using $||\cdot||_W$. Also define the expected value of a real-valued function to be $\mathbb{E}[f] := \int_\Omega f(\omega)p(\omega)\, d\omega$. With these spaces, the kinematic weak-form MHD equations are given by: find $\mathbf{J}_i \in \overline{V}, \mathcal{V} \in \overline{W}$ that satisfy

$$\mathbb{E}\left[\int_D \underline{\sigma}^{-1}\mathbf{J}\cdot\phi\right] - \mathbb{E}\left[\int_D \nabla\mathcal{V}\cdot\phi\right] = 0\ \forall\phi\in\overline{V}, \tag{1a}$$

$$-\mathbb{E}\left[\int_D \mathbf{J}_i\cdot\nabla\psi\right] = \mathbb{E}\left[\int_D (\underline{\sigma}(\mathbf{u}\times\mathbf{B}))\cdot\nabla\psi\right]\ \forall\psi\in\overline{W}. \tag{1b}$$

We denote the form (1) the *random weak form.* In the following work, to guarantee well-posedness, assume that $\forall\omega\in\Omega$, $\mathbf{u}(\omega), \mathbf{B}$ are in $(L^2(D))^3$ and bounded, and $\sigma(\omega), \beta_e(\omega)$ and $\beta_i(\omega)$ are positive, bounded, and real-valued on D. Then there exist unique solutions, $\mathbf{J}_i, \mathcal{V}$, to (1), that depend continuously on the parameters [4]. With the governing equations defined and well-posedness established, we turn to the focus of this paper, the numerical approximation of the solutions.

3. Polynomial Chaos

As is a standard approach when solving systems numerically, we would like to search for our solutions on some finite-dimensional representation of the solutions spaces. We do so by projecting onto the space of polynomials of random variables. We denote this projection a *polynomial chaos expansion.* However, we require that this polynomial space has finitely-many directions, which leads to the first major assumption required.

3.1. Finite-Dimensional Noise

To implement the SC method, we must first assume there are a finite number of random variables describing the noise [10]. To this end, we assume that the only random variables in the system are the real-valued random parameters μ_e, μ_i, σ, and $\mathbf{u}$. Furthermore, assume each are described by a finite number of independent random variables.

Let Γ be the tensor product of the image of the events under each random variable. Let ρ be the joint independent probability distribution. Then by the Doob-Dynkin's Lemma [11], we have that the solutions $\mathbf{J}_i, \mathcal{V}$ can be described by a finite number of random variables. The numerical problem then equates to approximating $\mathbf{J}_i(\mathbf{x}, y)$, $\mathcal{V}(\mathbf{x}, y)$ for $\mathbf{x}\in D$, $y\in\Gamma$. Analogous to the above problem, we attempt to find $\mathbf{J}_i \in V\times L_{2,\rho}(\Gamma)$, $\mathcal{V}\in W\times L_{2,\rho}(\Gamma)$ that satisfy

$$\int_D \underline{\sigma}^{-1}\mathbf{J}_i(\mathbf{x},y)\cdot\phi(\mathbf{x},y)\,d\mathbf{x} - \int_D \nabla\mathcal{V}(\mathbf{x},y)\cdot\phi(\mathbf{x},y)\,d\mathbf{x} = 0 \quad \forall\phi\in V,\ \text{for } \rho.a.e.\ y\in\Gamma \tag{2a}$$

$$-\int_D \mathbf{J}_i(\mathbf{x},y)\cdot\nabla\psi(\mathbf{x},y)\,d\mathbf{x} = \int_D \underline{\sigma}(\mathbf{x},y)\big(\mathbf{u}(\mathbf{x},y)\times\mathbf{B}(\mathbf{x})\big)\cdot\nabla\psi(\mathbf{x},y)\,d\mathbf{x} \quad \forall\psi\in W,\ \text{for } \rho.a.e.\ y\in\Gamma \tag{2b}$$

For notational convenience, we define $\widetilde{V} := V\times L_{2,\rho}(\Gamma)$ and $\widetilde{W} := W\times L_{2,\rho}(\Gamma)$ as the *new* random solution spaces for which we seek a numerical approximation. Note now that this is equivalent to (1), only with the alternative probability space (Γ, H, ρ), where H is the appropriately defined sigma algebra.

We now define the finite-dimensional (FD) random solution subspaces in which we search for our approximate solutions. We begin with the spatial dimension. Define $V_h\subset V$ to be the standard finite element approximation to V, with quadratic polynomials, on some Delauney triangular prism mesh t_h with max side length h. Similarly define $W_h\subset W$, on the same mesh t_h. For the random solution space, define $P_{N_k}(\Gamma_k)\subset L_{2,\rho}(\Gamma_k)$ for $k = 1,\ldots,M$ as the span of all polynomials on Γ_k of degree up to N_k, for $N_k\in\mathbb{N}$. In each direction Γ_k, we choose to

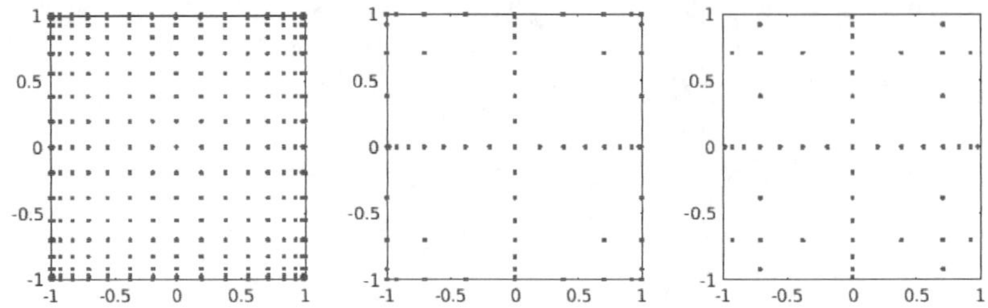

Figure 1: Quadrature grids with level = 5. (From left to right) CC Full: 289 nodes, CC Sparse: 65 nodes, Fejér Sparse: 55 nodes. Constructed using Burkhardt's repository [12].

use the basis of Legendre polynomials $\{r_k^j\}_{j=0}^{N_k}$ [7], which are orthogonal with respect to the uniform density function, e.g., $\int_{\Gamma_k} r_k^j r_k^l \, dy = \delta_{jl}$, where δ_{jl} is the Dirac delta function. Define $N = [N_1, \ldots, N_M]$. Then we approximate our random $L_2(\Gamma)$ space with a tensor product of these polynomial spaces, i.e. $P_N(\Gamma) := \prod_{k=1}^{M} P_{N_k}(\Gamma_k)$. Note that the dimension of P_N is clearly finite. The FD approximation to the random solution spaces are thus given by:

$$\widetilde{V}_{h,N} := V_h \otimes P_N(\Gamma) \text{ and } \widetilde{W}_{h,N} := W_h \otimes P_N(\Gamma).$$

With these spaces, we now describe the SC method in detail, defining the finite-dimensional polynomial chaos expansion.

3.2. Stochastic Collocation Method

We begin the collocation by solving the deterministic equivalent system, for fixed y, to (2). These will then used to build an interpolatant. By solving for fixed y, we treat $\mathbf{J}_i^h$ and $\mathcal{V}^h$ as mappings from Γ to their respective deterministic spaces, i.e. $\mathbf{J}_i^h : \Gamma \to V_h$, $\mathcal{V}^h : \Gamma \to W_h$. We perform the *collocation* by averaging the solutions sampled at the zeros of the polynomial basis of $P_{N_k}(\Gamma_k)$ for each $k = 1, \ldots, M$. By using the Legendre polynomials, we are able to use the Clenshaw-Curtis (CC) or Fejér method of numerical quadrature, which guarantees nesting of the zeros in each random direction [7]. Furthermore, we reduce the number of points, by constructing a Smolyak sparse grid [8]. Although there is not a closed-form method of giving the number of nodes required in each domain, we let $\tilde{N}_k$ denote the total number of points in the Γ_k direction. A more detailed discussion of the construction of such grids can be seen in [8] or [6], while the error analysis for the full-tensor Clenshaw-Curtis grid can be seen in [4]. Figure 1 depicts the difference between a full-tensor and sparse CC and Fejér grid. For the purposes of the grid construction, we assume a uniform max level in each random direction, i.e. an isotropic sparse grid.

With this in mind, we let $y_k^{m_k}$ for $m_k = 1, \ldots, \tilde{N}_k$, $k = 1, \ldots, M$ be the m_k^{th} unique zero in the direction Γ_k. To ease the notation, define $m = [m_1, \ldots, m_M]$ as an array of orders, and define $y_m = [y_1^{m_1}, \ldots, y_M^{m_M}]$, that is, a collection of zeroes in each direction of Γ. As well, define the product of the polynomials of a given order in each direction as $r_m(y) = \prod_{j=1}^{M} r_j^{m_j}$.

Then the polynomial chaos expansions of $\mathbf{J}_i$ and $\mathcal{V}$ are given by

$$\mathbf{J}_i^{h,N}(\mathbf{x}, y) = \sum_{m_1=1}^{\tilde{N}_1} \cdots \sum_{m_M=1}^{\tilde{N}_M} \mathbf{J}_i(\mathbf{x}, y_m) r_m(y), \text{ and } \mathcal{V}_i^{h,N}(\mathbf{x}, y) = \sum_{m_1=1}^{\tilde{N}_1} \cdots \sum_{m_M=1}^{\tilde{N}_M} \mathcal{V}(\mathbf{x}, y_m) r_m(y). \quad (3)$$

Table 1: Distribution parameters for the random parameters. Beta distribution shape parameters recovered from mean and standard deviation, and then translation maps from [0, 1] to the bounded space. Uniform distribution determines the bounds from the given mean and standard deviation.

Parameter	Distribution	Lower Bound	Upper Bound	Mean	Standard Deviation
μ_e	beta	8/6	12/6	10/6	0.5/6
$\mathbf{u}_x$	beta	1200	1800	1700	100
μ_e	uniform	9.13/6	10.86/6	10/6	0.5/6
$\mathbf{u}_x$	uniform	1527	1873	1700	100

Using this interpolation, we arrive at a deterministic form of estimating the expected values of the true solutions. Using Gaussian quadrature to approximate the integral yields

$$\mathbb{E}[\mathbf{J}_i^{h,N}] = \sum_{m_1=1}^{\tilde{N}_1} \cdots \sum_{m_M=1}^{\tilde{N}_M} \mathbf{J}_i(\mathbf{x}, y_m)\rho(y_m)w_k^m(y), \text{ and } \mathbb{E}[\mathcal{V}^{h,N}] = \sum_{m_1=1}^{\tilde{N}_1} \cdots \sum_{m_M=1}^{\tilde{N}_M} \mathcal{V}(\mathbf{x}, y_m)\rho(y_m)w_k^m(y)$$

where $w_k^m := \prod_{j=1}^{M} w_{k_j}^{m_j}$ and $w_{k_j}^{m_j} := \int_{\Gamma_{k_j}} (r_{k_j}^{m_j})^2 \, dy$, e.g. the weights of the polynomial in each direction. Note that we must include the density functions in the summations as well, to account for the transformation from the given joint density function to a strictly uniform one, which is required to use the CC or Fejér grid.

4. Numerical Experiments

With the collocation method established, and our choice of grid made, we now demonstrate the robustness of the method. For simplicity, we consider only $\mu_e, \mathbf{u}$ as the random parameters in the system. Furthermore, the fluid flow is assumed to be in one direction, $\mathbf{u} = (\mathbf{u}_x, 0, 0)$. We also assume that each is described by a single random variable and are spatially constant. For demonstration purposes, we assume both random variables are described by either a uniform or beta distribution, with a given mean and variance. The distributional choices are described in Table 1.

The results from these numerical experiments can be seen in Figure 2. In these, we examine a 1-D center line of the full 3-D model, running from channel inlet to outlet, of the electric potential $\mathcal{V}$. The random-grid and weights are implemented using Burkhardt's repository [12], while the deterministic solutions are computed with the COMSOL [13] model desribed in [5]. In Figure 2, we see a spatial dependence for the variation of the solution, with higher variances seen at the inlet of the channel. However, despite having the same mean and variance, the shape of the distribution clearly also had an impact on the distribution shape of the solutions, as seen in the right plot of Figure 2. Thus, the shape and variance of the random variables have a significant effect on the variance of the solutions, and inclusion of uncertain parameters within the model is necessary for reliable simulations.

5. Conclusion

In this paper, we explored the numerical propagation of uncertainty within an MHD system. We began by defining the governing equations for our model. We then introduced major assumptions required to implement the SC method, including the FD noise assumption. Next, we described the SC method itself, performing a FD polynomial chaos expansion of the solutions as an interpolation in the random space. We also defined the interpolating points by using the

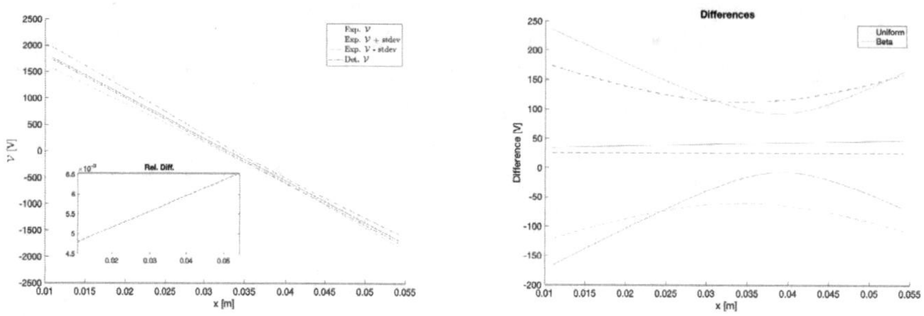

Figure 2: Results from numerical experiments. (Left) beta distributions. Exp. $\mathcal{V}$ is the expected value of $\mathcal{V}$, solved for using the SC method at a level of 5. Det. $\mathcal{V}$ is the deterministic $\mathcal{V}$, solved for with the means of each random parameter. (Inset plot) Relative error defined as the normalized difference between the Exp. $\mathcal{V}$ and Det. $\mathcal{V}$. (Right) Difference between Exp. $\mathcal{V}$ and Det. $\mathcal{V}$, with variances included. Solid line indicates uniform distribution results, dashed line indicates beta distribution results.

zeros of the orthogonal Lagrange basis of the random space, and further reducing the system complexity by utilizing a CC or Fejér sparse grid. Finally, we demonstrated the effectiveness of the SC method with two random variables within the system. It was shown that the inclusion of uncertainty is warranted. The efficiency of the SC method with a sparse grid enables the inverse problem to be considered under uncertainty, and allows for future work involving the optimal design problem of the MHD system.

Acknowledgments

The work was supported by the United States Department of Energy's Crosscutting Research Program, an appointment to the Science Education Programs at the National Energy Technology Laboratory, administered by ORAU through the United States Department of Energy Oak Ridge Institute for Science and Education, and by the National Science Foundation, grant number DMS-2012882.

References

[1] Rosa R 1968 *Magnetohydrodynamic Energy Conversion.* McGraw-Hill
[2] Sutton G and Sherman A 1965 *Engineering Magnetohydrodynamics.* McGraw-Hill series in mechanical engineering. McGraw-Hill
[3] Nacev A, Beni C, Bruno O, and Shapiro B 2011 The behaviors of ferro-magnetic nano-particles in and around blood vessels under applied magnetic fields. In *Journal of Magnetism and Magnetic Materials* **323** 6 651–668
[4] Rajbhandari E, Gibson N, and Woodside C Preprint Stochastic collocation error analysis for the full 3-d magnetohydrodynamic system. In
[5] Rajbhandari E, Gibson N, and Woodside C Preprint Parameter estimation for 3-d magnetohydrodynamics generator. In
[6] Johnson B, Gibson N, and Cotilla-Sanchez E 2021 A coupled Karheunen–Loeve and anisotropic sparse grid interpolation method for the probabilistic load flow problem. In *Electric Power Systems Research*
[7] Smith R 2014 *Uncertainty Quantification: Theory, Implementation, and Applications.* SIAM. ISBN 978-1-611973-21-1
[8] Burkhardt J 2012 The "combining coefficient" for anisotropic sparse grids. In *Interdisciplinary Center of Applied Mathematics And Information Technology*
[9] Bokil V, Gibson N, McGregor D, and Woodside C 2015 Toward estimating current densities in magnetohydrodynamic generators. In *Journal of Physics: Conference Series* **640**
[10] Babuska I, Nobile F, and Tempone R 2007 A stochastic collocation method for elliptic partial differential equations with random input data. In *SIAM Journal on Numerical Analysis* **45**
[11] Oksendal B 2003 Stochastic differential equations. In
[12] Burkhardt J 2021. J. Burkhardt's homepage
[13] Comsol multiphysics v. 5.4

Journal of Physics: Conference Series **2207** (2022) 012008 doi:10.1088/1742-6596/2207/1/012008

Multicanonical simulations of the 2D spin-1 Baxter-Wu model in a crystal field

Nikolaos G Fytas[1], **Alexandros Vasilopoulos**[1], **Erol Vatansever**[2 1], **Anastasios Malakis**[3 1] **and Martin Weigel**[4 1]

[1] Centre for Fluid and Complex Systems, Coventry University, Coventry, CV1 5FB, United Kingdom
[2] Department of Physics, Dokuz Eylül University, TR-35160, Izmir, Turkey
[3] Department of Physics, University of Athens, Panepistimiopolis, GR 15784 Zografou, Greece
[4] Institut für Physik, Technische Universität Chemnitz, 09107 Chemnitz, Germany

E-mail: nikolaos.fytas@coventry.ac.uk

Abstract. We investigate aspects of universality in the two-dimensional (2D) spin-1 Baxter-Wu model in a crystal field Δ using a parallel version of the multicanonical algorithm employed at constant temperature T. A detailed finite-size scaling analysis in the continuous regime of the $\Delta - T$ phase diagram of the model indicates that the transition belongs to the universality class of the 4-state Potts model. The presence of first-order-like finite-size effects that become more pronounced as one approaches the pentacritical point of the model is highlighted and discussed.

1. Introduction

The spin-1/2 Baxter-Wu model is defined by nearest-neighbor three-spin interactions on a triangular lattice. It was first conceptualized as a model that is not invariant under inversion by Wood and Griffiths [1] and subsequently analytically solved by Baxter and Wu who showed that its critical universality is identical to that of the 4-state Potts model [2, 3, 4]. The Hamiltonian of the BW model reads as

$$\mathcal{H} = -J \sum_{\langle ijk \rangle} \sigma_i \sigma_j \sigma_k, \quad (1)$$

where the spins take on the values ± 1, $J > 0$ is the ferromagnetic exchange interaction, and the sum extends over all elementary triangles of the lattice. Due to the fact that the triangular lattice can be decomposed into three sublattices, each spin of an elementary triangle of the lattice belongs to a different sublattice. Thus, inverting the spins of any two sublattices would result in no change in the energy. Hence, the ground state consists of one ferromagnetic state, where all spins are pointing up, and three ferrimagnetic states, where one sub-lattice is pointing up and the remaining two are pointing down.

An interesting extension of the Baxter-Wu model arises when one considers spin values $\sigma_x = \{-1, 0, 1\}$ and includes an extra crystal field (or single-ion anisotropy) Δ coupled to σ_x^2, so that the new Hamiltonian reads as

$$\mathcal{H} = -J \sum_{\langle ijk \rangle} \sigma_i \sigma_j \sigma_k + \Delta \sum_i \sigma_i^2 = E_J + \Delta E_\Delta, \quad (2)$$

Journal of Physics: Conference Series **2207** (2022) 012008 doi:10.1088/1742-6596/2207/1/012008

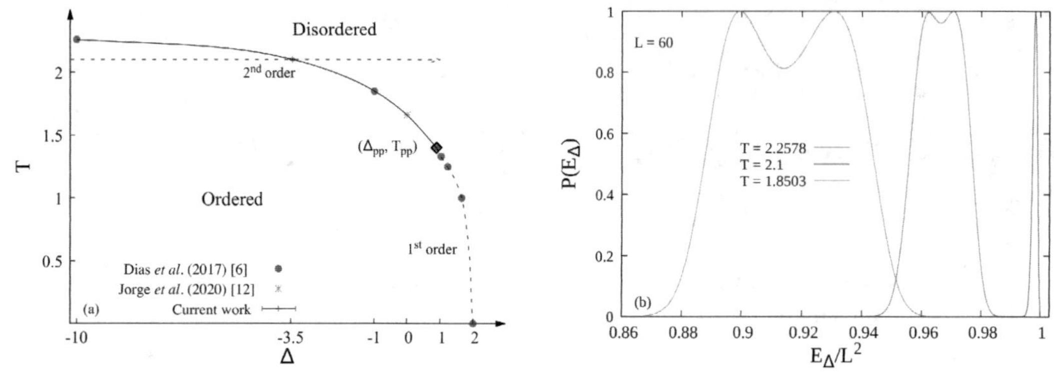

Figure 1: (a) Phase diagram of the 2D spin-1 Baxter-Wu model with the ferromagnetic and paramagnetic phases separated by a continuous transition for larger T (solid line) and a first-order transition for smaller T (dotted line). The line segments meet at the pentacritical point $(\Delta_{\rm pp}, T_{\rm pp}) \approx (0.8902, 1.4)$ [6] marked by the black rhombus, where three ferrimagnetic configurations and a ferromagnetic configuration, along with that of zero spins coexist [6, 7, 8]. The dashed arrow at $T = 2.1$ indicates the temperature choice of our simulations that corresponds to $\Delta \approx -3.5$, see also figure 2(a). (b) Probability density functions of the crystal-field energy $P(E_\Delta)$ for a system of linear size $L = 60$ at different temperatures in the second-order transition regime, as indicated. Note that for $T = 2.2578$: $\Delta \approx -10$, and for $T = 1.8503$: $\Delta \approx -1$. Finite-size effects of first-order-type (double-peak structure) appear as we lower the temperature and approach the pentacritical point.

where E_J and E_Δ denote the nearest-neighbor interaction and the crystal-field energies, respectively. Note that Δ controls the density of the zero spins: when $\Delta \to -\infty$ spins are confined to the values ± 1, retrieving the spin-1/2 Baxter-Wu model. Correspondingly, and in full analogy to the Blume-Capel case [5], one expects for the model defined in equation (2) the same kind of competition between the ordered and disordered phases (mediated by the crystal field), and therefore a similar phase diagram but a different universality class is expected, see figure 1(a).

Because dilution in the Potts model with four states [9] has the same effect as Δ in the Baxter-Wu model, one should also expect the critical behavior of both models to be the same. Surprisingly though, there is an ongoing debate with respect to the degree of universality obeyed by the spin-1 Baxter-Wu model. This is mainly due to contradicting numerical evidence suggested over the course of last years: The results of Ref. [7] via renormalization group, conventional finite-size scaling, and conformal invariance techniques indicated that the critical exponents vary continuously with Δ along the second-order transition line, differently from the expected behavior of the 4-state Potts model (with exponents $\nu = 2/3$, $\alpha/\nu = 1$, and $\gamma/\nu = 7/4$). A similar conclusion was drawn in Ref. [10], where using importance sampling Monte Carlo simulations for the special case with $\Delta = 0$ the values $\nu = 0.617(3)$, $\alpha = 0.692(6)$, and $\gamma = 1.13(1)$ were obtained. The complementary Monte Carlo work of Ref. [11] at $\Delta = -1$ and 1 further corroborated this hypothesis. Conversely, the renormalization-group work of Dias *et al.* [6] suggested that along the critical line, the conformal anomaly c and the exponents ν, η are the same as that of the pure spin-1/2 Baxter-Wu model (or the 4-state Potts model). Finally, the most recent works by Jorge *et al.* [12] used Wang-Landau entropic sampling simulations to probe the system at $\Delta = 0$. According to these authors the model exhibits an indeterminacy regarding the order of phase transition.

In this conflicting situation we present here an updated finite-size scaling analysis of the model in the second-order transition regime of the phase diagram. Using multicanonical simulations outlined in section 2, we scrutinize the critical properties of the system at the fixed temperature $T = 2.1$. Via a standard finite-size scaling analysis, as discussed in section 3, we recover estimates of critical exponents that are in good agreement to those of the 4-state Potts universality class [13]. Our work also features the existence of first-order-like finite-size effects in the second-order regime of the phase diagram that become more pronounced as one approaches the pentacritical point. A summary with a work-in-progress plan closes this short contribution in section 4.

2. Numerical approach

We now turn to a description of the multicanonical (MUCA) method [14]. In this approach, instead of using the canonical Boltzmann weight $e^{-\beta E}$, with the inverse temperature $\beta = 1/(k_{\rm B}T)$, a correction function is introduced, designed to produce a flat histogram. For the needs of the current work, the multicanonical method was applied with respect to the crystal-field energy E_Δ, fixing the temperature and allowing us to continuously reweight to arbitrary values of Δ. To this end, the partition function

$$\mathcal{Z} = \sum_{\{E_J, E_\Delta\}} g(E_J, E_\Delta) e^{-\beta(E_J + \Delta E_\Delta)} \tag{3}$$

is generalized to

$$\mathcal{Z}_{\rm MUCA} = \sum_{\{E_J, E_\Delta\}} g(E_J, E_\Delta) e^{-\beta E_J} \, W(E_\Delta), \tag{4}$$

where $g(E_J, E_\Delta)$ is the two-parametric density of states.

It follows that

$$P_{\rm MUCA}(E_J, E_\Delta) = \frac{g(E_J, E_\Delta) e^{-\beta E_J} W(E_\Delta)}{\mathcal{Z}_{\rm MUCA}}, \tag{5}$$

where $P_{\rm MUCA}$ is the equilibrium probability distribution. In order to produce a flat E_Δ histogram, by carrying out a summation with respect to E_J, the modified weight should be given by

$$W(E_\Delta) \propto \mathcal{Z}_{\rm MUCA} \left[\sum_{E_J} g(E_J, E_\Delta) e^{-\beta E_J} \right]^{-1}. \tag{6}$$

These weights can be calculated in an iterative fashion starting with an initial guess. At the $n^{\rm th}$ step spins are flipped using the weights $e^{-\beta E_J} W^{(n)}(E_\Delta)$ and the histogram $H^{(n)}(E_\Delta)$ of the energies E_Δ is sampled. After a specified number of spin-flip attempts the histogram is used to recalibrate the weights via $W^{(n+1)}(E_\Delta) = W^{(n)}(E_\Delta) / H^{(n)}(E_\Delta)$. The process is completed when a flat-enough histogram has been sampled, after which a series of production runs is carried out. At each step the normalized histogram $H^{(n)}_{\rm norm}(E_\Delta)$ satisfies the equation

$$\langle H^{(n)}_{\rm norm}(E_\Delta) \rangle = P^{(n)}(E_\Delta) = \frac{1}{\mathcal{Z}_{\rm MUCA}} \sum_{E_J} g(E_J, E_\Delta) e^{-\beta E_J} W^{(n)}(E_\Delta) \propto \frac{W^{(n)}(E_\Delta)}{W(E_\Delta)}, \tag{7}$$

justifying the scheme for updating the weights using sampled histograms.

We employed a parallel implementation of the multicanonical method [15, 16], guided by its already successful application in the study of the Blume-Capel model [5, 17, 18]. In this setup weights are distributed to parallel workers, each one producing a histogram. At the end all histograms are added and the resulting total histogram is then used to recalibrate the

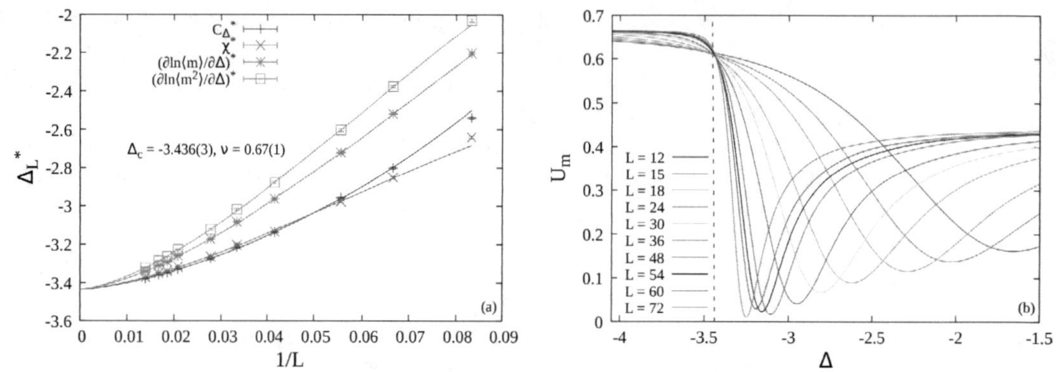

Figure 2: (a) Shift behaviour of several pseudocritical fields Δ_L^* as indicated in the legend. (b) Crossings of the fourth-order magnetization's Binder cumulant for different system sizes.

weights. Our simulations were implemented on an "Nvidia Tesla K80" Graphics Processing Unit boosting the numerical capacity to 26 624 simulations of uncorrelated systems in parallel. Finally, the histogram's flatness was tested using the Kullback-Leibler divergence as discussed in reference [16].

The numerical protocol described above was applied on triangular lattices with periodic boundary conditions. As the system has in addition to the ferromagnetic phase three different ferrimagnetic phases, the allowed values of the linear size of the lattice L must be a multiple of three. In this way, all ground states of the infinite system would fit on any finite lattice [11]. In the course of our simulations we considered linear sizes within the range $L = \{12-72\}$ respecting this constraint. The bulk of our simulations was performed at $T = 2.1$. Other values of T closer to the pentacritical point are currently being considered and will be reported elsewhere [19].

In the framework of the multicanonical approach it is natural to compute Δ-derivatives of observables rather than the usual T-ones. For instance, in place of the usual specific heat one may define a specific-heat-like quantity [17]

$$C_\Delta = \frac{1}{N}\frac{\partial E_J}{\partial \Delta} = -\left[\langle E_J E_\Delta\rangle - \langle E_J\rangle \langle E_\Delta\rangle\right]/(NT), \tag{8}$$

where $N = L^2$, the number of lattice sites. Other finite-size scaling observables include the partial derivative of the logarithm of the nth power of the magnetization m $(\partial \ln \langle m^n\rangle/\partial\Delta)$, the magnetic susceptibility (χ), as well as the fourth-order Binder cumulant of the magnetization (U_{m}) [12, 17, 20].

3. Results

We start with some typical illustrations of the (reweighted) probability density function $P(E_\Delta)$ (normalized to unity). It is well known that a single-peak structure in $P(E_\Delta)$ is characteristic of a continuous transition, whereas a double-peak structure signals the emergence of a first-order phase transition [17, 21, 22]. Figure 1(b) depicts $P(E_\Delta)$ for a system with $L = 60$ at several temperatures (see caption). The distributions were reweighted close to their respective critical crystal field according to the calculations of Dias *et al.* [6] and the current work. Some comments are in order: (i) For large-enough T a single peak is observed, but as we lower T (and increase Δ up to -1) first-order-like characteristics appear similar to the observations by Jorge *et al.* [12] at $\Delta = 0$. (ii) These effects become more pronounced as we approach the pentacritical point and lead to corrections in finite-size scaling in this vicinity. Clearly, a refined investigation of the

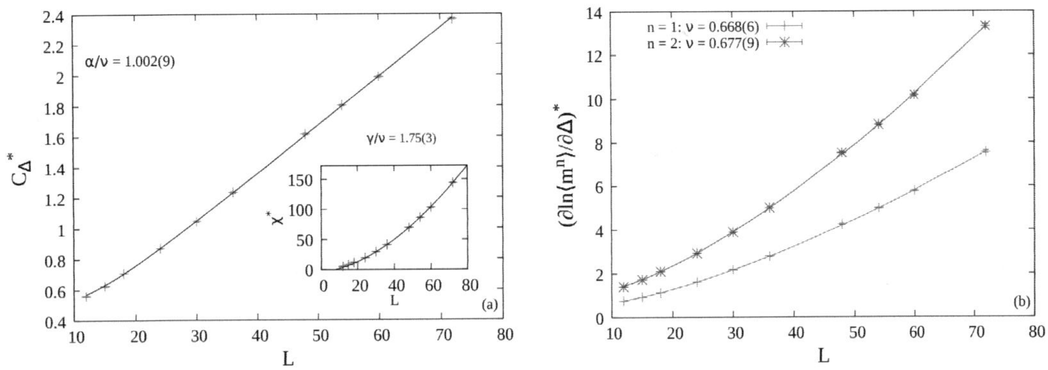

Figure 3: Finite-size scaling behaviour of C^*_Δ (main panel) and χ^* (inset) in panel (a) and of $(\partial \ln \langle m^n \rangle / \partial \Delta)^*$ in panel (b).

surface tension and latent heat of the transition is called for, as it is quite often the case that pseudo-first-order signatures emerge at moderate system sizes in many spin models with complex interactions, see [23] and references therein. Some first analysis along these lines indicates that these are indeed mere finite-size effects that disappear in the thermodynamic limit [19].

We proceed now with a finite-size scaling analysis at $T = 2.1$ in the second-order regime. For the fitting procedure discussed below we restricted ourselves to data with $L \geq L_{\rm min}$. As usual, to determine an acceptable $L_{\rm min}$ we employed the standard χ^2 test for goodness of the fit. Specifically, we considered a fit as being fair only if $10\% < Q < 90\%$, where Q denotes the probability of finding a χ^2 value which is even larger than the one actually found from our data [24].

To extract the critical crystal field $\Delta_{\rm c}$ and a first estimate of the correlation-length exponent ν we study the shift behaviour of several suitable pseudocritical fields, Δ^*_L, deduced from the peak locations of the relevant observables C_Δ, χ, and $\partial \ln \langle m^n \rangle / \partial \Delta$. Figure 2(a) features the joint fit [6, 20]

$$\Delta^*_L = \Delta_{\rm c} + bL^{-1/\nu}(1 + b' L^{-\omega}), \tag{9}$$

over four different data sets and with $L_{\rm min} = 15$. Note that b and b' are non-universal coefficients and the corrections-to-scaling exponent was set to the well-known value $\omega = 2$ [6, 11, 25, 26]. We obtain the estimates $\Delta_{\rm c} = -3.436(3)$ and $\nu = 0.67(1)$, the latter being in very good agreement to the value $\nu = 2/3$ of the 4-state Potts model [2, 3]. Typical curves of the Binder cumulant U_m are shown in figure 2(b) where the crossing point agrees (within some minor finite-size effects) with the value $\Delta = -3.436$, as marked by the vertical dashed line.

Further, in figure 3(a) we present the finite-size scaling behaviour of the maxima of the specific-heat-like quantity (main panel) and the magnetic susceptibility (inset). The solid lines are fits of the form [5]

$$C^*_\Delta \sim L^{\alpha/\nu}(1 + b' L^{-\omega}) \; ; \; \chi^* \sim L^{\gamma/\nu}(1 + b' L^{-\omega}), \tag{10}$$

using $L_{\rm min} = 18$ and 30 respectively. The resulting estimates $\alpha/\nu = 1.002(9)$ and $\gamma/\nu = 1.75(3)$ are clearly compatible to the exact values 1 and 7/4 of the 4-state Potts model [2, 3]. Finally, in figure 3(b) we provide an additional verification of the critical exponent ν via the maxima of the logarithmic derivatives of the order parameter. A fit of the form [20]

$$(\partial \ln \langle m^n \rangle / \partial \Delta)^* \sim L^{1/\nu}(1 + b' L^{-\omega}), \tag{11}$$

where $L_{\rm min} = 18$ gives $\nu = 0.668(6)$ and $0.677(9)$ for $n = 1$ and 2 respectively, in further support of the 4-state Potts universality class.

4. Summary and outlook

Using parallel multicanonical simulations and finite-size scaling analysis we studied the universality aspects of the 2D spin-1 Baxter-Wu model at the second-order transition regime of the $\Delta - T$ phase boundary. Our estimates for the critical exponents provide evidence that this ferromagnetic-paramagnetic transition belongs to the universality class of the 4-state Potts model. In order to capture possible finite-size effects of first-order type as already discussed recently [12], we monitored the structure of the crystal-field energy probability density function. Indeed, on lowering the temperature a double-peak structure in the energy probability density function is observed, which becomes more pronounced as the pentacritical point is approached. The actual role of these finite-size effects including simulation of larger system sizes and a systematic scaling analysis of the respective surface tension and latent heat of this pseudo-first-order transition are currently under study by our group [19]. Some future plans include a dedicated study of the location and universality principle of the pentacritical point, as well as a high-accuracy reproduction of the model's phase diagram.

Acknowledgments

N G Fytas would like to thank J A Plascak for useful communication on the topic. We acknowledge the provision of computing time on the parallel computer cluster *Zeus* of Coventry University.

References

[1] Wood D W and Griffiths H P 1972 *Journal of Physics C: Solid State Physics* **5** L253
[2] Baxter R and Wu F 1973 *Phys. Rev. Lett.* **31** 1294
[3] Baxter R 1974 *Aust. J. Phys.* **27** 357
[4] Baxter R and Wu F 1974 *Aust. J. Phys.* **27** 369
[5] Zierenberg J, Fytas N G, Weigel M, Janke W and Malakis A 2017 *Eur. Phys. J Special Topics* **226** 789–804
[6] Dias D A, Xavier J C and Plascak J A 2017 *Phys. Rev. E* **95** 012103
[7] Costa M L M, Xavier J C and Plascak J A 2004 *Phys. Rev. B* **69** 104103
[8] Jorge L, Martins P, DaSilva C, Ferreira L and Caparica A 2021 *Physica A* **576** 126071
[9] Nienhuis B, Berker A N, Riedel E K and Schick M 1979 *Phys. Rev. Lett.* **43**(11) 737–740
[10] Costa M L M and Plascak J A 2004 *Braz. J. Phys.* **34** 419–421
[11] Costa M L M and Plascak J A 2016 *J. Phys.: Conf. Ser.* **686** 012011
[12] Jorge L, Ferreira L and Caparica A 2020 *Physica A* **542** 123417
[13] Wu F Y 1982 *Rev. Mod. Phys.* **54** 235
[14] Berg B A and Neuhaus T 1992 *Phys. Rev. Lett.* **68** 9–12
[15] Zierenberg J, Marenz M and Janke W 2013 *Comput. Phys. Commun.* **184** 1155–1160
[16] Gross J, Zierenberg J, Weigel M and Janke W 2018 *Computer Physics Communications* **224** 387–395
[17] Zierenberg J, Fytas N G and Janke W 2015 *Phys. Rev. E* **91** 032126
[18] Fytas N G, Zierenberg J, Theodorakis P E, Weigel M, Janke W and Malakis A 2018 *Phys. Rev. E* **97** 040102
[19] Vasilopoulos A, Fytas N G, Vatansever E, Malakis A and Weigel M 2021 *(in preparation)*
[20] Ferrenberg A M and Landau D P 1991 *Phys. Rev. B* **44**(10) 5081–5091
[21] Binder K and Landau D P 1984 *Phys. Rev. B* **30** 1477
[22] Binder K 1987 *Rep. Prog. Phys.* **50** 783
[23] Fytas N G, Malakis A and Eftaxias K 2008 *J. Stat. Mech.* P03015
[24] Press W, Teukolsky S, Vetterling W and Flannery B 1992 *Numerical Recipes in C* 2nd ed 27 (Cambridge: Cambridge University Press)
[25] Alcaraz F C and Xavier J C 1997 *J. Phys. A: Math. Gen.* **30** L203
[26] Alcaraz F C and Xavier J C 1999 *J. Phys. A: Math. Gen.* **32** 2041

Efficient algorithms for computing ground states of the 2D random-field Ising model

Argyro Mainou[1], Nikolaos G Fytas[1] and Martin Weigel[2,1]

[1] Centre for Fluid and Complex Systems, Coventry University, Coventry, CV1 5FB, United Kingdom

[2] Institut für Physik, Technische Universität Chemnitz, 09107 Chemnitz, Germany

E-mail: mainoua@uni.coventry.ac.uk

Abstract. We investigate the application of graph-cut methods for the study of the critical behaviour of the two-dimensional random-field Ising model. We focus on exact ground-state calculations, crossing the phase boundary of the model at zero temperature and varying the disorder strength. For this purpose we employ two different minimum-cut–maximum-flow algorithms, one of augmenting-path and another of push-relabel style. We implement these approaches for the square and triangular lattice problems and compare their computational efficiency.

1. Introduction

The random-field Ising model (RFIM) is one of the simplest disordered systems [1–4]. Its study not only allows the investigation of a series of complex phenomena, but also is directly connected to a significant number of experiments in condensed-matter physics [5–7]. These facts have established it as one of the main model systems for the study of collective behaviour under quenched disorder. It is described by the Hamiltonian:

$$\mathcal{H} = -J \sum_{\langle x,y \rangle} s_x s_y - \sum_x h_x s_x, \tag{1}$$

where $s_x = \pm 1$ and $J > 0$ denotes the ferromagnetic nearest-neighbour interaction (we set $J = 1$ throughout this study). The disorder is introduced via the variables h_x, which are independent random magnetic fields acting on each spin s_x and following some distribution $\mathcal{P}(h_x)$. In the present work, we consider quenched random fields following the Gaussian distribution ($\mu = 0$, $\sigma^2 = h^2$), so that h defines the disorder strength.

A number of numerical techniques, mainly of Monte-Carlo type, have been employed for studying the critical (and other) properties of the RFIM [5]. However, the limitations of these methods due to the rough free-energy landscape of the model led the community to seek other more efficient methods. Fortunately, the renormalization group tells us that the random field is a relevant perturbation at the pure fixed point, and the random-field fixed point is at $T = 0$ (see figure 1) [8]. Hence the critical behavior is the same everywhere along the phase boundary and we can predict it simply by staying at $T = 0$ and crossing the phase boundary at the critical field point. This is a convenient approach because we can determine ground states of the system

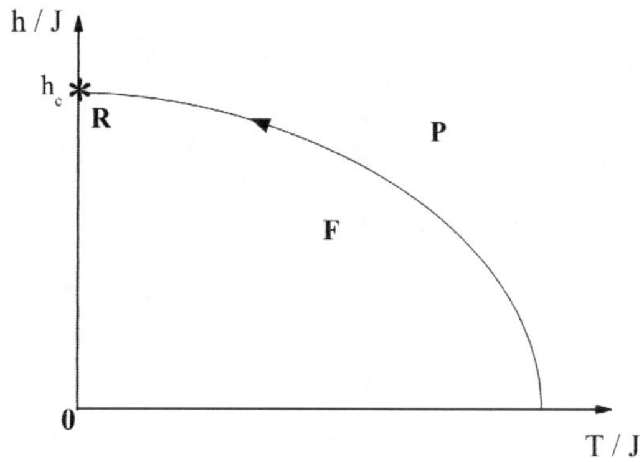

Figure 1. A sketch of the phase diagram of the RFIM. The solid line (phase boundary) separates the ferromagnetic (**F**) and paramagnetic (**P**) phases. The black arrow shows the renormalization-group flow to the random fixed point (**R**) at $T = 0$ and $h = h_c$, the critical disorder strength, as marked by an asterisk.

exactly using efficient optimization algorithms through an existing mapping of the ground-state problem to a maximum-flow optimization task [9–11].

In 1975, Picard and Ratliff introduced the idea of treating energy minimization problems of a class of systems, including the RFIM, as minimum-cut problems on a network [12], by mapping the system under consideration onto a network and calculating its maximum flow. The value of the maximum flow is then equal to the ground-state energy of the system. The orientations of the spins in the ground state can be constructed from the flow values of the corresponding network. These techniques have allowed the extensive employment of network theory for calculating ground states [13–16].

Motivated by the aforementioned results, we apply graph-cut theory in order to address such energy minimization problems for the Gaussian RFIM in two dimensions. For this purpose, we employ two distinct minimum-cut-maximum-flow algorithms. The most popular algorithms of this type are roughly divided into two main categories: augmenting-path-style and push-relabel-style. We chose an algorithm from each category and implement it for the study of the RFIM, as outlined in section 2. The applicability of the augmenting-path-style algorithm is examined for two lattice geometries of the model, namely square and triangular and the results are presented in section 3, where we also perform a comparison between the two algorithms in terms of their computational time for the square-lattice model. Finally, a summary of our results is provided in section 4.

2. Numerical Methods

In this section we provide an outline of our numerical implementation on various random-field systems with periodic boundary conditions. As a starting point, we point out that the mapping of the model to a network was carried out according to reference [17] and that we used the Mersenne Twister [18] random-number generator (RNG) for the production of the Gaussian local random-field values.

Let us start with the Boykov-Kolmogorov (BK) algorithm [19], which is a variation of standard augmenting-path algorithms [21–23] and was developed in order to improve their empirical performance on graphs in computer vision problems. Its input is a weighted, directed, residual graph, along with a source s and a sink t. BK involves the construction of two non-

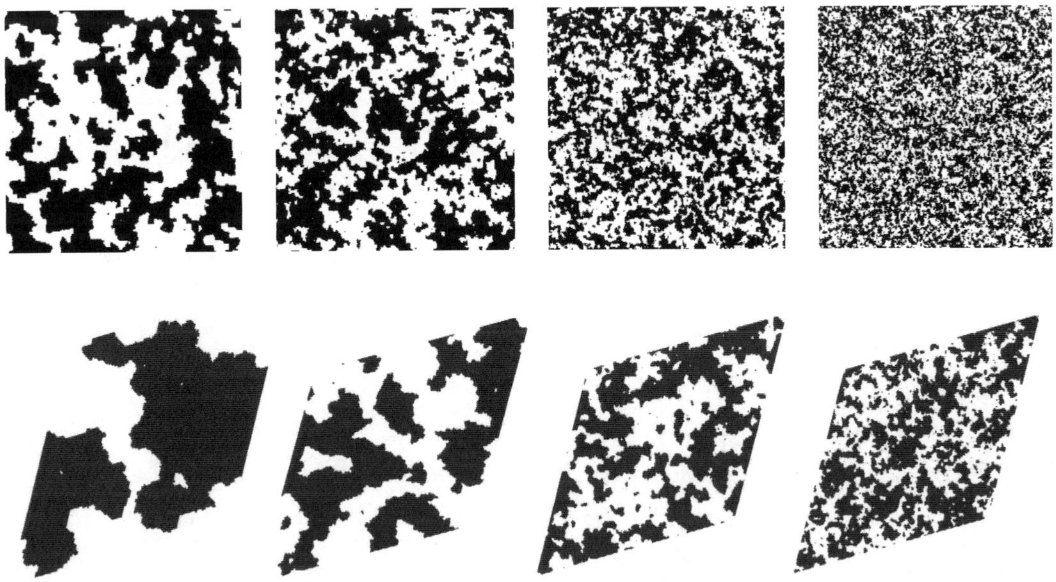

Figure 2. Example ground-state spin configurations of the RFIM on the square lattice (top row, $h = 1$) and the triangular lattice (bottom row, $h = 1.7$). System sizes are $L = 512$, 1024, 2048 and 4096 (top row), and $L = 128$, 256, 512 and 1024 (bottom row).

overlapping search trees, that is a source tree and a sink tree, which treats the terminals symmetrically and so outperforms an earlier version of the algorithm where a single tree rooted at the source was used [24]. What is more, instead of the usual process of building new augmenting paths at each new iteration, the above trees are reused. We refer the interested reader to reference [19] for further details regarding the algorithm. The worst case complexity of BK is $\mathcal{O}(mn^2|C|)$, where n is the number of edges of the network, m is the number of nodes and $|C|$ is the cost of a minimum cut, which is the value of the maximum flow. Boykov and Kolmogorov showed [19] with experimental tests in two dimensions and for lattice sizes of up to $L = 511$, that despite the fact that theoretically speaking this is worse than the time complexities of other standard minimum-cut–maximum-flow algorithms [21–23, 25, 26], the BK algorithm is significantly faster when applied to typical system sizes in real applications.

In the present work we validate their results for the case of the square-lattice RFIM. Before doing so, we examine the algorithm's efficiency in working out the model's ground-state energy and spin configuration, not only for the square but also for the triangular geometry. In order to achieve this goal, we adapt the code for BK provided in [19] (version 3.01) to the problem of computing RFIM ground states for square and triangular lattices. We pick $h = 1$ for the square lattice and $h = 1.7$ for the triangular lattice as test cases, since these values correspond to the same breakup length $\ell_b \approx 32$ for the two lattice geometries [20].

We now turn to the second category of minimum-cut-maximum-flow algorithms, namely algorithms of push-relabel-style. One of the most popular network flow algorithms used to solve the RFIM is the general push-relabel algorithm [25] with a computational time complexity of $\mathcal{O}(n^2m)$. Proofs and theorems regarding the algorithm have been provided in references [10, 11]. Two implementations of the push-relabel method are the H_PRF and Q_PRF techniques [26], which outrun many other minimum-cut-maximum-flow algorithms in real-size experiments, with computational time complexities of $\mathcal{O}(n^2\sqrt{m})$ and $\mathcal{O}(n^3)$, respectively. However, according

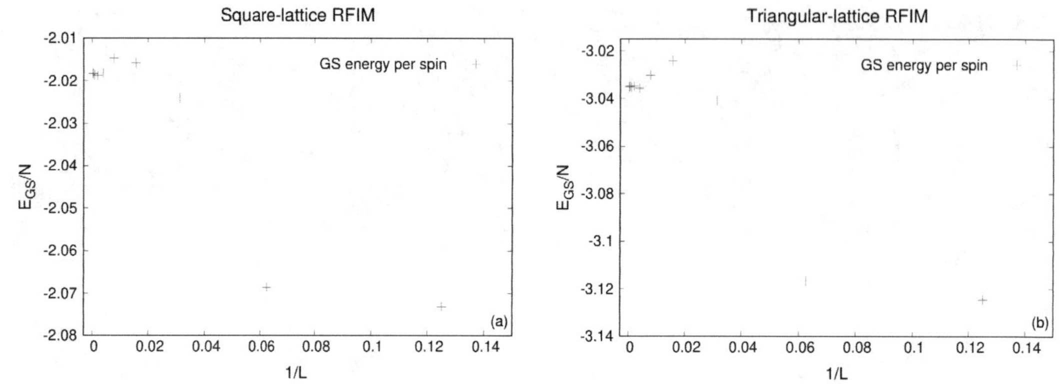

Figure 3. Ground-state (GS) energy per spin, E_{GS}/N (where $N = L \times L$), of single realizations of the RFIM (a) on the square lattice and (b) on the triangular lattice for linear system sizes $L \in \{8, 16, 32, 64, 128, 256, 512, 1024, 2048, 4096\}$, where $h = 1$ and $h = 1.7$ for the square and the triangular lattices, respectively.

to [19], this does not hold for the case of the BK algorithm. With the aim of confirming this, here we choose to focus on a Q_PRF-style modification [27, 28] that removes the source and sink nodes, reducing memory usage and clarifying the physical connection. In particular, we follow the implementation described in reference [29].

The implementation of the BK algorithm for creating multiple realizations of the quenched disorder can be carried out in two different ways, that is by either building the code's graph from scratch for every single disorder sample or building the graph once and reallocating the graph's edge capacities for every disorder realization. We expect that the latter constitutes an optimized version of the code, since the graph needs to be constructed outside of the loop for the many samples of the disorder. We therefore examine both versions and our results are provided in subsection 3.2, along with the ones referring to the Q_PRF-style algorithm. Note that in order to ensure that the timing measurements are not perturbed by concurrent jobs, we reserve a full node on the cluster while only using a single core at each point in time.

3. Results

3.1. Boykov-Kolmogorov algorithmic efficiency: square- and triangular-lattice RFIM

We examine the BK algorithm's suitability for the study of the ground-state problem of the two-dimensional RFIM with the use of graph-cut methods. For this purpose, we implement it in order to work out the maximum flow of a network corresponding to a Gaussian $L \times L$ RFIM, for various lattice sizes L, of either square- or triangular-lattice geometry. In figure 2 we show the spin configurations for some example disorder realizations on square lattices of linear lattice sizes $L = 512$, 1024, 2048 and 4096 and random-field strength $h = 1.0$ as well as triangular lattices of sizes $L = 128$, 256, 512 and 1024 and random-field strength $h = 1.7$, respectively. Note that for a given lattice size L, a larger disorder strength is required for the triangular case such that the sizes of the spin clusters are comparable between the two different geometries. For example, for $L = 1024$, the square- and triangular-lattice RFIMs exhibit similar behaviour for $h = 1$ and $h = 1.7$, respectively. This is expected, since the two extra bonds between the nearest neighbours of the triangular lattice contribute to the resistance of breaking the ground-state order. From the given spin configurations, it is possible to extract the ground-state energies, which we present (also for single disorder realizations) in figure 3.

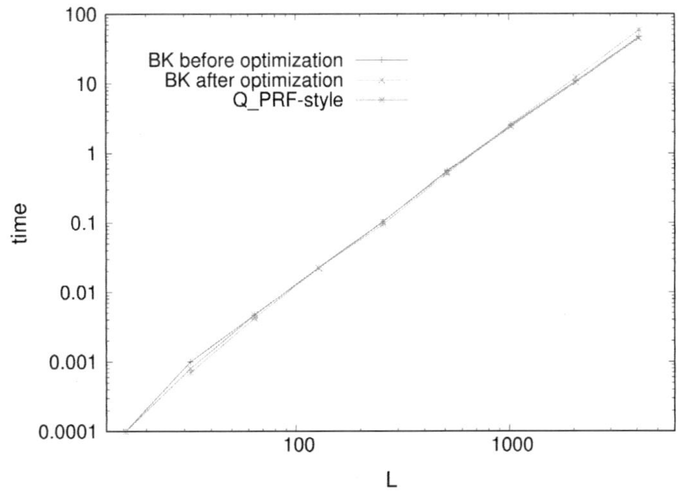

Figure 4. Average run-time in seconds for computing a single square-lattice ground state using the Q_PRF-style algorithm of [29] (blue line) as well as the BK algorithm before (violet line) and after (green line) optimization. The data are averaged over 100 disorder samples for $h = 1$.

3.2. Algorithmic comparison: square-lattice RFIM

We proceed to the inspection of both algorithms' time complexity for the square-lattice RFIM, for several lattice sizes of up to $L = 4096$. For this purpose we generate many realizations of the quenched disorder (exactly the same for the two algorithms) for some specific disorder strength (we pick $h = 1$) and work out the computational time of the codes for treating each one of them. Afterwards, the calculation of the mean value of those computational times will give us an estimate of the average computational time needed by each algorithm for working out a single sample's ground-state energy and corresponding spin configuration for $h = 1$.

In figure 4 we show the results regarding the computational time of the two versions of the BK algorithm compared to the ones we obtain by implementing the Q_PRF-style algorithm, using the same disorder samples. As expected, we find that the approach of building the graph just once outruns the method of building the graph for every realization of the disorder and therefore it may be preferred for large-scale simulations as an optimized version of the code. It is clear however, that both versions of the BK algorithm are somewhat faster than the Q_PRF-style algorithm. Performing fits of a power-law $t \sim L^a$ to the run-time data, we find $a \approx 2.3$ for all three codes. Note that this is a considerably slower increase than the worst-case scaling discussed above. Finally, one may notice that the difference between the two versions of the BK code implementation is relatively small. This is because the biggest portion of the computational time is used for the calculation of the maximum flow and not for setting up the graph structure.

4. Summary and Outlook

Motivated by the continuing use of graph-cut methods for the investigation of ground-state properties of the random-field magnets, we performed a comparative test of two different minimum-cut–maximum-flow algorithms on the Gaussian two-dimensional random-field Ising model. We implemented the particularly efficient Boykov-Kolmogorov algorithm [19] and applied it to the square-lattice and triangular-latice random-field Ising models. Additionally, an optimization of the code was developed and both versions were compared to a Q_PRF-style implementation of the push-relabel algorithm, currently reported as one of the fastest minimum-cut-maximum-flow algorithms. We showed that the Boykov-Kolmogorov method is more efficient than the Q_PRF approach when implemented for square lattice sizes up to $L = 4096$, a fact that verifies and extends the original claim of reference [19] that the Boykov-Kolmogorov algorithm outruns many of the most efficient minimum-cut-maximum-flow algorithms for lattice sizes of experimental interest.

Acknowledgments
We acknowledge the allocation of CPU time on the supercomputer *Zeus* of Coventry University. A. Mainou would like to thank Abhishek Kumar for his support at various stages of this work.

References

[1] Y. Imry and S.-K. Ma, Phys. Rev. Lett. **35**, 1399–1401 (1975).
[2] A. Aharony, Y. Imry and S.-K. Ma, Phys. Rev. Lett. **37**, 1364 (1976).
[3] A. P. Young, J. Phys. C **10**, 257 (1977).
[4] G. Parisi, Phys. Rev. Lett. **43**, 1754 (1979).
[5] H. Rieger, *Annual Reviews of Computational Physics II*, edited by D. Stauffer (1995).
[6] D. P. Belanger and T. Nattermann, *Spin Glasses and Random Fields*, edited by A. P. Young (1998).
[7] R. L. C. Vink, K. Binder and H. Löwen, Phys. Rev. Lett. **97**, 230603 (2006).
[8] A. J. Bray and M. A. Moore, J. Phys. C **18**, L927 (1985).
[9] H. Rieger, *Lecture Notes in Physics* (Springer-Verlag, Heidelberg, 1998).
[10] T. H. Cormen, C. E. Leiserson and R. L. Rivest, *Introduction To Algorithms*, 1st ed. (MIT Press, Cambridge, 1990).
[11] C. H. Papadimitriou, *Computational Complexity*, 1st ed. (Addison-Wesley, Reading, MA, 1994).
[12] J. C. Picard and H. D. Ratliff, Networks **5**, 357 (1975).
[13] M. R. Swift, A. J. Bray, A. Maritan, M. Cieplak and J. R. Banavar, EPL (Europhysics Letters) **38**, 273 (1997).
[14] E. T. Seppäla and M. J. Alava, Phys. Rev. E **63**, 066109 (2001).
[15] J. D. Stevenson and M. Weigel, EPL (Europhysics Letters) **95**, 40001 (2011).
[16] N. G. Fytas and V. Martín-Mayor, Phys. Rev. Lett. **110**, 22720 (2013).
[17] A. Hartmann and H. Rieger, *Optimization Algorithms in Physics* (John wiley & Sons, 2004).
[18] M. Matsumoto and T. Nishimura, ACM Trans. Model. Comput. Simul. **8**, 3–30 (1998).
[19] Y. Boykov and V.Kolmogorov, IEEE Transactions on PAMI **26**, 1124–1137 (2004).
[20] K. Binder, Z. Phys. B Condensed Matter **50**, 343–352 (1983).
[21] L. R. Ford and D. R. Fulkerson, Canadian J. Math **8**, 399–404 (1956).
[22] E. A. Dinic, Soviet Math. Dokl. **11**, 1277 (1970).
[23] J. Edmonds and R. M. Karp, J. ACM **19**, 248 (1972).
[24] Y. Boykov and V.Kolmogorov, in: International Workshop on Energy Minimization Methods in Computer Vision and Pattern Recognition (EMMCVPR), 359–374 (Springer-Verlag, 2001).
[25] A. V. Goldberg and R. E. Tarjan, J. ACM **35**, 921 (1988).
[26] B. V. Cherkassky and A. V. Goldberg, Algorithmica **19**, 390–410 (1997).
[27] A. A. Middleton, Phys. Rev. Lett. **88**, 017202 (2001).
[28] A. A. Middleton and D. S. Fisher, Phys. Rev. B **65**, 134411 (2002).
[29] N. G. Fytas and V. Martín-Mayor, Phys. Rev. E **93**, 06330 (2016).

Sharpening the Universality of Network Resilience Patterns using Motifs

Huixin Zhang, Sean P. Cornelius*

Department of Physics, Ryerson University, 350 Victoria Street, M5B 2K3, Toronto, Canada

E-mail: cornelius@ryerson.ca

Abstract. The resilience of networked systems to perturbations is a fundamental problem with applications to ecosystem management, financial system stability, and cell reprogramming. This key challenge is that in high-dimensional systems, there is no "oracle" that can predict, a priori, which changes to a nonlinear system's parameters will be harmless vs. which will cause a system-wide failure (bifurcation). Here, we present a proof of principle using the Florida Bay food web network, showing how one can use higher-order network structure to arrive at a reliable, universal scalar indicator of a system's proximity to a bifurcation. Our framework builds on and sharpens a recently introduced mean-field theory for nonlinear dynamics on networks. We find that by incorporating information on high-order network structure in the form of network motifs, the prediction of resilience is greatly improved, especially near a bifurcation point. Our results stress the key role of higher-order structure in driving a system's dynamics, offering new ways to anticipate and prevent the collapse of large networks raging from ecosystems to infrastructure networks.

1. Introduction

Most real systems possess some degree of *resilience*–an ability to inhabit a desirable dynamical state despite perturbations to one or more system parameters [1, 2]. This ability is not without limits, however. Whether caused by a terrorist attack in an infrastructure network [3], a new pathogen in a social network [4], or climate change in an ecological network [5], at some point a change in parameters can be sufficiently large to destabilize the system's "good" state as it crosses a bifurcation point [6]. The net result is a loss of resilience, manifesting as power blackouts [7], epidemics [8], and multi-species extinctions [9, 10]. In order to anticipate and prevent these catastrophic outcomes, we must be able to judge just how close a system is to one of these "tipping points" [11].

Unfortunately, most theoretical frameworks for detecting proximity to a bifurcation–such as critical slowing down [12, 13, 14]–are based on one-dimensional systems having a single control parameter. Their intuition does not necessarily extend to large networked systems, which have many dynamical degrees of freedom and a large number of parameters encoding nodal properties and the network's wiring diagram. Such systems can exhibit multiple bifurcation [15] surfaces coexisting in extremely high-dimensional parameter space. In general, there is no way to predict which sequences of perturbations are benign vs. which will lead to a loss of resilience [16].

Recent high-profile research [17] has made significant advances in bridging this gap. Specifically, the authors devise a way to map a general class of nonlinear dynamics on networks to an effective scalar system, with a single control parameter. This "scalarization" is based on averaging over nearest neighbors, thereby explicitly incorporating the detailed network structure into the mean-field representation. The authors show that in large part, this maps diverse behaviors onto a single, universal resilience function, with all bifurcations largely collapsing onto a single bifurcation point in 1D. Some ambiguity remains, however, in the vicinity of this point; indeed, near a bifurcation, the states of different nodes can vary dramatically [18]; in an ecological network, for example, some neighbors of a node may be extinct while others remain alive [4, 5, 19]. In this case, simply treating all network links equally may overlook other important structural information relevant to the system's dynamical fate.

In this paper, we propose a way to sharpen the mean-field prediction of network resilience by explicitly incorporating information on higher-order network structure—patterns of connectivity beyond individual links (dyads). We use ecological dynamics and the Florida Bay food web network as a case study. Food webs are an example *par excellence* of networked systems where higher-order structure matters, displaying large-scale patterns of connectivity patterns such as trophic communities [20, 21] and nestedness [22]. Here, we will focus on one of the simplest forms of higher-order structure: network motifs—that is, small, frequently-occurring subgraphs on $n > 2$ nodes [23]. Here, we study the case $n = 3$, which yields (up to isomorphism) a total of 13 weakly-connected, directed sub-graphs containing 3 species.

2. Results

Model

We start by considering a general network consisting of N entities (nodes), where the state x_i of node i follows:

$$\frac{dx_i}{dt} = F(x_i) + \sum_{j=1}^{N} A_{ij} G(x_i, x_j). \tag{1}$$

On the right hand of equation (1), the first term (F) describes the self-dynamics of a given node, while the second term (G) captures interactions between node i and its neighbors. Finally, the adjacency matrix A_{ij} encodes the network wiring diagram and the strength of these interactions. Through appropriate choices of F and G, equation (1) can capture many common nonlinear processes on networks, including mutualistic dynamics in ecological networks [24], signaling in gene regulatory networks [25], and the dynamics of power supply networks [26].

In this paper, we focus on the dynamics of food-web networks, by choosing logistic growth for F and saturating mutualistic dynamics for G. This results in the following rate equation for the dynamics of the population (x_i) of species i:

$$\frac{dx_i}{dt} = x_i \left(r_i - s_i x_i + \frac{\sum_{j=1}^{N} A_{ij} x_j}{1 + h \sum_{j=1}^{N} A_{ij} x_j} \right). \tag{2}$$

Here, the first term r_i describes the intrinsic reproductive rate of species i [27]; the second term describes intraspecific competition (with strength s_i) within species i [28]; and the third term captures mutualistic interactions between species, where $A_{ij} = 1$ if there is a network link from species i to species j, and $A_{ij} = 0$ otherwise. The sigmoidal form of G models the fact that, when the abundance of two mutualistic partners becomes too high, the beneficial effect conferred by the interaction saturates. The strength of this nonlinearity is captured by the exponent h. We explore the dynamics represented by equation (2) on the Florida Bay food web [29], consisting of $N = 128$ species (nodes) and $L = 2106$ directed links. For simplicity, we set $r_i = -1$, $s_i = 1$, and $h = 0.2$ identically for all species, in all simulations that follow.

The loss of network links representing trophic interactions is a key consequence of habitat destruction [30], and can serve as a harbinger of extinction risk. In figure 1(b), we show 100 realizations of a simulated link-loss process, highlighting one of them in black. In each realization, we initialize $\vec{x}$ as the unique co-existing stable fixed point, wherein $x_i > 0$ for all species. We then randomly remove 2% of the original network's links (setting the corresponding entries A_{ij} to zero) at each step, for a total of 50 steps. After each link removal, we simulate the dynamics of equation (2) for a total of $100,000$ time units, allowing the system to reach a new stable fixed point. We then record the average species population $\langle x_i \rangle$. Because each realization represents a different sequence of perturbations, the collapse of the system (as measured by $\langle x_i \rangle$ falling to near-zero) occurs at a different step in each. This illustrates the fundamental challenge we seek to overcome: namely, that the bifurcation point happens unpredictably.

Resilience Framework

Recent high-profile research [17] has introduced an effective mean-field theory for the dynamics of equation (1). Specifically, one can introduce an operator

$$\mathcal{L}(A, \mathbf{y}) = \frac{\mathbf{1}^{\mathrm{T}} A \mathbf{y}}{\mathbf{1}^{\mathrm{T}} A \mathbf{1}}, \tag{3}$$

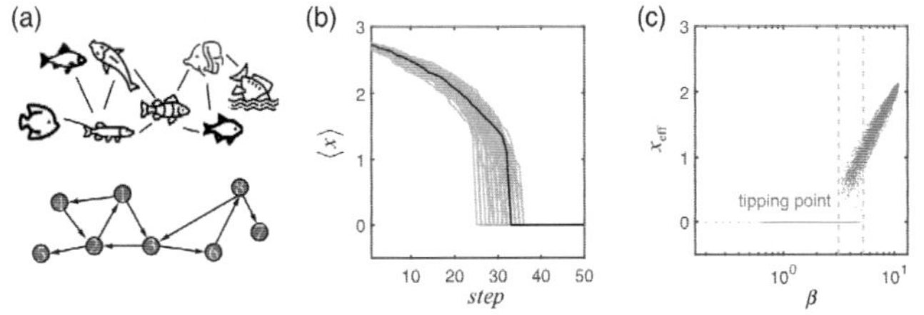

Figure 1. Demonstration of resilience universality in ecological dynamics. (a) Schematic illustration of a marine food-web network, in which nodes represent species and links represent predator-prey interactions. (b) The average value is a common scalar indicator of the system-level health, which is agnostic of the details of the system's detailed network structure. We ran the dynamics of equation (2) on the Florida Bay food web [29] and monitored the average equilibrium abundance of all species x_i across 100 realizations of the link-loss process, highlighting one in black. In each realization, we randomly delete 2% of the original number of links at each step. (c) The same data, viewed using the scalar mean-field theory of ref. [17], which summarizes the system state with a network-aware average (x_{eff}) and folds all network parameters (the N^2 entries of the adjacency matrix A) into a single effective control parameter (β). In this space, all sequences of link removals largely collapse onto a universal curve, with a one-to-one relationship between β and the qualitative nature of the average species abundance x_{eff} (high or low), *except* near the tipping point. Dashed lines show the range of the control parameter β for which a one-to-two relationship exists with x_{eff}.

which maps a vector $\mathbf{y} = (y_1, ..., y_N)^{\mathrm{T}}$ to a scalar, representing the average of $\mathbf{y}$ over the links (nearest neighbors) in a network. Here, A is the adjacency matrix from equation (1) and $\mathbf{1} = (1, \ldots, 1)^{\mathrm{T}}$. Using this, one can define an effective scalar order parameter x_{eff} as

$$x_{\text{eff}} = \mathcal{L}(A, \mathbf{x}) = \frac{\mathbf{1}^{\mathrm{T}} A\mathbf{x}}{\mathbf{1}^{\mathrm{T}} A\mathbf{1}}, \tag{4}$$

representing the *average nearest neighbour activity.* One can likewise define an effective interaction strength (control parameter) via

$$\beta = \mathcal{L}(A, A\mathbf{1}), \tag{5}$$

capturing the aggregate impact of node i's neighbors on its dynamics. Thus, perturbations to the network structure (A) translate to changes in both β and x_{eff}. In ref. [17], the authors find that the abstracted information contained in β can, except near the bifurcation, uniquely predict whether the system is resilient or has suffered a system-level collapse (figure 1(c)).

Network motifs

Higher-order connectivity patterns are essential to the fundamental mechanisms that govern the behavior of many complex systems [31, 32, 23]. A prominent example of higher-order structure in networks is network motifs—small network sub-graphs that are over- or under-represented relative to comparable random networks [23]. In different network systems, specific network motifs correspond to important structural or dynamical features of the system at hand. For example: 1) closed triangular motifs (figure 2, M1–M7) represent clustering of ties in social networks [33]; 2) in neuronal networks, open bidirectional wedges (figure 2, M13) represent the composition of brain hubs [34]. Finally, in food webs [21, 35] Motif M5 captures a hierarchical flow of energy considered being a building block of food webs [36], in which a higher trophic level species preys on different levels of a food chain. Likewise, Motif M6—modeling two species that prey on each other and then compete to feed on a common third species—is predicted to be prevalent within the *niche model* [21]—a widely-used generative model for food-web structure. Given the importance of specific network motifs to the dynamics of ecological networks, we hypothesize that "some links are more equal than others" in determining the system's fate.

Given a network, one way to emphasize the potential importance of a specific motif M_k is to weight the nonzero entries of the adjacency matrix according to how many times the corresponding link

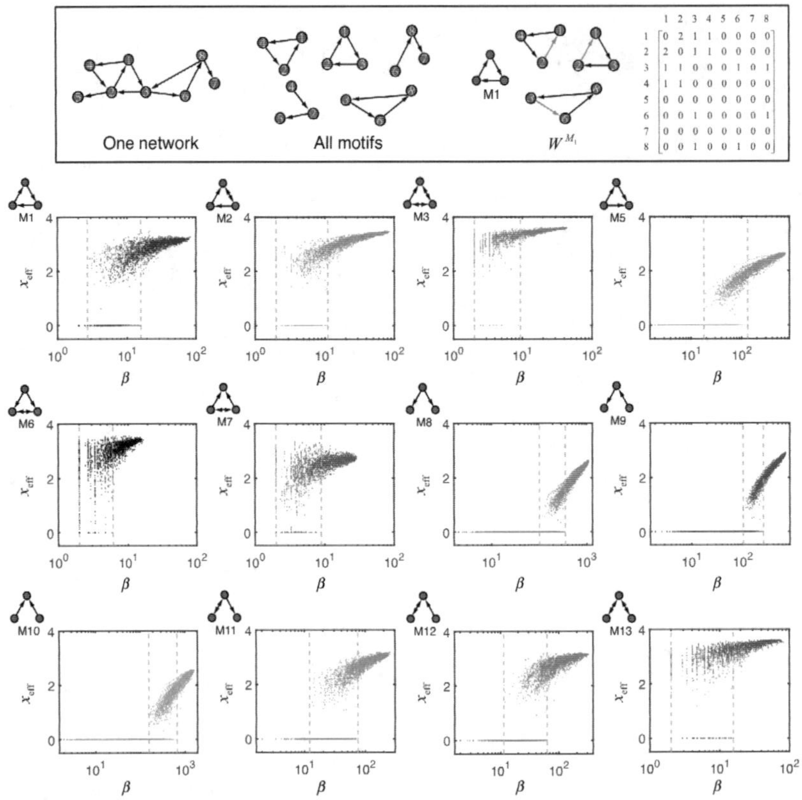

Figure 2. Predictive power of different motifs on the resilience. (top) There are 6 instances of different motifs on $n = 3$ nodes in this small network, and three of them are of motif M1 (a directed 3-cycle). We can build a *motif-weighted adjacency matrix* W for a specific motif by counting the number of times that each link appears in an instance of that motif. (top right) The motif-weighted adjacency matrix W^{M_1} for motif 1. For example, consider the link between node 1 and node 2 (red link). It participates in two instances of motif M1 (*i.e.*, involving nodes 1-2-3 and 1-2-4). As a consequence, $W_{12}^{M_1} = 2$. (bottom) We show the same 100 link loss processes from figure 1, using the the effective abundance $x_{\text{eff}} = \mathcal{L}(W, \mathbf{x})$ and the effective interaction strength $\beta = \mathcal{L}(W, W\mathbf{1})$ for each possible directed motif on $n = 3$ species (with the exception of Motif M4—representing a complete directed graph on 3 nodes—which never occurs in this network). Dashed lines show the upper and lower limits of β for which a one-to-two relationship exists with x_{eff}.

appears in an instance of that motif [23]. We denote this *motif-weighted adjacency matrix* by W^{M_k}. Figure 2 (top) shows the construction of the W^{M_1} for motif 1 (a directed 3-cycle) in a small example network. By substituting W^{M_k} in place of the adjacency matrix A in the operator $\mathcal{L}$, we can obtain *motif-weighted* versions of the effective system state and interaction strength as

$$\begin{aligned} \beta^{M_k} &= \mathcal{L}(W^{M_k}, W^{M_k}\mathbf{1}), \\ x_{\text{eff}}^{M_k} &= \mathcal{L}(W^{M_k}, \mathbf{x}). \end{aligned} \tag{6}$$

Figure 2 (bottom panels) shows the same 100 link-loss processes from figure 1, but viewed in $(\beta^{M_k}, x_{\text{eff}}^{M_k})$ space for each candidate directed motif k †. We see that there are mainly two qualitative

†In the Florida Bay food web, motif M4—a complete directed graph on 3 nodes—is absent, hence we do not consider it.

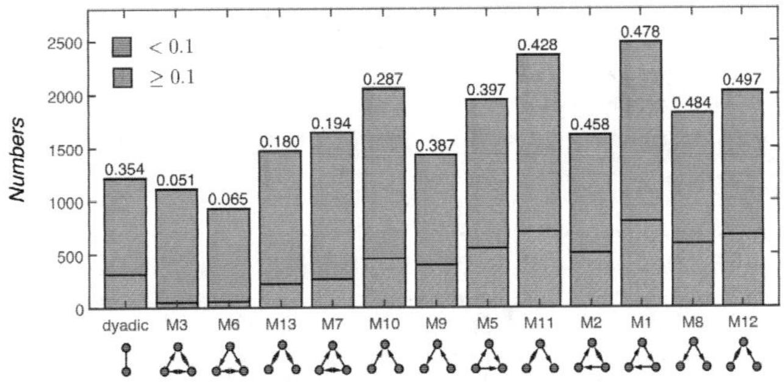

Figure 3. The prediction of resilience when approaching tipping points. For both the original ('dyadic') and the motif-aware ('M') versions of the mean-field resilience framework, we numerically identify the range of the control parameter β for which a one-to-two relationship exits with the effective system state (dashed lines in figure 1(c) and figure 2). For the simulation data falling in this range, we then calculate the fraction in the resilient state (x_{eff} is high) vs. the extinct/failed state ($x_{\text{eff}} = 0$). For concreteness, we use a threshold of $x_{\text{eff}} = 0.1$ to separate these outcomes. The total height of each bar indicates the number of data points falling into the "ambiguous region" of β, and the colors denote the portion of those corresponding to each outcome of x_{eff}. The bars corresponding to the motif-based analyses are sorted in ascending order by this probability. We see that two motifs (M3 and M6) offer a sharper picture of the system's resilience. In both cases, fewer perturbed networks fall into the ambiguous region (shorter bar), and even those that do overwhelmingly have one outcome for the system state (in these cases, x_{eff} = high). Three other motifs also share this latter property; in the dyadic resilience framework, the ambiguous values of β split roughly $65\% - 35\%$ into x_{eff} being high/low (respectively). In contrast, under motifs M13, M7, and M9, a given value of β implies an upward of 70% chance of a single outcome.

patterns for the relationship between β and x_{eff}:

- Case 1: $\beta_c > 0$. For motifs M1, M2, M3, M6, M7, M13.
- Case 2: $\beta_c \approx 0$.. For motifs M5, M8, M9, M10, M11, M12.

Here, β_c denotes the "tipping point" in each case, *i.e.*, the point where x_{eff} has collapsed to zero in all simulations. Notably, the motifs in Case 1 all have two-way links except M1. But M1 has a loop and seems to fall between the two cases. Among the motifs in Case 2, it is noteworthy that none contains a directed cycle. Besides, for motifs M11 and M12 possessing two-way links, they share similar patterns with Case 1.

As in the case of the original (non-motif) resilience framework (figure 1(c)), we find that regardless of the motif used, all sequences of perturbations largely fall onto a universal curve $x_{\text{eff}}(\beta)$. Knowledge of the individual species' populations or the exact sequence of perturbations is therefore not required to predict the system's resilience. Indeed, all that required is the value of the parameter β, *provided* that $x_{\text{eff}}(\beta)$ is one-to-one. But how often and for which motifs is this true?

With and without higher-order structure

As noted above, the ideal scenario for predicting system-level resilience is a one-to-one relationship between β and x_{eff}. However, there exists the possibility that for a single β there are multiple values of x_{eff}, especially near the tipping point. In this case, we would like the least amount of ambiguity possible; any given value of β should map to one outcome (x_{eff} = high or x_{eff} = low) with much higher probability than the other.

To quantify the ability of a given resilience framework to disambiguate "good" vs. "bad" system states, we first find the range of β for which a one-to-two relationship exits with x_{eff} in our simulations for each candidate motif. We set 0.1 as the threshold to separate x_{eff} as the resilient state ($x_{\text{eff}} > 0$) and the non-resilient state ($x_{\text{eff}} \approx 0$). We then tabulate the portion of the networks in this range of β that correspond to one outcome vs. the other.

In figure 3, we show the amount of ambiguity in resilience prediction, comparing the mean-field dynamics with weighting for each possible motif (figure 2, bottom panels), alongside that without motifs (figure 1(c)). It can be seen that the possibility of one-to-two is around 1/3 without accounting for a higher-order structure. On the other hand, for motifs M3 and M6, the possibility of a single value of β mapping to two qualitatively different values of x_{eff} is only around 5%. Accordingly, there is a sharp transition to failure ($x_{\text{eff}} \to 0$) at a well-defined value of β.

3. Conclusion and Discussions

In this case study, we demonstrate that specific network motifs can help to improve the predictability of network resilience, mapping different sequences of perturbations to a sharp bifurcation point in terms of a universal scalar parameter β. However, not all motifs confer this benefit. Why certain motifs perform better or worse in specific network dynamics is a topic for future study. Regardless, once the beneficial motifs are identified, they may be the key to reliable early warning signals of impending catastrophes in high-dimensional systems. This, too, is a promising direction for future work.

Acknowledgments

Huixin Zhang conceived of the project and performed all numerical simulations. Huixin Zhang and Sean P. Cornelius wrote the manuscript.

References

[1] Holling C S 1973 *Annual review of ecology and systematics* **4** 1–23
[2] Moritz C and Agudo R 2013 *Science* **341** 504–508
[3] Alderson D L, Brown G G and Carlyle W M 2015 *Risk Analysis* **35** 562–586
[4] Hodes G E, Pfau M L, Leboeuf M, Golden S A, Christoffel D J, Bregman D, Rebusi N, Heshmati M, Aleyasin H, Warren B L *et al.* 2014 *Proceedings of the National Academy of Sciences* **111** 16136–16141
[5] Malone E 2009 *Population Action International* 31
[6] Meyer K 2016 *Natural Resource Modeling* **29** 339–352
[7] Braun M, Hachmann C and Haack J 2020 *IEEE Power and Energy Magazine* **18** 54–63
[8] Massaro E, Ganin A, Perra N, Linkov I and Vespignani A 2018 *Scientific reports* **8** 1–9
[9] Solé R V, Alonso D and Saldaña J 2004 *Ecological complexity* **1** 65–75
[10] Walker R T 2020 *Environment: Science and Policy for Sustainable Development* **63** 15–25
[11] Dai L, Vorselen D, Korolev K S and Gore J 2012 *Science* **336** 1175–1177
[12] Veraart A J, Faassen E J, Dakos V, van Nes E H, Lürling M and Scheffer M 2012 *Nature* **481** 357–359
[13] van de Leemput I A, Wichers M, Cramer A O, Borsboom D, Tuerlinckx F, Kuppens P, van Nes E H, Viechtbauer W, Giltay E J, Aggen S H *et al.* 2014 *Proceedings of the National Academy of Sciences* **111** 87–92
[14] van de Leemput I A, Dakos V, Scheffer M and van Nes E H 2018 *Ecosystems* **21** 141–152
[15] Guckenheimer J 1984 *SIAM journal on mathematical analysis* **15** 1–49
[16] Kéfi S, Dakos V, Scheffer M, Van Nes E H and Rietkerk M 2013 *Oikos* **122** 641–648
[17] Gao J, Barzel B and Barabási A L 2016 *Nature* **530** 307–312
[18] Thompson J M T and Stewart H B 2002 *Nonlinear dynamics and chaos* (John Wiley & Sons)
[19] Steiner C F, Long Z T, Krumins J A and Morin P J 2006 *Ecology* **87** 996–1007
[20] Stouffer D B, Camacho J, Guimera R, Ng C and Nunes Amaral L 2005 *Ecology* **86** 1301–1311
[21] Stouffer D B, Camacho J, Jiang W and Nunes Amaral L A 2007 *Proceedings of the Royal Society B: Biological Sciences* **274** 1931–1940
[22] Bastolla U, Fortuna M A, Pascual-García A, Ferrera A, Luque B and Bascompte J 2009 *Nature* **458** 1018–1020
[23] Benson A R, Gleich D F and Leskovec J 2016 *Science* **353** 163–166
[24] Holland J N, DeAngelis D L and Bronstein J L 2002 *The American Naturalist* **159** 231–244
[25] Alon U 2019 *An introduction to systems biology: design principles of biological circuits* (CRC press)
[26] Zimmerman R D, Murillo-Sánchez C E and Thomas R J 2010 *IEEE Transactions on power systems* **26** 12–19
[27] Kondoh M 2003 *Science* **299** 1388–1391
[28] Spagnolo B, Fiasconaro A and Valenti D 2003 *Fluctuation and Noise Letters* **3** L177–L185
[29] Ulanowicz R E and DeAngelis D L 2005 *US Geological Survey Program on the South Florida Ecosystem* **114** 45
[30] Melián C J and Bascompte J 2002 *Ecology Letters* **5** 37–46
[31] Rosvall M, Esquivel A V, Lancichinetti A, West J D and Lambiotte R 2014 *Nature communications* **5** 1–13
[32] Shen-Orr S S, Milo R, Mangan S and Alon U 2002 *Nature genetics* **31** 64–68
[33] Holland P W and Leinhardt S 1972 Holland and leinhardt reply: some evidence on the transitivity of positive interpersonal sentiment
[34] Honey C J, Kötter R, Breakspear M and Sporns O 2007 *Proceedings of the National Academy of Sciences* **104** 10240–10245
[35] Bascompte J 2009 *Science* **325** 416–419
[36] Bascompte J, Melián C J and Sala E 2005 *Proceedings of the National Academy of Sciences* **102** 5443–5447

Journal of Physics: Conference Series **2207** (2022) 012011 doi:10.1088/1742-6596/2207/1/012011

Diffusional characteristics of a Newtonian event-chain Monte Carlo in hard disk systems

H Banno, D Mugita and M Isobe

Graduate School of Engineering, Nagoya Institute of Technology, 466-8555, Japan

E-mail: h.banno.968@stn.nitech.ac.jp, cmm12069@ict.nitech.ac.jp, isobe@nitech.ac.jp

Abstract. The equilibration and relaxation times of the physical properties of time correlation functions have historically been of fundamental importance in terms of molecular simulation. An efficient algorithm for the acceleration of equilibration, termed the 'event-chain Monte Carlo' has been investigated over the past decade. Recently, by adding the collision rule often used in event-driven molecular dynamics, a Newtonian event-chain Monte Carlo was developed. In this event-chain Monte Carlo and variants thereof, optimal performance strongly depends on chain length, physical properties, and system size; the microscopic mechanism in play remains elusive. To identify key factors affecting equilibration, we investigated the diffusional characteristics per event (collision) for each algorithm, focusing systematically on the diffusion coefficients and pair dispersions in two-dimensional hard disk systems.

1. Introduction

In molecular simulation, 'equilibration' refers to the preprocessing step prior to a production run; such a step yields the physical properties in the equilibrium state of the specific system. However, the computational cost is often high, especially for high-density and larger scale particle systems, and may in fact sometimes exceed the human lifetime. The two major standard molecular simulation methods for many-body systems are well known as the molecular dynamics (MD) using Newton's equations of motion and the Markov chain Monte Carlo (MC) [1]. Various novel algorithms focusing on equilibration, seeking to correlate the relaxation time with attainment of equilibrium, have played major roles in the history of molecular simulation.

One important milestone in the effort to improve the efficiency of equilibration and the relaxation time is the so-called event-chain MC (ECMC) [2]. This, and its variants and extensions, have been actively investigated over the past decade [3]. Recently, a ECMC variant termed the Newtonian ECMC (NEC) [4] was proposed. The velocity collision rule was added to dictate the direction of particle movement; this is often used in the field of event-driven molecular dynamics (EDMD) [5, 6] of hard-sphere systems. The NEC clearly improves the efficiencies of the diffusion coefficient, nucleation rate, and melting process [4]. This efficient algorithm has been extended to anisotropic hard particles (i.e., polygons) without the need for any approximations [7]. In the ECMC and variants thereof, the systems are driven by the sequential collisions of many particles that seem to behave as a chain; the optimal performance strongly depends on chain length, physical properties, and system size. However, the microscopic mechanism in play remains elusive. We searched for the key factors explaining why the rate of equilibration differed between the algorithms. We compared the diffusional characteristics per event between the ECMC and NEC, with a particular (systematic) focus on the diffusion coefficient and pair dispersion of two-dimensional hard disk systems. Sec. 2 summarizes the model and numerical settings and defines the diffusion coefficient we used. Sec. 3 presents the results of our comparison of the diffusional characteristics of the algorithms; we systematically changed the chain length and the system size. Sec. 4 provides our concluding remarks.

2. Model and simulation methods

We investigated two-dimensional (2D) mono-disperse systems consisting of N hard disks in the absence of any external force, friction, or dissipation among the disks. The disks of radius σ were placed in a $L_x \times L_y (= A)$ rectangular box (i.e., $L_y/L_x = \sqrt{3}/2$). Periodic boundary conditions were imposed in both directions. The basic units of the system are the mass of one disk m, the disk diameter d $(= 2\sigma)$, and the energy $1/\beta$, from which we derive the unit of time as $d\sqrt{\beta m}$, where β is $1/k_BT$ using the Boltzmann constant k_B and the temperature T. With regard to the initial states, we prepared a couple of equilibrated configurations generated by sufficiently long EDMD runs at a packing fraction $\nu = N\pi\sigma^2/A = 0.45$; the system is in a liquid state under such circumstances. Note that 2D hard disk systems are governed by two phase transitions between a fluid, hexatic and solid phases, at around $\nu \sim 0.700$ and 0.716, respectively [8, 9, 10]. The production runs proceeded using the three different Markov-chain MC sampling algorithms described below: The ECMC exhibits some displacement variants. Movements can be restricted

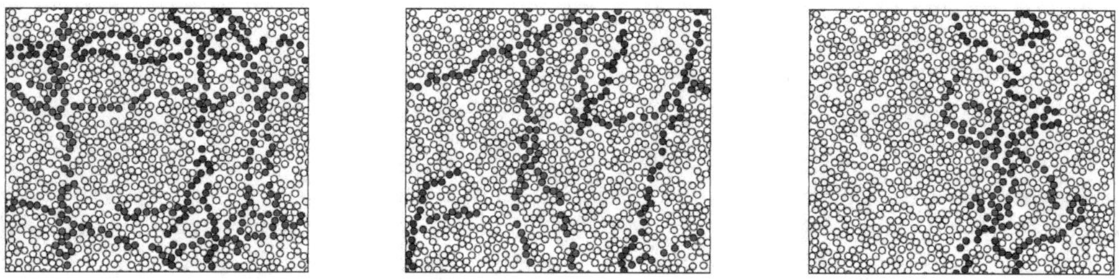

Figure 1. The typical trajectories of active particles involved in the events within the three chains (red, green, and blue) are displayed (from the left) for ECMC-xy, ECMC-all, and NEC when L_c/d (or $T_c v_{rms}/d$) $= 22.7$, respectively.

to only the positive x- and y- axes ('ECMC-xy') or may proceed in all possible directions that are uniformly sampled employing random numbers ('ECMC-all'). The NEC is also an ECMC variant; however, by introducing the Maxwell-Boltzmann velocity distribution of disks (thus the EDMD), the direction of displacement is determined by the collision rule of velocities. The basic algorithms for ECMC and its variants are as follows:

(i) A disk i in the system is selected by a random number and becomes the first active disk for the successive collisions (i.e., the starter of the event chain).

(ii) The direction of displacement (event) of the disk is determined by the corresponding rules for the variants of the ECMC. Note that in the NEC, the positions are updated as sequential collisions as for the ECMC; but the direction obeys the collision rule of velocities as for the EDMD. Disk i is moved until it collides with another disk j, and then the collided disk j is recognized as the active disk i for the subsequent displacement.

(iii) Repeat (ii) (i.e., a chain of collisions) until the total displacement of the active disk during one chain attains the predetermined chain length L_c. Note that in the NEC, the chain duration of the acculturated time T_c is introduced rather than the chain length L_c, because each disk has a velocity and its collision time affects the displacement length L_c via multiplication by the root mean square velocity $v_{rms} = \sqrt{\langle v^2 \rangle}$, where $\langle v^2 \rangle$ is the ensemble average of the squared velocity).

We define the self-diffusion coefficient in two dimensions in terms of $N^*_{disp} = N_{disp}/N$ (N_{disp}: displacement number), as is the case for three dimensions [4],

$$D^s_{disp} = \lim_{N^*_{disp} \to \infty} \frac{\sum_{i=1}^{N} |\Delta \boldsymbol{x}_i - \langle \Delta \boldsymbol{x} \rangle|^2}{4NN^*_{disp}}, \tag{1}$$

where $\Delta \boldsymbol{x}_i = |\mathbf{x}_i(t) - \mathbf{x}_i(0)|$ is the displacement of disk i at time t from the initial position $\mathbf{x}_i(0)$, and $\langle \Delta \boldsymbol{x} \rangle$ is the ensemble average. We also consider the pair separation of two disks $\Delta R^k(t) = |\mathbf{x}_{i'}(t) - \mathbf{x}_{j'}(t)|$ at time t, where k is the pair index of the two disks (i', j') that are initially the nearest neighbors ($t = 0$). Using $\Delta R^k(t)$, the pair diffusion coefficient D^p_{disp} is defined as:

$$D^p_{disp} = \lim_{N^*_{disp} \to \infty} \frac{\sum_{k=1}^{N_p} |\Delta R^k(t) - \Delta R^k(0)|^2}{4N_p N^*_{disp}}, \tag{2}$$

where N_p is the number of pairs for sampling.

3. Results

3.1. Chain length dependency

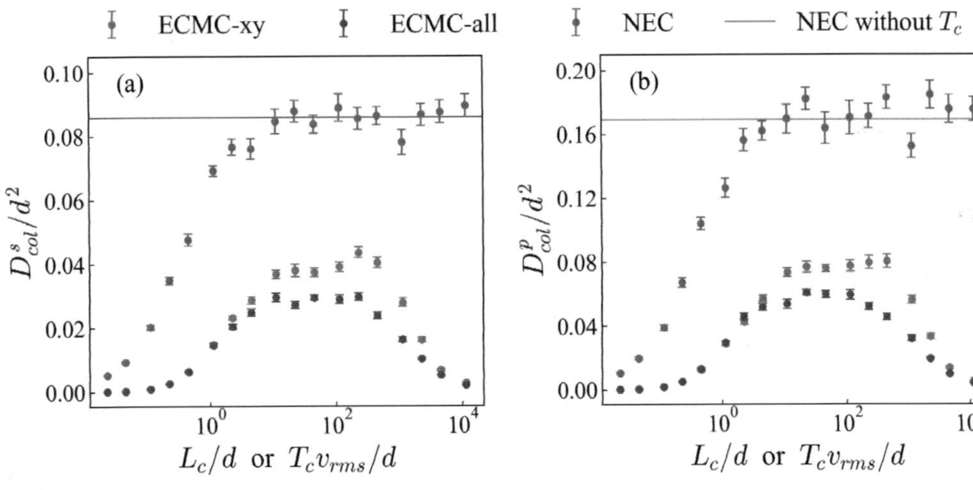

Figure 2. The chain length (or chain duration) dependencies of the three algorithms (ECMC-xy, ECMC-all, and NEC) in terms of (a) the self- D^s_{disp} and (b) the pair-diffusion coefficient D^p_{disp} figures in liquid hard disk systems at $(N, \nu) = (1,024, 0.45)$. Each datum was obtained by averaging 25 independent samples. The horizontal (green) lines in (a) and (b) indicate the values of the NEC without T_c (i.e., when the chain duration was infinite).

Figure 2 presents the chain length L_c (or chain duration T_c) dependencies of (a) the self- D^s_{disp} and (b) the pair-diffusion coefficient D^p_{disp} for the three algorithms (ECMC-xy, ECMC-all, and NEC). The chain length dependencies of (a) and (b) are similar, but the values of D^p_{disp} are approximately twice those of D^s_{disp} for all data yielded by both chain lengths (durations) and algorithms. Because the system is in a homogeneous liquid state, this work diffusion coefficient would be expected to be a standard Brownian. The factor of 2 might reflect how the diffusion coefficients of the one- or two-disk scenarios were calculated. For all three algorithms, the diffusion coefficients rise as the chain length (or duration) increases. As L_c further increases, the diffusion coefficients plateau at the highest values and then decrease only for ECMC-xy and ECMC-all. The L_c peak breadth differs markedly between ECMC (-xy, -all) and NEC, suggesting that ECMC is associated with an optimal chain length region. We also found that the ECMC-xy diffusion coefficient tended to be slightly larger than that of ECMC-all, around $L_c/d = 10 \sim 10^3$.

3.2. System size dependency

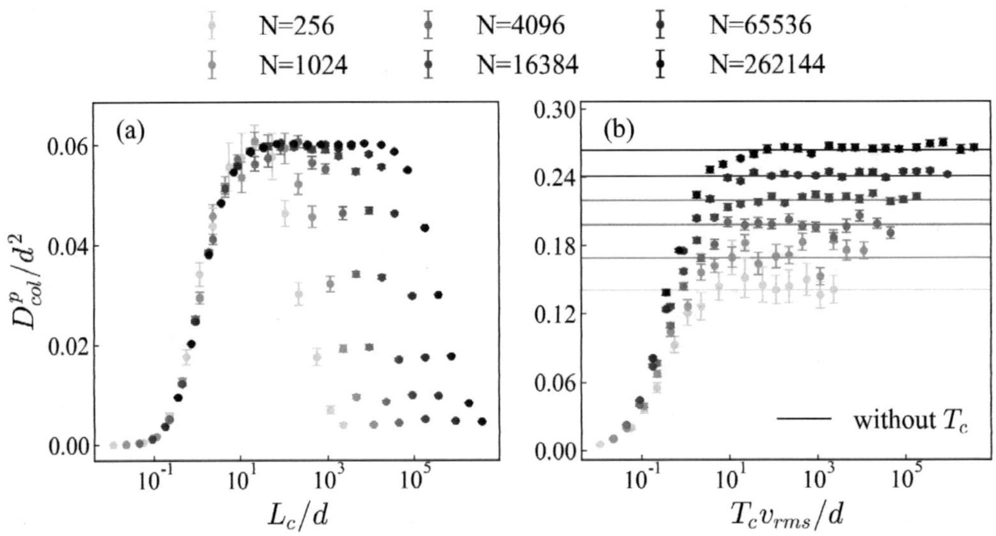

Figure 3. System size dependencies of the pair-diffusion coefficients D^p_{disp} in terms of the chain length (or duration) of (a) ECMC-all and (b) NEC, respectively. The system sizes differ remarkably.

Next, we focus on the pair-diffusion coefficient D^p_{disp} as a function of the chain length by changing the system size (particle number) in (a) ECMC-all and (b) NEC. In Fig. 3, the D^p_{disp} data exhibit the tendency described in Sec. 3.1. A remarkable difference emerges if the chain length is increased beyond the peak and then decreases in ECMC-all. For ECMC-all, the D^p_{disp} attains almost the same values as those of the universal curves for systems of all sizes. The persistence of the highest D^p_{disp} as L_c increases is longer in larger systems. Finally, all D^p_{disp} of systems of different sizes fall to near zero. In contrast, in the NEC, growth of the values of highest D^p_{disp} is observed by increasing the system size, which converges to the value of D^p_{disp} without the need for T_c, (i.e., an infinite chain duration).

3.3. Spatial distributions of displacements between algorithms

To explore why the algorithms differed in terms of the system size-dependence, we calculated the spatial distributions of the accumulated displacement (sampling) numbers per particle N^p_{disp}

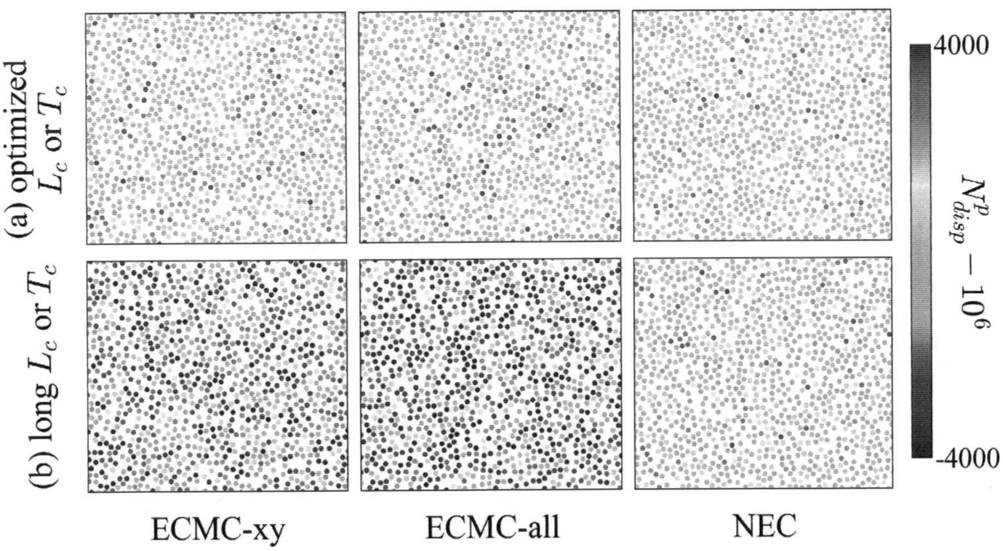

Figure 4. Spatial distributions of the accumulated displacement (sampling) numbers per disk N^p_{disp} within $N_{disp} = 10^6$, where the three algorithms are driven at (a) the optimal chain length (L_c/d or $T_c v_{rms}/d \sim 22.7$) or (b) a sufficiently long length (L_c/d or $T_c v_{rms}/d \sim 10^4$).

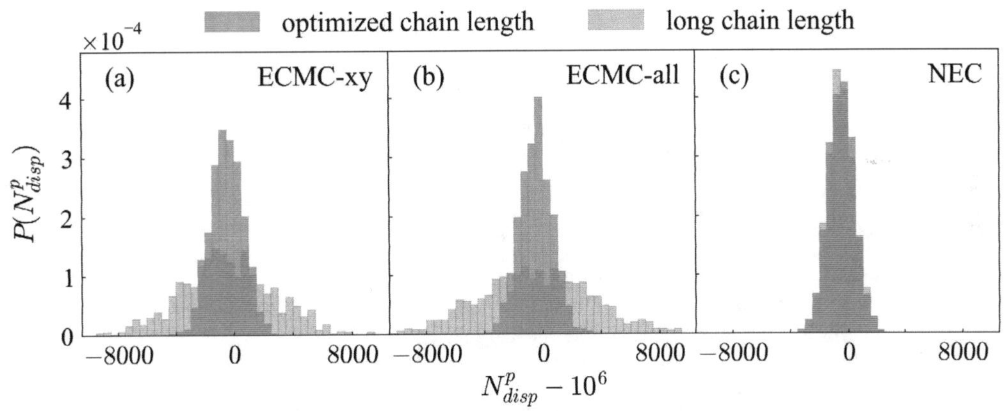

Figure 5. Probability density distributions (PDFs) $P(N^p_{disp})$ in terms of the displacement numbers in Fig. 4 for (a) ECMC-xy, (b) ECMC-all, and (c) NEC. The data are the averages of 10 independent samples. Spatial inhomogeneity is obvious in the PDFs of ECMC-xy and -all at long chain lengths, which may reduce the diffusion coefficients.

and the probability density distributions $P(N^p_{disp})$ in the $(N, \nu) = (1,024, 0.45)$ systems. Fig. 4 presents a snapshot of the spatial distributions of the accumulated displacement (sampling) number per disk within $N_{disp} = 10^6$ for each algorithm when the chain lengths were optimal or sufficiently long. ECMC-xy and -all exhibit strong inhomogeneities of displacement numbers per particle for the entire systems with long chain lengths. These results are confirmed by the probability distribution function $P(N^p_{disp})$ shown in Fig. 5. In the NEC, the direction of displacement is randomly sampled via elastic velocity collisions in liquid, ensuring a homogeneous displacement distribution and efficient (normal) diffusion. These results suggest that a specific correlation may exist between the spatial inhomogeneity of sampling and the highest value of D^p_{disp} by L_c as shown in Fig. 2 and 3.

In a smaller system, if the event chain exceeds the system size imposed by the periodic boundary conditions, some disks would repeatedly collide during a single event chain. The spatial inhomogeneity of N^p_{disp} may also explain the system size-dependence of the diffusion coefficient, thus the decrease in the highest diffusion coefficient D^p_{disp}.

4. Concluding remarks

We systematically investigated the diffusional characteristics per event (collisions) for ECMC-xy, ECMC-all, and NEC in liquid 2D hard disk systems. We defined self- and pair-diffusion coefficients based on the displacement number and compared the dependencies of both the chain length and the system size among the algorithms. We then visualized the spatial distributions of the accumulated numbers of displacements per particle and calculated probability density distributions. As the chain length increases, the diffusion coefficient grows to attain its highest value at a specific chain length. We assume that NEC has an advantage compared to ECMC in such a system because of the difference in the spatial homogeneity of sampling (displacement) given the random velocities of the disks after each displacement during the event chain. The disks are uniformly sampled in the liquid state.

We will further investigate the efficiencies of the diffusion coefficients in terms of CPU time. We will also evaluate the validity of sampling homogeneity in other situations, thus in co-existent and solid phases, and the equilibration processes commencing in the artificial (non-equilibrated) states.

Acknowledgments

We are grateful to Profs. Werner Krauth and Michael Engel for helpful discussions. M.I. was supported by JSPS KAKENHI Grant Number No. 20K03785. Part of the computations was performed using the facilities of the Supercomputer Center, ISSP, Univ. of Tokyo.

References

[1] W. Krauth, *Statistical mechanics: Algorithms and computations* (Oxford University Press, Oxford, 2006)
[2] Bernard E P, Krauth W and Wilson D B 2009 *Phys. Rev. E* **80** 056704
[3] Krauth W 2021 *Front. Phys.* **9** 229
[4] Klement M and Engel M 2019 *J. Chem. Phys.* **150** 174108
[5] Alder B J and Wainwright T E 1959 *J. Chem. Phys.* **31** 459
[6] Isobe M 2016 *Mol. Sim.* **42** 1317.
[7] Klement M, Lee S, Anderson J A and Engel M 2021 *J. Chem. Theor. and Comp.* **17** 4686
[8] Alder B J and Wainwright T E 1962 *Phys. Rev.* **127** 359
[9] Bernard E P and Krauth W 2011 *Phys. Rev. Lett.* **107** 155704
[10] Engel M, Anderson J A, Glotzer S C, Isobe M, Bernard E P and Krauth W 2013 *Phys. Rev. E* **87** 042134

Resampling schemes in population annealing – numerical results

Denis Gessert[1,2], Martin Weigel[1,3] and Wolfhard Janke[2]

[1] Centre for Fluid and Complex Systems, Coventry University, Coventry CV1 5FB, UK
[2] Institut für Theoretische Physik, Universität Leipzig, IPF 231101, 04081 Leipzig, Germany
[3] Institut für Physik, Technische Universität Chemnitz, 09107 Chemnitz, Germany

E-mail: gessertd@uni.coventry.ac.uk

Abstract. Population annealing (PA) is a population-based algorithm that is designed for equilibrium simulations of thermodynamic systems with a rough free energy landscape. It is known to be more efficient in doing so than standard Markov chain Monte Carlo alone. The algorithm has a number of parameters that can be fine-tuned to improve performance. While there is some theoretical and numerical work regarding most of these parameters, there appears to be a gap in the literature concerning the role of resampling in PA. Here, we present a numerical comparison of a number of resampling schemes for PA simulations of the 2D Ising model.

1. Introduction

Systems with a rough free energy landscape, such as spin glasses, structural glasses and some polymers, to name a few, appear in many fields of research and are notoriously hard to simulate [1]. Over the last decades a number of methods have been developed with the aim of making such systems more accessible such as, e.g., multicanonical simulations [2], the Wang-Landau algorithm [3], parallel tempering [4] and the population annealing (PA) algorithm [5, 6] which is the subject of the present work.

While there is no clear-cut answer to the question which of these methods is best and the answer may depend on the system considered, we motivate the use of PA as follows: First, PA is well suited for the highly parallel execution on supercomputers, as the number of replicas is only limited by the underlying computing architecture and not the method itself. In contrast, in parallel tempering the level of parallelization is limited by the number of temperature points considered which is bounded by increasing round-trip times as the number of replicas grows. Additionally, with increasing population size not just the statistical error but also the bias is systematically reduced [7]. Second, as part of the PA algorithm an equilibration routine is run at each temperature, which typically consists of a number of Markov chain Monte Carlo (MCMC) sweeps (MCS). However, the type of update is not specified and can freely be replaced by, e.g., a cluster algorithm, a rejection-free update or even by molecular dynamics [8]. This allows for an easy adaptation of PA to the system under study, and any algorithmic improvement in MCMC immediately translates to PA.

The fundamental idea of PA, first presented by Hukushima and Iba [5, 6], is to perform parallel simulated annealing (SA) simulations (replicas) subject to a population-control step at each temperature. It is worth mentioning that thanks to this control step PA is an equilibrium

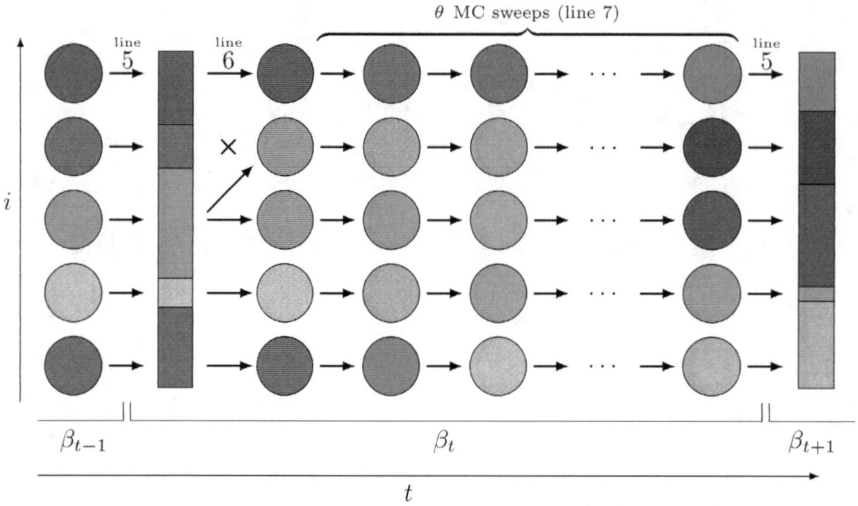

Figure 1. Schematic of the population annealing algorithm. Vertical axis corresponds to R replicas (here five) and horizontal axis to simulation time. Line numbers correspond to algorithm 1 in Appendix A.

method for calculating canonical averages although SA itself by default is not. The population control is implemented by a resampling phase employed at each temperature step (or in principle after M steps, but researchers have focused on $M = 1$) and is crucial to the performance of PA. Despite its importance to the success of the algorithm, it has received little attention in the literature – the effect of different resampling methods has not yet been systematically studied and particularly the fact that resampling introduces additional artificial noise into the simulations has not been discussed.

We use the two-dimensional ferromagnetic Ising model in zero field corresponding to the Hamiltonian

$$\mathcal{H} = -J \sum_{\langle ij \rangle} \sigma_i \sigma_j \tag{1}$$

where the sum is over nearest neighbors $\langle ij \rangle$ only. In our simulations we consider lattices of size 64×64 with periodic boundary conditions. Our implementation of the PA algorithm is based on the Open Source GPU code of Ref. [9] that is available online. The code uses a checkerboard decomposed Metropolis update as an equilibration routine. All resampling methods were also implemented on GPUs.

2. Simulation method

2.1. Population annealing

In PA a population of $R_0 = R$ replicas is initialized at the starting inverse temperature β_0, chosen to be zero if possible; here, R is the target population size. What follows is an iteration scheme (see Figure 1) in which in each iteration the inverse temperature β_t is increased to $\beta_{t+1} = \beta_t + \Delta\beta$. The step $\Delta\beta$ has to be chosen appropriately and can in principle be adaptive [9, 10].

At the end of iteration $t - 1$, the population consists of R_{t-1} replicas at β_{t-1}. Following the line of thought of histogram reweighting [11], if we were asked to estimate the energy histogram at β_t based on the population at β_{t-1}, we would attribute a weight $w_i \propto \exp(-\Delta\beta E_i)$ to each replica i with energy E_i where the weights are normalized such that they sum to one. Next, the population is resampled according to these weights w_i, while attempting to keep the population size close to the target size R. In practice, the number of descendants r_i of replica i is, on average,

$$\langle r_i \rangle = \hat{\tau}_i = w_i R. \tag{2}$$

Clearly, this only specifies the mean of r_i and the exact way how r_i is determined depends on the used resampling method. Note that it is common that $r_i = 0$ for some replicas, in which case they are culled while others are replicated multiple times.

Immediately after resampling, the population is potentially highly correlated as it may contain identical configurations. To reduce the correlation among replicas, an equilibration routine is used (here a checkerboard decomposed version of the Metropolis algorithm, see Ref. [9] for details). After θ MCS, population averages for observables of interest are calculated. The scheme then proceeds with the iteration for inverse temperature β_{t+1} until the final temperature is reached.

For a pseudocode implementation of the PA algorithm we refer to Appendix A.

2.2. Resampling methods

When the number of copies of one replica r_i is chosen only based on the value of $\hat{\tau}_i$ of the same replica i, the resampling method is completely described by a univariate distribution $P_{\hat{\tau}_i}(r_i = k)$. In this case the sum of all r_i's is a random variable and thus the population size fluctuates with time. In this group we consider "nearest integer" resampling [12] and Poisson resampling [13] given by

$$P_{\hat{\tau}_i}(r_i = k) = \begin{cases} \hat{\tau}_i - \lceil \hat{\tau}_i \rceil & \text{if } k = \lceil \hat{\tau}_i \rceil \\ 1 - (\hat{\tau}_i - \lceil \hat{\tau}_i \rceil) & \text{if } k = \lfloor \hat{\tau}_i \rfloor \\ 0 & \text{else} \end{cases} \tag{3}$$

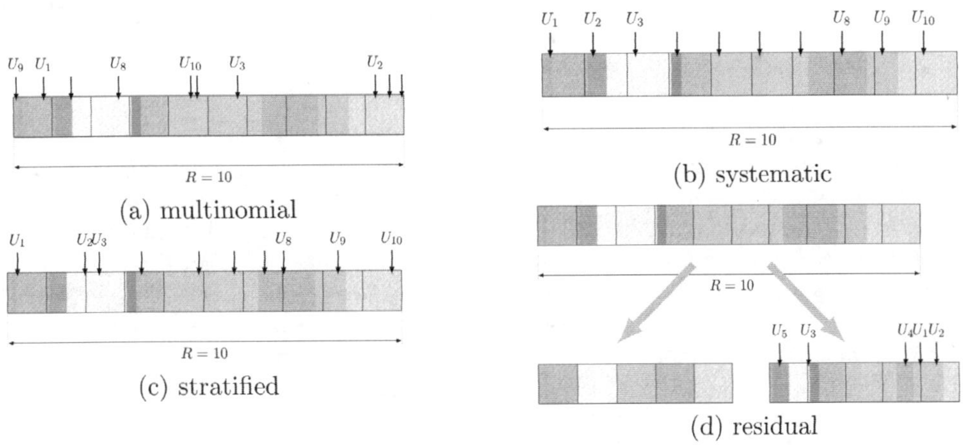

Figure 2. Visualization of various population size preserving resampling methods. Colored boxes correspond to different replicas i and their lengths are proportional to $\hat{\tau}_i$. The number of copies of replica i is determined by the number of arrows in box i. (a) - (c) R arrows are distributed over the interval $[0, R]$. (a) multinomial: Each arrow is uniformly distributed over the interval $[0, R]$. (b) systematic: The first arrow is uniformly distributed within $[0, 1]$. The remaining $R - 1$ arrows are placed with unit spacing. (c) stratified: In each square exactly one arrow is placed with uniform probability, i.e., $U_i \sim \mathcal{U}([i-1, i])$. (d) residual: At first each replica is copied $\lfloor \hat{\tau}_i \rfloor$ times. The population is brought to its original size by multinomially drawing from the residuals, i.e., performing (a) with $\tau_i - \lfloor \tau_i \rfloor$ replacing τ_i.

for nearest-integer and

$$P_{\hat{\tau}_i}(r_i = k) = \frac{\hat{\tau}_i^k}{k!} e^{-\hat{\tau}_i} \tag{4}$$

for Poisson, respectively. Here, $\lceil x \rceil$ denotes the smallest integer that is larger than or equal to x, i.e., rounding up, and similarly $\lfloor x \rfloor$ denotes the largest integer smaller than or equal to x, i.e., rounding down.

Vice versa, if one requires the population size to be fixed, the set $\{r_i\}$ is drawn at once and the individual r_i's may show some correlation. See Figure 2 for a visualization of the population size preserving resampling methods used here and the caption for explanations. These methods are quite well known in the field of particle filters [14] whereas within the PA community multinomial resampling is the only population size preserving method used so far.

In principle, one might come up with many more methods such as geometric resampling. Here we restrict ourselves to methods with sampling variance less than or equal to that of the widely used multinomial resampling.

2.3. Benchmarking quantities

The most obvious quantities to use for comparing the algorithmic performance are systematic and statistical errors. We looked at various observables and present here the errors measured for the specific heat C_v, as this is where we found the strongest differences among the methods.

Additionally, Ref. [12] introduced population quantities designed to identify the maximal correlation within the population introduced through the resampling step, i.e.,

$$\rho_t = R \sum_{i=1}^{R} \mathfrak{n}_i^2, \qquad \rho_s = R \exp\left(\sum_{i=1}^{R} \mathfrak{n}_i \ln \mathfrak{n}_i \right) \qquad \text{and} \qquad f = \sum_{i=1} \min\{1, R\mathfrak{n}_i\}, \tag{5}$$

where $\mathfrak{n}_i$ is the fraction of replicas descending from the initial replica i, ρ_t is known as the mean square family size, ρ_s as the entropic family size and f is the number of families. As the authors of Ref. [12] show, these quantities are closely related to each other.

3. Results

The following results were obtained from simulations of the Ising model on a square lattice with periodic boundary conditions of dimensions 64×64. The numerical data presented here was attained through independent PA simulations for each resampling method and was averaged over 5 000 simulations per method. The target population size is $R = 20\,000$ and the annealing schedule was chosen to be $\beta_k = k/300$ with $k \in \{0, \ldots, 299\}$. The choice of $\beta_0 = 0$ allows us to initialize each spin configuration simply by choosing every spin as up or down with equal probabilities which can easily be implemented. At each temperature $\theta = 5$ MCS were carried out.

Figure 3 shows bias and statistical error immediately after resampling (before the equilibration routine) of the specific heat. We calculate the bias by using the known exact solution [15, 16], whereas the statistical errors were obtained by estimating the standard deviation of independent runs. Both quantities peak around the critical temperature. The bias shows little deviation among the methods, although multinomial and Poisson resampling appear to have slightly stronger bias in the vicinity of the critical temperature. Around the critical temperature statistical errors are almost indistinguishable and away from criticality the statistical error using multinomial and Poisson resampling is significantly larger compared to the other methods. The inset, which shows all errors normalized to the nearest-integer errors, illustrates that Poisson and multinomial resampling have statistical errors of up to 25% higher than nearest-integer resampling and outside the depicted near-critical regime these even exceed 30%. We argue that near criticality fluctuations intrinsic to the model are very strong

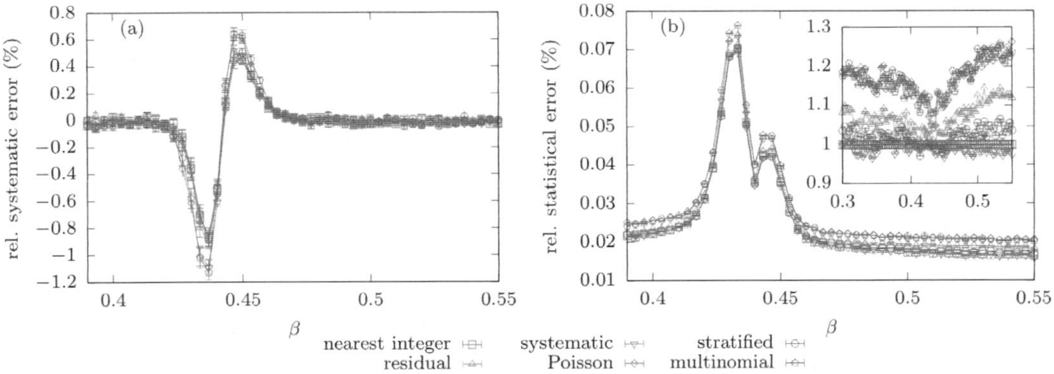

Figure 3. Relative (a) systematic and (b) statistical error of the specific heat measured after the resampling step for various resampling methods. (a) Systematic errors hardly differ between different resampling schemes. (b) The inset shows the statistical error relative to the error using nearest-integer resampling. Away from criticality, the curves differ significantly whereas around $\beta_c \approx 0.44$ the choice of the resampling method appears to have little effect on the statistical error.

and dominate the observed statistical errors whereas further away the resampling noise has a significant effect.

The family quantity ρ_t is shown in Figure 4(a). It increases rapidly around the critical point and slowly otherwise. It agrees with the previous observations in the sense that again multinomial and Poisson resampling stick out and ρ_t consistently takes its highest values for these two methods. Larger average family size typically corresponds to more correlation within the population and thus poorer statistics. The same observations are made for the entropic family size ρ_s and the inverse of the number of families R/f.

4. Conclusion

We have demonstrated that resampling has a significant effect on the data obtained through population annealing, and we see a difference in error bars of up to 30% using standard resampling methods. Furthermore, we identified that resampling matters most away from criticality when it contributes as much or more to the statistical error as the fluctuations intrinsic to the system. Through the study of family quantities we found that higher variance in r_i at

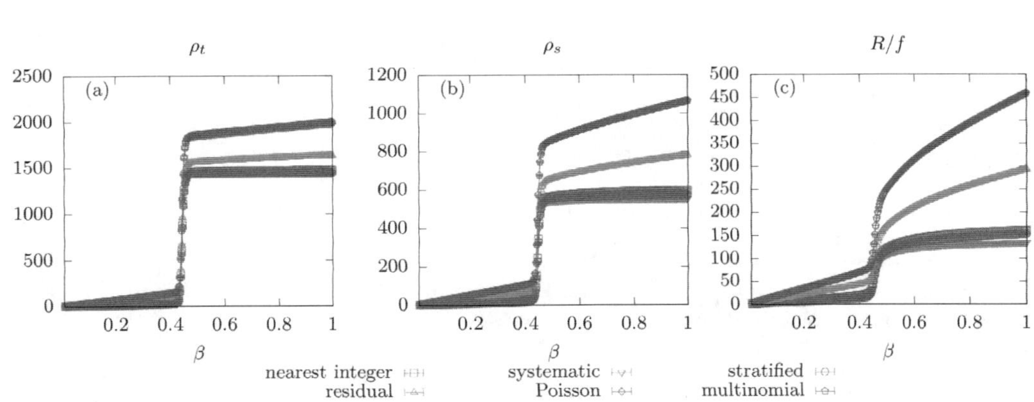

Figure 4. Family quantities (a) ρ_t, (b) ρ_s and (c) R/f for various resampling methods.

high temperatures causes fewer families of larger sizes surviving, thus causing more correlation in the population.

In a quest of finding the best resampling method nearest-integer resampling is the front runner due to it consistently outperforming the other methods and due to its simple implementation. Systematic resampling is also a good choice, in particular if a constant population size is desired, as it provides similar performance although its implementation is slightly more involved. In particular, we note that it performs significantly better than the multinomial method usually used for this purpose.

Acknowledgments
The project was supported by the Deutsch-Französische Hochschule (DFH-UFA) through the Doctoral College "$\mathbb{L}^4$" under Grant No. CDFA-02-07.

Appendix A. Population Annealing algorithm

Algorithm 1 Standard Population Annealing algorithm (introduced in Ref. [6])

Require: R,$\{\beta_k\}$,θ, M

$k \leftarrow 0$
Initialize population of R independent replicas at β_0
while $\beta_{k+1} < \beta_{\max}$ **do**
    $k \leftarrow t + 1$
    Calculate **Boltzmann weights** w_i of the replicas at β_t
    if $k \equiv 0 (\mathrm{mod}\ M)$ then **resample** population according to weights (on avg. $\tau_i = Rw_i$ copies of replica i)
    Monte Carlo update of the replicas (θ MC sweeps at β_k)
    Measure observables $\mathcal{O}$

Algorithm 1 shows the standard population annealing algorithm as originally devised by Hukushima and Iba [6]. The parameters are the target population size R (throughout the simulation R_i might fluctuate around R), the annealing schedule $\{\beta_k\}$, the number of equilibration updates θ (typically MCMC) between each temperature step and M, the resampling interval (usually $M = 1$).

References

[1] Janke W (ed) 2007 *Rugged Free Energy Landscapes – Common Computational Approaches to Spin Glasses, Structural Glasses and Biological Macromolecules* (*Lecture Notes in Physics* vol 736) (Berlin: Springer)
[2] Berg B A and Neuhaus T 1991 *Phys. Lett. B* **267** 249–53
[3] Wang F and Landau D P 2001 *Phys. Rev. Lett.* **86** 2050–3
[4] Hukushima K and Nemoto K 1996 *J. Phys. Soc. Jpn.* **65** 1604–8
[5] Iba Y 2001 *Trans. Jpn. Soc. Artif. Intell.* **16** 279–86
[6] Hukushima K and Iba Y 2003 *AIP Conf. Proc.* **690** 200–6
[7] Weigel M, Barash L, Shchur L and Janke W 2021 *Phys. Rev. E* **103** 053301
[8] Christiansen H, Weigel M and Janke W 2019 *Phys. Rev. Lett.* **122** 060602
[9] Barash L Y, Weigel M, Borovský M, Janke W and Shchur L N 2017 *Comput. Phys. Commun.* **220** 341–50
[10] Christiansen H, Weigel M and Janke W 2019 *J. Phys.: Conf. Ser.* **1163** 012074
[11] Ferrenberg A M and Swendsen R H 1988 *Phys. Rev. Lett.* **61** 2635–8
[12] Wang W, Machta J and Katzgraber H G 2015 *Phys. Rev. E* **92** 063307
[13] Machta J 2010 *Phys. Rev. E* **82** 026704
[14] Doucet A and Johansen A M 2011 *Handbook of Nonlinear Filtering* Crisan D and Rozovskii B (eds) (New York: Oxford University Press) chap 8.2, pp 656–704
[15] Beale P D 1996 *Phys. Rev. Lett.* **76** 78–81
[16] Kaufman B 1949 *Phys. Rev.* **76** 1232–43

Journal of Physics: Conference Series **2207** (2022) 012013 doi:10.1088/1742-6596/2207/1/012013

An agent-based modeling and simulation for the first stage of honeycomb construction

Takayuki Narumi[1], Takumi Akiyama[2], Maya Kageyama[2], Kenta Uemichi[2], Hisao Honda[3], Koichi Osaki[2]

[1] Department of Engineering, Yamaguchi University, Yamaguchi 755-8611, Japan
[2] Department of Mathematical Sciences, Kwansei Gakuin University, Hyogo 669-1337, Japan
[3] Graduate School of Medicine, Kobe University, Hyogo 650-0017, Japan

E-mail: `tnarumi@yamaguchi-u.ac.jp`

Abstract. Honeybees construct nests that consist of well-arrayed cells. In the first stage of honeycomb construction, they build a linear sequence of tetrapod structures that form the basis of the honeycomb. However, it is unknown how honeybees produce that initial pattern. To understand the mechanisms of honeycomb construction, we proposed an agent-based model, the attachment-excavation model, in which worker honeybees are classified into attachers and excavators. Simulation results showed that a tripod pattern emerges due to competition between the attachers and excavators. As time advances, isotropic wax growth causes the tripods to spread concentrically. We then employed anisotropic wax growth to obtain a linear sequence of constructed tripods. In the present paper, we discuss the origin of the anisotropy based on experimental results.

1. Introduction

Insects that live in colonies and divide up their tasks by class are called social insects. *Apis mellifera*, also known as the western honeybee, is a leading example of social insects, and there have been many studies on their behavioral characteristics [1, 2]. The structure of honeycomb has long been of interest to many scientists [3, 4]. Honeybees acquired the ability to create such structures through the process of evolution. However, it is still unclear how honeybees, which have limited abilities due to tiny brains, short lifespans, and so on, produce such elaborate structures. There are two primary viewpoints on how honeybees make hexagonal honeycomb structures. One is that the heat emitted by the bees melts the beeswax and the honeycomb structure appears naturally due to surface tension. The other is that honeybees create honeycomb structures as "dexterous architects". It should be noted that, with regard to the former theory, there have been no reports of direct observation of bees melting beeswax during nesting.

Although the construction mechanism of the hexagonal-frame cells on the comb surface is interesting, we do not address it here. Instead, we focus on how cells are aligned regularly. To understand how the cells are aligned on honeycomb, we made the hypothesis that honeybees construct their honeycombs by self-organization [5]. We then proposed an agent-based model based on this hypothesis, called the attachment-excavation model. This model reflects beeswax deposition and excavation, resulting in the inflow and outflow of beeswax at the system. This flow means that the honeycomb structure is considered a dissipative structure. We carried out a multi-agent simulation based on the model to reproduce the structure seen in the first stage of honeycomb construction on a substrate such as a ceiling [Fig. 1 (a)] . The results demonstrated the validity of our hypothesis for the initial process of nest building.

The next step in the development of the attachment-excavation model is to investigate whether the model can be applied to the later stages of nest building. The main problem is that the current model cannot describe the anisotropic form of nests, as shown in Fig. 1 (b). Our previous study introduced anisotropy into the beeswax attachment process [Fig. 1 (c)], but we believe that this point needs further verification. In the present paper, we review recent experimental results to improve the model, and then discuss the future development of our research.

The paper is organized as follows. In the next section, we briefly review the essential aspects of our proposed attachment-excavation model. Then we describe our recent experiment and present the observation results, followed by a discussion of the development of future research in the last section.

2. Model

2.1. A brief review of the attachment-excavation model

In this section, we examine the abilities used by the honeybee to create the regularity of honeycomb cells. Honeybees have limited abilities in general, so they are not thought to have the ability to perform complex tasks. Therefore, even if some techniques are partially inherited from previous generations, the ability to create the regularity is likely to be intuitive, simple, and short-range in its effect.

We first assumed that beeswax is randomly deposited on the outside of wax lump regardless of the lump shape. The complex structure will not appear if the wax is only enlarged by random adhesion. We next assume that honeybees excavate the existing beeswax. These functions—attachment and excavation—are the basis of our model. It should be emphasized that we were

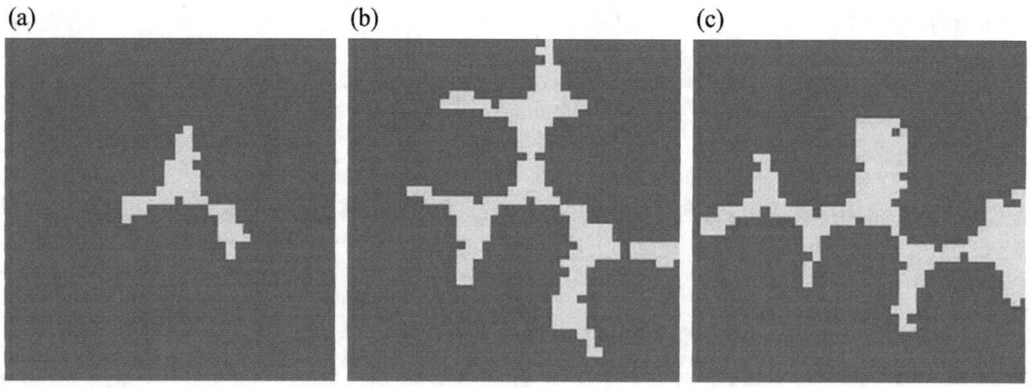

Figure 1. Wax patterns obtained from the attachment-excavation model, where the white region is wax on the substrate (brown). These results were previously reported in Ref. [5]. (a) A tripod structure, which corresponds to the tetrapod structure seen on the substrate at the first stage of honeycomb construction. (b) A tiled pattern of tripod structures, which is obtained from the simulation of isotropic wax growth. This structure is not observed in actual honeycomb construction. (c) A pattern obtained from the simulation of anisotropic wax growth. This pattern consists of tripod patterns that line up and is observed in the first stage of honeycomb construction.

not trying to create any complex structure in proposing either attachment or excavation, and that each of these steps has only a simple and localized effect. A characteristic structure appears in our model due to the competition between attachment and excavation.

2.2. *Wax growth*

We next describe how to implement in simulations the most critical assumptions of the agent-based model discussed in the previous section. Since nest building begins when beeswax is deposited on a substrate such as a ceiling, a two-dimensional (2D) system is sufficient to consider the initial process of nest building. The simulation domain is $l_x \times l_y$, where l_x and l_y are assumed to be relatively large compared to the body length of honeybees. In the following, a 2D system will be mainly described; the assumption is the same in a three-dimensional (3D) system, and the extensions to 3D will be mentioned as necessary.

Since we have assumed that beeswax deposition is random, we do not need to consider the presence of the bees that deposit beeswax. Instead, we concentrate on the growth of the wax that occurs as a result of beeswax attachment.

The simulation domain is divided into $N_x \times N_y$ small regions. We map the {presence, absence} of beeswax in the (i, j)-th cell to the {true, false} of the boolean variable wax(i,j). In other words, a change of the wax(i,j) variable from false to true in a cell realizes adhesion of beeswax. We assume that wax is deposited on one small region of the lump per unit time Δt (1 step). Since honeybees secrete beeswax from their bodies, the amount of beeswax deposited at a given time is small compared to the total volume of a honeybee. The situation where beeswax is deposited sequentially around the lump can be simulated by a model of bacterial colony growth called the Eden model [6]. The probability of beeswax spread in the x direction is denoted by p_x, and that in the y direction is denoted by $1-p_x$. Depending on the value of p_x, the direction of attachment is chosen probabilistically using a random number. Note that, in a 3D case, the probability of growth direction to x, y, and z direction are described by p_x, p_y, and $1 - p_x - p_y$. Next, one of the candidate cells, which are next to the existing (i.e., true) cells, is probabilistically selected using a random number again, and the variable wax(i,j) of that cell is set to true from false.

2.3. *Beeswax excavators*

We introduce entities into the model that play the role of beeswax excavators. There are N agents in the system, and they can move continuously in the system independently of the cells. Each agent's function is to remove all the beeswax from the regions overlapping its position (i.e., all wax(i,j) under the entities are set to false). Thus, we call an entity an excavation zone. The attachment and the excavation of beeswax are done independently, so the agent's location does not affect the selection of the attachment location. However, when beeswax is deposited in a cell overlapping an agent, the beeswax is excavated as fast as it is deposited, resulting in a local equilibrium state, and the amount of beeswax in the system remains unchanged.

The shape of an entity corresponds to the area where a honeybee can excavate. The honeybees excavate wax while moving back and forth. Thus, in a 2D system, a semicircle can move perpendicular to its diameter, and an excavation zone is a combination of a rectangle and a semicircle. In the same way, an agent in a 3D system is a combination of a cylinder and a hemisphere. The size of an excavation zone is characterized by its width, w, which is the diameter of the semicircle (2D) or hemisphere (3D), and the height, d, of the rectangle (2D) or cylinder (3D).

The area covered by the 2D entity oriented in a direction θ is divided into rectangular and semicircular portions and is represented by an equation. The coordinates (x, y) of the area covered by the rectangle satisfy

$$|(x - X_r)\cos\theta + (y - Y_r)\sin\theta| \ < \ \frac{d}{2}, \qquad (1)$$

$$|-(x - X_r)\sin\theta + (y - Y_r)\cos\theta| \; < \; \frac{w}{2}, \tag{2}$$

where (X_r, Y_r) define the position of the center of the rectangular. On the other hand, the coordinates (x, y) of the area covered by the semicircle satisfy

$$\left(x - X_r - \frac{d}{2}\cos\theta\right)\cos\theta + \left(y - Y_r - \frac{d}{2}\sin\theta\right)\sin\theta \; > \; 0, \tag{3}$$

$$\left(x - X_r - \frac{d}{2}\cos\theta\right)^2 + \left(y - Y_r - \frac{d}{2}\sin\theta\right)^2 \; < \; \left(\frac{w}{2}\right)^2. \tag{4}$$

Thus an agent's size is determined by the height d and the aspect ratio $r = w/d$. Note, however, that d and w do not directly correspond to the length of a honeybee, although they are comparable.

We next consider the motion of the excavation zones. We only need to consider translational and rotational motions because we treat the entity as a rigid body. For simplicity, it is assumed that the agent moves straight with speed v regardless of the presence or absence of beeswax. Thus, the translational motion of a 2D simulation is expressed as

$$X_r(t + \Delta t) \; = \; X_r(t) + v\cos\theta \cdot \Delta t, \tag{5}$$

$$Y_r(t + \Delta t) \; = \; Y_r(t) + v\sin\theta \cdot \Delta t. \tag{6}$$

Since all the beeswax touching the excavation zones is removed, a larger speed v will result in more beeswax excavation per unit time. On the other hand, it is necessary to decide where the center of rotation should be for the rotational motion. However, if the amount of rotation per unit time, $\Delta\theta$, is not large, the shape of the nest produced will not be affected by the position of the center of rotation. Therefore, the center of the semicircle is chosen as the center of rotation to reduce the computational cost. In this case, when the angle rotates from θ to $\theta + \Delta\theta$ at time Δt, the rotational motion is expressed as

$$X_r(t + \Delta t) \; = \; X_r(t) + \frac{d}{2}\left[\cos\theta - \cos(\theta + \Delta\theta)\right], \tag{7}$$

$$Y_r(t + \Delta t) \; = \; Y_r(t) + \frac{d}{2}\left[\sin\theta - \sin(\theta + \Delta\theta)\right]. \tag{8}$$

3. Experiment

Our attachment-excavation model successfully explains the tripod pattern that emerges as a basic building block during the initial stages of nest building. Later stages of the actual nest-building process involve multiple repetitions of the basic tripod structure, arranged in a straight-line configuration. Understanding this behavior forces us to consider the presence of some form of anisotropy in one or more sectors of the model. In our present analysis, we assume that anisotropy exists only on the attachment side—that is, we introduce anisotropy into the nest-growth process while assuming that excavation-zone motion remains entirely isotropic, attempting in this way to reproduce the observed straight-line pattern of nest building. However, the *origin* of the anisotropy in nest growth remains unclear. To investigate this point, we conducted an experiment to analyze the relationship between anisotropic nest shapes and the portals through which honeybees enter and exit the nest [7]. The remainder of this section briefly summarizes key aspects of our experiment, while the following section discusses its ramifications for the attachment-excavation model.

Figure 2. Honeycombs formed on the roof of our experimental chamber. (a) By the 8th day of the experiment, we observed the formation of three honeycombs surrounded by a swarm of honeybees. (b) By the 29th day of the experiment, a new (fourth) honeycomb had formed. (c) The roof of the experimental chamber after cutting away the four honeycombs. Each honeycomb left behind a fishbone-shaped discoloration on the roof of the chamber. The fishbone-like structure corresponds to the linear sequence of the tripods in our simulation [Fig. 1 (c)]. *Note:* The three images in this figure do not share a common length scale; we have separately adjusted the magnification of each image for maximal clarity.

3.1. Method

Honeybees in the wild build nests in tree hollows and at similar sites, while laboratory honeybees build nests in nest boxes. This suggests that the entrances and exits of nests may be related in some way to anisotropic nest shapes. To investigate this possibility, we eliminated entrance/exit anisotropy by creating a hole at the center of the lower surface of our experimental chamber, then observed the process of nest formation in this isotropic environment.

Our experimental nest chamber consisted of a hollow wooden cylinder (diameter 45 cm, height 50 cm) mounted on a block (height 10 cm) and covered above by a rectangular "roof" slab (with side lengths 52 cm and 45 cm) made from the same material as commercially available beehive. We partitioned the interior of the chamber by inserting a circular slab at a height of 17 cm above the chamber floor; the nest space was the region between this slab and the roof slab. The frame of the chamber was structurally reinforced by three support beams in the nest space. A circular hole of diameter 3 cm at the center of the partitioning slab furnished a portal through which honeybees could enter and exit the nest. We introduced approximately 4000 western honeybees and one queen bee into the nest space and observed the ensuing process of honeycomb formation.

3.2. Results

On the 8th day after initiating our experiment, we observed 3 honeycomb structures growing downward from the chamber roof [Fig. 2 (a)]. In descending order of proximity to the chamber side walls, we refer to these as honeycombs 1, 2, and 3. The three honeycombs were separated by roughly equal distances and grew in parallel to one another, with the direction of growth perpendicular to the radial direction of the cylinder. The tallest of the three structures was honeycomb 1, with a height of approximately 30 cm. The widest was honeycomb 2, which measured approximately 27 cm at its widest point. We also observed visual evidence of attempted nest-building efforts in the direction perpendicular to the growth of the three honeycombs. The side walls of the nest space were characterized by a high density of adhered pieces of beeswax. By the 29th day of the experiment, the continued progress of the nest-building process resulted in the formation of a new honeycomb structure—honeycomb 4 [Fig. 2 (b)], approximately 13 cm in height—while at the same time the original three honeycombs had continued to grow in size. In particular, honeycomb 1 reached a height of approximately 32 cm, around the same height as the entire nest space. Final observations on day 36 revealed no further changes in the sizes or number of honeycombs present.

4. Discussion

In the experiment, we observed anisotropic nest growth despite meticulous efforts to eliminate anisotropy in the horizontal direction at the nest entrance/exit portal. This demonstrates that asymmetry in the entrance and exit of bees to and from the nest space is *not* the mechanism responsible for anisotropic honeycomb formation. On the other hand, honeycombs were observed to form in the vicinity of the support beams providing structural reinforcement to the nest, indicating that the nest-building behavior of bees is sensitive to the local environment near the nest.

At the conclusion of our experiment, we removed all four honeycombs from the experimental chamber, leaving behind only trace portions of honeycombs stuck to the roof of the chamber. As seen in Fig. 2 (b), before the honeycombs were removed from the chamber, their outermost surfaces exhibited straight-line profiles and parallel alignment. In contrast, the discolored regions left on the chamber roof by the honeycomb-removal process did not define straight lines, but were instead slightly rounded [Fig. 2 (c)]. This finding suggests that, as honeycombs grow vertically downward, they tend toward a configuration resembling evenly spaced parallel lines. The slightly rounded shapes of the discolorations remaining after honeycomb removal were most likely due to the curvature of the cylindrical walls defining the boundaries of the nest space in this case. Clusters of western honeybees are known to form spherical swarms known as "bee balls", within which individual bees make measurements to update their sense of distance to other bees as they go about nest building. In these nest-building regions, anisotropy may also result from competition between externally applied pressure and internal repulsive forces. In view of this possibility, it may be necessary to enlarge our consideration of anisotropy to a phenomenon affecting not only beeswax growth—as in this paper—but excavation-zone motion as well.

Smooth regularity seen in Fig. 2 (c) has not been reproduced in our simulation as shown in Fig. 1 (c). The difference from the actual shape might suggest that the wax growth rate is too fast in simulation. If the rate is fast, there will be much wax around the excavation zone, and the thickness of the cells may vary from place to place. We have observed that real honeybees dig irregular holes when enough wax is placed in advance. It thus does not mean that there is a fatal defect in the model, and it may be possible to reproduce the actual shape depending on the parameter settings.

References

[1] von Frisch K 1964 *Bees: Their Vision, Chemical Senses, and Language* (Ithaca: Cornell Univ. Press)
[2] Tautz J 2008 *The Buzz about Bees - Biology of a Superorganism* (Berlin: Springer)
[3] Hepburn H R 1986 *Honeybees and Wax: An Experimental Natural History* (Berlin: Springer)
[4] Hepburn H R, Pirk C W W and Duangphakdee O 2014 *Honeybee Nests: Composition, Structure, Function* (Springer)
[5] Narumi T, Uemichi K, Honda H and Osaki K 2018 Self-organization at the first stage of honeycomb construction: Analysis of an attachment-excavation model *PLoS One* **13** e0205353
[6] Eden M 1961 A two-dimensional growth process *Berkeley Symp. Math. Stat. Probab.* **4** 223-39
[7] Akiyama T, Terai D, Kageyama M, Narumi T and Osaki K *submitted*

Journal of Physics: Conference Series **2207** (2022) 012014 doi:10.1088/1742-6596/2207/1/012014

Pursuit and Evasion: from singles to groups

Sota Yoshihara and Toru Ohira

Graduate School of Mathematics, Nagoya University, Furocho, Nagoya, 464-8602, Japan

E-mail: ohira@math.nagoya-u.ac.jp

Abstract. We provide a brief review of the current status of the traditional problem of pursuit and evasion. The new approach proposed by Eliezer and Barton is presented with an example of the one-to-one pursuit and evasion. Consequently, we describe recently proposed "Group Chase and Escape" and its characteristic behaviors. The directions of possible topics for future investigations and possible applications are also discussed.

1. Introduction

The original formulation of the problem of pursuit and evasion is said to be due to Leonardo da Vinci. He considered a situation of a cat chasing a mouse[1]. There have been mathematical interests in this topic[2]. Some specific problems have been solved analytically, but the problems in this topic generally remain challenging. Because of that, the cases where a single chaser pursues a single target have been traditionally investigated.

One of the authors has proposed a concept and a model of "Group Chase and Escape" in 2010[3, 4] by fusing this pursuit and evasion problem with collective behaviors of the self–moving units, such as animals, insects, birds, cars. In this model, one group chases another group. Thus, it is an extension of the original problem to multiple players. The model can also be viewed as a variant of collective motion[5] of self-moving units into a mixture of two groups with different objectives. In nature, such occurs in hunting situations by some animal groups.

In this paper, we give a brief review of the current status of the problem of pursuit and evasion. The new approach proposed by Eliezer and Barton is presented with an example of the one-to-one pursuit and evasion. Then, we describe group chase and escape and its characteristic behaviors. The directions of possible topics for future investigations are also discussed. We close the paper with the discussion of further fusion of the pursuit and evasion problem with other fields such as game theory[6] and optimization problems[7].

2. Pursuit and Evasion: Singles

First, we consider the traditional pursuit and evasion by singles. The representative problem is the one considered by Pierre Bouguer in 1732[8]. He obtained the pursuit curve for the one-to-one pursuit and evasion problem where the evader moves on a straight line. Though this is a classic problem, the general solution including the mouse moving on a straight inclined line was obtained only recently in 1991[9, 10, 11].

Also, another classic example is the problem where the evader moves on a circular path, which originates from the mid-18th century[12]. This problem was formulated in terms of coupled nonlinear differential equations. Hence, it is not possible to obtain the general closed-form solution.

It was in 2000 when Eliezer and Barton derived a new mathematical formulation [13]. Their approach has made it possible to treat problems beyond two-dimensional space. In three-dimensional space, this formulation is as follows.

We set the location of the evader and the pursuer by (X, Y, Z) and (x, y, z) respectively. The speeds of the evader v_e and pursuer v_p have a ratio $n = v_e/v_p$. Further, we impose the condition that the pursuer always points its direction of motion to the location of the evader.

$$(X, Y, Z) = (x + \lambda\dot{x}, y + \lambda\dot{y}, z + \lambda\dot{z}), \tag{1}$$

$$\dot{X}^2 + \dot{Y}^2 + \dot{Z}^2 = n^2(\dot{x}^2 + \dot{y}^2 + \dot{z}^2). \tag{2}$$

Given $X(t), Y(t), Z(t)$, and n, we want to solve the four first-order coupled differential equations to obtain $x(t), y(t), z(t), \lambda(t)$.

Because of the nonlinear nature of the above set of equations, its solution is generally difficult to obtain. There are only a handful of special explicit solutions known[13]. We note that they derived solutions for the case of constant λ.

We can generate mathematically different forms of solutions given one set of solutions. Let us assume that we have a set of solutions that satisfies (1) and (2) as $x(t), y(t), z(t), \lambda(t)$ and $X(t), Y(t), Z(t)$ with n as a constant. If we take an orthogonal 3×3 matrix $\hat{A}$. Then, since $\hat{A}^T\hat{A} = \hat{A}\hat{A}^T = \hat{E}$ with $\hat{E}$ as the identity matrix, it can be shown that $(\hat{x(t)}, \hat{y(t)}, \hat{z(t)}) = \hat{A}(x(t), y(t), z(t))$ and $(\hat{X(t)}, \hat{Y(t)}, \hat{Z(t)}) = \hat{A}(X(t), Y(t), Z(t))$ also satisfies (1) and (2). We note, however, that the orthogonal matrix $\hat{A}$ gives the same physical paths, as $\hat{A}$ simply changes one orthogonal basis to another.

Here, we consider the case of changing the 2-dimensional circular problem. In particular, we consider the evader moves in an ellipse with constant angular velocity.

$$X = a\cos t, \quad Y = b\sin t, \tag{3}$$

The pursuer starts from the center and points its velocity vector to the position of the evader. The velocities of the two stay constant. Some examples are shown in Figure 1 (A). We note that this numerically generated solution has associated time changing $\lambda(t)$. The oscillation of $\lambda(t)$ which has a period of approximately π is shown in Figure 1(B). The pursuer's trajectory approaches ellipsoidal shape with axis tilted. The analytical proof, however, remains to be explored.

We can also consider a slightly different condition. Instead of keeping the ratio n of speeds to be constant, we assume the λ in equation (1) as constant. In this case, we can find an asymptotic solution where the trajectories are both ellipses with the same angular velocity. The ratio n, however, oscillates. (Thus, it does not satisfy Eq.(2).) One example is described by the following equations for the pursuer (x, y, z) and the evader (X, Y, Z). The trajectories are shown in Figure 2.

$$x = \cos t, \quad y = \beta\sin t, \tag{4}$$

$$X = \cos(t + \phi)/\cos(\phi), \quad Y = \beta\sin(t + \phi)/\cos(\phi), \quad \phi = \arctan(\lambda). \tag{5}$$

We comment that having n as constant is the standard approach. As λ has a dimension of time, keeping it constant can have intricate effects on relations of initial positions and velocities in equation (1), requiring care in numerical simulations.

We also note that with the circular pursuit, these two constraints of either keeping n or λ as constant merge asymptotically. When the evader moves on a perfect circle, the pursuer's trajectory converges to a perfect circle where the radius ratio matches the speed ratio.

Extension to the 3-dimensional case and further analytical understanding of these ellipsoidal problems are still to be investigated.

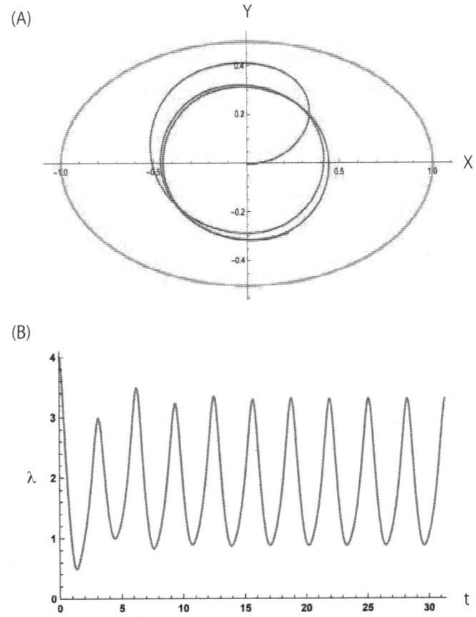

Figure 1. (A) An example of the pursuit of the evader with an elliptic trajectory as described in the equation (3): The red line is the path for the pursuer starting from $(0, 0)$, and the green line is the path for the evader starting from $(1, 0)$. The parameters are $a = 1.0, b = 0.5, n = 2.0$. (B) The associated oscillation of $\lambda(t)$.

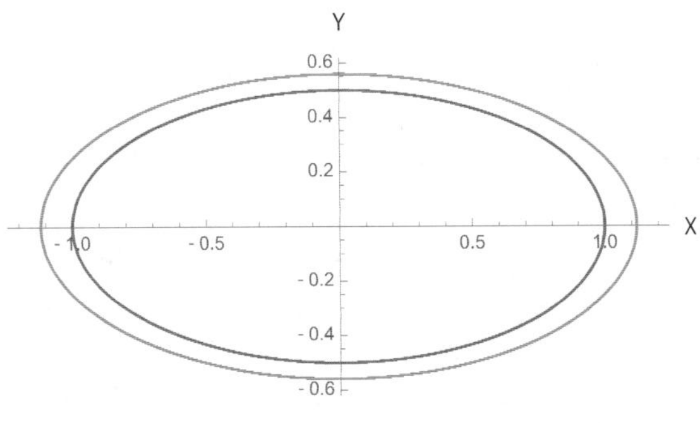

Figure 2. An example of asymptotic trajectories of the elliptic pursuit of the evader as described in the equation (4) and (5): The red line is the path for the pursuer starting, and the green line is the path for the evader. The parameters are $\beta = 0.5, \lambda = 0.5$.

3. Pursuit and Evasion: Groups

Here, we briefly discuss an extension of the problem of pursuit and evasion in groups. Essentially, this is a fusion of topics of "Chases and Escapes" and "Collective Motion" to propose a concept of "Group Chase and Escape" [3, 4] The concept can be viewed as a natural extension of the pursuit and evasion in singles. With this extension, it is an example of collective motion[14], and a simplified model of hunting a group of deer by a pack of wolves.

In the original model, a simple rule was implemented for pursuers and evaders: each pursuer moves closer to the nearest evader, while each evader moves away from the nearest pursuer. Albeit this simplicity, mathematical analysis is quite intricate and challenging, so it was mainly investigated by computational simulations.

The followings are the main observation of group chase and escape models.

- We observe aggregation of evaders even though no such interactions are implemented. These are mainly due to the exclusion volume effect.
- The pursuers tend to surround such aggregates of evaders and catch them at once. Thus, the number of evaders decreases rather intermittently and abruptly as a function of time.
- Given the initial number of evaders, there is typically an optimal number of pursuers to have the most efficient performance of catching.

A series of extensions of this basic model has been proposed and studied. They include extention to three groups, conversion of captured evaders to pursuers, adding interactions among pursuers, delays in pursuing, and so on.

There are still some open problems and topics to be explored relating to group chases and escapes. We list some possibilities below to close this section.

- Boundary conditions seem to affect pursuit and evasion rather strongly. We need to evaluate more quantitatively their effects.
- Also, the effect of changing initial configurations needs to be investigated more thoroughly.
- A possible application of the model is hunting in nature. Some animals such as lions, wolves, and dolphins, perform cooperative hunting of prey. It may be interesting to investigate the model from an ethological perspective.

4. Discussion

There are other directions where the problem of pursuit and evasion has expanded its horizon. "Discrete Search Games" [15] and "Differential Games" [6, 16] have been developed as a result of fusions with game theories. We do not introduce them in detail here, but there are many interesting topics in this new field.

For pursuit and evasion in groups, we have proposed an application to optimization problems such as Traveling Salesman Problem[7]. The pursuit and evasion process is used to search for the optimal state, which has shown some promise. We can view this application as a new example in the field called Swarm Intelligence.

References

[1] Guha A and S Biswas 1994 *Bulletin of Institute of Mathematics and its Application* **30** 12
[2] Nahin P J 2007 *Chases and Escapes: The Mathematics of Pursuit and Evasion* (Princeton: Princeton University Press)
[3] Kamimura A and Ohira T 2010 *New Journal of Physics* **12** 053013
[4] Kamimura A and Ohira T 2019 *Group Chase and Escape: Fusion of Pursuits-Escapes and Collective Motion* (Singapore: Springer)
[5] Vicsek T and Zafeiris A 2012 *Physics Reports* **517** 71
[6] Isaacs R 1965 *Differential Games* (New York: Wiley)
[7] Ohira T, Hosaka T and Nogawa T 2015 *Artificial Life and Robotics* **20** 257
[8] Bouguer P 1732 *Memories de l'Aacademie Royale des Sciences* 1
[9] Coleman W J A 1991 *Bulletin of Institute of Mathematics and its Application* **27** 45
[10] Eliezer C J and Barton J C 1992 *Bulletin of Institute of Mathematics and its Application* **28** 182
[11] Eliezer C J and Barton J C 1995 *Bulletin of Institute of Mathematics and its Application* **31** 139
[12] Archibald R C and Manning H P 1921 *American Mathematical Monthly* **28** 91
[13] Barton J C and Eliezer C J 2000 *Journal of the Australian Mathematical Society Series B* **41** 358
[14] Vicsek T 2010 *Nature*, **466** 43
[15] Ruckle W H 1991 *Stochastic Games and Related Topics* ed T E S Raghavan, T S Ferguson, T Parthasarathy and O J Vrieze (Dordrecht: Kluwer Academic), p 29
[16] Lewin J 1994 *Differential Games: Theory and Methods for Solving Game Problems with Singular Surface* (London: Springer-Verlag)

Phase diagram of the dodecahedron spin model on square lattice with diluted bonds

S. Yunita, R. Rahmat and T. Surungan

E-mail: tasrief@unhas.ac.id
Department of Physics, Hasanuddin University, Makassar, South Sulawesi 90245, Indonesia

Abstract. We study the effect of bond dilution of a dodecahedron spin model on ferromagnetic square lattice. The dodecahedron spin constitutes a discrete (polyhedral) counterpart of the continuous Heisenberg spin. Previous work found that, in the absence of dilution, the ferromagnetic dodecahedron system exhibits a finite temperature phase transition. We use Monte Carlo simulation with Wolff cluster algorithm to monitor how phase transitions are impacted by the bond dilution. We simulate the model with various bond concentrations and investigate the resulting phase diagram and critical behavior.

1. Introduction

As has long been appreciated, the study of phase transitions is of paramount importance. This is so since phase transitions are ubiquitous in nature. Amongst seemingly uncountable many other phase transitions, the melting of ice and the low temperature spontaneous magnetization of ferromagnets are notable examples [1]. A phase transition is commonly associated with a breaking of an underlining symmetry of a system. Most common transitions are of either the "first" or "second" order types, according to the Ehrenfest classification. The criterion of this classification is based on the derivative of the free energy function of the system with respect to its corresponding thermodynamic variable. A transition is termed "first order" (or discontinuous) if the derivative of its free energy relative to thermodynamic variable is discontinuous. Conversely, if the latter derivative is continuous, the phase transition is, naturally, termed "continuous" phase transition [2]. In most continuous phase transitions, the second derivative of the free energy is non-analytic; the latter are called "second order" phase transitions. A notable feature of systems exhibiting first order transitions is the presence of of latent heat and a region of phase co-existence.

Typically, at sufficiently high temperatures, various theories exhibit high symmetry. This is so since at high temperature (or, equivalently, high energy scales) the system can readily transition between various symmetry related configurations. As the temperature drops, the system breaks one or more of its symmetries. Typically, at these low temperatures, large free energy barrier need to be surmounted in order to transition from one state to another. The latter large energy barriers thwart transitions between symmetry related configurations and lead to "symmetry breaking". This can, e.g., be seen in the two dimensional (2D) Ising model in which, an extensive free energy barrier separating the existing two minima (having, respectively, positive and negative magnetization, $M \to -M$) in the energy landscape. These minima are associated with the underlining Ising Z_2 symmetry of the model. Most transitions in condensed matter systems are associated with such breaking of symmetries. There are additional transitions,

Journal of Physics: Conference Series **2207** (2022) 012015 doi:10.1088/1742-6596/2207/1/012015

Table 1. Characteristics of regular polyhedrons.

Name	Vertices	Faces	Edges	Group Symmetry
Tetrahedron	4	4	6	S_4
Octahedron	6	8	12	$O_h = S_4 \times C_2$
Hexahedron (cube)	8	6	12	O_h
Icosahedron	12	20	30	$A_5 \times C_2$
Dodecahedron	20	12	30	$A_5 \times C_2$

termed topological, in which no symmetry is lifted but rather the transition is intrinsically linked to topology.

Apart from the Ising model, other paradigmatic 2D systems featuring discrete symmetry include the q-state clock models (having discrete rotational symmetry) and Potts models (discrete permutational symmetry). The q-state clock models, with $q \geq 2$ being a natural number, form discrete counterparts to the continuous XY model [3]. (The $q = 2$ system is the above noted Ising model.) According to the Mermin-Wagner-Hohenberg theorem, models (of finite interaction range and strength) that are invariant under a continuous symmetry, such as the XY and Heisenberg models, cannot exhibit true long range order (TLRO) in 2D (and 1D) lattices [4]. In these systems, there is no conventional transition marked by a finite temperature symmetry breaking. Nonetheless, the 2D XY model exhibits a particular topological type of phase transition called Kosterlitz-Thouless (KT) transition[5]. Here, a finite temperature transition appears between a disordered high temperature paramagnetic phase and a low-temperature "KT phase" that features quasi-long range order (QLRO) in which the correlations between two XY spins decay algebraically with their spatial separation. At sufficiently low temperatures, the system does break its discrete rotational symmetry. At these temperatures, the q-clock systems exhibit TLRO [3, 6, 7]. At somewhat higher (intermediate) temperatures, the $q > 4$-state clock model features an algebraic phase. The effect of randomness, in the form of bond and site dilution, on the KT transition has been studied. It was the KT phase survives so long as the lattice allows for percolation of one side of the system to the other [8].

Analogous to the effect of dilution on KT phase of clock model, in the present study, we probe the effect of bond dilution on phase transition of 2D polyhedral spin models. These spin models form the discrete counterpart of the Heisenberg model of three component spins. As such, these are three component extensions of the discrete two spin component clock models. Here, polyhedra, on which the polyhedral spins are defined, are geometrical structures constructed by equally partitioning the (4π) solid angle of a sphere[9]. As known from time immemorial, there are five regular polyhedra (the Platonic solids)- i.e., the tetrahedron, octahedron, cube, icosahedron and dodecahedron, A possible orientation of spin is uniquely associated with corners of a polyhedron; the origin of the spin is situated at the center of a polyhedron. Tetrahedron and cubic spins will, for example, have four and eight possible spin orientations, respectively. Polyhedral spin models on the square lattice have been previously studied for pure (i.e., undiluted) ferromagnets. These models all exhibit a second order phase transition[10]. Alternatively, the possible orientation of spin can be prescribed to lie towards the center of a face or edge of the structure, rather than to a corner. The so-called edge-cubic spin model on square lattice has been studied previously and is known to exhibit a double phase transition[11].

Table 1 tabulates the characteristics of each polyhedral structure. Spin orientations are restricted to point to the vertices of each corresponding structure. In the current work, we aim to systematically study the effect of bond dilution on the critical properties of these models. In this paper, we report only for dodecahedron model. More complete report will be presented elsewhere. The remainder of the paper is organized as follows: Section II describes the model and the method. The result is discussed in Section III. Section IV is devoted to the summary and concluding remark.

2. Model and Simulation Method

The $(L \times L)$ square lattice dodecahedron spin model is defined by the following Hamiltonian

$$H = -\sum_{\langle ij \rangle} J_{ij}\ \vec{s}_i \cdot \vec{s}_j \tag{1}$$

The coupling interaction J_{ij} is 0 or 1, depending on whether the bond is depleted or not and $\{\vec{s}_i\}$ are the above mentioned dodecahedral spins. We assign bond density as $\kappa = 1.0$ for a pure (non-diluted) case, i.e., all $J_{ij} = 1$. The ground state energy for the pure case is $-2NJ$, N being the number of spins.

The ensemble average of a quantity Q is expressed as follows

$$\langle Q \rangle_\beta = \frac{\sum_j Q_j e^{-\beta E_j}}{\sum_j e^{-\beta E_j}} \tag{2}$$

which will be calculated here through Monte Carlo (MC) simulation. Q_j is the measurement of Q in j-th configuration. For a large real system, the number of configurations is huge (being infinite also for finite continuous models such as the XY model). Nonetheless, only a small number of configurations significantly contribute in the calculation of $\langle Q \rangle$. MC simulation is basically a sampling method which generates those significant configurations. Using a canonical ensemble MC algorithm, one performs a Markov process to generate a Markov chain, through a series of updating spin configurations based on the following probability

$$P_{E \to E'} = \min\left[1, \exp^{-\beta \Delta E}\right] \tag{3}$$

where E and E' are the energy of the old and the updated configuration, respectively; with $\Delta E = E' - E$. The update is automatically accepted if $\Delta E \leq 0$, or $E' \leq E$. With this rule, the ensemble average $\langle Q \rangle$ is then written as

$$\langle Q \rangle_M = \frac{\sum_j Q_j e^{-\beta E_j}}{M}; \tag{4}$$

where M is the number of configurations taken from the Markov chain [12]. In generating a new configuration, we implement a single cluster spin update introduced by Wolff [13].

A cluster for the dodecahedron spin is constructed by adopting the idea of Wolff where all spins are projected into a plane perpendicular to a randomly chosen existing spin orientation of the model, then divide them into two groups of Ising-like spins. This procedure is essential in implementing cluster algorithm for continuous spins in 2D and 3D. The usual steps of the cluster algorithm are performed after the projection [15]. A randomly chosen spin is connected to its nearest neighbors, next-nearest neighbors and so on, of similar group with suitable probability until no more spins to include; then flip them collectively. A Monte Carlo step (MCS) is defined as visiting once each spin randomly and perform cluster spin update in each visit. Measurement is performed after enough equilibration MCS's (10^4 MCS's). Each data point is obtained from an average over several parallel runs; each run is comprised of 4×10^4 MCS's. To evaluate the statistical error, each run is treated as a single measurement. For the accuracy in the estimate of critical temperatures (and exponents), data are collected up to more than 100 measurements for each system size, which is a standard procedure in doing simulation for random system. Taking into account the tendency of self-averaging data for larger system size, we took less samples for larger sizes.

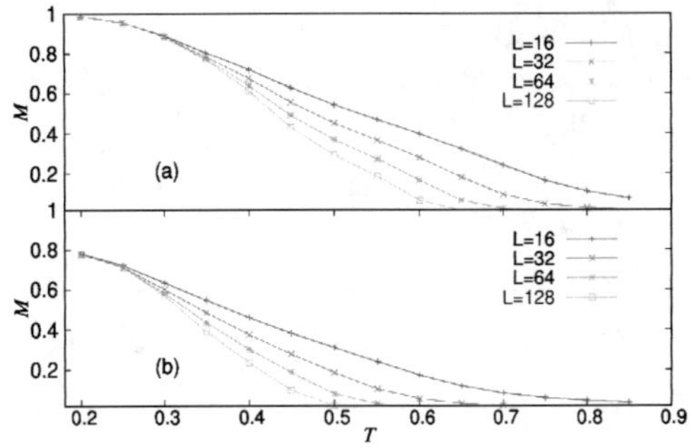

Figure 1. Magnetization vs temperature for system size L =16,32,64 and 128. Error bar is in the order smaller than the symbol size.

3. Results and Discussion

A phase transition separates the ordered and disordered phase. At critical transition points, the correlation length diverges. This divergence enables one to extract the critical temperature (T_c) and exponents from the data of finite system sizes. In the present study, we performed MC simulation and run several system sizes (L=16, 32, 64 and 128). To search for any existing transition, we measured the ensemble average of magnetization and the ratio of its correlation function.

The magnetization $\vec{M} = |\sum \vec{s}_i|$ is the natural order parameter for a ferromagnetic system. Fig. 1 shows the temperature dependence of magnetization for the case of $\kappa = 1.0$ and of $\kappa = 0.90$. The magnetization is normalized so that for the pure case it becomes unity at very low temperatures. A clear sign of a magnetic phase transition is indicated by the increasing magnetization at lower temperatures. The effect of dilution is already pronounced even when only 10% bonds are depleted (see Fig. 1(b)). The critical temperature decreases as dilution increases.

The estimate of critical temperatures for each corresponding value of κ is obtained from the finite size scaling (FSS) plot of the correlation ratio, which is the ratio of correlation functions of two different distances[14]. The correlation function and its ratio are written respectively as follows

$$g(\vec{R}) = \sum_{\vec{r}} \vec{s}(\vec{r}) \cdot \vec{s}(\vec{r} + \vec{R}) \tag{5}$$

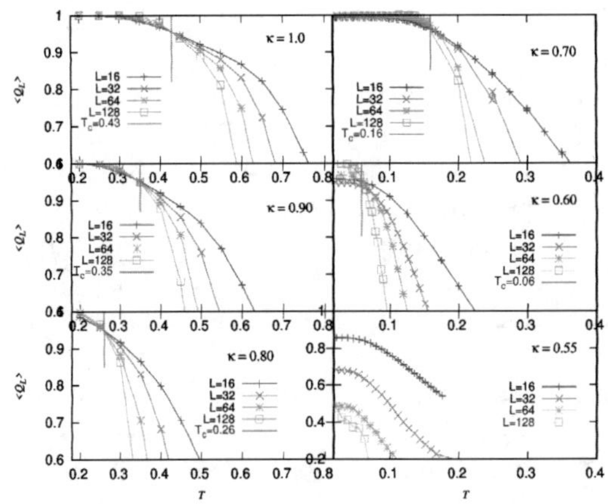

Figure 2. The plot of correlation ratio for each corresponding value of κ; The crossing point is a clear sign of a phase transition.

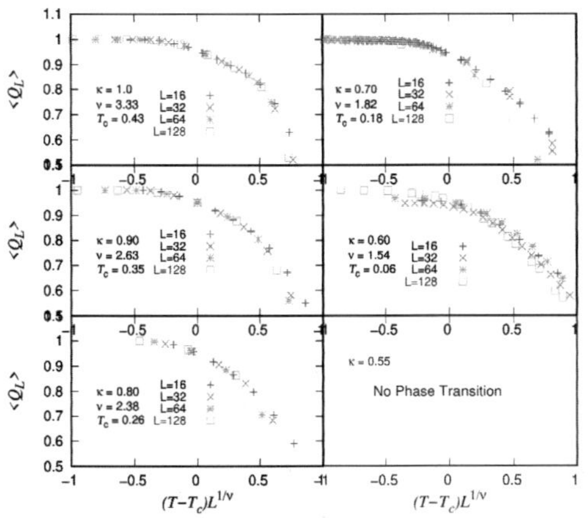

Figure 3. The FSS plot of correlation ratio and the estimate of T_c and ν for each corresponding value of κ.

Table 2. Table of critical temperature and exponent for each bond density. The number in bracket is the uncertainty for the last digit.

Bond Density (κ)	T_c	ν
1.0	0.43(1)	3.33(1)
0.90	0.35(1)	2.63(1)
0.80	0.26(1)	2.38(2)
0.70	0.18(1)	1.82(2)
0.60	0.06(2)	1.54(2)

$$Q_L = \frac{\langle g(L/2)\rangle}{\langle g(L/4)\rangle} \tag{6}$$

The displacement $\vec{R}$ is a vectorial quantity. For rotationally invariant correlations, $\langle g \rangle$ is a function of the separation $R \equiv |\vec{R}|$. In the second of the above equations, we consider two simple values of $\vec{R}$, i.e., $L/2$ and $L/4$, along the x- and y-directions.

The existence of a phase transition can be clearly observed from the temperature dependence plot of Q_L. As has been demonstrated in several previous studies, this quantity is a good estimator for determining the critical point of the second-order and Kosterlitz-Thouless transitions [6, 8, 11, 14]. We briefly describe the property of Q_L. At very low temperature where spins are highly correlated, Q_L is very close to unity. As temperature increases, Q_L decreases. Fig. 2 shows the temperature dependence of Q_L for various L where the existence of a phase transition is exhibited by the presence of a crossing point, which is due to the fact that the correlation length spans the whole system at critical temperature. We assign in Fig. 2 the associated temperature of the crossing point by a vertical line. Then, by means of finite size scaling plot of the correlation ratio, as shown in Fig. 3, we estimate the correlation length exponent ν for each corresponding κ. The results for both critical temperatures and exponents are listed in Table 2; from which we construct the phase diagram as shown in Fig 4.

Using the correlation ratio one can also extract the decay exponent η of the correlation function as has been reported in several previous studies [6, 11, 14]. Calculation of η and other exponents will be considered in the next study and the result will be reported elsewhere.

4. Summary and Concluding Remarks

In summary, we have studied critical properties of the dodecahedron spin model on a square lattice with diluted bonds. The dodecahedron spin is one of the five discrete counterparts of

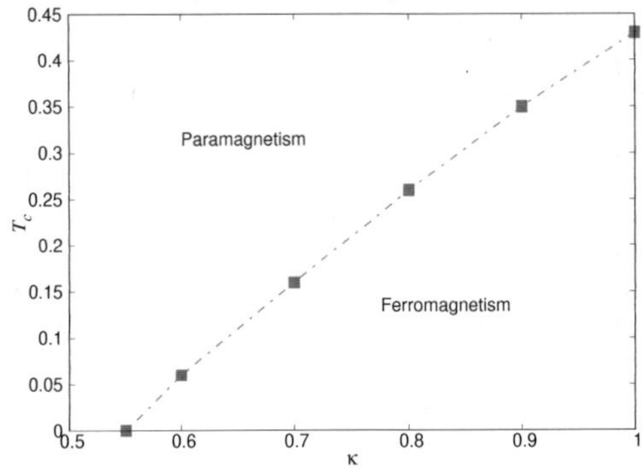

Figure 4. Phase diagram of the dodecahedron spin model on diluted square lattice. The plot suggests that the critical dilution for this model is above the the percolation threshold $p_c = 0.5$ of the square lattice.

the Heisenberg model, obtained by equally dividing the 4π solid angle of a sphere. Six different dilution parameters were considered i.e., $\kappa = 1.0, 0.9, 0.8, 0.7, 0.6$ and 0.55. For each dilution, the model exhibits a second order phase transition, except for $\kappa = 0.55$. The existing phase transition is probed by calculating the magnetization and the spin correlation ratio. The critical temperature and exponent for each corresponding value of κ are obtained from the finite size scaling analysis of the correlation ratio. We found a systematic decrease in critical temperatures as bond density decreases. The magnetic order disappears at the bond concentration above the percolation threshold $p_c = 0.5$ of a square lattice.

Acknowledgments

The authors wish to thank Z. Nussinov, P. G. Gareso, and Bansawang BJ for valuable discussions. The extensive computation was performed using GRID computational facility of the National Research and Innovation Agency (BRIN) of Indonesia and parallel computers at the Department of Physics, Hasanuddin University. The present work is financially supported by Hasanuddin University Research Grant 2021.

References

[1] H. Nishimori and G. Ortiz, *Elements of Phase Transitions and Critical Phenomena*, Oxford Univ. Press (2011).
[2] H. Kleinert and V. Schulte-Frohlinde, *Critical Properties of ϕ^4-Theories*, World Scientific (2001).
[3] J. V. José, L. P. Kadanoff, S. Kirkpatrick, and D. R. Nelson, Phys. Rev. B **16**, 1217 (1977).
[4] N. D. Mermin and H. Wagner, Phys. Rev. Lett. **17** 1133, (1966) ; P. C. Hohenberg, Phys. Rev. **158**, 383 (1967).
[5] J. M. Kosterlitz and D. Thouless, J. Phys. C **6**, 1181 (1973); J. M. Kosterlitz, J. Phys. C **7**, 1046 (1974).
[6] T. Surungan, Y. Okabe, and Y. Tomita, J. Phys. A **37**, 4219 (2004).
[7] T. Surungan et al., J. Phys. A: Math. Theor. **52**, 275002 (2019).
[8] T. Surungan and Y. Okabe, Phys. Rev. B **71**, 188428 (2005).
[9] K. J. M. MacLean, *A Geometric Analysis of the Platonic Solids and Other Semi-Regular Polyhedra: with an Introduction to Phi Ratio* ,The Big Pictures Press, Oxford (2002).
[10] T. Surungan and Y. Okabe, arXiv.org, cond-mat:1709.03720.
[11] T. Surungan, N. Kawashima and Y. Okabe, Phys. Rev. B77, 214401 (2008).
[12] M. E. J. Newman and G. T. Barkema, *Monte Carlo Methods in Statistical Physics*, Oxford Univ. Press (2001).
[13] U. Wolff, Phys. Rev. Lett. **62**, 361 (1989).
[14] Y. Tomita and Y. Okabe, Phys. Rev. B **66**, 180401(R) (2002).
[15] P. W. Kasteleyn and C. M. Fortuin, J. Phys. Soc. Jpn, Suppl. **26**, 11 (1969); C. M. Fortuin and P. W. Kasteleyn, Physica (Amsterdam) **57**, 536 (1972).

Characterizing flexibility and mobility in the natural mutations of the SARS-CoV-2 spikes

James Panayis[1], Navodya S. Römer[2], Dom Bellini[3], A. Katrine Wallis[4], Rudolf A. Römer[1]

[1] Department of Physics, University of Warwick, Coventry, CV4 7AL, UK
[2] School of Life Sciences, University of Lincoln, Brayford Pool Campus, Lincoln, LN6 7TS, UK
[3] MRC Laboratory of Molecular Biology, Cambridge CB2 0QH, UK
[4] School of Life Sciences, University of Warwick, Coventry, CV4 7AL, UK

E-mail: r.roemer@warwick.ac.uk

Abstract. We use *in silico* modelling of the SARS-CoV-2 spike protein and its mutations, as deposited on the Protein Data Bank (PDB), to ascertain their dynamics, flexibility and rigidity. Identifying the precise nature of the dynamics for the spike proteins enables, in principle, the use of further *in silico* design methods to quickly screen for existing and novel drug molecules that might prohibit the natural protein dynamics. We employ a recent protein flexibility modeling approach, combining methods for deconstructing a protein structure into a network of rigid and flexible units with a method that explores the elastic modes of motion of this network, and a geometric modeling of flexible motion. Our results thus far indicate that the overall motion of wild-type and mutated spike protein structures remains largely the same.

1. Introduction

In recent years, drug discovery supported by computer modelling has become a viable method to support experimental efforts in large-scale screening of prospective drug compounds [1]. This approach is part of the overarching theme of bioinformatics which aims to harness the power of modern computational techniques for "in silico" design of medicines [2]. The advantage of this lies not only in the increased screening speed, but also in its predictive power for novel drug development.

In early 2020, protein structures related to SARS-CoV-2 began to appear in the international Protein Data Bank (PDB). Such structures give the precise location of atoms in each protein, forming the basis for an understanding on how their biological mechanism might function. Chief among these structures were the protease (PDB code 6LU7 [3]), important in the life cycle of SARS-CoV-2, and the ectodomain of the spike glycoprotein (3 structures with PDB codes 6VSB [4], 6VXX and 6VYB [5]), a homo-trimer protein, protruding out of the viral capsid, which the virus uses in attaching itself to a human cell during infection. The spike protein is hence a main target for human antibodies and a key molecule for vaccine design and possible therapeutic treatments. The spike is a homo-trimer, protruding from the viral surface. The ectodomain of each monomer consists of an N-terminal subunit, S1, comprising two domains, S^A or N-terminal domain (NTD) and S^B or Receptor-Binding Domain (RBD), followed by an S2 subunit forming a stalk-like structure. Each monomer has a single membrane-spanning segment and a short C-terminal cytoplasmic tail [6].

In March 2021, Gobeil *et al.* [7] reported the first structures for the mutations of SARS-CoV-2 emerging widely in the United Kingdom (α-variant), South Africa (β-variant) and Brazil (γ-variant), as well as for the variant identified in the Danish mink population. All these variants either retain or improve binding to human cells [7].

In order to understand the biological function of a given structure such as the spike protein, it is important to know how such a molecule moves. Experimentally, this is very hard to ascertain and, even if achievable in some cases, may require months of careful experimentation. Computational modelling, on the other hand, can provide crucial information faster. However, in making such models, one has to make simplifying assumptions that can reduce the accuracy of the results that are obtained. In our work [8], we used an approach in which a protein structure is divided into rigid and flexible parts. This then allows us to quickly model the motion of the protein by keeping the rigid parts stable whilst allowing movement for the flexible regions.

In the following, we shall briefly explain the details of the computational modelling used and review the results obtained for the original "wild-type" spike proteins, which are only a small subset of the nearly 300 SARS-CoV-2 related protein structures studied in [8]. In addition, the same study was extended to the available PDB structures of variants, namely α, β and γ, which have emerged through various mutations [7].

2. Methods

Protein selection The wild-type structures of the spike protein were deposited into the PDB already in March 2020 with code 6VSB [4] followed shortly afterwards by 6VXX and 6VYB [5]. With RMS resolution of 3.5 Å, 6VSB has a slightly lower resolution than 6VXX at 2.8 Å and 6VYB at 3.2 Å and we shall mainly refer to 6VXX and 6VYB when mentioning the "wild-type" structures. These latter two structures are distinguished by 6VXX having all 3 S^B domains in a "closed" or "down" configuration, while 6VYB shows one of these in an "open" or "up" configuration [4]. S1 is involved in recognition of the human receptor angiotensin-converting enzyme 2 (ACE2) [6], eventually leading to viral entry [9].

PDB available structures of mutated spike proteins are the α-variant with 4 structures (codes 7LWS to 7LWV), the β-variant with 7 structures (codes 7LYK to 7LYQ) and the γ-variant (code 7LWW). The variants identified in the Danish mink population (9 PDB codes 7LWI to 7LWQ), giving broadly similar results, are not shown here due to space constraints.

Coarse-grained protein flexibility dynamics We start our rigidity, flexibility and mobility modelling with the given atomic coordinate files from the PDB in *.pdb* format. The details have been described previously in Refs. [10, 11] and the particulars for the spike proteins in Ref. [8]. Briefly, the PDB structures are cleaned and hydrogenated, their rigid and flexible parts identified [12] then the lowest vibrational modes computed [13, 14] and geometric simulation along the simplest non-trivial modes is performed [15]. This explores the flexible motion available to a protein within a given pattern of rigidity and flexibility. Typically, several thousand new conformers for each mode are generated unless or until prohibited by steric clashes and/or bonding constraints. We emphasize that the resulting trajectories do not represent actual stochastic motion in a thermal bath as a function of time, but rather the *possibility of motion* along the most relevant elastic modes. The usefulness of this approach has previously been shown in a variety of systems [11, 16–19]. Methods similar in spirit, exploring flexible motion using geometric simulations biased along "easy" directions, have also been implemented by other groups [20, 21].

3. Results

Rigidity and Flexibility The rigidity pattern of the homotrimer in the wild-type structures 6VXX and 6VYB is dominated by a large rigid cluster encompassing most of the trimer

Journal of Physics: Conference Series **2207** (2022) 012016 doi:10.1088/1742-6596/2207/1/012016

structure [8], except for the region in one of the monomers of 6VYB corresponding to the RBD in the "up" or "open" configuration (Fig. 1). Hence the rigidity clearly distinguishes open from closed structures [23]. This study has found that this pattern repeats itself for the mutated variant structures shown in Table 1.

Protein Mobility We have performed a FRODA mobility simulation for each protein at several selected values of E_{cut}. In a previous publication, we discussed the criteria for a robust selection of E_{cut} [15]. For each protein structure as given in Table 1, we identified a bespoke set of E_{cut} values from the rigidity plots corresponding to values of important loss of rigidity. In Fig. 1, we choose $E_{\text{cut}} = -1$ kcal/mol and show results for the simplest non-trivial mode, m_7, with 6VYB and 7LWW. Clearly, 6VYB shows good motion in the RBD domain for monomer B. This is similar to the mutated structure 7LWW. In addition, we also find for 7LWW [25] that

PDB	Ref.	rigidity class.	mutation	behaviour at $\lvert E_{\text{cut}}\rvert = 1$	u/d
6VXX	[5]	domain	wild-type	m_7: flexibility in NTD-A/NTD-C, m_8: flexibility in NTD-B	3 down
6VYB	[5]	domain	wild-type	m_7: motion in RBD-B, m_8: motion in RBD-B, NTD-C	1 up
7LWS	[7]	1st	α (UK)	m_7: flexibility in RBD-A+C, NTD-B, m_8: flexibility in NTD-A+C, RBD-A+C	3 down
7LWT	[7]	1st	α (UK)	m_7: motion in RBD-B, m_8: flexibility in RBD-B, NTD-C	1 up
7LWU	[7]	1st, domain	α (UK)	m_7: motion in RBD-B, m_8: motion in RBD-B	1 up
7LWV	[7]	1st, domain	α (UK)	m_7: motion in RBD-B, m_8: motion in RBD-B, flexibility in NTD-C	1 up
7LYK	[7]	1st, domain	β (SA)	m_7: motion in RBD-B, m_8: motion in RBD-B+A	2 up
7LYL	[7]	1st	β (SA)	m_7: flexibility in NTD-A, RBD-B+C, m_8: flexibility in RBD-A, NTD-B+C	3 down
7LYM	[7]	1st	β (SA)	m_7: flexibility in NTD-A, RBD-B+C, m_8: flexibility in RBD-A, NTD-B+C	3 down
7LYN	[7]	1st	β (SA)	m_7: motion in RBD-B, m_8: motion in RBD-B, NTD-C	1 up
7LOY	[7]	1st, domain	β (SA)	m_7: motion in RBD-B, m_8: flexibility in NTD-C	1 up
7LYP	[7]	1st	β (SA)	m_7: motion in RBD-B, m_8: motion in NTD-C, flexibility in RBD-B	1 up
7LYQ	[7]	2nd	β (SA)	m_7: motion in RBD-B, m_8: motion in RBD-B, NTD-C	1 up
7LWW	[7]	1st	γ (BR)	m_7: motion in RBD-B, m_8: motion in NTD-C, flexibility in RBD-B	1 up

Table 1. List of SARS-Cov-2 spike structures investigated in this study. In column 3, a rigidity dilution classification into solid "brick", 1st and 2nd order behaviour, and rigidity domains is given [8]. In column 5, the use of descriptor "flexibility" indicates largely elastic behaviour while "motion" denotes larger scale movements.

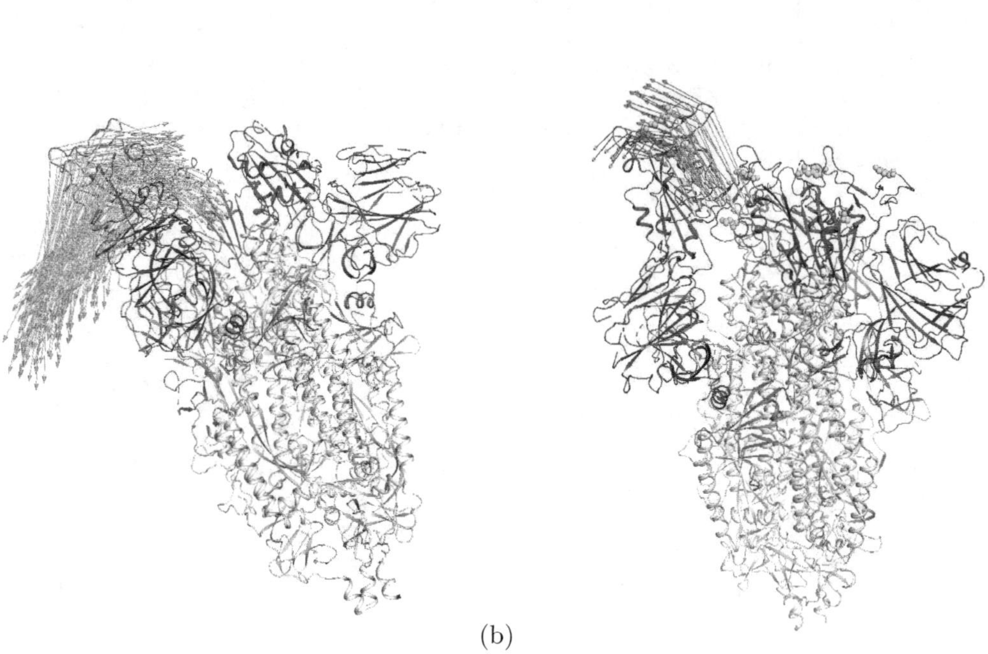

(a) (b)

Figure 1. Cartoon representation of motion along modes m_7 at $E_{\rm cut} = 1\,\rm kcal/mol$ in the (a) *open* wild-type spike ecto domain (6VYB) [5] and (b) the γ-variant 7LWW. Arrows indicate motion distances larger than 17.5Å. The RBDs are shown in red while the positions of mutations for 7LWW are given by the yellow spheres. Some other domains are also highlighted in color, i.e. the NTD is blue. An interactive animation of the motion is available for 6VYB [24] and 7LWW [25].

more motion is visible in the N-terminal domain (NTD) on monomer C than in the NTDs of 6VYB [24].

In Fig. 2 we quantify the motion by showing, for modes m_7 and m_8 of 6VYB and 7LWW, the root-mean-squared deviation (RMSD) compared to the initial (crystal) structure along the motion trajectories. For m_7, the motion of the RBD domain on monomer B can be clearly seen to emerge around conformer 50, corresponding to around 1000 steps in the motion simulation. Somewhat later, one sees the RMSD increase for NTD C on 7LWW. Mode m_8 has an even richer pattern of behaviour with nearly as much RMSD in NTD C as in RBD B in both structures.

Similar results have been obtained for the other mutations listed in Table 1. For modes 7 and 8 at $E_{\rm cut} = 1\,\rm kcal/mol$, the average RMSD across both modes and both directions (and the average of the largest observed RMSDs of each mode) for wild-types (6VXX, 6VYB), α (7LWS – 7LWV), β (7LYK – 7LYQ), and γ (7LWW) variants are respectively: 2.53Å (18.02Å), 2.95Å (22.99Å), 2.1Å (15.44Å) and 2.28Å (17.67Å). We find that, naturally, closed spike structures exhibit a smaller RMSD than the open structures. In addition, open structures seem to have roughly the same change in RMSD during the movement. Overall, this suggests that with respect to our modelling, there is no statistically relevant distinction between wild-type and mutated structures or between the mutated structures themselves.

4. Conclusions

The mutations of the SARS-CoV-2 virus that have emerged thus far in the human population are largely concerned with a few isolated mutations of residues on the spike protein. It seems entirely plausible that such isolated mutations should not dramatically change the overall flexibility of

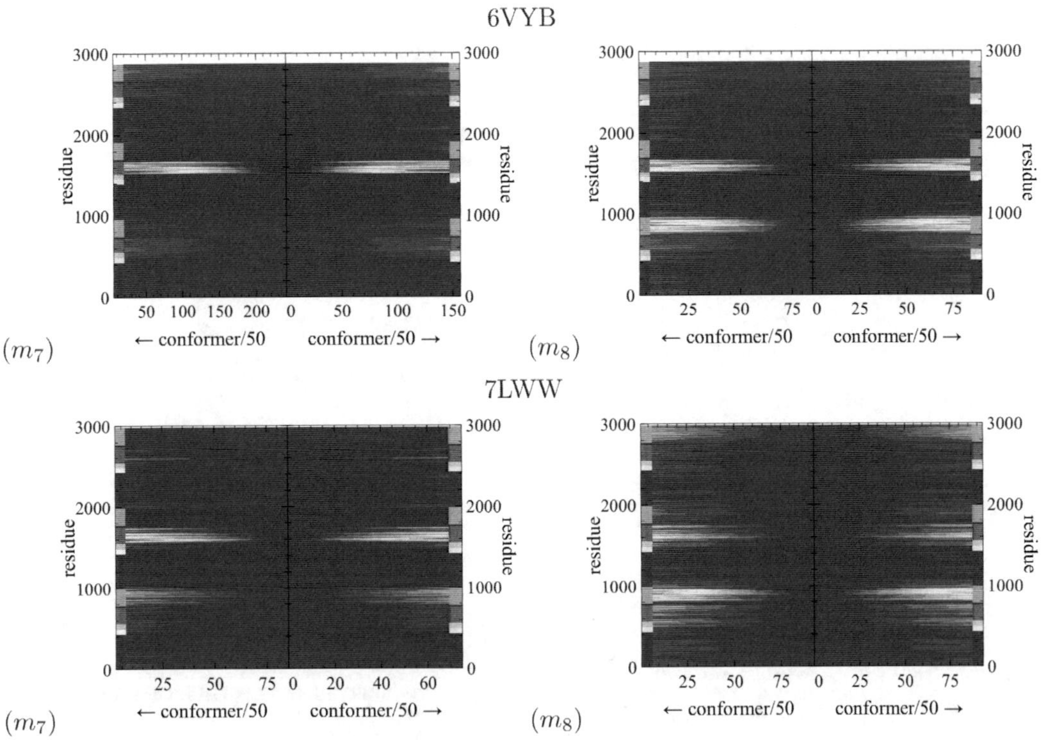

Figure 2. RMSD variation along modes m_7 (top) and m_8 (bottom) at $E_{\text{cut}} = 1\,\text{kcal/mol}$ in the (a) *open* spike ecto domain (6VYB) [5] and (b) the γ-variant 7LWW. The colored blocks at the left and right in each panel indicate domains with the RBD given in red while the NTD is cyan.

a protein with nearly 3000 residues. This is what we observe. Since the spikes are basically the prime functional target for all existing vaccines, our results provide reassurance that, at least from a geometric simulation approach, the geometry of the vaccines needs no major readjustment to continue to work also for the mutations of the spikes. However, since the mutations nevertheless are known to have somewhat different responses to vaccines, our result might also suggest that factors beyond simple geometric modelling are important – with Coulombic surface charges on proteins clearly being among the main suspects. On the other hand, the structures as used here are taken as deposited in the PDB. Perhaps an equilibration and thermalization process might lead to structural refinements that could also cause conformation changes in the motion trajectories as computed here. Investigations along these lines are proceeding.

Acknowledgements We thank Young-Jun Park for pointing us to Ref. [7]. JP and RAR thankfully acknowledge funding via Warwick's Undergraduate Research Support Scheme (URSS). We thank Warwick's Scientific Computing Research Technology Platform for computing time and support. UK research data statement: Data accompanying this publication are available for download at [22].

[1] Liu J, Li K, Cheng L, Shao J, Yang S, Zhang W, Zhou G, de Vries A A and Yu Z 2021 *International Journal of Infectious Diseases* **103** 300–304 ISSN 12019712 URL `https://doi.org/10.1016/j.ijid.2020.12.033`
[2] Glaser J, Vermaas J V, Rogers D M, Larkin J, Legrand S, Boehm S, Baker M B, Scheinberg A, Tillack A F, Thavappiragasam M, Sedova A and Hernandez O 2021 **35** 452–468
[3] Jin Z, Du X, Xu Y, Deng Y, Liu M, Zhao Y, Zhang B, Li X, Zhang L, Peng C, Duan Y, Yu J, Wang

L, Yang K, Liu F, Jiang R, Yang X, You T, Liu X, Yang X, Bai F, Liu H, Liu X, Guddat L W, Xu W, Xiao G, Qin C, Shi Z, Jiang H, Rao Z and Yang H 2020 *Nature* **582** 289–293 ISSN 14764687 URL `https://doi.org/10.1038/s41586-020-2223-y`

[4] Wrapp D, Wang N, Corbett K S, Goldsmith J A, Hsieh C L, Abiona O, Graham B S and McLellan J S 2020 *Science* **367** 1260–1263 ISSN 0036-8075 URL `https://www.sciencemag.org/lookup/doi/10.1126/science.abb2507`

[5] Walls A C, Park Y J, Tortorici M A, Wall A, McGuire A T and Veesler D 2020 *Cell* **181** 281–292 ISSN 00928674 URL `https://doi.org/10.1016/j.cell.2020.02.058`

[6] Li F 2016 *Annual Review of Virology* **3** 237–261 ISSN 2327-056X URL `http://www.annualreviews.org/doi/10.1146/annurev-virology-110615-042301`

[7] Gobeil S M C, Janowska K, Mcdowell S, Mansouri K, Parks R, Stalls V, Kopp M F, Manne K, Saunders K, Edwards R J, Haynes B F, Henderson R C and Acharya P 2021 *bioRxiv* 2021.03.11.435037 URL `https://doi.org/10.1101/2021.03.11.435037`

[8] Römer R A, Römer N S and Wallis A K 2021 *Scientific Reports* **11** 4257 URL `https://doi.org/10.1038/s41598-021-82849-2`

[9] Walls A C, Xiong X, Park Y J, Tortorici M A, Snijder J, Quispe J, Cameroni E, Gopal R, Dai M, Lanzavecchia A, Zambon M, Rey F A, Corti D and Veesler D 2019 *Cell* **176** 1026–1039 ISSN 10974172

[10] Jimenez-Roldan J E, Freedman R B, Römer R A and Wells S A 2012 *Physical Biology* **9** 016008 ISSN 1478-3967 URL `http://stacks.iop.org/1478-3975/9/i=1/a=016008?key=crossref.aef662ce28f5ce4956911a1e1a961e22`

[11] Römer R A, Wells S A, Emilio Jimenez-Roldan J, Bhattacharyya M, Vishweshwara S, Freedman R B, 2016 *Proteins: Structure, Function, and Bioinformatics* **84** 1776–1785 ISSN 08873585 URL `http://doi.wiley.com/10.1002/prot.25159`

[12] Thorpe M, Lei M, Rader A, Jacobs D J and Kuhn L A 2001 *Journal of Molecular Graphics and Modelling* **19** 60–69 ISSN 10933263 URL `http://linkinghub.elsevier.com/retrieve/pii/S1093326300001224`

[13] Suhre K and Sanejouand Y H 2004 *Acta Cryst D* **60** 796–799 URL `http://dx.doi.org/10.1107/S0907444904001982papers2://publication/doi/10.1107/S0907444904001982`

[14] Suhre K and Sanejouand Y H 2004 *Nucleic Acids Research* **32** W610–W614 URL `http://nar.oxfordjournals.org/content/32/suppl_2/W610.abstractpapers2://publication/doi/10.1093/nar/gkh368`

[15] Wells S A, Jimenez-Roldan J E and Römer R A 2009 *Physical Biology* **6** 046005 ISSN 1478-3975 URL `http://stacks.iop.org/1478-3975/6/i=4/a=046005?key=crossref.b21092b85c3ad5d79e5f7135756e84ae`

[16] Li H, Wells S A, Jimenez-Roldan J E, Römer R A, Zhao Y, Sadler P J and O'Connor P B 2012 *Protein Science* **21** 1269–1279 ISSN 09618368 URL `%http://doi.wiley.com/10.1002/pro.2111http://dx.doi.org/10.1002/pro.2111`

[17] Heal J W, Jimenez-Roldan J E, Wells S A, Freedman R B and Römer R A 2012 *Bioinformatics* **28** 350–357 ISSN 1460-2059 URL `https://academic.oup.com/bioinformatics/article-lookup/doi/10.1093/bioinformatics/btr683`

[18] Wells S A, van der Kamp M W, McGeagh J D and Mulholland A J 2015 *PLOS ONE* **10** e0133372 ISSN 1932-6203 URL `https://dx.plos.org/10.1371/journal.pone.0133372`

[19] Freedman R B, Desmond J L, Byrne L J, Heal J W, Howard M J, Sanghera N, Walker K L, Wallis A K, Wells S A, Williamson R A, Römer R A, Katrine Wallis A, Wells S A, Williamson R A and Römer R A 2017 *Biochimica et Biophysica Acta (BBA) - Proteins and Proteomics* **1865** 1383–1394 ISSN 15709639 URL `http://linkinghub.elsevier.com/retrieve/pii/S157096391730198X`

[20] Farrell D W, Speranskiy K and Thorpe M F 2010 *Proteins: Structure, Function, and Bioinformatics* **78** 2908–2921 ISSN 08873585 URL `http://doi.wiley.com/10.1002/prot.22810`

[21] Ahmed A and Gohlke H 2006 *Proteins: Structure, Function, and Bioinformatics* **63** 1038–1051 ISSN 08873585 URL `http://doi.wiley.com/10.1002/prot.20907`

[22] Römer R A, Römer N S and Wallis A K 2020 Flex-Covid19 data repository URL `https://www.warwick.ac.uk/flex-covid19-data`

[23] With nearly 3000 residues in the structure of the trimers, is it impossible to capture the behaviour in a single figure that at the same time would allow the reader to see enough detail. The reader is referred to Ref. 22 for access to the electronic images for 6VXX and 6VYB.

[24] Visualization of the m_7 movements at $E_{\mathrm{cut}} = 1\,\mathrm{kcal/mol}$ for 6VYB, using a simple cartoon representation and rainbow coloring from N to C terminals, URL `https://pdb2movie.warwick.ac.uk/animol/6VYB`

[25] As in [24], but now for 7LWW, URL `https://pdb2movie.warwick.ac.uk/animol/7LWW`

Nonequilibrium dynamics of a fluid vesicle: Turing patterns and traveling waves

Hiroshi Noguchi and Naoki Tamemoto

Institute for Solid State Physics, University of Tokyo, Kashiwa, Chiba 277-8581, Japan

E-mail: noguchi@issp.u-tokyo.ac.jp

Abstract. In living cells, the shapes of biomembranes are dynamically regulated by proteins. We briefly review our recent studies on the membrane dynamics coupled with reaction-diffusion dynamics of curvature-inducing proteins. We simulated a vesicle in a fluid phase using a dynamically triangulated membrane model combined with a modified Brusselator model. The mechanochemical feedback of curvature-inducing proteins changes the condition of Turing patterns relative to those that occur on a non-deformable curved surface. Budding and multi-spindle shapes are also stabilized by Turing patterns. The speed of traveling waves increases or decreases with increasing local membrane curvature depending on the condition. Moreover, the reaction-diffusion waves can induce the self-oscillation of vesicle shapes.

1. Introduction

Many types of proteins are known to generate a local curvature of biomembrane. They regulate dynamic processes such as endocytosis and vesicle transport. Although the thermal equilibrium membrane shape stabilized by proteins has been well studied, nonequilibrium dynamics have been much less understood. Experimentally, spatio-temporal patterns of membrane shapes has been observed during chemical reaction propagation in plasma membranes [1]. Thus, the reaction-diffusion dynamics in deformable curved membranes is an important problem.

We studied nonequilibrium pattern formation on a fluid vesicle induced by mechanochemical feedback between membrane deformation and chemical reactions, by using a coarse-grained membrane simulation. In Sec. 2, the simulation models are briefly described. In Sec. 3 and Sec. 4, Turing patterns and traveling waves accompanied by vesicle deformation are presented; The details are described in Ref. [2] and Ref. [3], respectively. In Sec. 5, the summary is described.

2. Simulation model

We employ a dynamically triangulated membrane model [4] combined with a Brusselator model [5] that is one of the simplest reaction-diffusion models. The surface area S and volume V of a vesicle are fixed by constraint potentials, such that the reduced volume $V^* = V/(4\pi R^3/3)$ is fixed, where $R = \sqrt{S/4\pi}$. The details of the model and method are described in Refs. [2, 3].

Two proteins (a curvature-inducing protein and a regulatory protein) bind to the membrane. The bending energy of the membrane is expressed as

$$F_{\rm cv} = \int f_{\rm cv} dS \tag{1}$$

$$f_{\rm cv} = 2\kappa_0 H^2(1-u) + \frac{\kappa_1}{2}(2H - C_0)^2 u, \tag{2}$$

where u is the concentration of the curvature-inducing protein in the membrane ($u = 1$ at the maximum coverage) and H is the mean curvature of the membrane [6]. The membrane bound by the curvature-inducing protein has a high bending rigidity κ_1 and spontaneous curvature C_0.

The reaction-diffusion equations are given by

$$\tau\frac{\partial u}{\partial t} = f(u,v) + D_1\nabla^2 u, \tag{3}$$

$$\tau\frac{\partial v}{\partial t} = g(u,v) + D_2\nabla^2 v, \tag{4}$$

where v is the concentration of a regulatory protein that does not directly modify the membrane properties, and D_1 and D_2 are diffusion constants. We employed two types of the reaction function for u,

$$f(u,v) = A - G\frac{\partial f_{\rm cv}}{\partial u} - (B+1)u + u^2 v, \tag{5}$$

$$f(u,v) = A\exp\Big[-\frac{1}{k_{\rm B}T}\frac{\partial f_{\rm cv}}{\partial u}\Big] - \Big(B + \frac{u}{1-u}\Big)u + u^2 v, \tag{6}$$

(7)

in Ref. [2] and Ref. [3] to mainly investigate Turing patterns and traveling waves, respectively, where $k_{\rm B}T$ is the thermal energy. We use the original Brusselator function for v,

$$g(u,v) = Bu - u^2 v. \tag{8}$$

Although we consider $0 < u < 1$, the original Brusselator model does not has a maximum value of u. Hence, we scaled the value of u as $u' = k_u u$ in Ref. [2] and modified the second term as in Eq. (6) in Ref. [3] to maintain $u < 1$. The relation of the binding rate and the energy change by protein binding/unbinding is not uniquely determined, since the binding energy barrier can be modified differently [7]. In the latter reaction (Eq. (6)), the energy barrier is varied following the bending energy change. In the former reaction (Eq. (5)), it varies linearly. The former can be interpreted as the leading term of the latter for small $\partial f_{\rm cv}/\partial u$.

3. Turing patterns

The conditions for the Turing pattern and temporal oscillation mode on a non-deformable curved surface were determined by linear stability analysis [2]. Membrane deformation modifies these conditions; The region of Turing patterns increases in the phase diagram so that the Turing patterns are stabilized by the membrane deformation. According to vesicle shapes and spontaneous curvature of proteins, different shapes and patterns appear as shown in Fig. 1. Budding and multi-spindle shapes are also induced by Turing patterns. For budded vesicles, a Turing domain boundary separating two phases with higher and lower values of u is formed at the connective neck, because the diffusion of the proteins is reduced at the narrow neck. It is also observed that a temporal oscillation of the protein concentration changes into a Turing pattern.

4. Traveling waves

When the membrane deformation is coupled with a reaction-diffusion system, traveling wave patterns can spontaneously occur as shown in Fig. 2 [3]. In the top and middle snapshots, the waves start at highly curved regions of the vesicles. These waves disappear when they collide.

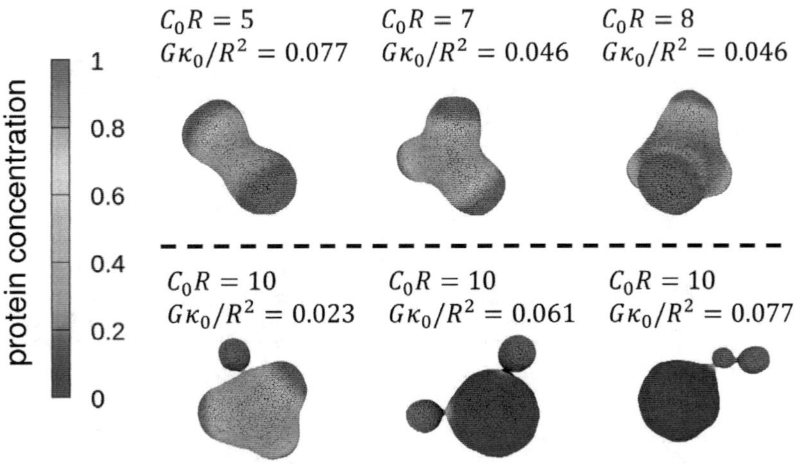

Figure 1. Snapshots of vesicles with Turing patterns at $V^* = 0.8$. Top panel: multi-spindle shapes. Bottom panel: budded vesicles. The concentration u of the curvature-inducing protein is indicated by color.

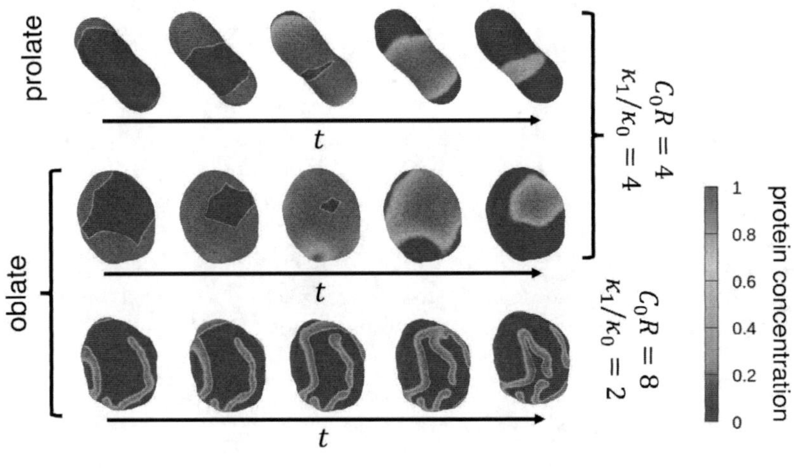

Figure 2. Sequential snapshots of propagating wave patterns. Top panel: an axisymmetric propagating wave pattern on prolate vesicle at $V^* = 0.8$. Middle panel: an axisymmetric propagating wave pattern on oblate vesicle at $V^* = 0.65$. Bottom panel: unordered and spiral wave patterns on oblate vesicle at $V^* = 0.65$. The concentration u of the curvature-inducing protein is indicated by color.

The bottom snapshots demonstrate that a propagating wave changes into spiral wave patterns and non-propagating patterns.

Moreover, the wave speed is changed by the local membrane curvature. It is positively or negatively correlated depending on the bending rigidity and spontaneous curvature.

Furthermore, the vesicle shape spontaneously oscillates, associated with the reaction-diffusion waves of curvature-inducing proteins as shown in Fig. 3 [3]. Similar oscillations have been

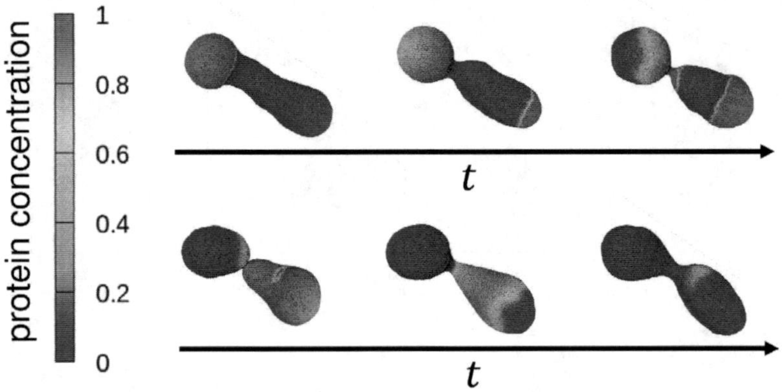

Figure 3. Sequential snapshots of a shape oscillatory vesicle at $V^* = 0.65$, $C_0R = 4$, and $\kappa_1/\kappa_0 = 4$. Budding repeatedly occurs accompanied by the traveling wave. The concentration u of the curvature-inducing protein is indicated by color.

experimentally observed for liposomes with a reconstituted Min system [8]. Min proteins play a crucial role in the cell division of Escherichia coli.

5. Summary

We have reviewed our recent studies [2, 3] on the coupling between a reaction-diffusion dynamics and membrane deformation. Membrane deformation stabilizes Turing patterns and modifies traveling waves. Moreover, the chemical patterns induce membrane deformations. Hence, the mechanochemical feedback plays an important role in pattern formation on deforming membranes.

We have examined two types of the modified Brusselator models. It is an interesting problem how the coupling of the membrane deformation and reaction-diffusion dynamics changes when other feedback relations with the bending energy and other reaction models are used.

Acknowledgments

This work was supported by JSPS KAKENHI Grant Number JP21K03481.

References

[1] Wu Z, Su M, Tong C, Wu M and Liu J 2018 *Nat. Commun.* **9** 136
[2] Tamemoto N and Noguchi H 2020 *Sci. Rep.* **10** 19582
[3] Tamemoto N and Noguchi H 2021 *Soft Matter* **17** 6589–6596
[4] Noguchi H and Gompper G 2005 *Phys. Rev. E* **72** 011901
[5] Prigogine I and Lefever R 1968 *J. Chem. Phys.* **48** 1695–1700
[6] Noguchi H 2021 *Phys. Rev. E* **104** 014410
[7] Goutaland Q, van Wijland F, Fournier J B and Noguchi H 2021 *Soft Matter* **17** 5560–5573
[8] Christ S, Litschel T, Schwille P and Lipowsky R 2021 *Soft Matter* **17** 319–330

A simple method to accelerate configurational sampling for a generalized hybrid Monte Carlo method

D Suzuki, T Hori and S Miura

Graduate School of Natural Science and Technology, Kanazawa University
Kakuma, Kanazawa 920-1192, Japan

E-mail: smiura@mail.kanazawa-u.ac.jp

Abstract. In this paper, a simple method to improve sampling efficiency of the generalized hybrid Monte Carlo (GHMC) method is presented. Compared to the standard GHMC method, our method is found to allow us to safely increase the time increment for solving an equation-of-motion in the GHMC calculations by a factor of 4. We have demonstrated various algorithmic parameter dependence on the sampling efficiency of a hydrated alanine dipeptide.

1. Introduction

The hybrid Monte Carlo (HMC) method [1-3] is an equation-of-motion guided Monte Carlo method to generate the canonical ensemble. The method has originally been developed to solve sampling problems related with a non-ergodicity found in numerical simulations of quantum field theory [1]. Then, the method has been extended to handle condensed matters such as liquids [2,3], including quantum many-body systems [3-9]. The HMC method has been generalized by introducing an additional algorithmic parameter [10]. The generalized HMC (GHMC) has been applied to simulations of lattice field theory [10,11] and then to molecular systems [12]. The GHMC method has also been extended to generate the multicanonical ensemble [13,14].

In the present study, we have tested two methods, the reversible reference system propagator algorithm (r-RESPA) [15,16] and the hydrogen mass repartitioning (HMR) [17,18] to improve sampling efficiency of the GHMC method. The r-RESPA is a general and systematic method to introduce the multiple time step integrator of the equation-of-motion. The HMR method enables us to set larger time increments in the MD simulation by increasing the mass of the hydrogen atoms in the system. We have applied our GHMC combined with the above methods to a hydrated alanine dipeptide to examine the computational efficiency.

2. Method

2.1. A generalized hybrid Monte Carlo method

A hybrid Monte Carlo algorithm is a method that combines molecular dynamics (MD) and Monte Carlo (MC) techniques. Unlike the standard MC, whole system coordinates are simultaneously updated by equations of motion. The trial configuration generated by several molecular dynamics steps is then accepted or rejected by an appropriate Metropolis criterion as in MC. The HMC algorithm has been proved to yield the canonical distribution as long as a time-reversible and volume-preserving numerical integration algorithm is employed to solve the equations of motion [2,3]. In the standard HMC method, particle momenta are randomly sampled from the Maxwell distribution at each

HMC step [1]. This condition can be relaxed so that the particle momenta are partially refreshed [10]. This method is called the generalized hybrid Monte Carlo method. To our experience, the complete refreshment of the momenta in the standard HMC requires many molecular dynamics steps in a single HMC step to achieve good computational efficiency for some systems [4,9,13]. The partial mixing or partial reuse of the momenta provides a possibility to achieve good computational efficiency by much smaller number of molecular dynamics steps in a single HMC step.

A unit GHMC step is outlined as follows:

- Start a GHMC step from an initial state $(\{\boldsymbol{r}_i\}, \{\boldsymbol{p}_i\})$.
- Mix a random momentum $\boldsymbol{u}_i$ drawn from the Maxwell distribution to the momentum $\boldsymbol{p}_i$ according to the following equation with a given angle ϕ in the range of $0 < \phi \le \pi/2$
$$\boldsymbol{p}_i \leftarrow \boldsymbol{p}_i \cos\phi + \boldsymbol{u}_i \sin\phi. \tag{1}$$
- Perform MD calculations of n_{MD} steps with a time increment Δt by
$$U_{\Delta\tau}\colon (\{\boldsymbol{r}_i\}, \{\boldsymbol{p}_i\}) \to (\{\boldsymbol{r}_i'\}, \{\boldsymbol{p}_i'\}) \tag{2}$$
where $U_{\Delta\tau}$ is a map from the initial to the final state by an appropriate integrator with $\Delta\tau = n_{\mathrm{MD}} \times \Delta t$.
- The trial state is accepted according to the following Metropolis criterion:
$$P_{\mathrm{A}} = \min\left\{1, \exp\left(-\frac{\Delta H}{k_{\mathrm{B}}T}\right)\right\} \tag{3}$$
where $\Delta H = H(\{\boldsymbol{r}_i'\}, \{\boldsymbol{p}_i'\}) - H(\{\boldsymbol{r}_i\}, \{\boldsymbol{p}_i\})$. Here, H denotes the total energy of the system.
- When the trial state is rejected, negate the momentum in the initial state,
$$(\{\boldsymbol{r}_i\}, \{\boldsymbol{p}_i\}) \to (\{\boldsymbol{r}_i\}, \{-\boldsymbol{p}_i\}). \tag{4}$$

It is worthwhile to note that when we set $\phi = \pi/2$, the GHMC method recovers the standard HMC method where the momenta of the system are fully refreshed at each HMC step. In the standard GHMC method, the velocity Verlet method is used to perform MD calculation by equation (2). In the present study, we have adopted the r-RESPA to improve the sampling efficiency of the GHMC method.

2.2. *The reversible reference system propagator algorithm*

The reversible reference system propagator algorithm (r-RESPA) is a general and systematic method to derive a time-reversible and volume-preserving numerical integration algorithm that achieves highly stable time evolution; the method has successfully been applied to various systems [16,19-21]. In the present study, we have used the r-RESPA to perform MD calculations in the GHMC method, equation (2). As in the previous study [16], we decompose force of the system $\boldsymbol{F}$ into two parts, $\boldsymbol{F}_{\mathrm{rapid}}$ and $\boldsymbol{F}_{\mathrm{slow}}$; $\boldsymbol{F}_{\mathrm{rapid}}$ arises from the rapidly varying interactions including stretch, bending and torsion interactions and $\boldsymbol{F}_{\mathrm{slow}}$ from the slowly varying interactions including van der Waals and electrostatic interactions and solvent effects. A unit MD step is outlined as follows:

- Update momenta using F_{slow} with a large time increment Δt according to the following equation,
$$\boldsymbol{p}_i \leftarrow \boldsymbol{p}_i + \boldsymbol{F}_{\mathrm{slow}} * \Delta t/2. \tag{5}$$
- Update positions and momenta using F_{rapid}. Iterate n_{r} times on the basis of the following three equations:
$$\boldsymbol{p}_i \leftarrow \boldsymbol{p}_i + \boldsymbol{F}_{\mathrm{rapid}} * \delta t/2, \tag{6}$$
$$\boldsymbol{r}_i \leftarrow \boldsymbol{r}_i + \boldsymbol{p}_i * \delta t, \tag{7}$$
$$\boldsymbol{p}_i \leftarrow \boldsymbol{p}_i + \boldsymbol{F}_{\mathrm{rapid}} * \delta t/2, \tag{8}$$
where $\delta t = \Delta t/n_{\mathrm{r}}$ is a small time increment and n_{r} is a natural number chosen so that δt is small enough.

- Update momenta using equation (5) again.

In a unit MD step, F_{rapid} is computed n_{r} times, while F_{slow} is computed only once. As F_{rapid} needs significantly less computational time than F_{slow}, the total computational cost of Δt time evolution can be reduced with keeping numerical stability.

2.3. *Hydrogen mass repartitioning*

In standard molecular systems, the equilibrium properties are determined independently of the kinetic part of their Hamiltonian. Thus, in an artificial system with converted masses, the dynamics cannot be reproduced, but physical quantities can be accurately evaluated. In previous studies [22,23,24], MD simulations of some polypeptides were accelerated by simply increasing masses of the hydrogen atoms (or masses of all atoms). On the other hand, Feenstra, Hess and Berendsen pointed out that increasing the total mass of the system causes the increase in viscosity and the decrease in diffusivity [17]; they proposed the repartitioning algorithm to keep the total mass of the system constant, which is the basis of the hydrogen mass repartitioning (HMR). In a recent study [18], the HMR method has been applied to some biomolecules and the time increment of the simulation to be increased by up to twice. In the present study, we have applied the same repartitioning algorithm presented in a previous study [18] to an alanine dipeptide; the masses of the hydrogen atoms were increased by a factor of 3, and the masses of the other heavy atoms decreased so that the total mass is equal to that before the repartition. Details can be found in the reference 18.

3. Results

In the present study, we have performed GHMC calculations for a hydrated alanine dipeptide at 300 K. The following three methods have been tested: the standard GHMC method using the velocity Verlet algorithm for MD calculations (called "GHMC-1"), the GHMC method with the r-RESPA (called "GHMC-2") and the GHMC method with the r-RESPA and HMR (called "GHMC-3"). The dipeptide was modeled by ff14SB [25] as a force field. The effect of the hydration is described by an empirical generalized Born and surface area (GBSA) continuum model [26,27]. For the GB model, we adopted the Hawkins, Cramer, Truhlar model [28]; we set the values of dielectric constant for the peptide and water solvent to be 1 and 78.5, respectively. It is noted that the use of the ff14SBonlysc forcefield [25] with the GB-Neck2 model [29] is recommended to obtain better accuracy on peptide properties using implicit solvent models; however, results regarding computational efficiencies presented in this paper would be nearly unaffected by the choice of the forcefields. In MD calculations using the r-RESPA, the small time increment δt was fixed to be 1.0 fs, and the large time increment Δt was set as $n_{\mathrm{r}} \times \delta t$.

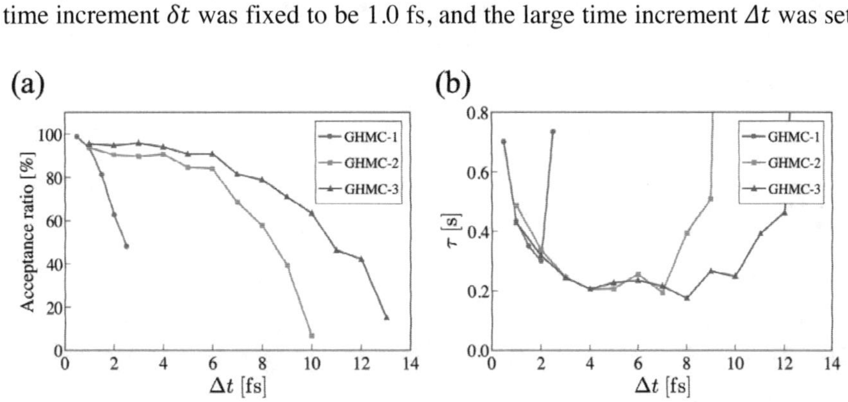

Figure 1. The acceptance ratios (a) and the correlation times τ for the potential energy (b) presented as a function of Δt. These results are obtained by three GHMC calculations (GHMC-1, GHMC-2 and GHMC-3) at 300 K in 2.0×10^5 steps using the following parameters: $\phi = \pi/4$ and $n_{\mathrm{MD}} = 10$.

Our simulation code has been developed using the OpenMM toolkit for molecular simulation [30].

We show results of GHMC-1, GHMC-2 and GHMC-3 calculations to examine computational efficiency. Each calculation was done by 2.0×10^5 GHMC steps for various Δt. To measure the sampling efficiency, we employ a quantity called a statistical inefficiency [31,32]. This quantity expresses the number of correlated steps needed to obtain independent sampling for a physical quantity. In the present study, we have chosen the potential energy as a quantity to measure the efficiency. We calculate the statistical inefficiency in units of the number of GHMC steps. We define the correlation time τ as the CPU time taken to compute the correlated GHMC steps. In figure 1, we present the correlation time and the associated acceptance ratio as a function of the time increment Δt; here, ϕ and n_{MD} are fixed to be $\pi/4$ and 10. If the equations of motion are accurately integrated, corresponding to the high acceptance ratio, the movement in the phase space is limited; this results in the long correlation. On the other hand, if we adopt large Δt corresponding to low acceptance ratio, the system moves widely in phase space; however, many of the trial configurations are rejected due to the large Hamiltonian error, resulting in the long correlation again. Thus, the correlation time τ has a minimum value between high and low acceptance ratio. As seen in figure 1, GHMC-2 and GHMC-3 show better computational efficiency than GHMC-1. The minimum correlation times for GHMC-1, GHMC-2 and GHMC-3 are given by $\Delta t = 2.0$, 7.0, and 8.0 fs, respectively and corresponding acceptance ratios are 63 %, 68 %, and 79 %. In the three GHMC simulations, the shortest correlation time is given by GHMC-3. This result shows that our new GHMC algorithm can sample

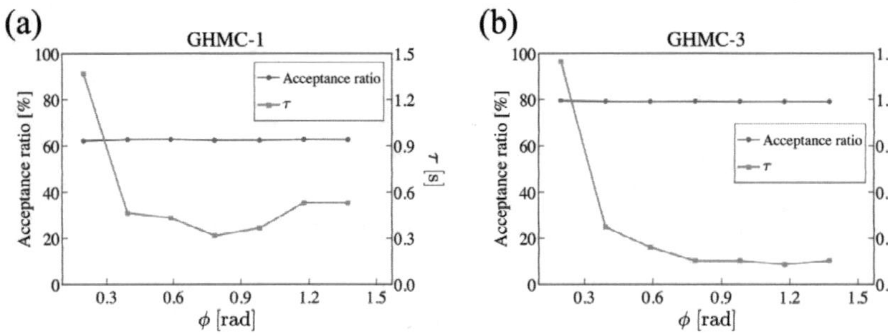

Figure 2. The acceptance ratios and the correlation times τ for the potential energy presented as a function of ϕ. These results obtained by two GHMC calculation at 300 K in 2.0×10^5 steps. (a) The result obtained by GHMC-1 using $\Delta t = 2.0$ fs and $n_{\mathrm{MD}} = 10$. (b) The result obtained by GHMC-3 using $\Delta t = 8.0$ fs and $n_{\mathrm{MD}} = 10$.

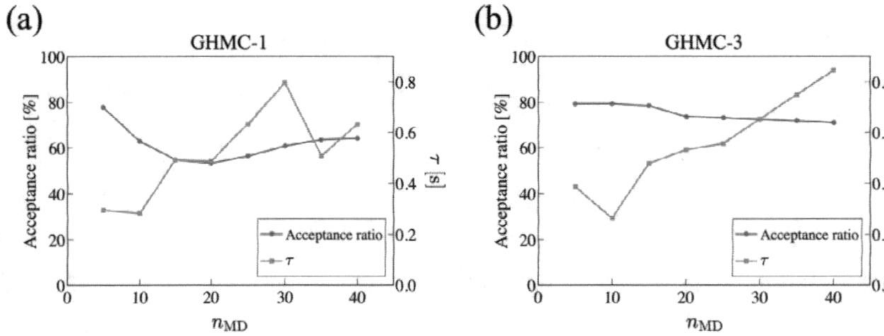

Figure 3. The acceptance ratios and the correlation times τ for the potential energy presented as a function of n_{MD}. These results obtained by two GHMC calculation at 300 K in 2.0×10^5 steps. (a) The result obtained by GHMC-1 using $\Delta t = 2.0$ fs and $\phi = \pi/4$. (b) The result obtained by GHMC-3 using $\Delta t = 8.0$ fs and $\phi = 3\pi/8$.

configurations more efficiently than the standard method.

We also performed GHMC-1 and GHMC-3 calculations of 2.0×10^5 steps for various ϕ and n_{MD}; here, Δt was fixed to be the best value found in figure 1 for each method, 2.0 fs for GHMC-1 and 8.0 fs for GHMC-3. Figure 2 shows ϕ dependence of the correlation time τ and the associated acceptance ratio; n_{MD} is fixed to be 10. As seen in figure 2, the minimum correlation is given by $\phi = \pi/4$ for GHMC-1 and $\phi = 3\pi/8$ for GHMC-3. Similarly, figure 3 shows n_{MD} dependence of τ and the best value of n_{MD} is given to be 10 for both GHMC-1 and GHMC-3; here, ϕ is fixed to be $\pi/4$ for GHMC-1 and $3\pi/8$ for GHMC-3. From these results, the optimal parameters of GHMC method are determined to be $\Delta t = 2.0$ fs, $\phi = \pi/4$ and $n_{\mathrm{MD}} = 10$ for GHMC-1, $\Delta t = 8.0$ fs, $\phi = 3\pi/8$ and $n_{\mathrm{MD}} = 10$ for GHMC-3. With the help of the r-RESPA and HMR, we can use four times longer time increment for the GHMC calculations. As in our previous study [13], it was found that good computational efficiency can be gained in the case of the acceptance ratio around 60 - 80 %, and small n_{MD} can be used to achieve good efficiency with small ϕ for the isolated system.

We finally show results of long GHMC-1 and GHMC-3 calculations. Each calculation was done by 1.0×10^7 GHMC steps using the optimal parameters obtained above. Figure 4 shows dihedral angles distribution $P(\varphi, \psi)$ obtained by each calculation. Compared with the previous study [33], overall distribution is well described by both methods. While a weak peak is found around $(\varphi, \psi) = (1.0\ \mathrm{rad}, 0.5\ \mathrm{rad})$ for the GHMC-3 result, which is consistent with the previous study, the corresponding peak is missed by the GHMC-1 calculations. This indicates that the standard GHMC calculation needs to be extended to properly sample the configurations around the weak peak. This result demonstrates that our new GHMC method can sample configurations more efficiently than the standard one in the simulation of the hydrated alanine dipeptide.

4. Concluding remarks

In the present study, we have tested two methods, the reversible reference system propagator algorithm (r-RESPA) and hydrogen mass repartitioning (HMR) to improve the computational efficiency of the generalized hybrid Monte Carlo (GHMC) method. We have applied our method to the hydrated alanine dipeptide. For this system, we can increase the time increment by a factor of 4 while maintaining numerical stability. Dihedral angles distribution is well described by our GHMC methods, although a weak peak in the distribution is missed by the standard method within a limited simulation time. This demonstrates that more efficient calculations can be realized by our GHMC accelerated by the r-RESPA and HMR methods. Finally, we comment on calculations of hydrated biomolecules using explicit water models by our GHMC method. For the explicit solvent models, the r-RESPA and HMR methods must be carefully used to achieve high computational efficiency; a recent

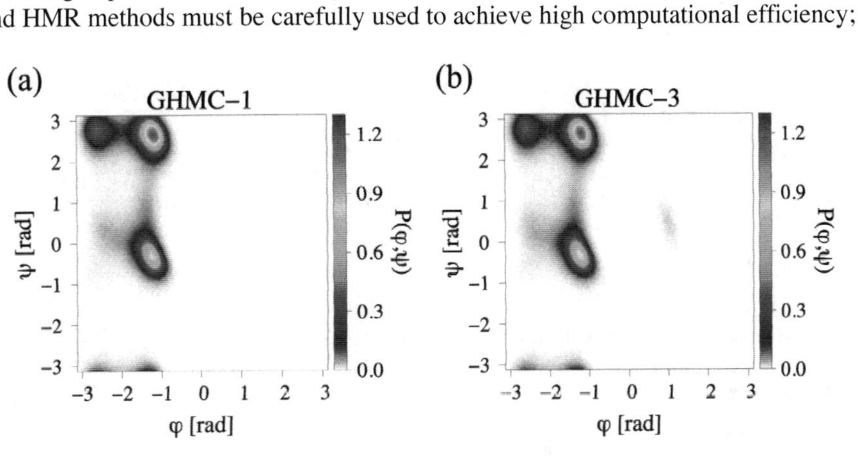

Figure 4. Dihedral angles distribution $P(\varphi, \psi)$ of the hydrated alanine dipeptide at 300 K. (a) The distribution obtained by GHMC-1. (b) The distribution obtained by GHMC-3.

attempt on this issue can be found in reference 34.

Acknowledgements
This work was supported by JSPS KAKENHI (Grant Nos. 20K03879 and 20H04464; Tokyo, Japan).

References

[1] Duane S, Kennedy A D, Pendleton B J and D. Roweth 1987 *Phys. Lett.* B **195** 216
[2] Mehlig B, Heermann D W and Forrest B M 1992 *Phys. Rev*. B **45** 679
[3] Tuckerman M, Berne B J, Martyna G J and Klein M L 1993 *J. Chem. Phys.* **99** 2796
[4] Miura S and Tanaka J 2004 *J. Chem. Phys.* **120** 2160
[5] Miura S 2005 *J. Phys.: Condens. Matter* **17** S3259
[6] Miura S 2007 *J. Chem. Phys.* **126** 114308
[7] Miura S 2007 *J. Chem. Phys.* **126** 114309
[8] Miura S 2012 *Advances in Quantum Monte Carlo* (*The ACS Symposium Series* vol 1094) ed Tanaka S *et al* (Washington DC: American Chemical Society) Chapter 14 pp 177–186
[9] Kamibayashi Y and Miura S 2006 *J. Chem. Phys.* **145** 074114
[10] Horowitz A M 1991 *Phys. Lets.* B **268** 247
[11] Kennedy A D and Pendleton B 2001 *Nucl. Phys.* B **607** 456
[12] Akhmatskaya E, Bou-Rabee N and Reich S 2009 *J. Comput. Phys.* **228** 2256
[13] Mukuta N and Miura S 2018 *J. Chem. Phys.* **149** 072322
[14] Mukuta N and Miura S 2020 *Biophys. Physicobiol*. **17** 14
[15] Tuckerman M E and Berne B J 1991 *J. Chem. Phys*. **95** 8362
[16] Tuckerman M E, Berne B J and G J Martyna 1992 *J. Chem. Phys*. **97** 1990
[17] Feenstra K A, Hess B and Berendsen H J C 1999 *J. Comput. Chem*. **20** 786
[18] Hopkins C W, Grand S L, Walker R C and Roitberg A E 2015 *J. Chem. Theory Comput*. **11** 1864
[19] Tuckerman M E, Berne B J and Martyna G J 1991 *J. Chem. Phys*. **94** 6811
[20] Humphreys D D, Friesner R A and Berne B J 1994 *J. Chem. Phys*. **98** 6885
[21] Zhou R and Berne B J 1995 *J. Chem. Phys*. **103** 9444
[22] Mao B and Friedman A R 1990 *Biophys. J*. **58** 803
[23] Mao B, Maggiora G M and Chou K C 1991 *Biopolymers* **31** 1077
[24] Mao B 1991 *Biophys. J*. **60** 611
[25] Maiser J A, Martinez C, Kasavajhara K, Wickstrom L, Hauser K E and Simmerling C 2015 *J. Chem. Theory Comput*. **11** 3696
[26] Schaefer M, Bartels C and Karplus M 1998 *J. Mol. Biol.* **284** 835
[27] Onufriev A, Bashford D and Case D A 2004 *Proteins* **55** 383
[28] Hawkins G D, Cramer C J and Truhlar D G 1996 *J. Phys. Chem*. **100** 19824
[29] Mongan J, Simmerling C, McCammon J A, Case D A and Onufriev A 2006 *J. Chem. Thoery Comput.* **3** 156
[30] Eastman P, Swails J, Chodera J D, McGibbon R T, Zhao Y, Beauchamp K A, Wang L-P, Simmonett A C, Harrigan M P, Stern C D, Wiewiora R P, Brooks B R and Pande V S 2017 *PLoS Comput. Biol*. **13** e1005659
[31] Friedberg R and Cameron E 1970 *J. Chem. Phys*. **52** 6049
[32] Allen M P and Tildesley D J 1987 *Computer Simulation of Liquids* (Oxford: Clarendon Press)
[33] Rubio-Martinez J, Tomas M S and Perez J J 2017 *J. Mol. Graph. Model*. **78** 118
[34] Jaewoon J, Kasahara K, Kobayashi C, Oshima H, Mori T, and Sugita Y 2021 *J. Chem. Theory Comput*. **17** 5312

Optical Kerr Effect and Structural Tetrahedrality of Supercooled Water at Ambient Pressure

Ping-Han Tang[1,#], Yu-Wei Kuo[1] and Ten-Ming Wu[1,*]
[1]Institute of Physics, National Yang Ming Chiao Tung University, Hsinchu 30010, Taiwan, R.O.C.
[#]Present address: Molecular Sciences and Digital Innovation Center, Genetics Generation Advancement Corp., Taipei 11493, Taiwan, R.O.C.

[*]Corresponding author Email: tmw@nycu.edu.tw

Abstract. The correlation between the optical Kerr effect (OKE) spectroscopy of supercooled water at ambient pressure and its structural tetrahedrality was investigated by contrasting simulation results of two non-polarizable water models modified with the same collective polarizability, which involves intrinsic molecular polarizability and induced polarizability arising from interactions between molecular dipoles. The tetrahedrality of water structure was typified with the second-peak maximum in the pair distribution function of oxygens and the fraction of molecules, which and their neighbours up to the second hydration shell all have four H-bond coordinators. Our results indicate that the intermolecular vibrational band in the OKE spectrum of supercooled water is considerably correlated to its structural tetrahedrality.

1. Introduction

The thermodynamic properties of water display many anomalies on cooling into its supercooled states [1]. The temperature dependences of the anomalies at ambient pressure (AP) appear to diverge in a power law near 228 K [2]. It is subtle to understand the origin of these anomalies, because of the nucleation of water to ice near 235 K at AP. A widely accepted explanation for water anomalies is the liquid-liquid critical point (LLCP) scenario [3], where the anomalies were hypothetically resulted from a critical point imbedded in the "no-man's land" region in the phase diagram of metastable water [4].

In the LLCP scenario, water is considered as a mixture of two types of local structure typified with tetrahedrally ordered H-bonds and strongly distorted H-bonds [5], where the associated molecules are referred as low-density liquid (LDL) and high-density liquid (HDL), respectively. In the deeply supercooled regime, LDL and HDL are two distinct phases divided with a transition line, which has been illustrated by a recent experiment [6]. The transition line is a continuation of the coexistence line between low-density-amorphous (LDA) and high-density-amorphous (HDA) ices existing at low and high pressures, respectively [7]. As increasing temperature, the phase transition line ends at a critical point, above which the two types of local structure mix together and exchangeable and the thermodynamic response functions of the mixture are expected to show a maximum. The loci of these maxima merge on a line starting from the critical point into the single phase region; this line is referred as the Widom line (WL) [8], where the fractions of LDL and HDL are almost equal. Recently, the WL of supercooled water has been evidenced to exist by the experiment of water droplet below the nucleation temperature [9], where the measured isothermal compressibility, correlation length, and isobaric heat capacity all display a maximum near 229 K [10,11], indicating that the structural tetrahedrality of a water molecule extends rapidly to its second hydration shell as near the WL [10].

The coexistence of two types of local structure in supercooled water was evidenced by the optical Kerr effect (OKE) spectrum [12], the Fourier-sine transform of the nuclear response function (NRF) measured by the time-resolved pump-probe femtosecond spectroscopy. The NRF observed by the OKE measurement decays in a relaxation time with a power-law temperature variation divergent near 227 K in accord with the behaviors of thermodynamic response functions. More significances related to water structure come from the intermolecular vibrational band in the OKE spectrum, which can be fitted by two constituents with a relative shift in frequency: The low-frequency constituent barely depends on temperature but the high-frequency constituent has an amplitude increasing noticeably in the supercooled regime [12]. The two constituents were attributed to two types of local structure in supercooled water, comparable to the structures of HDA and LDA ices [13].

In this paper, we investigated the correlation between the intermolecular vibrational band in the OKE spectrum of supercooled water and its tetrahedral structure, by comparing simulation results of TIP4P/2005 [14] and SPC/E [15] water models at AP. The TIP4P/2005 water shows stronger tetrahedral order in structure and has a LLCP, which is comprehensible by recent simulations [16], whereas the SPC/E water shows weaker tetrahedral order and is absent from a convincing evidence for the existence of a LLCP [17]. In our study, the two non-polarizable water models were modified with the same collective polarizability model [18] to manifest that the distinctions between their NRFs and OKE spectra are resulted from water structure. This paper is organized as follows: In Sec. II, the globally and locally tetrahedral structures of water are addressed. In Sec. III, the background for the OKE of water is described. In Sec. IV, the results of the two models are presented. Our conclusions are given in Sec. V.

2. Tetrahedral Structure of Supercooled Water

MD simulations distinctly with TIP4P/2005 and SPC/E force fields are described in Ref. [19] and [20], respectively, where the temperature variation of the pair distribution function (PDF) $g(\mathrm{r})$ of oxygens in each model liquid is displayed. The tetrahedrality of global water is depicted with the second-peak maximum g_2 of the PDF. For each water model at 1 atm, the temperature variation of g_2 is presented in Fig. 1 (a), indicating that the tetrahedrality of water structure gradually develops upon cooling, where at each temperature the g_2 of TIP4P2005 water is higher than that of SPC/E water.

The tetrahedrality of local structure in water is described with molecules, which and their neighbors up to the second hydration shell all have four H-bond coordinators [21]. In this paper, these molecules are LDL and the others are HDL. For the two model waters, the molecular fraction of LDL and HDL are shown in Fig. 1 (b). For TIP4P/2005 water, the equivalent of the two fractions occurs near 225 K; however, for SPC/E liquid, the temperature at equal fraction is lower than 220 K. At each temperature, the LDL fraction f_L of TIP4P/2005 liquid is higher than that of SPC/E water, revealing that the former model possesses a higher degree in locally structural tetrahedrality than the latter one.

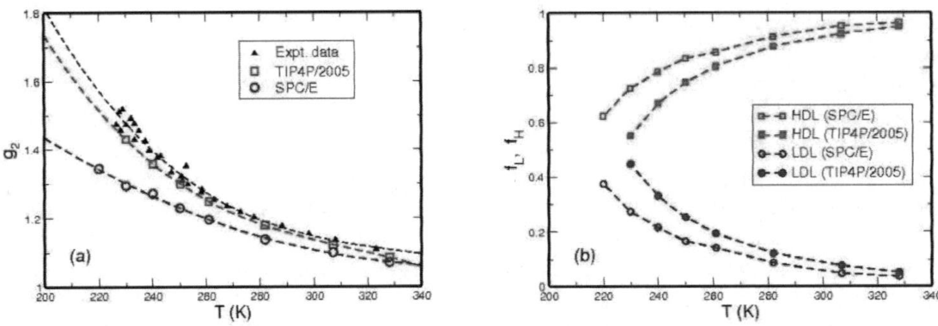

Figure 1 Globally and locally tetrahedral structure of water at 1 atm. (a) Temperature dependence of g_2. Squares and circles are simulation data of water model, and triangles are experimental data referred from Ref. [9]. The dash lines are the least-squares fit for each data set. (b) Molecular fraction f_L and f_H. The symbols are the results calculated with each water model, with each set of data guided by a dash line.

3. The OKE of Water

The time-resolved OKE spectroscopy of a molecular liquid is resulted from modifications caused by molecular motions on off-diagonal elements of its collective polarizability tensor $\boldsymbol{\Pi}$ [22], so that the spectroscopy provides information related to collective dynamics of the liquid. In the classical limit, the OKE NRF of a liquid with N anisotropic molecules at temperature T is given as [23]

$$R(t) = -\frac{\theta(t)}{k_B T}\frac{d\Psi(t)}{dt}, \qquad (1)$$

with k_B the Boltzmann constant, $\theta(t)$ a unit step function, and $\Psi(t)$ the polarizability anisotropy time correlation function (PATCF), which is a TCF of off-diagonal elements of $\boldsymbol{\Pi}$.

In our calculations, the collective polarizability model of water is given as [18]

$$\boldsymbol{\Pi} = \sum_{i=1}^{N} \boldsymbol{\alpha}_i^M + \sum_{\substack{i,j=1 \\ i \neq j}}^{N} \boldsymbol{\alpha}_i^M \cdot \boldsymbol{T}_{ij} \cdot \boldsymbol{\alpha}_j^M, \qquad (2)$$

where $\boldsymbol{\alpha}_i^M$ is the intrinsic polarizability of molecule i viewed in the Lab. frame and $\boldsymbol{T}_{ij}$ is the interaction tensor between molecular dipoles located at the centres of mass of molecule i and j. For each molecule, $\boldsymbol{\alpha}_j^M$ is obtained through a similarity transformation with the intrinsic polarizability tensor $\boldsymbol{\alpha}^M$ in the body frame of a water molecule. In Eq. (2), the first term is the molecular polarizability (MP) $\boldsymbol{\Pi}^M$, which is a sum of intrinsic polarizability of each molecule, and the second term is the induced polarizability (IP) $\boldsymbol{\Pi}^I$ caused by the dipole-induced-dipole (DID) interactions between molecular dipoles. For a rigid-molecule liquid, the MP is only associated with reorientations of molecule but the IP has more dependence on molecular translations.

In the frequency domain, the OKE reduced spectral density (RSD) $\chi''(\omega)$ is the Fourier-sine transform of the NRF after detached from its long-time decay due to the diffusion of molecular reorientations [19]. By dividing $\boldsymbol{\Pi}$ into $\boldsymbol{\Pi}^M$ and $\boldsymbol{\Pi}^I$, the NRF $R(t)$ can be decomposed into the MM, II, and MI components, with the first two resulted from $\boldsymbol{\Pi}^M$ and $\boldsymbol{\Pi}^I$ individually and the last one from the cooperation of $\boldsymbol{\Pi}^M$ and $\boldsymbol{\Pi}^I$. The OKE RSD $\chi''(\omega)$ is also a sum of three corresponding components.

4. Results and Discussion

Evaluated with the collective polarizability model in Eq. (2), the NRF up to 0.5 ps and $\chi''(\omega)$ of SPC/E water at 1 atm are presented in Fig. 2 (a) and (b), respectively. The results of TIP4P/2005 water are displayed in Ref. [19]. For both models, $R(t)$ shows two peaks near 15 and 50 fs, which are generally insensible to temperature. After the two peaks, the time profile of $R(t)$ displays a dip and then a damped oscillatory behaviour before converging to a monotonical decay. As decreasing temperature, the amplitude of the dip increases progressively and the damped oscillations extend to a longer time.

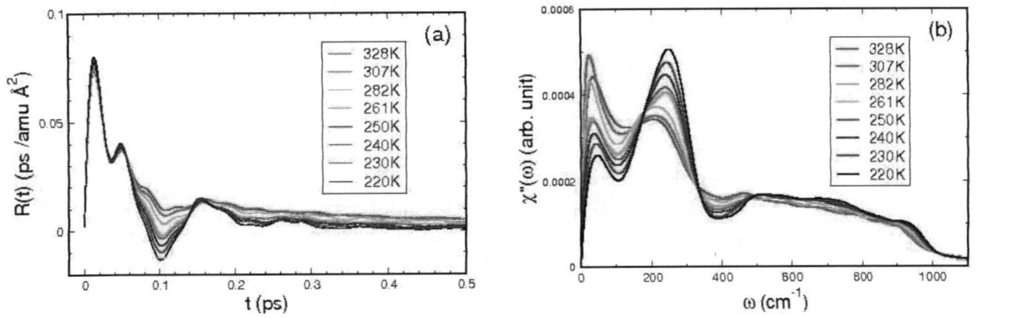

Figure 2 (a) NRF and (b) OKE RSD of SPC/E water.

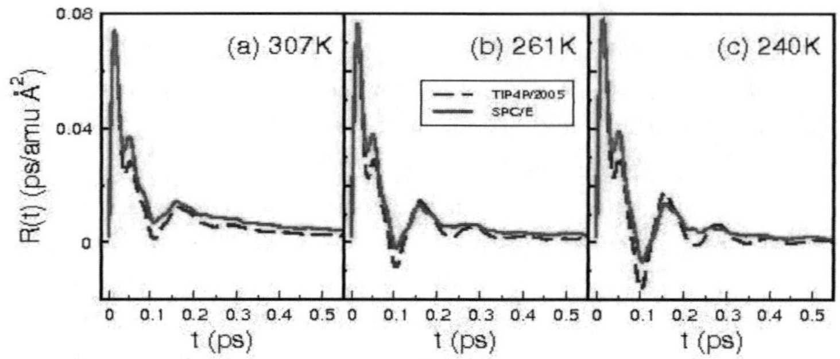

Figure 3 Comparison of NRF between SPC/E (red solid lines) and TIP4P/2005 (blue dash lines) waters.

Shown in Fig. 3 are comparisons of NRF between the two model waters at the same pressure and temperature. Relatively, the overall $R(t)$ of TIP4P/2005 water delays slightly in time, where the dip is somewhat enhanced and the subsequent oscillatory behaviour has larger amplitudes and persists for a longer time. The difference in $R(t)$ between the two model liquids becomes more prominent on cooling to low temperatures, where the TIP4P/2005 water has a higher degree of tetrahedral order, which makes its structure more persistent in time.

As shown in Fig. 2 (b), $\chi''(\omega)$ of each model water displays three bands. The broad band in $\chi''(\omega)$ at frequencies over 400 cm^{-1} is related to molecular librations, or hindered rotations [24], where the band amplitudes have a slight increase on cooling. For SPC/E water, the two other low-frequency bands in $\chi''(\omega)$ are centred around 35 and 200 cm^{-1} at 328 K, and shift to 45 and 250 cm^{-1}, respectively, as decreasing temperature to 220 K. The band around 200 cm^{-1} in $\chi''(\omega)$ is related to intermolecular vibrations, with the O-atom motions along connecting H-bonds [25]. The peak magnitude I_{max} of the vibrational band in $\chi''(\omega)$ increases considerably upon cooling, where the band at low temperatures is similar as the stretching band in Raman spectrum of LDA ice [13]. The band around 30 cm^{-1} in $\chi''(\omega)$ is related to the O-O-O bending of triplet water molecules, with the motions of O-atom perpendicular to the H-bonds connecting them [25,26], where the intensity of the band decreases with temperature.

Shown in Fig. 4 are comparisons of OKE RSD between the two model waters at the same pressure and temperature. Relative to SPC/E liquid, the vibrational band in $\chi''(\omega)$ of TIP4P/2005 water has a higher peak and a narrower bandwidth in the deeply supercooled regime and the portion around 500 ~ 600 cm^{-1} is enhanced, where the related water dynamics is mainly associated with molecular reorientations about the axis perpendicular to the molecular plane [24]. The bending band of TIP4P/2005 water has a reduced amplitude. The differences in $\chi''(\omega)$ between the two waters are attributed to the distinction of their liquid structures, where the distinction can be quantified by the values of g_2 and f_L.

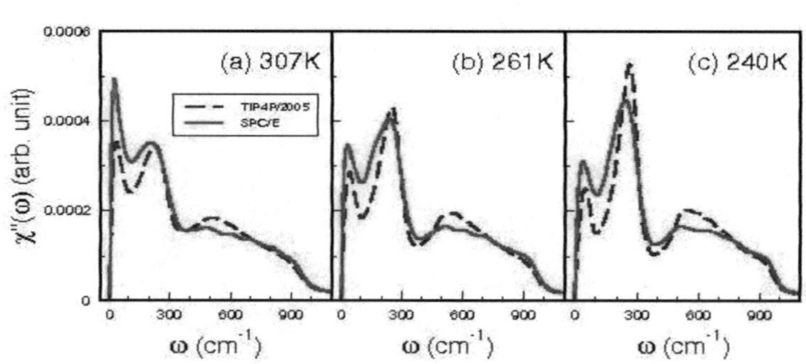

Figure 4 Comparison of OKE RSD between SPC/E (red solid lines) and TIP4P/2005 (blue dash lines) waters.

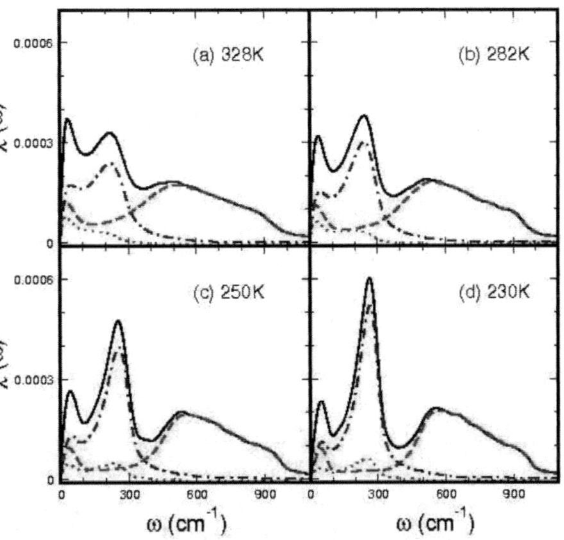

Figure 5 Contributions to the OKE RSD of TIP4P/2005 water. The black solid lines are the total spectrum calculated with both the MP and the IP. The red-dash and blue-dot-dash lines indicate the MM and II components, respectively. The green-dot lines are the MI component.

Shown in Fig. 5 are the MM, II and MI contributions to the OKE RSD of TIP4P/2005 water at 1 atm. Generally, the librational band is resulted from the MM component, the vibrational band is governed by the II component, and the O-O-O bending band is caused by all three components.

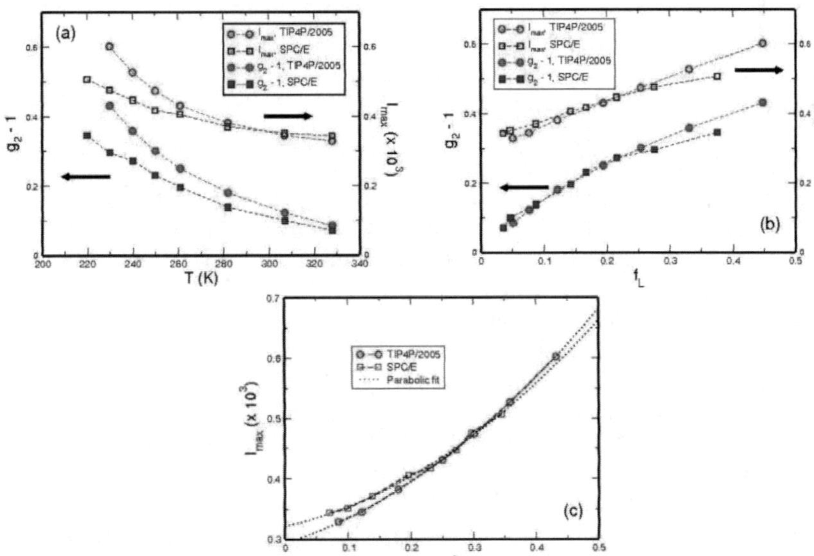

Figure 6 Correlation of vibrational band maximum I_{max} in OKE spectrum with $g_2 - 1$, where g_2 is the second-peak maximum of the PDF. Panel (a) presents the dependences of I_{max} (open) and $g_2 - 1$ (filled) on temperature. Panel (b) presents their dependences on the LDL fraction f_L, with the scales of I_{max} and $g_2 - 1$ referred to right and left vertical axes, where I_{max} has been amplified by a factor of 10^3. Panel (c) displays the direct correlation between I_{max} and $g_2 - 1$, with the black-dot lines showing a fitting by using a parabolic function. In all panels, circles and squares, guided with the dash line, are the data of the two model waters at 1 atm.

We have addressed that the vibrational band in $\chi''(\omega)$ is associated with the tetrahedrality of water structure. Here, we manifest quantitatively the correlation between its band maximum I_{max} and $g_2 - 1$, which specifies the tetrahedral order of global water structure with respect to a completely random arrangement of molecules as an ideal gas. The variations of I_{max} and $g_2 - 1$ with temperature are shown in Fig. 6 (a), and their dependences on the LDL fraction f_L are displayed in Fig. 6 (b). Our results reveal that I_{max} and $g_2 - 1$ increase in a similar manner with decreasing temperature and they have a similar ascending trend with f_L. In Fig. 6 (c), the direct correlation between I_{max} and $g_2 - 1$ is presented, where the correlation is described by a parabolic function of each model water. Hence, we have given a numerical correlation between the vibrational band amplitude in the OKE RSD and the tetrahedral order of water structure. This result suggests a possible way to explore the structural transformation in supercooled water through the temperature variation of vibrational band intensity in its OKE spectrum.

5. Conclusions

We have scrutinized the correlation between the OKE spectrum and the tetrahedrality of water structure. Our results indicate that the intermolecular vibrational band in the OKE spectrum of supercooled water at AP is dominated by the induced polarization caused by the DID interactions between molecules and the intensity of vibrational band has a significant correlation with the tetrahedrality of water structure.

Acknowledgement

TMW acknowledge supports from Ministry of Science and Technology, Taiwan R.O.C. under Grant No. MOST 109-2112-M-009-010.

References

[1] Debenedetti P G 2003 *J. Phys., Condens. Matter* **15** R1669
[2] Speedy R J and Angell C A 1967 *J. Chem. Phys.* **65** 851
[3] Poole P H, Sciortino F, Essmann U and Stanley H E 1992 *Nature* **360** 324
[4] Amann-Winkel K, Böhmer R, Fujara F, Gainaru C, *et al* 2016 *Rev. Mod. Phys.* **88** 011002
[5] Soper A K and Ricci M A 2000 *Phys. Rev. Lett.* **84** 2881
[6] Kim K H, Amann-Winkel K, Giovambattista N, Späh A, *et al* 2020 *Science* **370** 978
[7] Mishima O and Stanley H E 1998 *Nature* **396** 329
[8] Xu L, Kumar P, Buldyrev S V, Chen S H, *et al* 2005 *Proc. Natl. Acad. Sci. U.S.A.* **102** 16558
[9] Sellberg J A, Huang C, McQueen T A, Loh N D, *et al* 2014 *Nature* **510** 381
[10] Kim K H, Späh A, Pathak H, Perakis F, Mariedahl D, *et al* 2017 *Science* **358** 1589
[11] Pathak H, Späh A, *et al* 2021 *Proc. Natl. Acad. Sci. U. S. A.* **118** e2018379118
[12] Taschin A, Bartolini P, Eramo R, Righini R and Torre R 2013 *Nature Commun.* **4** 2401
[13] Suzuki Y, Takasaki Y, Tominaga Y and Mishima O 2000 *Chem. Phys. Lett.* **319** 81
[14] Abascal J L F and Vega C 2005 *J. Chem. Phys.* **123** 234505
[15] Berendsen H J C, Grigera J R and Straatsma T P 1987 *J. Phys. Chem.* **91** 6269
[16] Debenedetti P G, Sciortino F and Zerze G H 2020 *Science* **369** 289
[17] Giovambattista N, Loerting T, Lukanov B R and Starr F W 2012 *Sci. Rep.* **2** 390
[18] Tang P H and Wu T M 2020 *J. Mol. Liq.* **301** 112363
[19] Tang P H and Wu T M 2018 *J. Mol. Liq.* **269** 38
[20] Kuo Y W, Tang P H, Wang H, Wu T M and Saito S 2021 *J. Mol. Liq.* **341** 117269
[21] Saito S, Bagchi B and Ohmine I 2018 *J. Chem. Phys.* **149** 124504
[22] Bartolini P, Taschin A, Eramo R and Torre R 2008 *Time-Resolved Spectroscopy in Complex Liquids,* edited by Torre R (Springer, New York, 2008) Chapter 2
[23] Cho M, Du M, Scherer N F, Fleming G R and Mukamel S 1993 *J. Chem. Phys.* **99** 2410
[24] Chang S L, Wu T M and Mou C Y 2004 *J. Chem. Phys.* **121** 3605
[25] Walrafen G E, Chu Y C and Piermarini G J 1996 *J. Phys. Chem.* **100** 10363
[26] Tsai K H and Wu T M 2006 *Chem. Phys. Lett.* **417** 389

Effect of Coulomb interaction between secondary electrons on plasma formation due to heavy-ion irradiation

Kengo Moribayashi,

National Institutes for Quantum and Science and Technology (QST), 8-1-7, Umemidai, Kizugawa City, 619-0215, Japan

e-mail: moribayashi.kengo@qst.go.jpe

Abstract. Plasma formed by heavy-ion irradiation has been investigated in this study. In this study, a simulation model that effectively uses recent computers is developed. The effect of Coulomb interaction between secondary electrons on the motion of secondary electrons and plasma formation is demonstrated, where the secondary electrons are the electrons produced by incident ion impact ionisation.

1. Introduction

Heavy ion beams contribute to cancer therapy [1]; however, the treatment mechanism for this therapy is unclear. Plasma produced near the incident ion path may be important for understanding this mechanism. It has been found that the calculation of the survival rates of cells using radial dose distribution, including the effect of the plasma [2], is closer to the experimental results than that obtained when excluding the effect of plasma [1]. Here, the radial dose is the dose as a function of r and the radial dose distribution is employed in the treatment planning system for cancer therapy to calculate the survival rates [1], where r denotes the distance from the incident ion path. We developed a novel simulation model [3,4] to effectively study the plasma using recently manufactured supercomputers. In this study, we examine the effect of $\boldsymbol{F}_{ee}^{(a)}$ on plasma formation, where $\boldsymbol{F}_{ee}^{(a)}$ represents the Coulomb force of the other secondary electrons acting on the a-th secondary electron [see Eq. (2)]. Here, the secondary electrons are the electrons produced by incident ion impact ionisation.

Approximately a century ago, theoretical treatment was proposed for the interaction between an ion and material using the free electron gas model [2,5]. In this model, plasma is formed by an electric field caused by the change in the positions of free electrons near the incident ion path. However, large-scale simulations have been performed using the binary collision model [5-9]; nevertheless, these simulations do not result in the appearance of plasma. The theoretical treatment and simulations produce different radial dose distributions; this difference is caused by the presence or absence of plasma, as mentioned earlier.

Recently, computers have become significantly advanced, and a new factor to determine the secondary electron yield has been established [10]. Although the new simulation model developed in this study [3,4] is based on the binary collision model, plasma appearance is based on the effect of the electric field produced by heavy-ion irradiation on the motion of secondary electrons (see details in Section 2). These simulations [4] suggest that plasma appears for large values of σ or stopping power, but disappears with decreasing σ, where σ represents the incident ion impact ionisation cross-section [3,4]. Furthermore, it has been shown that the plasma theory can reproduce these simulation results [11].

Based on the measurement of the secondary electron yield [10], it can be concluded that slow secondary electrons are trapped by the electric field created near the incident ion path, and that P_{es} depends on the stopping power, which is almost proportional to σ when E_{ion} is fixed, where P_{es} denotes the probability of the secondary electrons escaping this electric field, and E_{ion} represents the incident ion energy. Because secondary electrons are not trapped by the electric field in the conventional binary collision model [5-9], P_{es} is almost zero, independent of σ. In the free electron gas model [5], P_{es} is assumed to be a constant value independent of σ, because the rate at which the free electrons move in plasma energy is a constant value regardless of σ, that is, the measurement [10] cannot be explained by the conventional models. Our simulations successfully reproduced the trend of this measurement [10] as a function of τ, where τ denotes the mean path of the incident ion impact ionisation events and is inversely proportional to σ [4].

We incorporate $\boldsymbol{F}_{ei}^{(a)}$ and $\boldsymbol{F}_{ee}^{(a)}$ into our model to consider the effect of the electric field on the motion of the secondary electrons, where $\boldsymbol{F}_{ei}^{(a)}$ [see Eq. (1)] denotes the Coulomb force of the molecular ions acting on the a-th secondary electron. Notably, $\boldsymbol{F}_{ee}^{(a)}$ is not considered in the conventional models. As no study has been conducted on the effect of $\boldsymbol{F}_{ee}^{(a)}$ on plasma formation thus far, to the best of our knowledge, our study shown here is the first to investigate this effect.

2. Simulation model

To clarify the physical phenomena produced by heavy-ion irradiation, we considered individual molecules in the target. Additionally, we considered individual molecular ions and secondary electrons produced by incident ion impact ionisation for these molecules; however, continuum targets were employed in conventional simulation models [6–9]. By incorporating $\boldsymbol{F}_{ei}^{(a)}$ and $\boldsymbol{F}_{ee}^{(a)}$ into our model, plasma was formed, as mentioned in Section 1. Here, $\boldsymbol{F}_{ei}^{(a)}$ and $\boldsymbol{F}_{ee}^{(a)}$ are given by

$$\boldsymbol{F}_{ei}^{(a)} = -\sum_{c} \frac{Q_c e \boldsymbol{\rho}_{ac}}{4\pi\varepsilon_0 \rho_{ac}^3}, \tag{1}$$

and

$$\boldsymbol{F}_{ee}^{(a)} = \sum_{b \neq a} \frac{e^2 \boldsymbol{\rho}_{ab}}{4\pi\varepsilon_0 \rho_{ab}^3}, \tag{2}$$

where ε_0 denotes the dielectric constant in vacuum, $\boldsymbol{\rho}_{ab}$ and $\boldsymbol{\rho}_{ac}$ denote the lengths from the a-th secondary electron to the b-th secondary electron and c-th molecular ion, respectively, and Q_c denotes the charge of the c-th molecular ion. It is impossible to simulate $\boldsymbol{F}_{ei}^{(a)}$ and $\boldsymbol{F}_{ee}^{(a)}$ using computers produced in the 1970s, that is, the age when the conventional models were proposed, because of the associated computational effort because much simulation time is required even using recent computers.

In our model, only σ and E_{ion} were used as input data. For each molecule, σ was assigned. When the incident ion passed through σ, a molecular ion and secondary electron were produced. The initial energy and initial emission angle of the secondary electrons were determined by the distribution obtained from the experimental data [12,13]. After the initial states of the secondary electrons were determined, only the simulations of the motions of the secondary electrons were executed.

Using a random number, the water molecule positions were randomly determined. Furthermore, we prepared a few 100 targets arranged at different molecular locations. For each target, we simulated a heavy ion passing through; thus, a few 100 simulations were performed. The average values of these results are presented in Section 3. The details of the simulation model are presented in Ref. [4].

3. Results and discussion

Figure 1 depicts N_e ($=2\pi r n_e$) obtained from simulations vs. r at $\sigma = 8 \times 10^{-16}$ and 2×10^{-15} cm^2 when $E_{ion} \sim 500$ keV/u with and without the consideration of $\boldsymbol{F}_{ee}^{(a)}$ [see Eq. (2)], where n_e denotes the electron density. In this figure, the approximation solution of the one-dimensional Poisson's equation [11,14] is also shown.

$$N_e = 2\pi r n_e \sim A \exp\left(-\frac{r}{\lambda}\right) \tag{3}$$

where A and λ denote the constant value and Debye length [11,14], respectively, and are obtained by fitting the simulation results. Our simulation results agree with this solution except when $r \sim 0.25$ nm and when $\boldsymbol{F}_{ee}^{(a)}$ is not considered. Notably, A and λ are obtained without the consideration N_e at $r \sim 0.25$ nm when $\boldsymbol{F}_{ee}^{(a)}$ is not considered. The velocities of the secondary electrons become so high that some electrons may not be able to stay here, because $\boldsymbol{F}_{ei}^{(a)}$ near the ion path is significantly strong. However, as the electric field becomes weak because of the screening of other secondary electrons when $\boldsymbol{F}_{ee}^{(a)}$ is considered, electrons may be easily trapped. It is also confirmed that N_e, when considering $\boldsymbol{F}_{ee}^{(a)}$, is larger and smaller than that when $\boldsymbol{F}_{ee}^{(a)}$ is not considered at $r < 1.25$ nm and $r > 1.25$ nm, respectively. This implies that $\boldsymbol{F}_{ee}^{(a)}$ plays a role in the swelling of plasma.

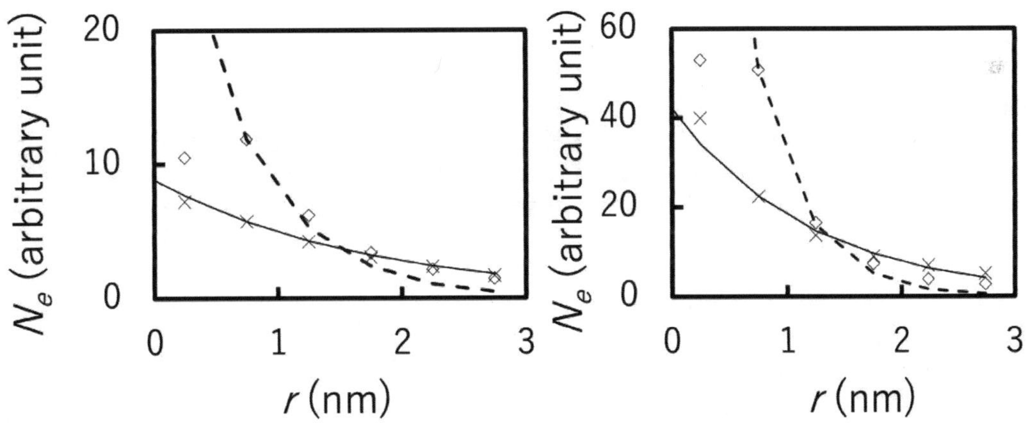

Fig. 1 N_e vs. r at σ = (a) 8×10^{-16} and (b) 2×10^{-15} cm^2 when E_{ion} is approximately 500 keV/u with and without the consideration of $\boldsymbol{F}_{ee}^{(a)}$, indicated by symbols × and ◇, respectively. Straight and dashed lines show the results obtained from Eq. (3) with and without the consideration of $\boldsymbol{F}_{ee}^{(a)}$, respectively.

Figure 2 plots λ vs. σ when $E_{ion} \sim 500$ keV/u with and without the consideration of $\boldsymbol{F}_{ee}^{(a)}$. In this figure, the function of $\lambda \propto \sigma^{1/2}$, which is obtained from the plasma theory {see Eq. (9) of Ref. [11]}, is also shown. When $\boldsymbol{F}_{ee}^{(a)}$ is not considered, the relationship of $\lambda \propto \sigma^{1/2}$ is almost satisfied. Because λ expresses the size of the plasma, it is determined that the plasma size obtained when $\boldsymbol{F}_{ee}^{(a)}$ is considered becomes approximately 2.7 times larger than that when $\boldsymbol{F}_{ee}^{(a)}$ is not considered.

Figure 3 shows P_{es} as a function of τ with and without the consideration of $\boldsymbol{F}_{ee}^{(a)}$ when $E_{ion} \sim 500$ keV/u. Notably, τ determines the electric field produced by molecular ions, which are created by incident ion impact ionisation near the ion path. Based on our simulations, it can be found that the trend of $P_{es} \propto (\tau)^{1/2}$ is almost satisfied regardless of target materials when $\boldsymbol{F}_{ee}^{(a)}$ is considered [4]. This trend is also observed in the measurement of the insular target KCl [4,10]. However, the trend of P_{es} becomes approximatcly proportional to τ when $\boldsymbol{F}_{ee}^{(a)}$ is not considered. $\boldsymbol{F}_{ee}^{(a)}$ significantly contributes to the increase of the electron density in plasma, because P_{es}, which is related to the plasma density, is proportional to $(\tau)^{1/2}$ and τ with and without the consideration of $\boldsymbol{F}_{ee}^{(a)}$, respectively.

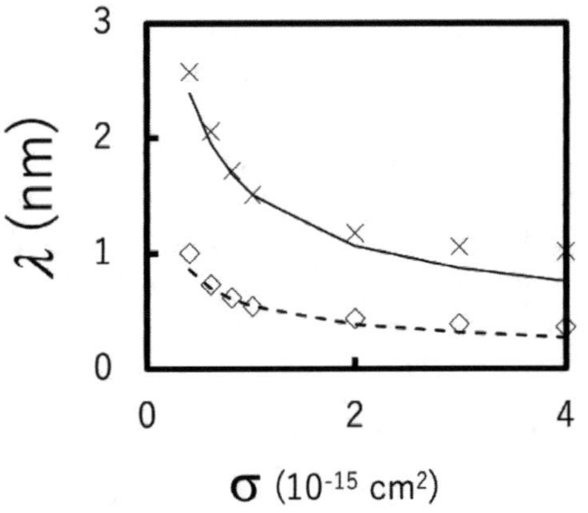

Fig. 2 Same as Fig. 1 for λ vs. σ.

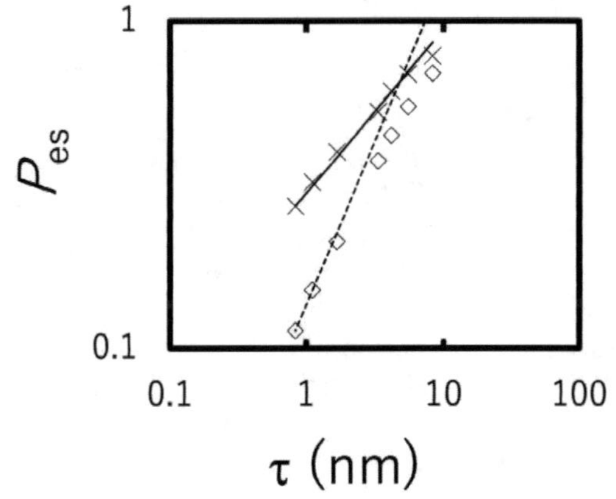

Fig. 3 Same as Fig. 1 for P_{es} vs. τ.

4. Summary

To clarify the biological effects of heavy ion beams, a simulation study was conducted. The simulation model used in this study enables the investigation of the plasma formed by heavy-ion irradiation. The effect of the Coulomb interaction between secondary electrons on the formation of plasma was demonstrated. This interaction plays a role in plasma swelling and decreases the probability of secondary electrons escaping from the electric field formed. This probability is proportional to $(\tau)^{1/2}$ and τ with and without the consideration of this interaction, respectively, where τ denotes the mean path of the incident ion impact ionisation event.

Acknowledgments
This work was supported by JSPS KAKENHI, Grant Numbers 16H02959, 18K11935, and 20K12489. For the simulations executed here, the supercomputer 'PRIMERGY BX900' in JAEA was employed.

References

[1] Kase Y et al. 2008 *Phys. Med. Biol.* **53** 37
[2] Chatterjee A and Schaefer H J 1976 *Radiat. Environ. Biophys.* **13** 215
[3] Moribayashi K 2011 *Phys. Rev. A* **84** 012702
[4] Moribayashi K 2020 *Jpn. J. Appl. Phys.* **59** SH0801
[5] Ziegler J F et al. 1985 *The Stopping and Range of Ions in Matt*er. Vol. 1. Pergamon Press, New York
[6] Katz R and Kobetich E J 1969 *Phys. Rev.* **186** 344
[7] Wang H and Vassiliev O N 2017 *Phys. Med. Biol.* **62** N219
[8] Nikjoo S et al. 1998 *Int. J. Radiat. Biol.* **73** 355
[9] Uehara S and Nikjoo H 2002 *J. Phys. Chem. B* **106** 11051
[10] Kimura K et al. 2002 *Nucl. Instrum. Methods Phys. Res. B* **193** 661
[11] Moribayashi K 2021 *Radiat. Phys. Chem.* **184** 109436
[12] Dal Cappello C D et al. 2009 *Nucl. Instrum. Meth. Phys. Res. B* **267** 781
[13] Rudd M E et al. 1992 *Rev. Mod. Phys.* **64** 441
[14] Chen F F 1974 Introduction to Plasma Physics. Plenum Press, New York

Theoretical Study of Dissociation Process of Plastocyanins by PaCS-MD Simulation

D Fitrasari[1,3], M S Arwansyah[1,2], K Kawaguchi[5], A Purqon[3], Suprijadi[3,4] and H Nagao[5]

[1] Graduate School of Natural Science and Technology, Kanazawa University, Japan
[2] Departement of Chemistry, Faculty of Science, Universitas Cokroaminoto Palopo, Palopo, Indonesia
[3] Departement of Physics, Bandung Institute of Technology, Indonesia
[4] Research Center of Nanoscience, Bandung Institute of Technology, Indonesia
[5] Institute of Science and Engineering, Kanazawa University, Japan

E-mail: nagao@wriron1.s.kanazawa-u.ac.jp,dian@wriron1.s.kanazawa-u.ac.jp

Abstract. We present a procedure of calculation of free energy landscape of two proteins by using parallel cascade molecular dynamics (PaCS-MD) and multiple independent umbrella sampling (MIUS). The free energy landscape of two plastocyanins for association/dissociation process is investigate by using the present method. We find that binding free energy is around 3 kcal/mol and that the barrier energy at around the middle range between the equilibrium point and the dissociation state becomes about 1 kcal/mol from the association process. The present results suggest that the energy barrier may arise from hydrogen bonds between two plastocyanins. We also find that the effective interaction between two plastocyanins already vanishes at the distance of 2 Å from equilibrium state. The equilibrium point of the complex around 27.9 Å is a good agreement with the experimental result 27.8 Å.

1. Introduction

Plastocyanins are small type I blue copper proteins consisting of 97 to 105 amino acid residues. They play an essential role in oxygenic photosynthesis, where they act as mobile electron carriers, of receiving an electron from cytochrome f (Cyt-*f*) which is a subunit of the membrane-bound cytochrome b_6f complex and donating. The electron is transferred to the $P700^+$ in photoreaction center of photosystem I (PS I)[1],[2],[3]. The multi-subunit complex plastocyanin is known as carriers of a single PEG. PEG is polyethylene glycol, a synthetic polymer who has wide application specially in increasing stability and half-life of therapeutic proteins[4].

In the previous works, we have investigated the electronic structure of active site type I copper protein by quantum calculation [5] and free energy profile of plastocyanin to cytochrome-f by using molecular dynamic simulations [6]. In our research for electronic structure we have been found the force field parameters for active cite structure especially in plastocyanin.

In this work, we investigate free energy profile of plastocyanin by using PaCS-MD simulation. The advantage of PaCS-MD simulation is to shorten the simulation time to find the conformational pathways. In this study, we investigate the dissociation process of the complex plastocyanin using the Parallel Cascade Molecular Dynamics (PaCS-MD) simulation method. PaCS-MD simulation is one of the molecular simulation methods use to generate conformational

Journal of Physics: Conference Series **2207** (2022) 012021 doi:10.1088/1742-6596/2207/1/012021

transition pathways. This method is one of the effective computational calculations which is using short cycle multiple independent molecular dynamics simulation to find the pathway. We discuss dissociation processes of plastocyanins in relation to the electron transfer in the photosystem and to find the mechanism of two plastocyanin in this system.

2. Method

Plastocyanin from X-ray diffraction which is solved by Schmidt, et al. (PDB ID: 2GIM) [7] is used as the initial structure for the simulation of dissociation process. Plastocyanin consists of 105 amino acid residues, and two copper ions in its active site.

In this calculation, we use NAMD2.12 software program package[8]. The force field parameters used are CHARMM22[9],[10] and TIP3P water model [11]. The timestep that we perform during this simulation is 2fs/step. The box size is 70 $\AA\times$59 $\AA\times$100 $\AA$. NPT ensemble is applied to this simulation at 310 K and 1 atm. The cutoff distance is 12 $\AA$.

We perform Parallel Cascade Molecular Dynamics (PaCS-MD) to find dissociation pathways of complex Plastocyanins. The All-atom model is used to build a complex Plastocyanin structure. The premelinary MD is used to generate initial structure for PaCS-MD. This initial structure, then divide in to ten independent simulation as known as short Multiple Independent Molecular Dynamics (MIMD) simulation [12]. The ten independent simulation differentiate by the number of seed (random seed) from 1 to 10. The center of mass distance between two plastocyanin use as the ranking measure. The furthest distance choose as reactive trajectories for dissociation pathways. This reference structure will be use as initial structure in free energy landscape (FEL) analysis. Multiple independent umbrella sampling (MIUS) with weighted histogram analysis method (WHAM) is used to analyze the free energy landcape [12].

3. Results and Discussion

All-atom molecular dynamic simulations of two plastocyanins are performed by using the initial configuration of the crystal structure with PDB ID 2GIM[7]. The equilibrium state of the complex system is obtained shown in Figure 1. In this Figure 1 the red solid line corresponds to RMSD of chain A in complex plastocyanin. The blue dashed line corresponds to RMSD of chain C. The yellow dashed line corresponds to all chain in complex plastocyanin. In Figure 2, the average distance of two centers of mass becomes around 34 $\AA$ for 70ns simulation time. Meanwhile in experimental data the average center of mass is 27.847 $\AA$. The red solid line corresponds to the center of mass distance (COM) between two plastocyanins and the yellow dashed line corresponds to the distance between two copper ions in complex plastocyanin.

The equilibrium state of the complex system which is obtained from reactive trajectories PaCS-MD All-atom molecular dynamic shown in Figure 3. The red solid, blue dashed and yellow dotted lines correspond to RMSD of the one plastocyanin, another plastocyanin, and the complex system, respectively. Figure 4 shows the dependence of distance between center of mass of two plastocyanins coloured in red and of distance between two copper ions in blue dashed line. From Figure 4 the average of two center of mass becomes around 31 $\AA$ for 1.7ns time simulation.

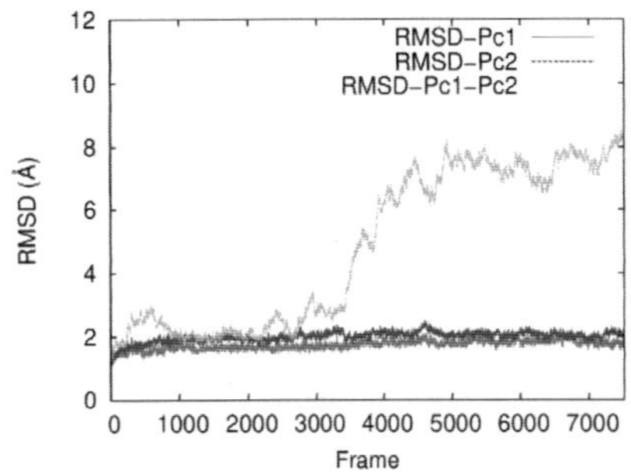

Figure 1: RMSD of complex Plastocyanin obtained from All-Atom simulation each frame step correspond to 10ps.

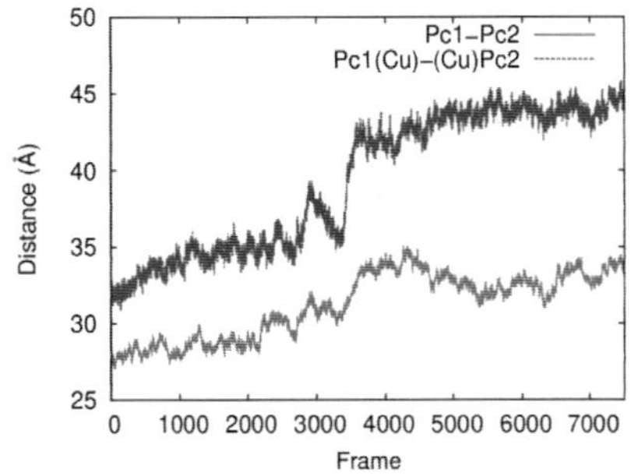

Figure 2: Center of Mass Distance between two Plastocyanin obtained from All-Atom simulation each frame step correspond to 10ps.

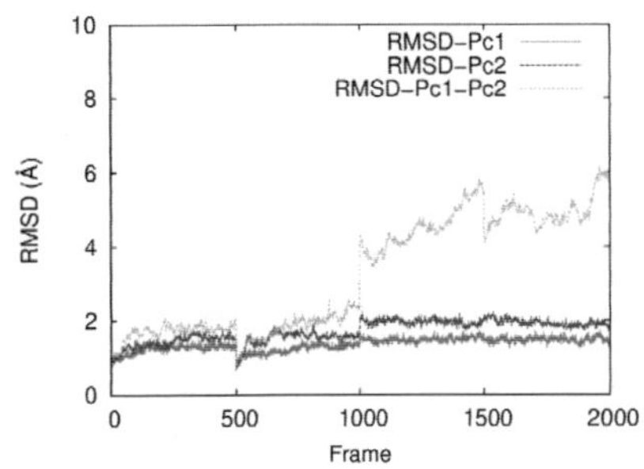

Figure 3: RMSD of complex Plastocyanin obtained from reactive trajectories PaCS-MD All-Atom simulation each frame step correspond to 10ps.

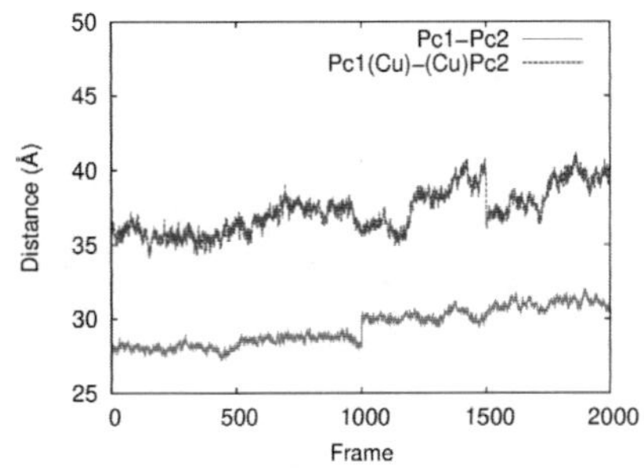

Figure 4: Center of Mass Distance between two Plastocyanin obtained from reactive trajectories PaCS-MD All-Atom simulation each frame step correspond to 10ps.

Figure 5 shows a free energy landscape of dissociation process of the two plastocyanin proteins. We can find that the free energy landscape equilibrium point become 27.9 $\AA$ which has good agreement with the experimental equilibrium point. We can observe the free energy barrier around 29 $\AA$, which may arise from hydrogen bonds between two plastocyanins. The barrier energy for association process is around 1 kcal/mol. The binding free energy of two plastocyanins becomes around 3 kcal/mol. From the free energy landscape (FEL) shown in Figure 5, we observe the convergence of free energy around 30 $\AA$ to 32 $\AA$. We also suggest that the dissociation state of two plastocyanin can be found at those distance which is shown in Figure 6. The short distance of dissociation state may be arising from the characteristic structure of plastocyanin which means the small and hard protein. In Figure 6 there are 8 β-sheet in yellow arrow and 1 α-helix in purple in one plastocyanin, and the gray strings correspond to the turn structure.

Journal of Physics: Conference Series **2207** (2022) 012021 doi:10.1088/1742-6596/2207/1/012021

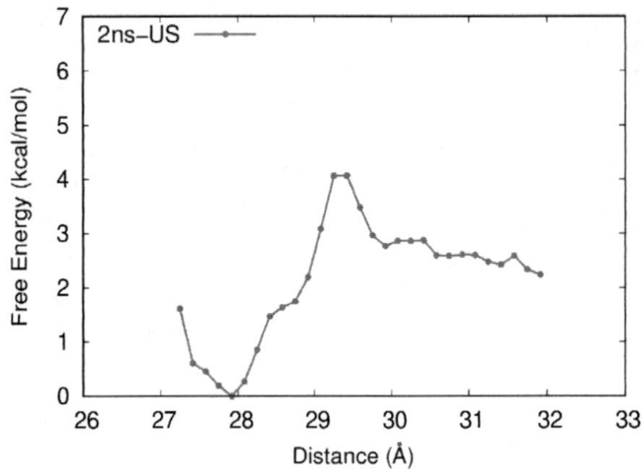

Figure 5: Free energy dissociation (obtained from reactive trajectories PaCS-MD All-Atom simulation with force constant=5 kcal/mol.

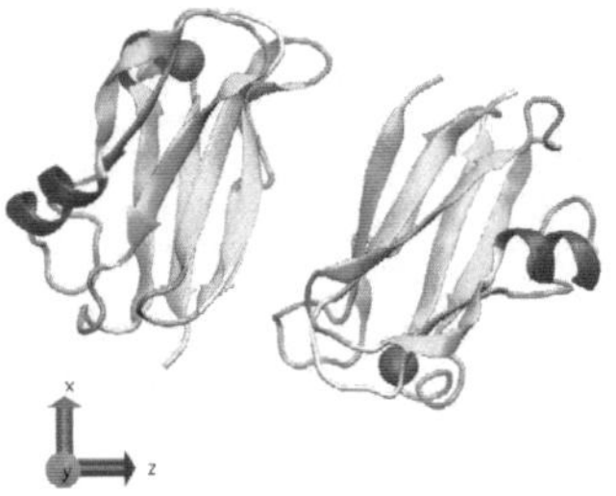

(a) COM distance between two Plastocyanin ≈ 28 Å.

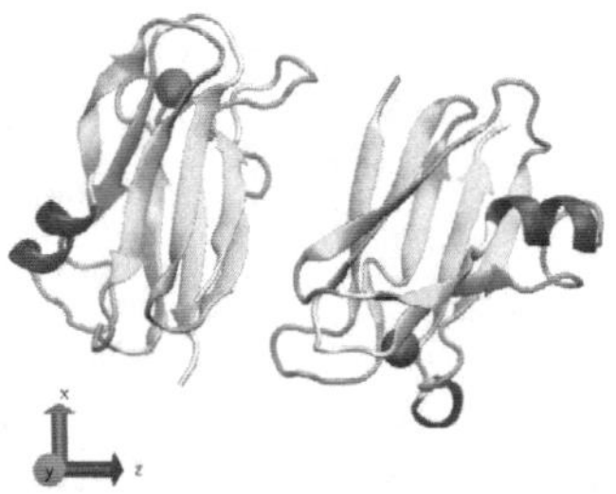

(b) COM distance between two Plastocyanin ≈ 29Å.

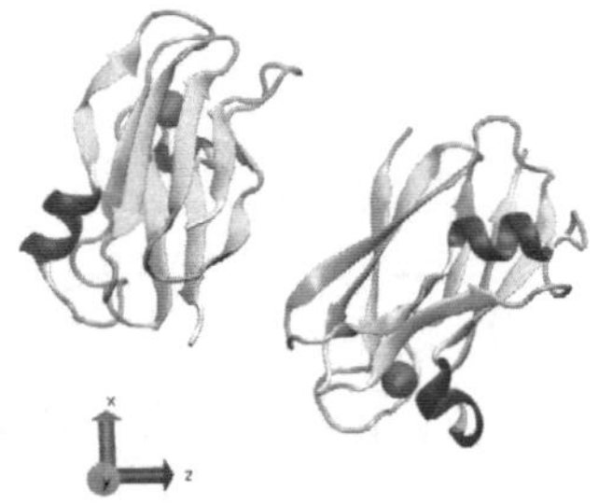

(c) COM distance between two Plastocyanin ≈ 31Å.

Figure 6: Snapshot structure of two Plastocyanin

4. Concluding Remarks

We have presented a procedure of calculation of free energy landscape of two proteins by using parallel cascade molecular dynamics (PaCS-MD) and multiple independent umbrella sampling (MIUS). The free energy landscape of two plastocyanins for association/dissociation process has been investigated by using the present method. We have found that binding free energy is around 3 kcal/mol and that the barrier energy at around the middle range between the equilibrium point and the dissociation state becomes about 1 kcal/mol from the association process. The present results suggest that the energy barrier may arise from hydrogen bonds between two plastocyanins. We have also found that the effective interaction between two plastocyanins already vanishes at the distance of 2 Å from equilibrium state. The equilibrium point of the complex around 27.9 Å is a good agreement with the experimental result 27.8 Å.

References

[1] Davis D J, Krogmann D W and Pietro A S 1980 *Plant Physiology* **65**(4) 697-702.
[2] Hope A B 2000 *Biochimica et Biophysica Acta (BBA) - Bioenergetics* **1456**(1) 5-26.
[3] Redinbo M R, Yeates T O and Merchant S 1994 *J Bioenerg Biomembr* **26** 49-66.
[4] Cattani G, Vogeley L, Crowley P B 2015 *Nature chemistry* **7**(10) 823-828.

[5] Kurniawan I, Kawaguchi K, Sugimori K, Sakurai T and Nagao H 2019 *Sci. Rep. Kanazawa Univ.* **63** 1-13. 012026.

[6] Kurniawan I, Matsui, T, Nakagawa S, Kawaguchi K and Nagao H 2018 *IOP Conf. Series: Journal of Physics: Conf. Series* **012026**(1136) 1-12.

[7] Schmidt L, Christensen, H E M and Harris P 2006 *Acta Crystallographica Section D* **62** 1022-1029.

[8] Phillips J C, Hardy D J, Maia J D C, Stone J E, Ribeiro J V, Bernardi R C, Buch R, Fiorin G, Henin J, Jiang W, McGreevy R, Melo M C R, Radak B K, Skeel R D, Singharoy A, Wang Y, Roux B, Aksimentiev A, Luthey-Schulten Z, Kale L V, Schulten K, Chipot C, and Tajkhorshid E 2020 *Journal of Chemical Physics* **153** 044130.

[9] Brooks B R, Brooks III C L, MacKerell, Jr. A D, Nilsson L, Petrella R J, Roux B, Won Y, Archontis G, Bartels C, Boresch S, Caflisch A, Caves L, Cui Q, Dinner A R, Feig M, Fischer S, Gao J, Hodoscek M, Im W, Kuczera K, Lazaridis T, Ma J, Ovchinnikov V, Paci E, Pastor R W, Post C B, Pu J Z, Schaefer M, Tidor B, Venable R M, Woodcock H L, Wu X, Yang W, York D M, and Karplus M 2009 *J. Comput. Chem.* **30** 1545-1614.

[10] Lee J, Cheng X, Swails J M, Yeom M S, Eastman P K, Lemkul J A, Wei S, Buckner J, Jeong J C, Qi Y, Jo S, Pande V S, Case D A, Brooks III C L, MacKerell Jr A D, Klauda J B and Im W 2016 *J. Comput. Chem.* **12** 405-413.

[11] Jorgensen W L, Chandrasekhar J, Madura J D, Impey R W and Klein M L 1983 *The Journal of Chemical Physics.* **79**(2) 926-935.

[12] Harada R and Kitao A 2013 *The Journal of Chemical Physics* **139**(3) 0351103.

Theoretical Study of Complex Aspirin and Hydroxypropyl-β-cyclodextrin in Solvent Phase

H Jayyinunnisya[1], L S Solihat[1], K Sugimori[2], K Kawaguchi[3], H Nagao[3]

[1]Graduate School of Mathematical and Physical Science, Kanazawa University, Kanazawa, Japan
[2]Center for the Advancement of Higher Education, Hokuriku University, Kanazawa, Japan
[3]Institute of Science and Engineering, Kanazawa University, Kanazawa, Japan

E-mail: nagao@wriron1.s.kanazawa-u.ac.jp, helmia@ wriron1.s.kanazawa-u.ac.jp

Abstract. We estimate the configuration of complex ASA/HPCD in the gas phase by using PM3 and B3LYP methods. Some physical properties in the gas phase, such as complexation energy, deformation energy, dipole moment, etc., are calculated by using PM3 and B3LYP methods. We investigate the dependence of physical properties such as complexation energy, etc., on solvation by using PM3 and B3LYP methods with the PCM method. Complex ASA/HPCD becomes more stable than the dissociation state. The present results suggest that the effect of solvent can be increasing the energy complexation, LUMO energy, HOMO energy, HOMO-LUMO gap energy, electronic potential, and chemical hardness, whereas the electrophilicity decrease.

1. Introduction

Aspirin (ASA) is the family of the Non-Steroidal Anti-Inflammatory Drugs (NSAIDs). Besides the benefits of ASA therapy as an analgesic, antiplatelet, anti-inflammatory, anti-pyretic, antithrombotic, and anticancer actions, this drug causes severe and undesired side effects such as upset or/and stomach bleeding, hematemesis, gastrointestinal ulcers, tinnitus, thrombocytopenia, agranulocytosis, and leucopenia. Several methods, such as pro-drug or micro-encapsulation, etc., have been used to surmount these inconveniences. Or, more accurately, its inclusion in the cyclodextrins or in their derivatives permits the limitation of these effects [1].

ASA is used for anti-inflammatory property that reacts by inhibition of prostaglandin synthesis and COX-II inhibitor. Long-term consumption of ASA is associated with severe gastrointestinal side effects. The gastrointestinal side effects of ASA become reduced by increasing solubility and bioavailability. [2]. Cyclodextrin has a lipophilic central cavity that is able to form inclusion complexes with many drugs by taking up some lipophilic part of the drug molecules into its cavity [3].

Due to the peculiar chemical structure of cyclodextrins, they are constituted of an external hydrophilic surface and a relatively hydrophobic cavity. Given their particular structure and their dual polarity, cyclodextrins are used to improve the solubility, stability, physicochemical and biological characteristics, electrochemical, and spectroscopic properties of various drug molecules by forming inclusion complexes [1]

Generally, β-cyclodextrin has been widely used for inclusion complexes because of its availability and appropriate cavity size for a variety of drug molecules such as ASA, paracetamol, and so on. In contrast to β-cyclodextrin, α-cyclodextrin has a small cavity size, and γ-cyclodextrin is expensive. However, they are inadequate for the widest range of guest molecules, and they also reduce the inclusion complex formation [1].

The properties such as binding energy, deformation energy, etc. of cyclodextrins to form inclusion complexes have become the subject of intense interest in theoretical studies using molecular mechanics [4], molecular dynamics [5,6], and quantum mechanics [7] methods because of the combination of experimental and theoretical leads to successful results in solving structural, energetic, and dynamic problems. In our previous paper, the solvent effects of proteins have been studied by quantum calculations [8].

In this work, we report computational results of the complexation of hydroxypropyl-β-cyclodextrin (HPCD) and ASA with the solvent effect. We use ASA as a guest model and HPCD as a host model and calculate their complex structure by using PM3 and B3LYP methods. We discuss the stability of the complex in relation to the dependence of the solvation by using the implicit solvent method with a common organic solvent such as Chloroform, Acetone, Methanol, Ethanol, and Water as a solvent.

2. Computational method

Physical properties such as free energy etc., of the inclusion complex model of the guest molecule, host molecule, and complex molecule are calculated by using Gaussian 09 software [9] using the semiempirical PM3 method under vacuum. After that, the calculation results from the PM3 method are refined and validated with a higher calculation method, using full DFT method with XC functional of B3LYP and 6-31g (d,p) basis set. The polarizable continuum model (PCM) is used in DFT calculation to understand the solvent effect in this complex. The solvents used in this calculation are chloroform, acetone, ethanol, methanol, and water, where the dielectric constant obtained from the gaussian package: 4.7113 F/m, 20.493 F/m, 24.852 F/m, 32.613 F/m, and 78.3553 F/m, respectively.

The complexation energy ($E_{complexation}$) is defined as the difference between the sum of energy of individual host and guest molecule and the energy of inclusion complex given by the following formula:

$$E_{complexation} = E_{complex} - (E_{free\ guest} + E_{free\ host}). \quad (1)$$

The deformation energy of the guest or the host molecule can be obtained by Eqs. (2) and (3):

$$E_{def} (guest) = E_{sp}^{opt}(guest) - E_{opt} (guest), \quad (2)$$

$$E_{def} (host) = E_{sp}^{opt}(host) - E_{opt} (host), \quad (3)$$

where E_{def}(guest) stands for the deformation energy of the guest, E_{sp}^{opt}(guest) is the single point energy of the guest using its geometry in the optimized complex, and E_{opt} (guest) is the energy of the optimized geometry of the guest.

In this study, we analyze the chemical reactivity with some global properties such as electronic potential (μ), hardness (η), and global electrophilicity index (ω). Then, the global properties are calculated using the following equations:

$$\mu = \frac{1}{2}(E_{HOMO} + E_{LUMO}), \quad (4)$$

$$\eta = -\frac{1}{2}(E_{HOMO} - E_{LUMO}), \quad (5)$$

$$\omega = \mu^2/2\eta, \quad (6)$$

where E_{HOMO} stands for the HOMO energy, and E_{LUMO} is the LUMO energy.

3. Results and Discussion

3.1. Complexation and Deformation Energy

The optimized structure of ASA, HPCD, and the complex ASA/HPCD is shown in Figure 1. From figure 1(c), we can understand that ASA formed a stable inclusion complex with HPCD, with ASA is entirely inside the HPCD. This structure is obtained from molecular docking using Autodock4 [10], where the

ASA is restricted with the rigidity of its torsion. In the case of the PM3 method, the complexation energy ($E_{complexation}$)of this process becomes -12.6535 kcal/mol and -4.8294 kcal/mol for gas phase and water solvent respectively. The deformation energy (E_{def}) of ASA are 3.1567 kcal/mol and 3.3306 kcal/mol for gas phase and water solvent, respectively. On the other hand, in the case of HPCD, they become 0.2503 kcal/mol and 2.1936 kcal/mol for gas phase and water solvent, respectively.

Table 1 shows the dependence of complexation energy and deformation energy used PM3 and B3LYP methods on the solvents. In the gas phase, ASA is most stable, both in the PM3 and B3LYP calculations. Based on the energy complexation in PM3 calculation in Table 1, the stability system can be sorted as gas > chloroform > acetone > ethanol > methanol > water. In water solvation, ASA has higher energy complexation than the other solvent, which indicates that ASA is most unstable in HPCD.

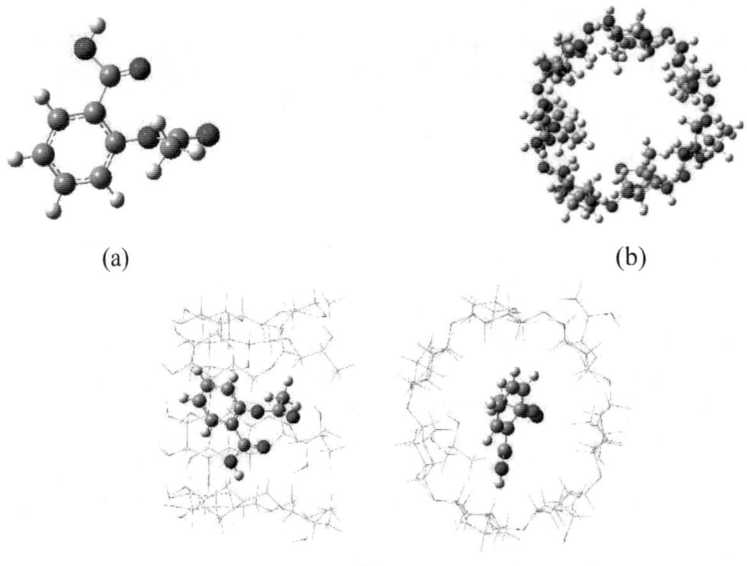

(a) (b)

(c)

Figure 1. Optimization structure used B3LYP of (a) ASA, (b) HPCD, and (c) complex ASA/ HPCD, where the left is the sideway, and the right is the front way.

Table 1. The energy of ASA, HPCD, and complex ASA/HPCD

	ε (F/m)	Energy of ASA (kcal/mol)	Energy of HPCD (kcal/mol)	Energy of ASA/HPCD (kcal/mol)	$E_{complexation}$ (kcal/mol)	E_{def} of ASA (kcal/mol)	E_{def} of HPCD (kcal/mol)
(a) PM3							
Gas		-138.1053	-1783.6635	-1934.4223	-12.6535	3.1567	0.2503
Chloroform	4.7113	-143.2726	-1803.4937	-1955.0619	-8.2957	3.2499	0.3296
Acetone	20.4930	-145.1377	-1811.7633	-1963.6666	-6.7656	3.3368	0.5137
Ethanol	24.8520	-145.2485	-1812.3211	-1964.2445	-6.6748	3.3376	0.5350
Methanol	32.6130	-145.3738	-1812.9724	-1964.9165	-6.5703	3.3369	0.5689
Water	78.3553	-145.6139	-1815.818	-1966.2612	-4.8294	3.3306	2.1936
(b) B3LYP							
Gas		-431725.19	-3531245.75	-3962977.02	-6.0817	2.1703	6.1341
Chloroform	4.7113	-431730.56	-3531270.23	-3963002.59	-1.7979	1.9949	4.7458
Acetone	20.4930	-431732.51	-3531279.73	-3963013.22	-0.9845	1.9442	3.8867
Ethanol	24.8520	-431732.63	-3531280.35	-3963013.92	-0.9458	1.9203	3.8671
Methanol	32.6130	-431732.76	-3531281.02	-3963014.74	-0.9630	1.8279	3.8330
Water	78.3553	-431733.02	-3531282.48	-3963018.46	-2.9617	1.6799	2.4030

The instability may make it will be easier to dissociate than in the other solvent. The deformation energy of ASA is relative the same in the solvation phase, and it is a bit smaller in the gas phase. In the water solvent, HPCD has the highest deformation energy.

B3LYP calculation gave a different result than in PM3 calculation. the stability system can be sorted as gas > water > chloroform > acetone > ethanol > methanol. The water solvent is more stable than the other solvents, and the most unstable is in the methanol solvent from the energy complexation. The deformation energy of ASA and HPCD has a smaller value in the water solvent. The computed deformation energies given in Table 1 indicate that the deformation of HPCD is greater than that of ASA in the systems during the inclusion process allowing the guest molecule to accommodate easily. From these results, the HPCD deformation energy seems to be one of the driving factors leading to the inclusion complex formation.

However, even if the gas phase has the most stable complex between the two methods. It is well known that generally, the formation of the inclusion complexes involving cyclodextrin takes place in an aqueous solution. The binding behavior of ASA and HPCD in solution seems to be more important than in the gas (vacuum).

In the PM3 method, the energy complexation is significantly increased with the increase of the dielectric constant. While in the B3LYP method, it suddenly decreases from methanol to water. PM3 calculation is not suitable for providing accurate relative conformer energy, so the B3LYP model is more reliable in this study. It is agreed with the literature where the inclusion complex is more stable in the water [11,12]. However, since the two methods give the opposite result, the effect of solvent by the energy complexation cannot be validated. It will be better to add another higher method to get a better result.

The increasing dielectric constant makes the deformation energy of ASA decrease, which means that ASA will use smaller energy to adapt in the HPCD cavity with more polar solvent. The same result is also shown in HPCD with B3LYP calculation. Otherwise, on the deformation energy of HPCD, the same explanation may be applied the same as the complexation energy case.

3.2. Dipole moments

The dipole moment, which also defines the dipole-dipole interactions in non-bonded systems is calculated, to give information about asymmetry in the molecular charge distribution in the molecule [13]. In Table 2, the dipole moment of ASA in the water phase has the largest value for both methods. While for the HPCD, the largest value dipole moment in methanol and water for PM3 and B3LYP methods, respectively. All the inclusion complexes showed dipole moment values higher than the isolated ASA. Compared to HPCD, the dipole moment values are lower in the PM3 method and higher for the solvent phase in the B3LYP method, indicating that the HPCD cavity's polarity changed after the drug entered the HPCD cavity [14]. From these results, it can be concluded that the dipole moment values show a strong correlation with the complexation behavior of the molecules. The polarity of the complex increases with the increase of the dielectric constant.

Table 2. The dipole moment (D) of ASA, HPCD, and complex ASA/HPCD

	(a) PM3			**(b) B3LYP**		
	ASA	HPCD	ASA/HPCD	ASA	HPCD	ASA/HPCD
Gas	4.4428	14.3887	12.0638	5.3620	13.0058	12.5611
Chloroform	5.1976	16.8359	14.4344	6.6461	16.6182	17.0333
Acetone	5.4033	17.8925	15.2903	7.1449	18.2491	18.5285
Ethanol	5.4186	17.9642	15.3337	7.1758	18.3653	18.6126
Methanol	5.4330	18.0554	15.3811	7.2111	18.9352	18.7321
Water	5.4627	16.6372	15.4606	7.2792	19.2390	19.8031

Table 3. The HOMO–LUMO energy calculations of complex ASA/HPCD

Properties	Gas	Chloroform (ε=4.711 F/m)	Acetone (ε=20.493 F/m)	Ethanol (ε=24.852 F/m)	Methanol (ε=32.613 F/m)	Water (ε=78.355 F/m)
(a) PM3						
E_{HOMO} (eV)	-10.3858	-10.4228	-10.3444	-10.3360	-10.3254	-10.3009
E_{LUMO} (eV)	-1.0710	-0.9499	-0.8678	-0.8593	-0.8487	-0.8234
E_{HOMO}-E_{LUMO} (eV)	9.3148	9.4729	9.4766	9.4767	9.4767	9.4775
μ (eV)	-5.7284	-5.6864	-5.6061	-5.5977	-5.5871	-5.5622
η (eV)	4.6574	4.7365	4.7383	4.7384	4.7384	4.7388
ω (eV)	3.5228	3.4134	3.3164	3.3064	3.2939	3.2643
(b) B3LYP						
E_{HOMO} (eV)	-6.7762	-6.7892	-6.7811	-6.7814	-6.7814	-6.7841
E_{LUMO} (eV)	-2.3277	-2.0926	-1.9361	-1.9195	-1.9013	-1.7579
E_{HOMO}-E_{LUMO} (eV)	4.4485	4.6966	4.8450	4.8619	4.8801	5.0262
μ (eV)	-4.5520	-4.4409	-4.3586	-4.3505	-4.3414	-4.2710
η (eV)	2.2243	2.3483	2.4225	2.4310	2.4401	2.5131
ω (eV)	4.6578	4.1991	3.9210	3.8928	3.8621	3.6293

3.3. HOMO-LUMO parameters

The energy gap (E_{HOMO}-E_{LUMO}) is an important stability scale with large energy gap values, indicating higher stability [15]. As shown in Table 3, the HOMO-LUMO energy gap in the water solvent is the most stable and less reactive conformation with a value of about 9.4775 eV for the PM3 method and 5.0262 eV for the B3LYP method. It has different results with the complexation energy where the most stable is in the gas phase. If we compare in the solvent phase, this result is agreed with the complexation energy for the B3LYP method. By the dependence of the dielectric constant, we can assume that increasing the dielectric constant can make the complex more stable.

The electronic chemical potential is represented by μ, where the compounds with larger chemical potential values are more reactive than those with small electronic chemical potentials [16]. From Table 3, we have noticed that all chemical potential (μ) values of complex ASA/HPCD are negative for all models in gas and solvation. This indicates that the inclusion process of ASA into the HPCD cavity is spontaneous, where the more negative value indicates, the more spontaneous. In the water solvent, the complex becomes the most reactive. The electronic potential of the complex is slightly increased with increasing dielectric constant, which means that increasing dielectric constant can make the complex less spontaneous and become more reactive.

Hard molecules have a large HOMO-LUMO energy gap, and soft molecules have a small HOMO-LUMO gap. The qualitative definitions of hardness were closely related to polarizability, where a small energy gap leads to easy polarization and a large gap to difficult polarization. Hardness also measures resistance to deformation or change [17]. Based on Table 3, in water solvent has larger η. If we compared the complex with the isolate ASA and HPCD, we could see that the chemical hardness of the complex is similar to ASA, yet in the HPCD, it decreases after the inclusion complex. The chemical hardness slightly increases with increasing dielectric constant. We also can assume that increasing the polarity can increase the chemical hardness of the complex.

Global electrophilicity index (ω) is an important index of electrophilicity, where the large values of ω have characteristics of most electrophilic systems. The gas phase is more electrophilic than the other system in this study. From Table 3, the complex has strong electrophilic ($\omega > 1.5$ eV), Parr defined the electrophilicity ω index [18]. The electrophilicity is slightly decreased with the increase of the dielectric constant. We can assume that increasing the polarity can decrease the electrophilicity of the complex.

4. Concluding Remarks

We have estimated the configuration of complex ASA/HPCD in the gas phase by using PM3 and B3LYP methods. Some physical properties in the gas phase, such as complexation energy, deformation energy, dipole moment, etc., have been calculated by using PM3 and B3LYP methods. We have investigated the dependence of physical properties such as complexation energy, etc., on solvation by using PM3 and B3LYP methods with the PCM method.

Complex ASA/HPCD becomes more stable than the dissociation state. The present results suggest that the effect of solvent can be increasing the energy complexation, LUMO energy, HOMO energy, HOMO-LUMO gap energy, electronic potential, and chemical hardness, whereas the electrophilicity decrease.

References

[1] Shan D K 2012 *Int. J. Pharm. Sci. Res* **3** 2831
[2] Szetli J 1988 *Cyclodextrin Technology* (Dordecht: Springer)
[3] Bezzina B, Djemil R and Bensouilah N 2019 *Theor. Chem. Acc.* **138** 43
[4] Ding H J, Chao J B, Zhang G M, Shuang S M and Pan J H 2003 *Spectrochim. Acta. Part. A.* **59** 3421
[5] Zheng Y, Haworth I S, Zuo Z, Chow M S S and Chow A H L 2005 *J. Pharm. Sci.* **94** 1079
[6] Kríž Z, Koča J, Imberty A, Charlot A and Auzély-Velty R 2003 *Org. Biomol. Chem.* **1** 2590
[7] Wen X H, Liu Z Y, Zhu T Q, Zhu M Q, Jiang K Zand Huang Q 2004 Bioorg. Chem. **32** 223
[8] Sugimori K, Shuku T, Sugiyama A, Nagao H, Sakurai T and Nishikawa K 2005 *Polyhedron.* **24** 2671
[9] Frisch M J *et al.* 2013 *Gaussian 09, Revision D.01* (Wallingford CT: Gaussian, Inc.)
[10] Morris G M, Huey R, Lindstrom W, Sanner M F, Belew R K, Goodsell D S and Olson A J 2009 *J. Comp. Chem.* **16** 2785
[11] Xu X Q, Hong N, Zheng J W, Guo B B and Cheng L 2015 *Adv Mat Res* **1120-1121** 882
[12] Niether D, Kawaguchi T, Hovancova J, Eguchi K, Dhont J K G, Kita R and Wiegand S 2017 *Langmuir.* **33** 8483
[13] Mariappan G 2014 *Spectrochim. Acta. A. Mol. Biomol. Spectrosc.* **117** 604
[14] Lui L, Song K S, Li X S and Guo Q X2001 *J. Incl. Phenom. Macrocycl. Chem.* **40** 35
[15] Bouchemela H, Madi F and Nouar L 2019 *J. Incl. Phenom. Macrocycl. Chem.* **95** 247
[16] Miar M, Shiroudi A, Pourshamsian K, Olliaey A R and Hatamjafari F 2020 *J. Chem. Res.* **45** 147
[17] Parr R G and Pearson R G 1893 *J. Am. Chem. Soc.* **105** 7512
[18] Parr R G, Szentpály L v and Liu S 1999 *J. Am. Chem. Soc.* **121** 1922

Collective pattern formation in a binary mixture of self-propelled particles

Sagarika Adhikary and S. B. Santra

Department of Physics, Indian Institute of Technology Guwahati, Guwahati-781039, Assam, India.

E-mail: a.sagarika@iitg.ac.in

Abstract. We have studied the collective behaviour of a binary mixture of self-propelled particles (SPPs) with non-identical motile properties. The two types of SPPs with low and high velocity is modelled in a two-dimensional system with periodic boundary conditions (PBCs) at a fixed particle density. The two types have two widely different velocities. The alignment interaction rule and angular noise similar to Vicsek Model (VM) is implemented in this model. Both inter and intra-particle interactions are considered among the SPPs. The system exhibits various self-organised pattern formation and phase segregation which mainly depends on the angular noise present in the system. Without any restriction of interaction or repulsion among the particles, the SPPs phase-separated into dense structures. The orientational order-disorder phase transition is studied, and nontrivial results are obtained. Cluster size distribution (CSD) for the entire system is also studied at the transition region. The CSD exhibits a power-law behaviour at the transition, with an exponent that of the standard percolation.

1. Introduction

Self-organised collective patterns of active or self-propelled particles (SPPs) are observed in different length scales such as cells [1], bacterial swarms [2, 3], fish schools [4], flocking birds [5], traffic flow etc. [6]. Motility induced phase separation is also observed in several such active systems [7, 8]. Collective motion of SPPs was studied in the seminal Vicsek model (VM) [9]. In this model, a large number of point-like particles move together at a constant speed, and they interact locally with the neighbours and align their direction of motion. However, an angular noise (η) is also present in the system. For a given density of particles, an orientational order-disorder transition occurs at $\eta = \eta_c$. One of the main criteria of the VM is that all the SPPs should have the same magnitude of velocity. Collective behaviour of SPPs of common motile properties is extensively studied following the VM in many other models [10, 11, 12, 13, 14]. However, in natural systems, the velocity need not be the same during the collective motion of SPPs. For example, fast-moving (active) and slow-moving (dormant) bacteria in a bacterial population or slow and fast-moving vehicles in a daily traffic scenario. A mixture of two types SPPs is proposed in this work, where the two have widely different motile properties. The interaction rules are kept similar to the VM. First, we describe the model. The model exhibits several nontrivial results, which will be discussed in the next section.

2. The Model

A binary mixture of SPPs is modeled over a two-dimensional square box of linear size L with periodic boundary conditions. Two widely different velocities are assigned to the two types. The fast-movers move with a velocity v_f and the slow-movers move with a velocity v_s and $v_f \gg v_s$. The two types of SPPs are kept in equal proportion. If N_f is the number of fast-movers and N_s is the number of slow-movers, then $N_f = N_s = N/2$. Where, the total number of particles is N with a fixed density $\rho = N/L^2$. The initial positions $\vec{r}$ of all the particles are randomly distributed over the two-dimensional square box. Whereas, the velocity orientations θ of all particles are randomly distributed over the range $[-\pi, +\pi]$. A particle interacts locally with other particles within a region of radius R. It aligns its velocity orientation with that of the average orientation of all the neighbours. The time evolution of the orientations of a fast-mover $\theta_{f,i}$ ($i = 1, 2, 3, \cdots, N/2$) and that of a slow-mover $\theta_{s,i}$ are determined by the following equations

$$\theta_{f,i}(t + \Delta t) = \langle\theta(t)\rangle_R + \Delta\theta \tag{1}$$
$$\theta_{s,i}(t + \Delta t) = \langle\theta(t)\rangle_R + \Delta\theta \tag{2}$$

where $\langle\cdots\rangle_R$ denotes the average of $\theta(t)$ over the particles within the interaction region of radius R around the ith particle. Both inter-particle and intra-particle interactions are considered in estimating the average orientation. It should be noted that the velocity magnitude of individual SPPs are igonred and only the orientations are taken into account in estimating $\langle\theta(t)\rangle_R$. The term $\Delta\theta$ is a random orientation chosen with a uniform probability from the interval $[-\eta\pi, +\eta\pi]$. The angular noise η acts as a control parameter in this model. The strength of the angular noise η typically varies from 0 to 1. The position of the fast movers $\vec{r}_{f,i}$ and for the slow movers $\vec{r}_{s,i}$ are updated following the forward update rules as given below

$$\vec{r}_{f,i}(t + \Delta t) = \vec{r}_{f,i}(t) + \vec{v}_{f,i}(t)\Delta t \tag{3}$$
$$\vec{r}_{s,i}(t + \Delta t) = \vec{r}_{s,i}(t) + \vec{v}_{s,i}(t)\Delta t \tag{4}$$

where Δt is the time between two successive updates, and it is chosen as $\Delta t = 1$. The terms $\vec{v}_{f,i}$ and $\vec{v}_{s,i}$ have the velocity amplitude v_f and v_s in the direction $\theta_{f,i}$ and $\theta_{s,i}$ respectively. In a special situation if $v_s = v_f = v$, the model will boil down to the VM with the same velocity amplitude v for all the SPPs [9].

3. Results and Discussion

For a given initial random distribution of SPPs over a square box of size L, the system's time evolution is studied following Eq.1, 2, 3 and 4 by performing Monte Carlo simulations. One Monte Carlo time step corresponds to the up-gradation of position and orientation of all the SPPs. Initial 7×10^5 Monte Carlo steps are neglected to achieve the steady-state. An ensemble of size 48×10^5 (2×10^5 samples at different times for 24 different initial configurations) is taken for statistical averages. System of sizes $L = 128$ and $L = 256$ are considered with a fixed density $\rho = N/L^2 = 0.5$ for the simulation. The velocities of the fast and slow-movers are considered as $v_f = 1.0$ and $v_s = 0.01$ respectively. First, we discuss the collective pattern formation that occurs at different angular noise regions for this model. Then other system properties such as orientational order-disorder phase transition, segregation analysis and cluster-size distribution study will be discussed accordingly.

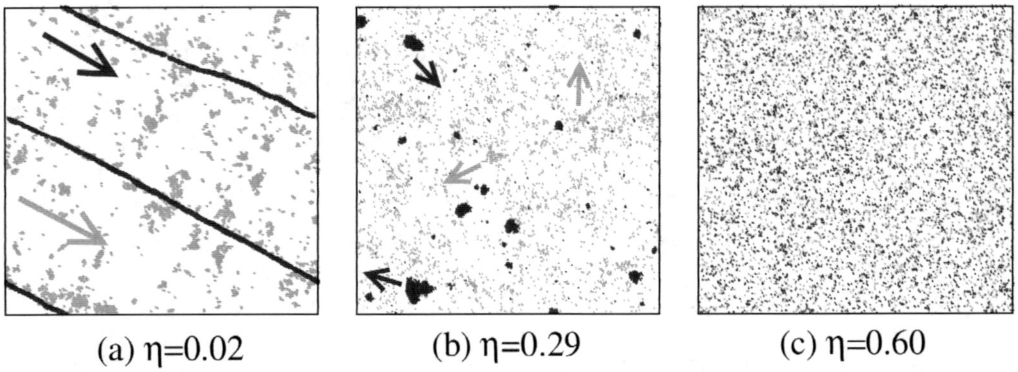

Figure 1. Morphologies for (a) $\eta = 0.02$, (b) $\eta = 0.29$ and (c) $\eta = 0.60$ with system size $L = 128$ and velocities $v_f = 1.0$ and $v_s = 0.01$. Orange colour represents the fast-movers and maroon represents the slow-movers. Arrows indicate the moving direction of a flock.

First, the system morphologies for different values of angular noise η are shown in fig.1 with system of size $L = 128$ and velocities $v_f = 1.0$ and $v_s = 0.01$. Orange colour represents fast-movers, and maroon colour represents slow-movers. At a low angular noise with $\eta = 0.02$, shown in fig.1(a), dense clusters of fast-movers are observed where all the clusters move in the same direction (indicated by the orange arrow). Whereas the dense lane pattern of slow movers has been observed with the same moving direction as the fast-movers (maroon arrow). These patterns do not form in the systems where all the particles have the same velocity. At an intermediate angular noise with $\eta = 0.29$, shown in fig.1(b), less dense clusters of fast-movers are observed in the system. Whereas dense clump structures of different sizes formed by the slow movers appear in the system in this noise region. However, the clusters and the clumps move in random directions. At a relatively high angular noise with $\eta = 0.60$, shown in fig.1(c), a homogeneously mixed phase of the two types of SPPs is observed in the system.

As the angular noise strength, η is tuned from 0 to 1, the system undergoes a transition from an orientationally ordered state to a disordered state. The order-parameter for the entire system (ϕ) for all SPPs and also for partial systems ϕ_p ($p \in \{f, s\}$) consisting only one type of SPPs can be defined as

$$\phi(\eta) = \frac{1}{N}\left|\sum_p \sum_{i=1}^{N_p} \frac{\vec{v}_{p,i}}{|\vec{v}_{p,i}|}\right|, \quad \phi_p(\eta) = \frac{1}{N_p}\left|\sum_{i=1}^{N_p} \frac{\vec{v}_{p,i}}{|\vec{v}_{p,i}|}\right| \tag{5}$$

where N_p is the number of p-type ($p \in \{f, s\}$) SPPs. The susceptibility (χ) for the entire system and that of the partial systems (χ_p) can be estimated from the fluctuation in their respective order parameters as

$$\chi = L^2\left[\langle\phi^2\rangle - \langle\phi\rangle^2\right], \quad \chi_p = L^2\left[\langle\phi_p^2\rangle - \langle\phi_p\rangle^2\right] \tag{6}$$

Now, we present data for the order parameters and susceptibilities for $v_f = 1.0$ and $v_s = 0.01$ ($v_f = 100v_s$) on a system of size $L = 256$. The order parameter ϕ of the entire system and ϕ_p, that of the partial systems, are plotted against η in fig.2. The values of ϕ and ϕ_p decrease from a positive value to zero as η increases. Thus, an orientational order-disorder transition is observed for the entire system as well as for the partial system at a particular η. In fig.3, the fluctuations in order parameters χ and χ_p are plotted against η. It is expected that the susceptibility will diverge at $\eta \approx \eta_c$. From the susceptibility plots shown in fig.3, the same peak

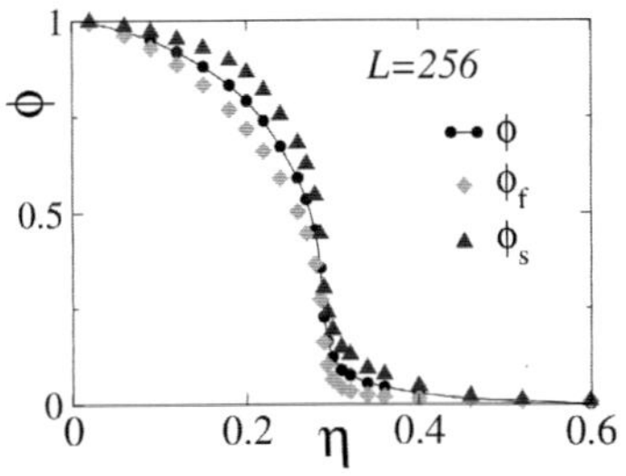

Figure 2. Plot of ϕ, ϕ_f and ϕ_s against η for the binary mixture of SPPs with $v_f = 1.0$ and $v_s = 0.01$.

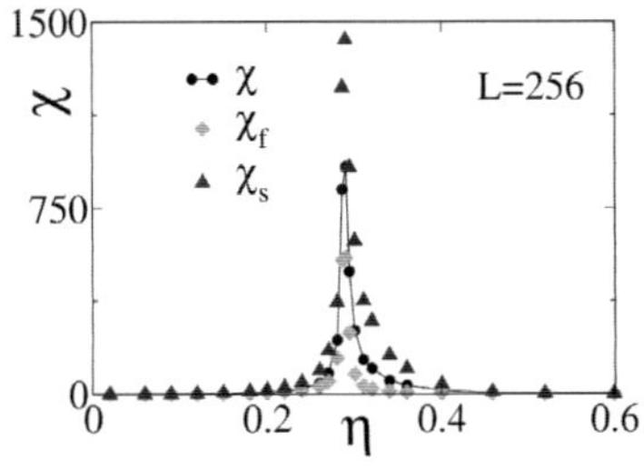

Figure 3. Plot of χ, χ_f and χ_s against η for the binary mixture of SPPs with $v_f = 1.0$ and $v_s = 0.01$.

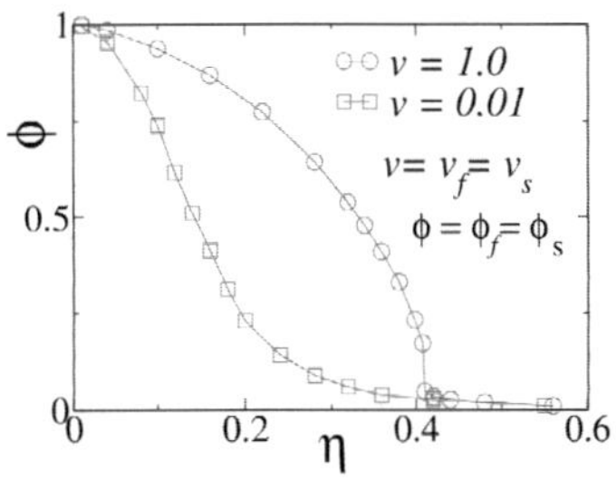

Figure 4. Plot of ϕ against η for the VM. Two cases with $v = 1.0$ and $v = 0.01$.

position is observed for the entire system as well as the partial systems, which indicates the transition point $\eta_c \approx 0.29$. The fluctuation for the slow-movers at η_c is higher than that of the fast-movers. It is interesting to note that the value of the order parameter of the fast-movers (ϕ_f) is less than the order parameter of the slow-movers (ϕ_s) at all values of η (fig.2). However, in the VM, the SPPs always have more polar order at a higher velocity than at a lower velocity [15]. The order parameter ϕ versus η plot for the VM for a system of size $L = 256$ is also shown in fig.4 separately for the two velocities, $v = 1.0$ and $v = 0.01$. It can be seen that the polar order with $v = 0.01$ is significantly less than the case with $v = 1.0$. Thus, the order parameter of a particular type of SPPs is highly influenced by the motion of other types of SPPs in this present binary model.

As various noise ranges, segregated patterns occur in the system. Hence, the segregation coefficient [16] is calculated to study demixing between the two types of SPPs in this model. The square box of area L^2 is divided into Q square sub-regions of linear size l and area L^2/Q each. The segregation coefficient is defined as

$$S = \frac{1}{N}\sum_{i=1}^{Q}|n_{fi} - n_{si}| \tag{7}$$

Here i is the index for the sub-regions. n_f and n_s are the number of the fast and slow-movers, respectively, present in the sub-region area. The length $l = 8$ is similar to the length scale of the average cluster size near the transition. The value S is plotted against η shown in fig.5 The coefficient S is high with smaller η where phase-separated patterns occur in the system. Whereas S decreases with higher η, as they mix together and form a homogeneous gas-like phase.

Next, the cluster size distribution (CSD) is obtained at the transition point for the entire system. The distribution, n_m is defined as $n_m = N_m/N$, where N_m is the number of m-sited clusters. It corresponds to the number of a cluster of size m per particle. The CSD for the entire system is determined at the transition point $\eta_c \approx 0.29$ shown in Fig.6. It is plotted in the double-logarithmic scale, and the behaviour is almost like a power-law distribution. However, a slight curvature is found. A straight line with a slope of 2.05 is found to fit the data points approximately. It is worth mentioning that n_m behaves like a power-law distribution at the critical point as $n_m \sim m^{-\tau}$ with $\tau = 2.05$ for a two-dimensional standard percolation [17, 18]. Thus, the exponent value is almost similar to the percolation threshold, where different cluster sizes occur in the system.

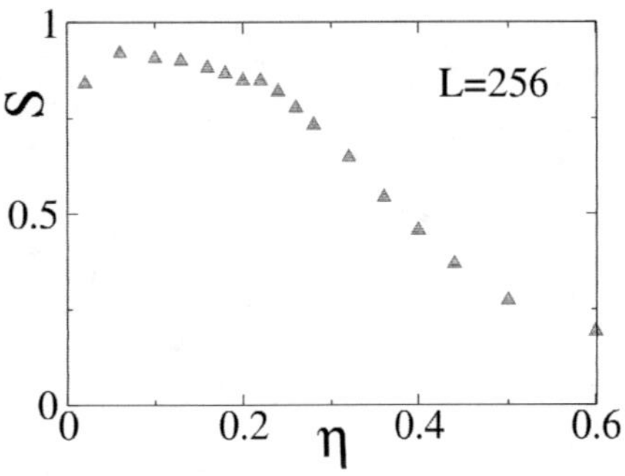

Figure 5. Plot of S against η.

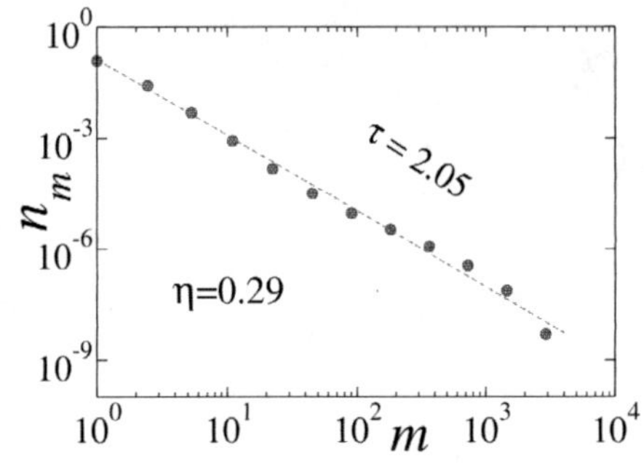

Figure 6. CSD plot at $\eta \approx \eta_c$.

4. Summary and Conclusion

A simple numerical model for a binary mixture of SPPs with different motile properties is proposed to study their collective motion. Formation of the dense clusters, lane pattern, clump-like structure and homogeneous gas-like phase in different noise regions are observed in this model. Interestingly, they are self-organised into coherent patterns, starting from a random initial state, only with the alignment interaction on a small scale. In this model, a large difference in speeds and mutual interaction among the species results in phase-separated collective patterns. Furthermore, an orientational order-disorder transition is observed, varying the angular noise. The order of the slow-movers is observed higher than the fast-movers for a given angular noise. However, in the VM, the order is higher for larger velocities for a particular angular noise. The mutual interaction between the two species seems to be responsible for such nontrivial behaviour. Another interesting feature is the power-law behaviour in the cluster size distribution at the order-disorder transition point. The related exponent is found similar to the standard percolation exponent value. Further study of these systems, varying the different parameters, can obtain several nontrivial results that are intriguing to explore in the future.

References

[1] Schaller V, Weber C, Semmrich C, Frey E and Bausch A R 2010 *Nature* **467** 73–77
[2] Sokolov A, Aranson I S, Kessler J O and Goldstein R E 2007 *Phys. Rev. Lett.* **98** 158102
[3] Zhang H, Be'Er A, Smith R S, Florin E L and Swinney H L 2009 *Europhys. Lett.* **87** 48011
[4] Makris N C, Ratilal P, Symonds D T, Jagannathan S, Lee S and Nero R W 2006 *Science* **311** 660–663
[5] Ballerini M, Cabibbo N, Candelier R, Cavagna A, Cisbani E, Giardina I, Lecomte V, Orlandi A, Parisi G, Procaccini A *et al.* 2008 *Proc. Natl. Acad. Sci. U.S.A.* **105** 1232–1237
[6] Vicsek T and Zafeiris A 2012 *Phys. Rep.* **517** 71–140
[7] Cates M E and Tailleur J 2015 *Annu. Rev. Condens. Matter Phys.* **6** 219–244
[8] Bechinger C, Di Leonardo R, Löwen H, Reichhardt C, Volpe G and Volpe G 2016 *Rev. Mod. Phys.* **88** 045006
[9] Vicsek T, Czirók A, Ben-Jacob E, Cohen I and Shochet O 1995 *Phys. Rev. Lett.* **75** 1226
[10] Romensky M, Lobaskin V and Ihle T 2014 *Phys. Rev. E* **90** 063315
[11] Chaté H, Ginelli F, Grégoire G and Raynaud F 2008 *Phys. Rev. E* **77** 046113
[12] Adhikary S and Santra S B 2021 *Europhy. Lett.* **135** 48003
[13] Ginelli F and Chaté H 2010 *Phys. Rev. Lett.* **105** 168103
[14] Shaebani M R, Wysocki A, Winkler R G, Gompper G and Rieger H 2020 *Nat. Rev. Phys.* **2** 181–199
[15] Cambui D S, de Arruda A S and Godoy M 2016 *Physica A* **444** 582–588
[16] Ai B q, Shao Z g and Zhong W r 2018 *Soft Matter* **14** 4388–4395
[17] Stauffer D and Aharony A 2018 *Introduction to percolation theory* (CRC press)
[18] Kyriakopoulos N, Chaté H and Ginelli F 2019 *Phys. Rev. E* **100** 022606

Simulation study of the function of domain swapping in the HSP90 chaperone cycle

L Matsukura*and N Miyashita**

Department of Biological Systems Engineering, Graduate School of Biology-Oriented Science and Technology, KINDAI University, 930 Nishimitani, Kinokawa, Wakayama 649-6493, Japan

*liz@miyashita-lab.net, **miya@waka.kindai.ac.jp

Abstract. HSP90 is one of the molecular chaperones, and it is known as an anti-cancer drug target. In the HSP90 chaperone cycle, when the ATP binds to the HSP90NTD, the HSP90 dimer forms a domain swapping in NTD. The stability of the HSP90 dimer is essential to the HSP90 chaperon cycle process. We have performed several molecular dynamics simulations of the HSP90 dimer to investigate how ATP binding and domain swapping have affected the stability. We analysed the fluctuation of critical residues for ATP binding and ATP hydrolysis in the HSP90NTD dimer conformation. As a result, we found that domain swapping is essential to the stability of the HSP90 dimer, while ATP is not so essential for the stability of the dimer conformation.

1. Introduction

Molecular dynamics simulation is a useful tool for understanding the fluctuations and characteristics of biomolecules. Recently, it has been possible to perform the calculations of large-scale systems and long-time calculations. Heat Shock Protein 90 (HSP90), one of the molecular chaperons, has been known as a target protein for anti-cancer drugs. It has the essential function of folding and stabilizing various client cancer-related proteins. Geldanamycin, an inhibitor of HSP90, binds to an ATP binding site of HSP90 N-terminal domain (NTD), and it inhibits the chaperone cycle of HSP90. Recently, a Cryptand-like binder has been developed as a new peptide inhibitor that binds to the β-sheet region of the NTD, which is not ATP binding site [1].

HSP90 forms a homo-dimer which consists of two chains, chain A and B. The HSP90 has three domains, NTD, Middle domain (MD), and C-terminal domain (CTD), and the ATP binding site is located on the NTD. In the HSP90 chaperone cycle, when the ATP and the client proteins bind to HSP90, HSP90 changes to a close state from the open state. At the same time, the N terminus regions of each NTD are swapped. It is called domain swapping (DS) (figure 1 (a)) [2]. After that, the co-chaperones such as p23 and Aha1 bind to the HSP90 dimer, and the chaperone cycle will be proceeded. Finally, the mature client protein and the co-chaperone are released from the HSP90 dimer complex [3]. The interface of the NTDs of the HSP90 dimer is consists of two helices, H_A and H_B. Each interfacial helix is connected to the DS region in the HSP90. In this study, we focused on the stability of HSP90 dimer. The stability of the HSP90 dimer would be related to the ATP binding and DS, and it is important to the HSP chaperon cycle. To investigate how ATP binding and DS are affected to the stability of HSP90 dimer system, we have performed several molecular dynamics simulations.

2. Materials and Methods

2.1. Molecular dynamics simulations of HSP90 dimer
We used the complex structure of the HSP90NTD dimer and co-chaperone p23 (PDB ID: 2CG9 [4]) for the simulations. We removed the co-chaperon p23 from the complex, and modeled the NTD dimer. To understand the function of domain swapping, we also modeled the NTD dimer without DS, which is eliminated the N terminus region of NTD. To investigate the effects of the domain swapping and ATP, we prepared four systems, HSP90NTD, HSP90NTD without DS, ATP bound HSP90NTD, ATP bound HSP90NTD without DS. In this study, the all-atom molecular dynamics simulations of these four systems were carried out 500 ns in the isothermal-isobaric (NPT) ensemble (300 K and 1 bar) with SPCE water molecules and 150 mM of K+ and Cl- ions using CHARMM36m force field [5] on GROMACS 2021 software [6].

2.2. The trajectory map of the elements (TME) of HSP90
To figure out the fluctuation of the elements in a complex system visually, we introduced the trajectory map of the elements (TME) on the two dimensional surface. The center of the mass of HSP90NTD in the dimer denotes A and B, respectively, and we define the vector, $\vec{AB}$ (figure 1 (b)). A vector $\vec{H_A H_B}$ was defined as the vector between the center of the mass of the interfacial helices H_A and that of H_B. The normal vector $\vec{C}$ is the perpendicular of both $\vec{AB}$ and $\vec{H_A H_B}$. Here we prepared a plane surface with the normal vector $\vec{C}$ is included in the center of the mass of A and B (figure 2).

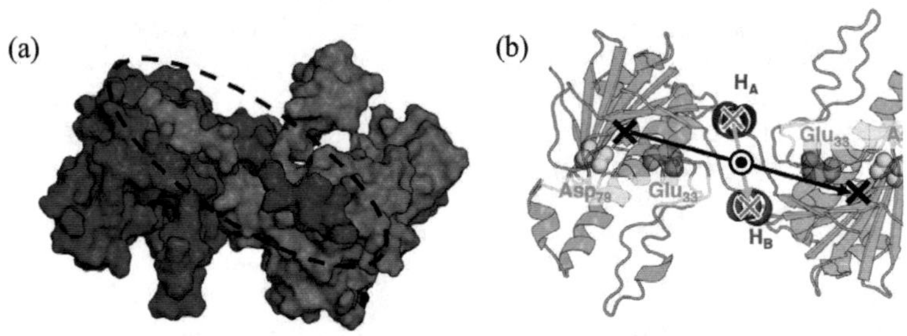

Figure 1. (a) The HSP90NTD dimer structure. The dashed open circle indicates the domain swapping region. Blue and red are chain A and B of HSP90NTD dimer. (b) The structure of HSP90NTD dimer (light blue an red). Green and Yellow indicate the Glu_{33} and Asp_{79}. H_A and H_B are the interfacial helices of HSP90. The four crosses indicate two of the center

3. Results and discussions
The stability of the HSP90 dimer would be important to the HSP chaperon cycle. ATP binding and the DS of each HSP90 would be critical to the stabilization of the HSP90 dimer conformation. To investigate how ATP binding and DS contribute to the stabilization of the HSP90 dimer, we analyzed the fluctuations of the interfacial helices of H_A and H_B and Glu_{33} and Asp_{79}. The Glu_{33} and Asp_{79} are key residues for the ATP binding, and the process of ATP hydrolysis [7]. Figure 2 indicates the TME of HSP90. The trajectories of the center of mass of H_A and H_B helices are denoted by blue and red dots. The trajectories of C_α of Asp_{79} and Glu_{33} were plotted by yellow and green dots, respectively. The dots were plotted every 1 ns. We also calculated the dispersions of H_A, H_B, Glu_{33}, and Asp_{79} (table.1).

The Glu_{33}, Asp_{79}, and the interfacial helices in the system without DS (figure 2 (b) and the right upper panel of table 1) have broader distributions than that with DS (figure 2 (a) and the left upper panel of table 1). It means that the DS affects the stabilization of HSP90. The ATP binding also contribute to stabilize the Glu_{33}, Asp_{79}, and interfacial helices, H_A and H_B, because the distributions of these elements without ATP binding (figure 2 (a) and upper panels of table 1) were broader than that with ATP binding (figure 2 (c) and lower panels of table 1). On the other hand, the ATP bound HSP90NTD without DS system is unstable, even if ATP bounds to the HSP90 (figure 2 (d)). It means that the DS is critical to stabilizing the interfacial helices, Glu_{33} and Asp_{79}. Thus, the DS is essential for stabilizing the HSP90 dimer conformation, ATP binding, and ATP hydrolysis.

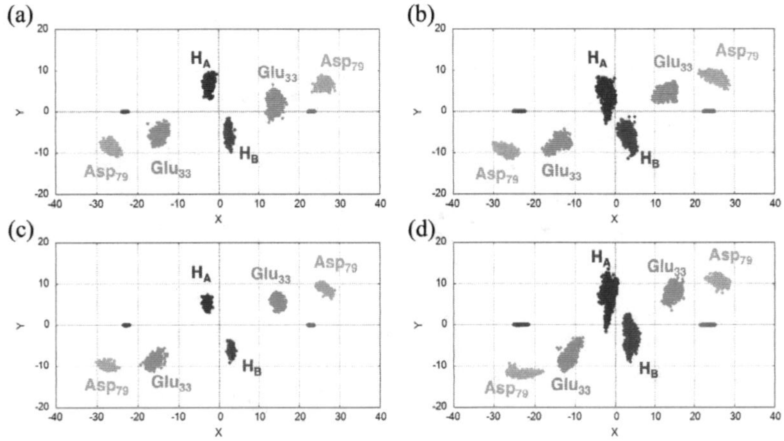

Figure 2. The trajectory maps of the element (TME) of (a) HSP90NTD, (b) HSP90NTD without DS, (c) ATP bound HSP90NTD, (d) ATP bound HSP90NTD without DS. Blue, red, green, and yellow dots indicate the trajectories of the interfacial helices (H_A, H_B), Glu_{33}, and Asp_{79}, respectively.

Table 1. The dispersions of the position of the interfacial helices of H_A and H_B, and Glu_{33} and Asp_{79}.

HSP90NTD

	x	y	z
H_A	0.320	2.606	1.360
H_B	0.127	2.314	1.998
GLU_A	0.702	0.900	0.985
GLU_B	0.780	0.488	1.322
ASP_A	0.603	0.451	0.615
ASP_B	0.644	0.285	0.520

HSP90NTD without DS

	x	y	z
H_A	0.958	2.404	4.344
H_B	0.894	3.022	2.804
GLU_A	2.065	0.941	1.266
GLU_B	2.079	3.168	0.908
ASP_A	1.840	0.484	0.486
ASP_B	2.237	0.350	0.724

ATP bound HSP90NTD

H_A	0.110	1.285	0.469
H_B	0.094	1.048	0.573
GLU_A	0.630	0.291	0.615
GLU_B	0.465	0.269	0.634
ASP_A	0.658	0.282	0.243
ASP_B	0.663	0.211	0.415

ATP bound HSP90NTD without DS

H_A	0.774	1.277	7.085
H_B	0.707	1.779	3.598
GLU_A	1.307	1.802	3.036
GLU_B	1.123	1.431	1.508
ASP_A	3.160	1.328	0.201
ASP_B	1.345	0.979	0.459

4. Summary

In this study, to investigate the cause of the stability of the HSP90 close state, we performed several molecular dynamics simulations of the HSP90NTD dimer. We found that the DS is essential for stabilizing the HSP90 close state, ATP binding, and hydrolysis. However, ATP binding is not necessary for stabilizing the HSP90 dimer conformation, though ATP binding also stabilizes the conformation of the HSP90 dimer.

References

[1] Mochizuki K, Matsukura L, Ito Y, Miyashita N and Taki M 2021 Org. Biomol. Chem. **19** 146-50
[2] Prodromou C, Roe S M, Piper P W and Pearl L H 1997 Nat. Struct. Mol. Biol. 4 477-82
[3] Butler L M, Ferraldeschi R, Armstrong H K, Centenera M M and Workman P 2015 Mol. Cancer Res. **13** 1445-51
[4] Ali M M U, Roe S M, Vaughan C K, Meyer P, Panaretou B, Piper P W, Prodromou C and Pearl H 2006 Nature **440** 1013-17
[5] Vanommeslaeghe K, Hatcher E, Acharya C, Kundu S, Zhong S, Shim J, Darian E, Guvench O, Lopes P, Vorobyov I and Mackerell Jr A D 2009 J. Comput. Chem. **31** 671-90
[6] Berendsen H J C, Spoel D V D, and Drunen R V 1995 Comp. Phys. Comm. **91** 43–56
[7] Panaretou B, Prodromou C, Roe S M, O'Brien R, Ladbury J E, Piper P W and Pearl L H 1998 EMBO J. **17** 4829-36

Interaction between PET tracer and the specific residues around the gate of the open form of Monoamine Oxidase B (MAO-B)

M Ottawa and N Miyashita*

Biological System Engineering, Graduate School of Biology-Oriented Science and Technology, KINDAI University, 930 Nishimitani, Kinokawa, Wakayama 649-6493, Japan.
Email: *miya@waka.kindai.ac.jp

Abstract. Monoamine Oxidase B (MAO-B) is the enzyme that metabolizes a monoamine neurotransmitter. Recently, a PET tracer targeting MAO-B, SMBT-1, has been developed as the biomarker of neurodegenerative diseases. However, the detailed binding mode of SMBT-1 has not been clear. To clarify the binding mode of SMBT-1 on MAO-B, we performed the molecular dynamics (MD) simulations of MAO-B in the outer mitochondrial membrane and the complex of MAO-B to SMBT-1, and the docking simulation of MAO-B with SMBT-1. We found that the Leu_{88}, Leu_{171}, Ile_{199}, and Tyr_{326} around the substrate-binding site interact with SMBT-1. These hydrophobic residues mainly support the two aromatic rings of the center of SMBT-1, making the stable binding of SMBT-1.

1. Introduction

Monoamine oxidase B (MAO-B) is the enzyme that metabolizes a monoamine neurotransmitter, such as dopamine and histamine [1]. The MAO-B is expressed in an astrocyte that observed in the basal forebrain, substantia nigra, basal ganglia, thalamus, and hippocampal relative to the cerebellar cortex [2]. Elevation of MAO-B in humans is related to neurodegenerative diseases, such as Parkinson's disease (PD), and amyotrophic lateral sclerosis (ALS) [3][4]. Thus, it is known as an essential biomarker for the neurodegenerative diseases. Recently, R. Harada. et al. has developed a PET tracer targeting MAO-B, SMBT-1 [5].

MAO-B, which forms homodimer structures in vivo, is bound to the outer mitochondrial membrane. Furthermore, MAO-B has expressed between the outer and inner mitochondrial membranes. The ligand-binding site consists of many hydrophobic residues and a part of Flavin Adenine Dinucleotide (FAD), MAO-B coenzyme. However, the entrance of the ligand-binding site has been unclear. Thus, is also not clear that the detailed binding mode of PET tracers.

To clarify the binding mode of SMBT-1 on MAO-B, we performed the molecular dynamics (MD) simulations of the MAO-B with the outer membrane of the mitochondria and the complex of MAO-B and SMBT-1, and the docking simulation.

2. Materials and Methods

2.1. MD simulation of SMBT-1
The structure of SMBT-1 is constructed by ChemDraw [6], and the force field was created using CGenFF [7]. We performed the MD simulation of SMBT-1 in solutions with 150 mM of K^+ and Cl^- ions. The most favorable structure in the solution was estimated by the classifications of the structures of MD trajectories using the principal component analysis (PCA).

2.2. MD simulation of MAO-B in the membrane.
We modeled the MAO-B dimer structure with the transmembrane domain. The transmembrane regions of MAO-B have been modeled by replica-exchange molecular dynamics simulations [8]. The soluble domain of MAO-B is obtained from PDB (PDB ID: 1GOS [9]), and we combined it with the helical transmembrane domains. We also prepared the membrane with similar content of the lipids in the mitochondrial outer membrane [10]. The MAO-B dimer model is embedded into the membrane using the membrane builder on CHARMM-GUI [11][12]. We set 150mM of K^+ and CL^- ions.

We performed a 2μs MD simulation of MAO-B in the membrane system. The NPT ensemble simulation was set to a temperature of 310 K and a pressure of 1 bar. We used Nose-Hoover [13][14] thermostat, Parrinello-Rahman [15] barostat, and Velocity-Verlet [16] integrator. We used GROMACS [17] for the MD simulation, and the force field was Charmm36m [18]. The suitable structure of MAO-B with the large entrance of a ligand docking was observed.

2.3. Docking Simulation.
We performed a docking simulation using the docking software AutoDock Vina [19]. We used the most favorable structure of SMBT-1 in the solution and the structure of the larger entrance of MAO-B, a target protein. This study used the MAO-B dimer structure as a rigid model, and the ligand was a partly flexible model.

2.4. MD simulation of the complex structure of MAO-B and SMBT-1.
We also performed MD simulations of a complex model of MAO-B bound SMBT-1 (figure 1). We used the structure obtained from the docking simulation for the initial structure of the complex model. We performed a 100 ns MD simulation of this complex model.

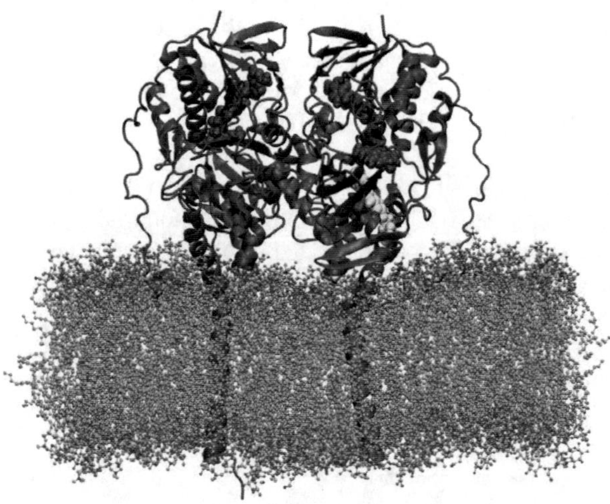

Figure 1. This figure shows the complex structure of the membrane-bounded MAO-B with SMBT-1. The blue and yellow indicate the MAO-B structure and the PET tracer, SMBT-1, respectively. The ball-and-sticks indicate the lipids.

2.5. Binding Free Energy using MM-PBSA.
We calculated the binding free energy using the trajectory of the MD simulation of the complex of MAO-B and SMBT-1 using the MM-PBSA method. The software g_mmpbsa [20] was used for binding free energy calculation.

3. Results
The MD simulations of the membrane-bound structure of MAO-B suggest that the open-close motion was observed around the ligand entrance region that is in the water-soluble domain of MAO-B. Therefore, we chose the largest entrance structure of MAO-B obtained from the MD simulations to use the docking simulations.

We performed the docking simulations with SMBT-1 using the structure and found many docking sites. In the docking simulation, MAO-B was set to rigid, but SMBT-1 was set to partially flexible. Since it has been known that the PET tracer targeting MAO-B binds to the substrate-binding site, we selected the most reliable model structure of MAO-B with PET tracer binding. We then performed the MD simulation of the MAO-B and SMBT-1 complex structure.

As the results, the SMBT-1 bound to the Leu$_{171}$, Ile$_{198}$, Val$_{294}$, and Tyr$_{326}$ in the MAO-B in the case of the initial structure, which is the most reliable model structure obtained by the docking simulations (figure 2 (a)(b)), while the SMBT-1 stably interacted with Leu$_{88}$, Leu$_{171}$, Ile$_{199}$, and Tyr$_{326}$ after 30 ns of the MD simulation (figure 2 (b)(d)). It seems that the hydrophobic residues catch two aromatic rings of the SMBT-1, and these residues support the stable binding of the SMBT-1 in the substrate-binding site.

Furthermore, we calculated the binding free energy of the SMBT1 and MAO-B using the MM-PBSA method. We found that the binding free energy is -18.9 kcal/mol and suggested that the binding of SMBT-1 and the MAO-B seems strong.

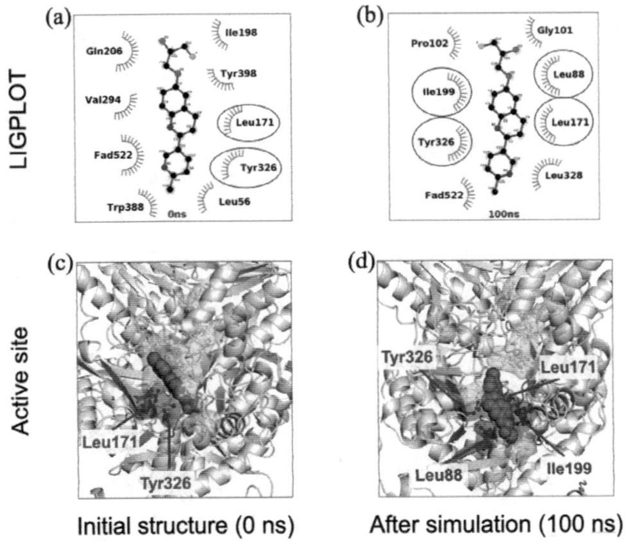

Figure 2. (a)(b) show a schematic picture of the essential residues for the SMBT-1 binding. (a) is for the initial structure (at 0 ns), and (b) is for the structure at 100 ns. We calculated these by LIGPLOT [21]. Blue circles indicate the favorable residues for SMBT-1 binding. (c) The binding site of SMBT-1 at 0 ns, and (d) that of 100 ns in the MD simulation. The red space fills, blue ball-and-sticks, and yellow regions indicate the SMBT-1, the essential residues, and other related hydrophobic residues.

4. Summary
We performed MD simulations and the binding energy calculation of MAO-B and the PET tracer, SMBT-1, to investigate the interaction between MAO-B and SMBT-1. This study modeled an outer mitochondrial membrane-bound dimer structure of MAO-B in vivo. We also performed an MD simulation, 2μs, using this structure. In addition, we performed docking simulations of MAO-B with SMBT-1 using the structure with the open ligand binding gate. As a result, we found that the SMBT-1 is bound to the substrate-binding site of MAO-B.

Journal of Physics: Conference Series **2207** (2022) 012025 doi:10.1088/1742-6596/2207/1/012025

In addition, we performed MD simulations for 100 ns on the complex structure of MAO-B bound to SMBT-1. Then, the analysis by LIGPLOT suggests that SMBT-1 interacts with Leu_{88}, Leu_{171}, Ile_{199}, and Tyr_{326} of MAO-B specifically (figure 2) and it has the lower binding free energy. These hydrophobic residues mainly support the two aromatic rings of the center of SMBT-1, making the stable binding of SMBT-1.

References

[1] Youdim M B H and Weinstock M 2004 *Neurotoxicology* **25** (1-2) 243–50

[2] Tong J, Meyer J H, Furukawa Y, Boileau I, Chang L, Wilson A A, Houle S and Kish S J 2013 *J. Cereb. Blood. Flow. Metab.* **33** (6) 863-71

[3] Mallajosyula J K, Kaur D, Chinta S J, Rajagopalan S, Rane A, Nicholls D G, Di Monte D A, Macarthur H and Andersen J K 2008 *PLoS One* **3** (2) 1-16

[4] Ekblom J, Jossan S S, Bergstrüm M, Oreland L, Walum E and Aquilonius S 1993 *GLIA* **8** (2) 122-32

[5] Harada R, Hayakawa Y, Ezura M, Lerdsirisuk P, Du Y, Ishikawa Y, Iwata R, Shidahara M, Ishiki A, Kikuchi A, Arai H, Kudo Y, Yanai K, Furumoto S and Okamura N 2021 *J. Nucl. Med.* **62** (2) 253-8

[6] Mills N 2006 *J. Am. Chem. Soc.* **128** (41) 13649-50

[7] Vanommeslaeghe K and MacKerell Jr A D 2012 *J. Chem Inf. Model.* **52** (12) 3144-54

[8] Sugita Y and Okamoto Y 1999 *Chem. Phys. Lett.* **314**(1-2) 141-51

[9] Binda C, Newton-Vinson P, Hubalek F, Edmondson D E and Mattevi A 2002 *Nat. Struct. Biol.* **9** 22-6

[10] Aufschnaiter A, Kohler V, Diessl J, Peselj C, Carmona-Gutierrez D, Keller W and Büttner S 2017 *Cell. Tissue. Res.* **367** (1) 125-40

[11] Jo S, Kim T and Im W 2007 *PLoS ONE* **2** (8) e800

[12] Jo S, Kim T, Lyer V G and Im W 2008 *J. Comput. Chem.* **29** 1859-67

[13] Nosé S 1984 *Mol. Phys.* **52** 255-68

[14] Hoover W G 1985 *Phys. Rev. A* **81** 1695-97

[15] Parrinello M and Rahman A 1981 *J. Appl. Phys.* **52** 7182-90

[16] Swope W C, Andersen H C, Berens P H and Wilson K R 1982 *J. Chem. Phys.* **76** 637-49

[17] Berendsen H J C, van der Spoel D and van Drunen R 1995 *Comput. Phys. Commun.* **91** (1-3) 43-6

[18] Huang J, Rauscher S, Nawrocki G, Ran T, Feig M, Groot B L, Grubmüller H and MacKerell Jr A D 2017 *Nat. Methods* **14** 71-3

[19] Trott O and Olson A J 2010 *J. Comput. Chem.* **31** 455-61

[20] Kumari R, Kumar R, Open Source Drug Discovery Consortium and Lynn A 2014 *J. Chem. Inf. Model.* **54** (7) 1951-62

[21] Wallace A C, Laskowski R A and Thornton J M 1996 *Protein Eng.* **8** (2) 127-34

Enhancing efficient computation of long-wavelength relaxation dynamics in a 2D liquid involving millions of particles

Hayato Shiba

Information Technology Centre, University of Tokyo, Kashiwa, Chiba 277-0882, Japan

E-mail: shiba@cc.u-tokyo.ac.jp

Abstract. Recently, a two-dimensional liquid cooled toward the glass transition was found to exhibit a t^{-1} long-time tail in the velocity autocorrelation function (VACF) owing to the presence of long-wavelength fluctuations. To directly observe this power-law behaviour, it is necessary to simulate a large system with millions of particles, which is a challenging task from the computational viewpoint. In this study, to address this difficulty, I first show that this power-law tail can be reproduced by differentiating the finite-time diffusivity with respect to time. In addition, the feasibility of another direction, a direct on-the-fly computation of the VACFs utilizing GPGPUs, wherein VACFs are evaluated as the simulation runs, is also demonstrated. A performance benchmark was executed on Wisteria/BDEC-01 (Aquarius subsystem) supercomputer using a simulation code developed by the author, which enabled the direct computation of the VACF of 4 million particlesx for as long as the 10^8 simulation steps within 10 days.

1. Introduction

Glass transition of a molecular liquid is characterised by the freezing of constituent particles induced by a decrease in temperature T or an increase in density n. The movement of particles at the microscopic level can be visualized in molecular dynamics (MD) simulations, which is their strong merit in comparison with the experiments. For many years, the MD simulations have become an indispensable tool for studying glass transition [1, 2]. The MD simulations have revealed that the movement of particles becomes inhomogeneous with correlation lengths significantly larger than the particle size.

Many MD simulations for glass transition have been two-dimensional (2D) [3, 4]. Because visualising particle trajectories in two-dimension is relatively easy, such studies have improved our understanding of the nature of slow glassy dynamics. The rationale for using two-dimensional (2D) systems is that the essential feature of glassy dynamics, i.e., the intermittent cage-jump of particles on the microscopic scale is similar to those in three-dimensional (3D) systems. Recently, in contrast to these ideas, the dynamics of a 2D glass-forming liquid were found to be different from its 3D counterpart [5, 6, 7, 8, 9, 10]. Similar to other 2D liquids, 2D glass-forming liquids exhibit long-wavelength density fluctuations, which affects the behaviour of various observable quantities. Particularly, as the system size increases, the plateau in the mean-squared displacement begins to disappear [6, 7, 11]. Moreover, the structural relaxation in terms of rearranging dynamics on the microscopic scale was found to be similar, given the introduction of quantities that characterise the relative motions between neighbouring pairs is independent of the long-wavelength fluctuations [5, 7, 8, 9, 10, 11]. These quantities include the relative displacement between neighbouring pairs (i.e., the "cage-relative" displacement) [8, 9, 10, 11] and bond-breakage functions representing the change in the coordination number of each particle [5, 7, 10]. In these contexts, the long-wavelength fluctuations are irrelevant in the vitrification process because such collective fluctuations do not directly give rise to particle rearrangement on the microscopic scale.

In a recent study by the present author, such long-wavelength fluctuations in two dimensions were further investigated by computing the velocity autocorrelation function (VACF) in the long time [11]. Beyond the time region wherein negative velocity autocorrelations appear, re-entrant positive correlations are found to exhibit a long-time power-law tail of the velocity autocorrelation [12]

$$Z(t) = \langle \boldsymbol{v}(t) \cdot \boldsymbol{v}(0) \rangle \sim \frac{k_{\mathrm{B}}T}{8\pi nm} \left[\left(D + \frac{\eta}{mn} \right) t \right]^{-1} \simeq \frac{k_{\mathrm{B}}T}{8\pi\eta} t^{-1}, \tag{1}$$

where $\boldsymbol{v}(t)$, t, k_{B}, η, and m denote the velocity, time, Boltzmann constant, viscosity, and particle mass, respectively. The diffusivity D is neglected in the final expression because $D \ll \eta/mn$ near the glass transition. Note that total momentum conservation is required for the power-law tail, which can be simulating the NVE ensemble. Although we succeeded in calculating this power-law tail at a high temperature (i.e., $T = 1.0$) (the method failed at lower temperaturew due to a technical difficulty mentioned later), quantitative agreement with the analytical expression of long-time tail suggests that this behaviour will be exhibited even at lower temperatures. Importantly, it is shown by a simple consideration of a breakdown of the Stokes-Einstein relation that this long-time correlation is not sufficiently weak to be considered as irrelevant in the slow dynamics near the glass transition [11].

In this study, I address this long-time power-law asymptotic behaviour in the VACF. After briefly describing the models we used in Sec. 2, the relationship between the VACF and the mean-squared displacement are clarified. In Sec. 3, the diffusion processes altered by this VACF will be clearly demonstrated. In order to efficiently compute the VACF of a large-scale system comprising millions of particles, the present author is now developing a simulation code. The current development status and benchmark data of this code are described in Sec. 4.

2. Model

The simulation and benchmark results presented in this study are for a 2D binary Kob-Andersen Lennard-Jones mixture with modified interaction parameters, which has already been employed in previous studies [6, 11, 13, 14]. The ratio between the two different species is altered from 80:20 to 65:35 to prevent crystallisation in two dimensions. The pairwise potential is defined as a function of the interparticle distance r as

$$v_{\alpha\beta}(r) = 4\epsilon_{\alpha\beta} \left[\left(\frac{\sigma_{\alpha\beta}}{r} \right)^{12} - \left(\frac{\sigma_{\alpha\beta}}{r} \right)^{6} \right] \tag{2}$$

and is truncated and shifted at $r = 2.5\sigma_{\alpha\beta}$. $\sigma_{\alpha\beta}$ and $\epsilon_{\alpha\beta}$ represent pairwise diameter and interaction energy, respectively, for arbitrary combinations of particle species (i.e., $\{\alpha, \beta\} = \{1, 2\}$). Here, we have $\sigma_{12} = 0.88\sigma_{11}$, $\sigma_{22} = 0.8\sigma_{11}$, $\epsilon_{12} = 1.5\epsilon_{11}$, and $\epsilon_{22} = 0.5\epsilon_{11}$; the number density is set at $n = 1.2$. Furthermore, the length, temperature, and time are presented in units reduced by σ_{11}, $\epsilon_{11}/k_{\mathrm{B}}$, and $\sigma_{11}\sqrt{m/\epsilon_{11}}$, respectively. We employed the typical annealing procedure, wherein the liquid at temperature $T = 2.5$ was cooled to the lower target temperature at a sufficiently slow rate. In this paper, we used a system size of $N = 1,024,000$ and 4,096,000.

3. Long-time tail and nonlinear diffusivity

The VACF can be computed by adding the product of velocities of each particle i at time t_0 and $t_0 + t$, i.e.,

$$Z(t) = \overline{\sum_{i=1}^{N} \boldsymbol{v}_i(t_0) \cdot \boldsymbol{v}_i(t_0 + t)}, \tag{3}$$

where the overline ($\bar{\cdot}$) represents the statistical average taken over initial times $t = t_0$ and independent simulations. Despite its simple definition, this VACF is technically difficult to calculate for a large system, because it requires the computation of the correlations over a large data that keep the history of the velocity of each individual particle of the entire system. In my previous paper [11], this quantity was computed and was shown to agree with the t^{-1}-tail in the long-time region; after re-entrance from negative to positive correlations [11].

It is also shown in Ref. [11] that, at a high temperature of $T = 1$, the time-dependent diffusivity can be evaluated by integrating the VACF with respect to time (i.e.,$D'(t) = \int_0^t Z(t)\, dt$), wherein its long-time limit $\lim_{t\to\infty} D'(t)$ coincides with the diffusivity D evaluated from the long-time mean-squared displacement (MSD). It is worth noting that, in the time region where the t^{-1}-tail of the VACF emerges, the time-dependent diffusivity slightly increases as a function of time. In other words, the MSD is not strictly linear but rather it becomes slightly nonlinear as it is incremented by logarithmic corrections originating from the long-wavelength fluctuations.

From the nonlinear behaviour of MSD, the VACF can be evaluated by differentiating the time-dependent diffusivity with respect to time, i.e.,

$$Z'(t) = \frac{dD'(t)}{dt}. \tag{4}$$

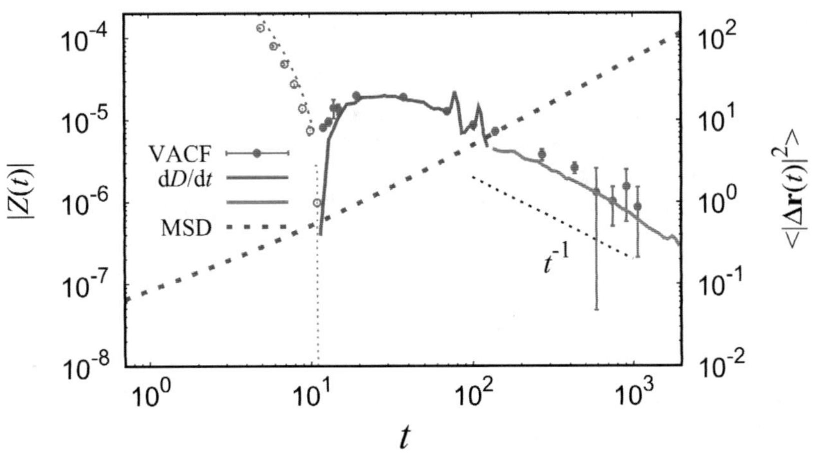

Figure 1. Power-law tail in the VACF evaluated by different methodologies. All the data are shown in absolute values. Red circles (with error bars, the left axis) show the VACF directly computed from the history data for the velocity (at $T = 1$), where open and filled ones show negative and positive correlations. Blue and Green lines (the left axis) are the evaluation of VACF (left axis) from the time-dependent MSDs by using Eqs. (5) and (6), where dotted and solid ones show negative and positive correlations, respectively. In both, the t^{-1} power-law tail is clearly observed. The purple dotted line shows the MSD (the right axis).

Journal of Physics: Conference Series **2207** (2022) 012026 doi:10.1088/1742-6596/2207/1/012026

In the short-time region, the time-dependent diffusivity $D'(t)$ can be evaluated by taking the second time derivative of MSD; however, such an evaluation becomes challenging in the long-time region because it becomes too small as time proceeds. Therefore, it is more appropriate to evaluate the time derivative of the time-dependent diffusivity $D'(t)$ from the MSD by taking a different way in the long-time region as follows

$$\frac{dD'(t)}{dt} = \frac{d^2}{dt^2}\frac{\langle|\Delta \boldsymbol{r}(t)|^2\rangle}{4} \qquad (t < t_a) \tag{5}$$

$$\frac{dD'(t)}{dt} = \frac{d}{dt}\frac{\langle|\Delta \boldsymbol{r}(t)|^2\rangle}{4t} \qquad (t \geq t_a). \tag{6}$$

Figure 1 shows the data for the VACF at $T = 1$ with $N = 4,096,000$ particles. The directly computed VACF from Eq. (3) is similar to that in our previous paper [11], wherein the statistical errors are reduced by augmenting the simulation data for longer duration by using my simulation code as described in Sec. 4. The time-dependent diffusivity was also analysed with $N = 1,024,000$ particles at $T = 1$, where its derivative in Eqs. (5) and (6) is plotted. We observed the coincidence in between, which indicates that the the weak power-law t^{-1}-tail in the VACF is embedded in the nonlinear diffusion in the MSD.

4. Accelerated computing of the VACF with GPGPUs

From a computational point of view, it is difficult to directly compute the VACF, because velocity correlation needs to be summed over independent particles and a long time. It is virtually impossible to directly compute the VACF of a million-particle (or more) system by post-processing, because storing a huge amount of velocity data with a fine resolution becomes necessary. An alternative is to store the velocity history of independent particles on main memories at appropriate time intervals. To minimize complex internode communications, storing the velocity histories on a shared memory space is desirable.

I have recently developed a simulation code that performs such on-the-fly computation of the VACF by using General-Purpose computing on Graphics Processing Units (GPGPUs). In the following, I present the benchmark results of the proposed code. For the benchmark, I employed one node from the Aquarius subsystem of the Wisteria/BDEC-01 supercomputer, installed at the Information Technology Center, University of Tokyo. Each computational node on the Aquarius subsystem is equipped with eight GPGPUs (NVIDIA A100 Tensor Core GPUs with 40 GB device memories), two CPUs (Intel Xeon Platinum 8360Y with 72 cores in total), and a total of 512 GB DDR-4 main memory.

Figure 2 presents a schematic illustration of the resource allocation. My simulation code was developed with the intent that the computation of forces are offloaded to the GPGPUs; other calculations such as the neighbour-list construction are performed on the host CPUs. Further, this allows the CPUs to process the velocity history accumulated in the main memory, while GPGPUs are computing interaction forces between the particles.

The set of particles was divided into eight groups to allow the inter-particle interaction of a group to be computed in its respective GPGPU, while the CPUs process the pair interactions between different groups. Because particles are less frequently exchanging their positions as the temperature decreases, the workload for force computation can be reduced by labelling the particle indices in accordance with the spatial positions after restarting the simulation from the dumped data. Parallelization on the host CPUs was implemented using OpenMP. Although one node of the Aquarius subsystem is equipped with 72 cores, the presented benchmark test was run with up to 64 threads, wherein 1 thread was processed on 1 CPU core because the hyperthreading was disabled.

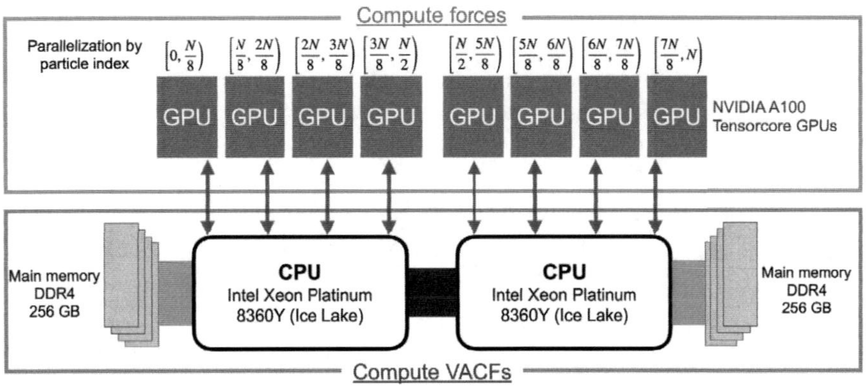

Figure 2. Schematic illustration of resource allocation for on-the-fly computation of the VACF on Wisteria/BDEC-01, Aquarius subsystem, installed at the University of Tokyo.

Table 1 presents a comparison of the benchmark results between simulations performed with and without GPGPUs for 10^4 time steps and $N = 4,096,000$ particles. The result was not severely affected by changing the simulated temperature. Data I/O routines were excluded from the benchmark executions. For the simulations that used eight GPGPUs, the neighbour-list construction and time integration of Newton equations were processed on the host CPUs. During the neighbour-list construction, the data layout is modified to allow the unit stride access to the neighbour-list data on the GPGPUs. At each time step, the host CPUs send the position data, and inversely retrieves the force data. While these communications consume the largest portion of time, the force computation was shortened to 3.4 s by using GPGPUs, which is approximately 80 times faster than without GPGPU. The total simulation time was 81.9 s, which enables a simulation of 10^8 steps in less than 10 days.

The VACF can be computed on the host CPUs by loading the previous velocity data for each particle from the main memory. We may choose an appropriate time interval for saving the velocity history when computing the VACF, in accordance with the time scale we

Table 1. Workload analysis result for $N = 4,096,000$ over 10^4 time steps. On the column of "host/device", "H" and "D" represent the host CPUs and GPU device respectively, where "H to D" means the data transfer time from host to device, and vice versa for "D to H".

	functions	host/device	elapse	
8 GPU devices	send position data	(H to D)	7.0 s	(8.5%)
– 81.9 s (total)	send neighbour-list	(H to D)	5.1 s	(6.2%)
	recv force data	(D to H)	16.1 s	(19.6%)
	neighbour-list construction	(H)	9.8 s	(12.0%)
	force computation	(D)	3.4 s	(4.1%)
	force computation (inter-GPUs)	(H)	< 20.0 s	(<25%)
64 CPU cores	neighbour-list construction	(H)	43.9 s	(9.5%)
without GPUs	force computation	(H)	295.7 s	(64.1%)
– 340.6 s (total)				

want to address. With the present 512 GB main memory on one computational node, we can accommodate approximately 10^4 previous time points for a system with $N = 4,096,000$. Therefore, we can correlate velocities separated by 10^7 steps at maximum by storing velocity correlations at every 10^3 time-steps. Typically, this computation takes less than 10 ms per step.

The present system is equipped with a high-bandwidth direct communication architecture (NVIDIA NVLink) between the GPGPUs. Although we have not tried to implement it, the communication bottleneck between the host and the device may be removed by offloading the computations of the equations of motion and the neighbour-list construction. This is one possible direction for further accelerating the computation of VACFs.

5. Summary

I demonstrated two independent computational techniques for the effective evaluation of the power-law relaxation of long-wavelength fluctuations that is characteristic of 2D liquids involving millions of particles. The first strategy involves computing the time derivatives of the finite-time diffusivity, which can be evaluated from the MSDs by utilizing the fact that the MSDs in two dimensions slightly deviate from linear diffusion. The other strategy involves the direct computation of VACF with the help of GPGPUs, which enable the computation that effectively utilises a large shared-memory. I performed the calculations of the VACFs at lower temperatures following the latter strategy. I am currently preparing a separate paper in which these results will be reported.

Acknowledgments

I thank Yohei Miki and Takashi Shimokawabe for the stimulating discussions on GPGPU computations. I am also grateful to Ludovic Berthier for his suggestions. This study was financially supported by the JSPS KAKENHI Grant No. JP18H01188. I also acknowledge "Joint Usage/Research Center for Interdisciplinary Large-scale Information Infrastructures" and "High Performance Computing Infrastructure" in Japan (Project ID: jh210035-NAH), for supports and for providing computational resources of Wisteria/BDEC-01 (FUJITSU Server PRIMERGY GX2570 M6, equipped with NVIDIA A100 Tensor Core GPUs 40GB SXM) at Information Technology Center, University of Tokyo. For numerical computations, I also used MASAMUNE-IMR (Cray CS-Storm 500, equipped with NVIDIA Tesla V100 for PCIe) at Institute for Materials Research, Tohoku University and Dell EMC PowerEdge (System B) at Institute for Solid State Physics, University of Tokyo.

References

[1] Binder K and Kob W 2011 *Glassy Materials and Disordered Solids* (World Scientific)
[2] Berthier L and Biroli G 2011 *Reviews of Modern Physics* **83**(2) 587–645
[3] Tanaka H, Kawasaki T, Shintani H and Watanabe K 2010 *Nature Materials* **9**(4) 324–331
[4] Widmer-Cooper A, Perry H, Harrowell P and Reichman D R 2008 *Nature Physics* **4** 711–715
[5] Shiba H, Kawasaki T and Onuki A 2012 *Physical Review E* **86** 041504 (Erratum: 2017 *ibid.* **95**, 069901)
[6] Flenner E and Szamel G 2015 *Nature Communications* **6** 7392
[7] Shiba H, Yamada Y, Kawasaki T and Kim K 2016 *Physical Review Letters* **117**(24) 245701
[8] Illing B, Fritschi S, Kaiser H, Klix C L, Maret G and Keim P 2017 *Proceedings of the National Academy of Sciences U. S. A.* **114**(8) 1856–1861
[9] Vivek S, Kelleher C P, Chaikin P M and Weeks E R 2017 *Proceedings of the National Academy of Sciences U. S. A.* **114**(8) 1850–1855
[10] Shiba H, Keim P and Kawasaki T 2018 *Journal of Physics: Condensed Matter* **30**(9) 094004
[11] Shiba H, Kawasaki T and Kim K 2019 *Physical Review Letters* **123**(26) 265501
[12] Ernst M H, Hauge E H and van Leeuwen J M J 1970 *Physical Review Letters* **25**(18) 1254–1256
[13] Flenner E and Szamel G 2019 *Proceedings of the National Academy of Sciences U. S. A.* **116**(6) 2015–2020
[14] Li Y W, Mishra C K, Sun Z Y, Zhao K, Mason T G, Ganapathy R and Ciamarra M P 2019 *Proceedings of the National Academy of Sciences U. S. A.* **116**(46) 22977–22982

Role of temperature and alignment activity on kinetics of coil-globule transition of a flexible polymer

Subhajit Paul, Suman Majumder, and Wolfhard Janke

Institut für Theoretische Physik, Universität Leipzig, IPF 231101, 04081 Leipzig, Germany

E-mail: subhajit.paul@itp.uni-leipzig.de, suman.majumder@itp.uni-leipzig.de, wolfhard.janke@itp.uni-leipzig.de

Abstract. We study the nonequilibrium kinetics during the coil-globule transition of a flexible polymer chain with active beads after a quench from good to poor solvent condition using molecular dynamics simulation. Activity for each bead is introduced via the well-known Vicsek-like alignment rule due to which the velocity of a bead tries to align towards the average direction of its neighbors. We investigate the role of quenching temperature with varying activity during collapse of this polymer. We find that although for lower activities the kinetics remains qualitatively similar for different temperatures, for higher activity noticeable differences can be identified.

1. Introduction

The typical conformation of a polymer undergoes changes when quenched from a good to a poor solvent condition [1]. In a good solvent or at high temperature the equilibrium conformation of the polymer is a coil. On the other hand, in a poor solvent or at low temperature the equilibrium conformation becomes a globule. The equilibrium aspects of this transition have been studied since many years and are quite well understood [1]. Despite efforts using computer simulations of coarse-grained as well as all-atom models, a concrete theoretical framework for the nonequilibrium properties is still missing. Among different phenomenological descriptions regarding the pathways of globule formation, the 'pearl-necklace' picture proposed by Halperin and Goldbart [2] is quite well accepted and observed in various polymeric systems. In recent years, the kinetics and associated scaling laws have been explored by using the analogy with usual coarsening in a spin or particle system [3, 4, 5].

Understanding of relaxation across the collapse transition for a polymer has relevance in many systems, e.g., protein folding, chromatin dynamics and conformational changes of other bio-polymers, to name a few. Such polymer chains can also be 'active' in nature. Their 'activity' can be modeled using frameworks borrowed from 'active' particle systems [6]. Active particles have the property of being motile either by using their own internal energy or by taking energy from the environment. Similarly, for an active polymer, the monomers can be active by themselves or can be activated by external forces. Like in a particle system, for a polymer also one should expect significant differences in its dynamics due to activity compared to that of a passive polymer. Whereas notable progress has been made for a passive polymer [3, 4, 5], studies regarding the kinetics of an 'active' polymer are rather recent [7, 8, 9, 10].

Journal of Physics: Conference Series **2207** (2022) 012027 doi:10.1088/1742-6596/2207/1/012027

In this paper, we consider a flexible homopolymer chain with each monomer as an active element. The activity is implemented via the well-known Vicsek-like alignment rule [9, 10, 11, 12, 13]. In this context a polymer can be visualized as a system in which the beads are moving with some constraints. Due to this activity, the velocity of each bead gets modified towards the average direction of its neighbors. In our earlier works [9, 10], we have investigated the effects of such an activity on the pathway across the coil-globule transition and compared them with those for the passive case for a particular quench temperature. Here we consider a lower quench temperature and investigate whether similar features can be observed for the conformations and kinetics.

2. Model and methods

We consider a bead-spring polymer chain with N linearly connected beads. The bonded interaction between two successive beads is modeled via a finitely extensible non-linear elastic (FENE) potential [3] in the form $V_{\text{FENE}}(r) = -0.5KR^2 \ln[1-((r-r_0)/R)^2]$ with $K = 40$, $r_0 = 0.7$ and $R = 0.3$. The non-bonded interaction is modeled as $V_{\text{nb}}(r_{ij}) = V_{\text{LJ}}(r) - V_{\text{LJ}}(r_c) - (r-r_c)\frac{dV_{\text{LJ}}}{dr}$, where $V_{\text{LJ}}(r) = 4\epsilon[(\sigma/r)^{12} - (\sigma/r)^6]$ is the standard Lennard-Jones (LJ) potential. Here $\sigma = r_0/2^{1/6}$ is the diameter of the beads, $\epsilon = 1$ the interaction strength and $r_c = 2.5\sigma$ is the cut-off distance.

Dynamics of this polymer has been studied using molecular dynamics (MD) simulation with velocity-Verlet integration scheme using the Langevin thermostat. Thus for each bead our working equation is [9, 10]

$$m\frac{d^2\vec{r}_i}{dt^2} = -\vec{\nabla}U_i - \gamma\frac{d\vec{r}_i}{dt} + \sqrt{2\gamma k_B T}\vec{\Lambda}_i(t)\,, \tag{1}$$

where m is the mass, γ the drag coefficient, k_B the Boltzmann constant, T the quench temperature measured in units of ϵ/k_B, and U_i is the interaction potential which contains both V_{FENE} and V_{nb}. For convenience we set m, γ and k_B to unity. $\Lambda_i(t)$ represents Gaussian white noise with zero mean and unit variance, and Delta-correlations over space and time. We have used the integration time step $\delta t = 0.0005$ in units of the timescale $\tau_0 = \sqrt{m\sigma^2/\epsilon}$.

At each MD step the activity is introduced in the following way. The direction of the velocity of the i-th bead obtained from Eq. (1) is modified by the active force [9, 10, 13]

$$\vec{f}_i = f_A\hat{n}_i\,; \quad \text{with } \hat{n}_i = \frac{(\sum_j \vec{v}_j)_{r_c}}{|(\sum_j \vec{v}_j)_{r_c}|}\,, \tag{2}$$

where f_A is the strength of the activity and $\hat{n}_i$ represents the average direction of the neighboring beads within a sphere of radius r_c. More technical details regarding the implementation of the active force are discussed in Refs. [9, 10].

In the rest of the paper, the activity strength will be expressed via the ratio of active ("ballistic") and thermal energy, i.e., the Péclet number [7]

$$Pe = \frac{f_A\sigma}{k_B T}. \tag{3}$$

As we will focus on comparing the kinetics of the polymer collapse with activity for different temperatures, it is more convenient to use the dimensionless parameter Pe for which we consider the values 0, 0.62, 1.25 and 5.0. $Pe = 0$ corresponds to the passive polymer case. The temperatures are chosen as $T = 0.25$ and 0.5 which are both well below the coil-globule transition temperature for the passive polymer of length $N = 512$ [4]. In the figures of the next section, $\langle \ldots \rangle$ represents an average over different initial configurations and thermal noise, using 200 independent realizations for $T = 0.25$ and 500 for $T = 0.5$.

3. Results

We start our discussion by presenting in Fig. 1 typical representative conformations of the polymer chain during its evolution with time for the two temperatures at low and high activity, i.e., $Pe = 0.62$ and 5.0, respectively. Comparative time evolution snapshots for different activities as well as for the passive case are discussed in Ref. [10]. Although in all four cases the final conformations are globules, one notices differences in the pathways. For $Pe = 0.62$ the conformations follow the three different stages of the 'pearl-necklace' picture [2]. For $Pe = 5.0$ and $T = 0.25$ the time evolution does not look very much different, although it takes longer time for the globule to form. For $T = 0.5$, however, already the early time conformations look somewhat different. It appears that the polymer becomes more elongated than its starting conformation and then starts collapsing. Later the two-cluster conformations, i.e., the dumbbell state, persists quite long and it takes very long to reach the collapsed state. Even though the value of Pe is the same, the conformations as well as the time required for globule formation at these two different temperatures are clearly different.

Before going into the quantification of conformational properties or the cluster growth, we first look at the effect of increasing activity on the alignment of the velocities of the beads. Thus

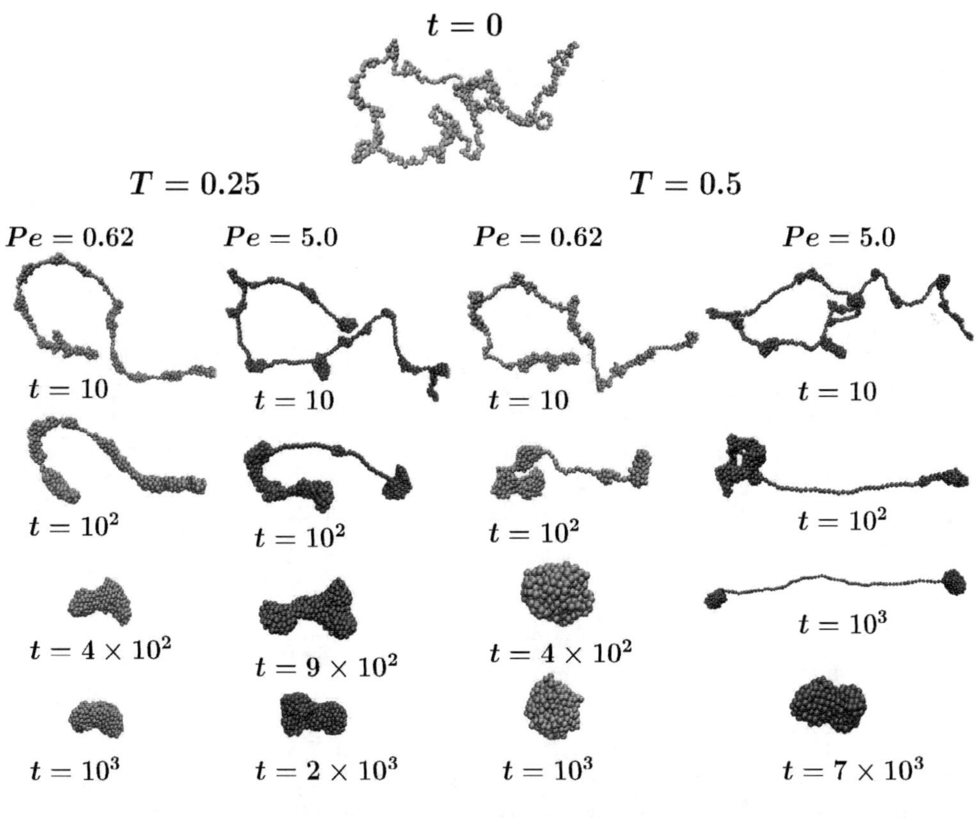

Figure 1. Typical snapshots representing the time evolution during collapse of an active polymer of length $N = 512$ at two different temperatures for two different values of the Péclet number Pe.

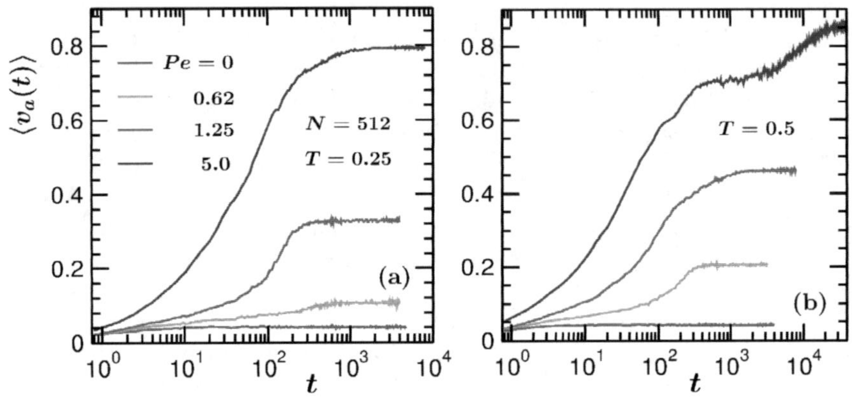

Figure 2. Semi-log plots of the average order parameter $\langle v_a(t) \rangle$ versus t for two different quench temperatures T of a polymer of length $N = 512$. For both temperatures, data are shown for different values of the Péclet number Pe.

we define a velocity order parameter as

$$v_a = \left| \sum_{i=1}^{N} \vec{v}_i \right| \Big/ \sum_{i=1}^{N} \left| \vec{v}_i \right|, \tag{4}$$

where $\vec{v}_i$ is the velocity of the i-th bead. In Figs. 2(a) and (b) we show plots of $\langle v_a \rangle$ vs time for different values of Pe for the two temperatures. For both of them, in the passive case, $\langle v_a \rangle$ always remains close to 0. Then for $Pe > 0$, $\langle v_a \rangle$ saturates at non-zero values which increase with increasing Pe. But for the same Pe the saturation values differ for the two temperatures. In fact, with $Pe = 5.0$ one finds a notable difference. For $T = 0.25$, $\langle v_a \rangle$ increases smoothly and reaches its saturation at ≈ 0.8, whereas for $T = 0.5$ the saturation occurs in two steps. The slower growth after the first plateau is due to the longer persistence of the dumbbell conformation for which the two clusters move slowly towards each other before they finally merge and $\langle v_a \rangle$ reaches up to 0.9 [10]. Even though the Pe values for the two temperatures are the same, this does not lead to the same degree of alignment of the beads.

The conformational changes during the collapse can be quantified via the squared radius of gyration R_g^2 of the polymer defined as

$$R_g^2 = \frac{1}{N} \sum_{i=1}^{N} (\vec{r}_{\rm cm} - \vec{r}_i)^2, \tag{5}$$

where $\vec{r}_{\rm cm} = \frac{1}{N} \sum_{i=1}^{N} \vec{r}_i$ defines the center-of-mass of the polymer. To show a comparative picture, in Figs. 3(a) and (b) we plot $\langle R_g^2 \rangle$ versus t for $T = 0.25$ and $T = 0.5$, respectively. For $T = 0.5$ with low values of Pe the decay is faster than that for the passive case. For $Pe = 5.0$ we see that $\langle R_g^2 \rangle$ initially increases from its starting point and then follows a much slower decay. For $T = 0.25$ and lower Pe we observe a similar trend of the decay of $\langle R_g^2 \rangle$, but here it even becomes faster with increasing Pe. For larger activity with $Pe = 5.0$ we do not observe the initial increase of $\langle R_g^2 \rangle$ but as for lower Pe a faster decay than in the passive case. Only asymptotically for large t, similar to $T = 0.5$, the decay appears to become slower. For a better visualization we plot the data for $Pe = 5.0$ for both T values over a much longer time on a semi-log scale in Fig. 3(c). This clearly shows that even though Pe is kept at the same value, the conformational changes

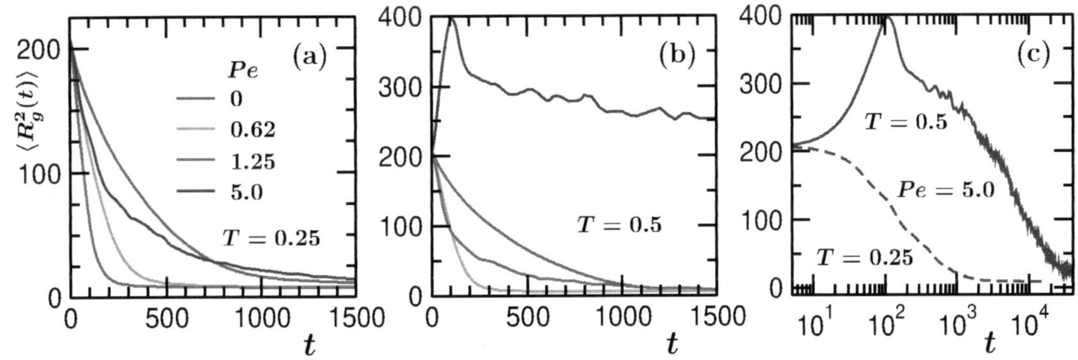

Figure 3. Plots of the average squared radius of gyration $\langle R_g^2(t)\rangle$ versus time t for passive as well as active polymers with $N = 512$ for (a) $T = 0.25$ and (b) $T = 0.5$. In both plots the values of Pe are the same. (c) Semi-log plot of $\langle R_g^2(t)\rangle$ versus t for the higher activity, i.e., with $Pe = 5.0$ for both the temperatures.

during the kinetics of globule formation are not similar. This has also been observed from the corresponding snapshots in Fig. 1. From this one can conclude that the Péclet number is not the only determining parameter for the kinetics.

Finally we investigated whether there exist any differences in the cluster growth at these two temperatures. For this we have identified the number of clusters $n_c(t)$ at time t along the chain and calculated their average size as $C_s(t) = (\sum_{k=1}^{n_c(t)} m_k)/n_c(t)$ where m_k counts the number of monomers in the k-th cluster. More technical details can be found in Refs. [3, 4, 10]. In Figs. 4(a) and (b) we plot $\langle C_s(t)\rangle$ versus t for different values of Pe for both temperatures. In general, $\langle C_s(t)\rangle$ follows a power-law behavior with time

$$\langle C_s(t)\rangle \sim t^{\alpha_c}, \tag{6}$$

where α_c is the growth exponent. We see that for both temperatures the qualitative behavior of $\langle C_s(t)\rangle$ with increasing Pe remains quite similar. For the passive case, in the scaling regime data look consistent with $\alpha_c \approx 1/2$ for both values of T. In both cases, with lower activities the growth is faster and the globular state is reached earlier than in the passive case. But the exponent for $T = 0.5$ appears to be lower than for $T = 0.25$. As guide to the eyes, we plot power-law lines with exponents 1 and 3/4 for $T = 0.25$ and 0.5, respectively. For $Pe = 5.0$ we see an opposite trend as the growth becomes slower and it takes a much longer time to reach the final globular state. In fact, similar to the lower activities, for this higher Pe also, it appears that the growth is slower for $T = 0.5$ compared to $T = 0.25$.

4. Conclusion

In this paper we have presented results for the kinetics during the coil-globule transition of a flexible polymer following a quench from a good to a poor solvent condition. The beads for the polymer are considered to be 'active' which is implemented in a Vicsek-like alignment manner. For a comparison among two different quench temperatures we keep the values of the Péclet number Pe the same. For the passive case as well as for low Pe values, the conformations and kinetics look more or less similar. Noticeable differences in the structural behavior and the kinetics appear for the higher activity. Our observations indicate that Pe cannot be the only dimensionless parameter determining the effect of activity. Rather, one may need to consider also

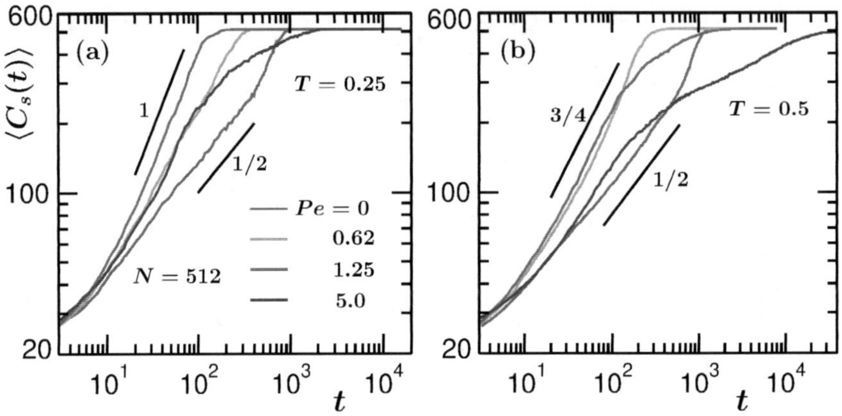

Figure 4. Log-log plots of $\langle C_s(t)\rangle$ versus t for a polymer with $N = 512$ governed by different values of Pe for (a) $T = 0.25$ and (b) $T = 0.5$. In both plots the exponents of the corresponding power laws are mentioned close to the curves.

another energy scale, namely the interaction energy ϵ. Investigation of the interplay of different energy scales in detail is beyond the scope of this paper and will be presented elsewhere in the future.

Acknowledgments
This project was funded by the Deutsche Forschungsgemeinschaft (DFG, German Research Foundation) under Grant No. 189 853 844–SFB/TRR 102 (Project B04). It was further supported by the Deutsch-Französische Hochschule (DFH-UFA) through the Doctoral College "$\mathbb{L}^4$" under Grant No. CDFA-02-07, the Leipzig Graduate School of Natural Sciences "BuildMoNa", and the EU COST programme EUTOPIA under Grant No. CA17139.

References
[1] Doi M and Edwards S 1986 *The Theory of Polymer Dynamics* (Clarendon Press, Oxford)
[2] Halperin A and Goldbart P 2000 *Phys. Rev. E* **61** 565-73
[3] Majumder S, Christiansen H and Janke W 2020 *Eur. Phys. J. B* **93** 1-19
[4] Majumder S, Zierenberg J and Janke W 2017 *Soft Matter* **13** 1276-90
[5] Christiansen H, Majumder S and Janke W 2017 *J. Chem. Phys.* **147** 094902
[6] Elgeti J, Winkler R G and Gompper G 2015 *Rep. Prog. Phys.* **78** 056601
[7] Winkler R G and Gompper G 2020 *J. Chem. Phys.* **53** 040901
[8] Bianco V, Locatelli E and Malgaretti P 2018 *Phys. Rev. Lett.* **121** 217802
[9] Paul S, Majumder S and Janke W 2021 *Soft Materials* **19** 306-15
[10] Paul S, Majumder S, Das S K and Janke W 2022 *arXiv:*2108.00342 to appear in *Soft Matter* (in print)
[11] Vicsek T, Czirôk A, Ben-Jacob E, Cohen I and Shochet O 1995 *Phys. Rev. Lett.* **75** 1226-29
[12] Das S K, Egorov S A, Trefz B, Virnau P and Binder K 2014 *Phys. Rev. Lett.* **112** 198301
[13] Paul S, Bera A and Das S K 2021 *Soft Matter* **17** 645-54

Numerical simulations of spectroscopic properties in two-dimensional Mott insulator

Takami Tohyama[1], Kazuya Shinjo[1], Shigetoshi Sota[2], and Kenji Tsutsui[3]

[1] Department of Applied Physics, Tokyo University of Science, Tokyo 125-8585, Japan
[2] Computational Materials Science Research Team, RIKEN Center for Computational Science (R-CCS), Kobe, Hyogo 650-0047, Japan
[3] Synchrotron Radiation Research Center, National Institutes for Quantum and Radiological Science and Technology, Hyogo 679-5148, Japan

E-mail: tohyama@rs.tus.ac.jp

Abstract.
We numerically investigate spectroscopic properties for a two-dimensional Mott insulator described by a half-filled Hubbard model on the square lattice. Firstly, we examine the spectral shape of the optical conductivity by using time-dependent density-matrix renormalization group implemented by the Legendre polynomial expansion. We find a shape peak resembling to an excitonic peak at absorption edge, whose origin is attributable to antiferromagnetic correlation in the spin background. We next investigate momentum-dependent transient charge and spin dynamics after photoirradiation by using time-dependent Lanczos-type exact diagonalization. We find a temporal oscillation in the intensity of the dynamical spin structure factor, which exhibits an antiphase behavior between, for example, two orthogonal momentum directions parallel and perpendicular to the electric field of a pump pulse. Observing such antiphase oscillations will be a big challenge for time-resolved resonant-inelastic x-ray scattering experiments.

1. Introduction

Mott insulators exhibit characteristic photoexcited states different from band insulators. However, our understanding of photoexcited states in two-dimensional (2D) Mott insulators on square lattice is far from complete. This is due to theoretical difficulty in treating photoexcited carriers induced by photoirradiation, which strongly couple to localized spins in the background. Since clarifying the nature of photoexcited states is a typical many-body problem, large scale numerical simulations are useful.

In this paper, we focus on two topics related to photoexcitation in a 2D Mott insulator described by a half-filled Hubbard model on the square lattice. We first examine the spectral shape of the optical conductivity by using time-dependent density-matrix renormalization group (tDMRG). Whether an excitonic peak exists or not in absorption spectra for cuprate Mott insulators such as Nd_2CuO_4 is an unsettled question [1, 2]. From our tDMRG calculation of the optical conductivity, we find a shape peak resembling to an excitonic peak at absorption edge. The origin of the peak is attributed to antiferromagnetic (AF) correlation in the spin background. We next investigate momentum-dependent transient charge and spin dynamics after photoirradiation by using a time-dependent Lanczos-type exact diagonalization (ED) technique. This work is motivated by the recent development of time-resolved resonant-inelastic x-ray scattering (trRIXS) in photoexcited AF Mott insulators [3]. We find that the intensity of the dynamical spin structure factor temporally oscillates with frequencies determined by the energy of bimagnons in the AF Mott insulator. We also find an antiphase behavior in the oscillation between, for example, two orthogonal momentum directions parallel and perpendicular to the electric field of a pump pulse. Observing the antiphase oscillations will be a big challenge for trRIXS experiments.

This paper is organized as follows. The half-filled Hubbard model on the square lattice is introduced in Sec. 2. In Sec. 3, we calculate the optical conductivity of the Hubbard model by using tDMRG. Time-resolved dynamical structure factors after photoirradiation by a pump pulse is calculated by the Lanczos technique in Sec. 4. Finally, a summary is given in Sec. 5.

2. Hubbard model on square lattice

In order to describe Mott insulating state in cuprates, we take a single-band Hubbard model on the square lattice at half filling given by

$$\hat{H}_0 = -t_\mathrm{h} \sum_{i\delta\sigma} \hat{c}_{i\sigma}^\dagger \hat{c}_{i+\delta\sigma} - t_\mathrm{h}' \sum_{i\delta'\sigma} \hat{c}_{i\sigma}^\dagger \hat{c}_{i+\delta'\sigma} + U \sum_i \hat{n}_{i\uparrow}\hat{n}_{i\downarrow}, \tag{1}$$

where $\hat{c}_{i\sigma}^\dagger$ is the creation operator of an electron with spin σ at site i, number operator $\hat{n}_{i\sigma} = \hat{c}_{i\sigma}^\dagger \hat{c}_{i\sigma}$, $i+\delta$ $(i+\delta')$ represents the four first (second) nearest-neighbor sites around site i, and t_h, t_h', and U are the first nearest-neighbor hopping, the second nearest-neighbor hopping, and on-site Coulomb interaction, respectively.

We incorporate an external spatially homogeneous electric field applied along the x direction via the Peierls substitution in the hopping terms, $\hat{c}_{i\sigma}^\dagger \hat{c}_{i+\delta(\delta')\sigma} \to e^{-iA_x(t)\delta_x(\delta_x')} \hat{c}_{i\sigma}^\dagger \hat{c}_{i+\delta(\delta')\sigma}$, leading to the time-dependent Hamiltonian $H(t)$. Here, $\delta_x(\delta_x')$ is the x component of $\delta(\delta')$ and $A_x(t)$ is the x component of vector potential given by

$$A_x(t) = A_0 e^{-(t-t_0)^2/(2t_\mathrm{d}^2)} \cos[\omega_\mathrm{p}(t-t_0)], \tag{2}$$

where a Gaussian-like envelope centered at t_0 has a temporal width t_d and a central frequency ω_p. Hereafter we use $t_\mathrm{h} = 1$ as the energy unit and $1/t_\mathrm{h}$ as the time unit.

3. Optical conductivity calculated by tDMRG

We apply a pulse of an electric field along the x direction. The time evolution of an electric current along the x direction induced by the pulse is given by $j_x(t) \equiv -\langle\psi(t)|\frac{\partial H(t)}{\partial A_x(t)}|\psi(t)\rangle$ with the time-dependent wave function $|\psi(t)\rangle$. The optical conductivity $\sigma(\omega)$ defined by a linear-response coefficient between the applied electric field and the induced electric current reads [4]

$$\sigma(\omega) = \frac{1}{L}\frac{j_x(\omega)}{i(\omega+i\eta)A_x(\omega)}, \tag{3}$$

where $A_x(\omega)$ and $j_x(\omega)$ are the Fourier transforms of $A_x(t)$ and $j_x(t)$, respectively, η is a small positive number determining a broadening width of $\mathrm{Re}\sigma(\omega)$, and L is the total number of sites in a $L_0 \times L_0$ lattice $(L = L_0^2)$. In order to guarantee a linear response, we use a weak vector potential with $A_0 = 0.001$, $t_d = 0.02$ together with $t_0 = 1$ and $\omega_p = 10$.

We briefly explain our tDMRG implemented by the Legendre polynomial, which is used for obtaining the time evolution of the wave function of large clusters to which the Lanczos method cannot apply. The dynamics of the wave function $|\psi(t)\rangle$ is described by the time-dependent

Schrödinger equation. When a time step Δt between two wave functions is as small as $\Delta t = 0.02$ used in our calculations, the solution of the Schrödinger equation is given by

$$|\psi(t+\Delta t)\rangle = e^{-i\hat{H}(t)\Delta t}|\psi(t)\rangle. \tag{4}$$

One of the efficient techniques to calculate $\exp[-i\hat{H}(t)\Delta t]$ by using DMRG is the Suzuki-Trotter decomposition scheme [5]. However, this approach is restricted to one-dimensional systems. Another approach that we employ is the use of the Legendre polynomial $P_l(x)$;

$$e^{-i\hat{H}(t)\Delta t} = \int_{-\infty}^{\infty} d\omega \delta(\omega - \hat{H}(t)) e^{-i\omega\Delta t} \tag{5}$$

$$= \int_{-1}^{1} d\omega \sum_{l=0}^{\infty} \frac{2l+1}{2} P_l(\omega) P_l(\hat{H}_s(t)) e^{-i[\lambda_s(t)+\omega/w_s(t)]\Delta t} \tag{6}$$

$$\simeq C(t) \sum_{l=0}^{M} (-i)^l (2l+1) j_l\left(\frac{\Delta t}{w_s(t)}\right) P_l(\hat{H}_s(t)). \tag{7}$$

The δ function in (5) is expanded by the Legendre polynomial. $\hat{H}_s(t)$ in (6) is a scaled Hamiltonian whose eigenvalues are confined within the interval of $[-1, 1]$ by using $\hat{H}_s(t) \equiv w_s(t)[\hat{H}(t) - \lambda_s(t)]$ with scaling parameters $w_s(t)$ and $\lambda_s(t)$. The integration in (6) with respect to ω leads to the spherical Bessel function $j_l(x)$ of the first kind. The summation with respect to l in (7) is truncated by M, since dozens of terms are enough for a good convergence of the summation. $C(t)$ is a normalization factor.

$j_l(x)$ is obtained by a recurrence relation $j_{l+1}(x) = \frac{2l+1}{x} j_l(x) - j_{l-1}(x)$ with $j_0(x) = \frac{\sin x}{x}$ and $j_1(x) = \frac{1}{x}[-\cos x + \frac{\sin x}{x}]$. Since $\exp[-i\hat{H}(t)\Delta t]$ is applied to $|\psi(t)\rangle$, we can use the following recurrence relation for $P_l(\hat{H}_s(t))$:

$$\left[P_{l+1}(\hat{H}_s(t)) - \frac{2l+1}{l+1} x P_l(\hat{H}_s(t)) - \frac{l}{l+1} P_{l-1}(\hat{H}_s(t))\right] |\psi(t)\rangle = 0 \tag{8}$$

with $P_0(\hat{H}_s(t)) = 1$ and $P_1(\hat{H}_s(t)) = \hat{H}_s(t)$.

We use two target states $|\psi(t)\rangle$ and $|\psi(t+\Delta t)\rangle$ in the DMRG procedure for a given t to construct a basis set that can express wave functions in the time-dependent Hilbert space. With the two-target time-dependent DMRG procedure, we can calculate time-dependent physical quantities with high accuracy even when the Hamiltonian varies rapidly with time.

Figure 1 shows the real part of the optical conductivity for a 6×6 Hubbard lattice with open boundary conditions (BC) with $U = 5$ and $t'_h = 0$. A sweeping path in the DMRG process is denoted by a broken zigzag line in the inset of Fig. 1. We keep 4500 density-matrix eigenstates in our tDMRG and use a parameter set for the vector potential (2) with $A_0 = 0.001$, $t_d = 0.02$, $t_0 = 1$, and $\omega_p = 10$. There is a sharp peak at the absorption edge $\omega = 3t_h$ followed by broad continuum. The peak height decreases with increasing U accompanied by a shift toward higher energy (not shown here) [6]. The decrease of the height seems to be correlated to the reduction of AF exchange interaction $J = 4t_h^2/U$. In addition, introducing t'_h in the lattice reduces the height, which seems to be a consequence of weakening of AF correlation due to frustration effect by the second nearest-neighbor AF interaction $J' = 4t'^2_h/U$ [6]. Judging from these observations, we can conclude that the sharp peak in Fig. 1 is predominantly originated from AF correlation in the system. In other words, the peak is produced by an excitonic effect due to a bound state of a holon composed of unoccupied site and a doublon composed of doubly occupied site, as a consequence of magnetic energy gain. We note that the excitonic state is weekly bound since there is no clear gap between the peak and continuum in $\mathrm{Re}\sigma(\omega)$ (see Fig. 1).

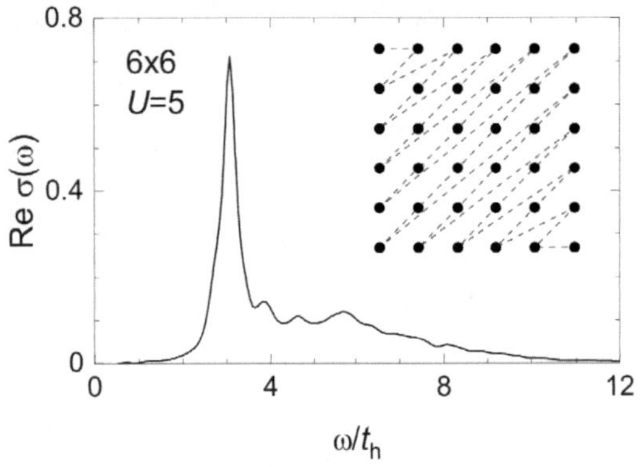

Figure 1. Real part of the optical conductivity of the half-filled Hubbard model for the 6×6 lattice with open BC. $U = 5$ and the broadening factor $\eta = 0.2$. The inset shows the 6×6 lattice whose snake-like sweeping path in the DMRG process is denoted by a broken zigzag line.

4. Time-resolved dynamical structure factors calculated by Lanczos method

Using eigenvalue $\epsilon_n(t)$ and eigenvector $|\phi_n(t)\rangle$ of $\hat{H}(t)$, we can express the time-dependent wave function (4) as

$$|\psi(t+\Delta t)\rangle = \sum_n e^{-i\epsilon_n(t)\Delta t}|\phi_n(t)\rangle\langle\phi_n(t)|\psi(t)\rangle. \tag{9}$$

In an ED technique based on the Lanczos method, we can generate eigenvalue $\tilde{\epsilon}_n(t)$ and eigenvector $|\tilde{\phi}_n(t)\rangle$ for M-dimensional Krylov subspace by diagonalizing a tridiagonal matrix generated in the Lanczos iteration applied for $\hat{H}(t)$ starting from an initial state $|\psi(t)\rangle$. It is known that if one replaces $\epsilon_n(t)$ and $|\phi_n(t)\rangle$ in (9) by $\tilde{\epsilon}_n(t)$ and $|\tilde{\phi}_n(t)\rangle$, respectively, a good convergence is achieved for $M \sim 30$ [7].

Another technique to calculate (4) based on the ED-type scheme is the use of a Taylor expansion [8]:

$$|\psi(t+\Delta t)\rangle = \sum_{n=0}^{\infty} \frac{(-i\hat{H}(t)\Delta t)^n}{n!}|\psi(t)\rangle \equiv \sum_{n=0}^{\infty} |\varphi_n(t)\rangle. \tag{10}$$

Each term is iteratively obtained by $|\varphi_{n+1}(t)\rangle = -i\hat{H}(t)\Delta t/(n+1)|\varphi_n(t)\rangle$ starting from $|\varphi_0\rangle = |\psi(t)\rangle$. The summation can be truncated if the norm $\langle\varphi_n(t)|\varphi_n(t)\rangle$ is smaller than a critical value such as 10^{-14}.

In calculating the dynamical charge and spin structure factors and the K-edge RIXS spectrum in the two-dimensional Hubbard model, a real-time representation of time-dependent cross section has been used [9, 10]. Instead, here we use the spectral representation of dynamical structure factors by regarding $|\psi(t)\rangle$ as an initial state, which is easily formulated for the systems without time-dependent terms in the Hamiltonian. Applying this formalism to a time region after turning off the pump pulse, $t > t_{\rm off}$, where the Hamiltonian is time-independent, i.e., $\hat{H}_0$, we obtain the time-resolved dynamical structure factor with momentum $\mathbf{q}$ and frequency ω for a physical quantity O as

$$O(\mathbf{q},\omega;t) = \mathrm{Im}\int_0^{\infty} ds\, ie^{i(\omega+i\eta)s}\langle\psi(t)|\left[\hat{O}_{\mathbf{q}}(s),\hat{O}_{-\mathbf{q}}\right]|\psi(t)\rangle$$
$$= \mathrm{Im}\sum_{m,n}\frac{\langle\psi(t)|\hat{O}_{-\mathbf{q}}|m\rangle\langle m|\hat{O}_{\mathbf{q}}|n\rangle\langle n|\psi(t)\rangle - \langle\psi(t)|m\rangle\langle m|\hat{O}_{\mathbf{q}}|n\rangle\langle n|\hat{O}_{-\mathbf{q}}|\psi(t)\rangle}{\omega+\varepsilon_m-\varepsilon_n+i\eta}, \tag{11}$$

where $\hat{O}_{\mathbf{q}}(s) = e^{i\hat{H}_0 s}\hat{O}_{\mathbf{q}}e^{-i\hat{H}_0 s}$, $\hat{H}_0|m\rangle = \varepsilon_m|m\rangle$, and $|\psi(t)\rangle = e^{-i\hat{H}_0(t-t_{\rm off})}|\psi(t_{\rm off})\rangle$. We note that replacing $|\psi(t)\rangle$ by the ground state $|0\rangle = |\psi(-\infty)\rangle$ in (11) formally gives the equilibrium dynamical structure factor $O(\mathbf{q},\omega)$. We choose $\hat{O}_{\mathbf{q}} = \hat{S}^z_{\mathbf{q}} = \sum_i e^{-i\mathbf{q}\cdot\mathbf{R}_i}\hat{S}^z_i$ for the time-resolved dynamical spin structure factor $S(\mathbf{q},\omega;t)$ and $\hat{O}_{\mathbf{q}} = \hat{N}_{\mathbf{q}} = \sum_i e^{-i\mathbf{q}\cdot\mathbf{R}_i}\hat{N}_i$ for the time-resolved

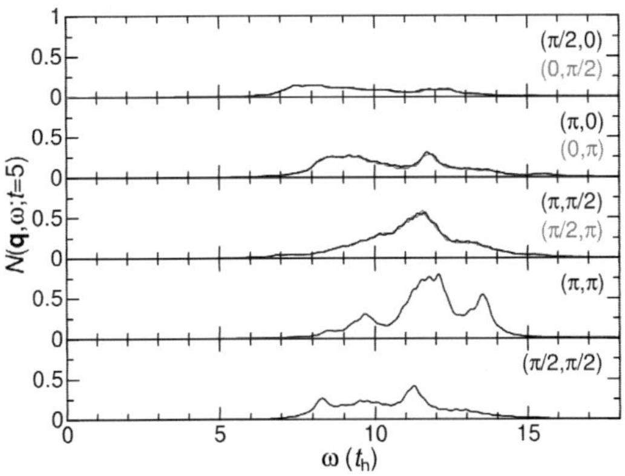

Figure 2. Dynamical charge structure factor at $t = 5$ after pumping for the half-filled 4×4 periodic Hubbard lattice with $t'_{\rm h} = -0.25$ and $U = 10$. Momenta $\mathbf{q}$ are shown in each panel and their color is the same as that of solid lines.

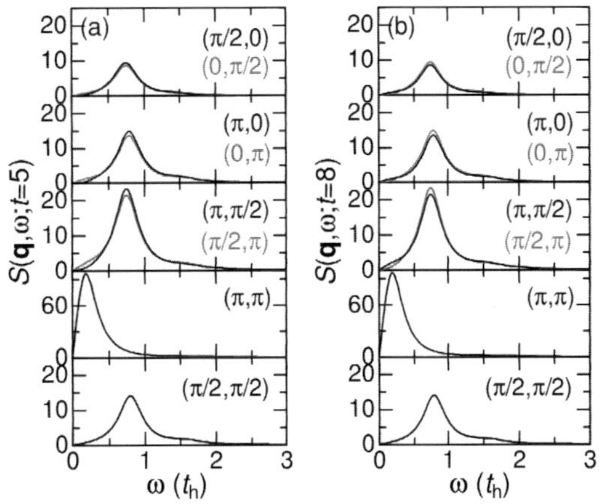

Figure 3. Dynamical spin structure factor at $t = 5$ (a) and $t = 8$ (b) after pumping for the half-filled 4×4 periodic Hubbard lattice with $t'_{\rm h} = -0.25$ and $U = 10$. Momenta $\mathbf{q}$ are shown in each panel and their color is the same as that of solid lines.

dynamical charge structure factor $N(\mathbf{q}, \omega; t)$, where $\hat{S}_i^z$ is the z component of a spin operator and $\hat{N}_i$ is an electron-number operator at site i.

We consider $\hat{H}(t)$ for a 4×4 lattice with periodic BC. We take $t'_{\rm h} = -0.25$ and $U = 10$, which are realistic values for cuprates. For the pump pulse, we set $A_0 = 0.5$, $t_d = 0.5$, $t_0 = 0$, and $\omega_{\rm p} = 20$. Since $\omega_{\rm p}$ is larger than the upper bound of optical absorption spectrum, which is around $\omega = 16$ for the same 4×4 Hubbard lattice [1], the energy absorbed into the system is small. This means that the change of the dynamical charge and spin structure factors before and after pumping is expected to be very small.

Figure 2 shows the dynamical charge structure factor at $t = 5$, $N(\mathbf{q}, \omega; t = 5)$. The spectra are almost the same as $N(\mathbf{q}, \omega)$ before pumping as expected from large $\omega_{\rm p}$, exhibiting momentum-dependent charge excitation across the Mott gap around $\omega = 7t_{\rm h}$. A small difference of spectral weight appears at originally equivalent momenta before pumping, for example, $\mathbf{q} = (\pi, 0)$ and $(0, \pi)$. Since the electric field is applied along the x direction, the appearance of such a difference makes sense even after the field is turned off. We note that if $\omega_{\rm p}$ is set within finite absorption region, photoinduced spectral weights appear inside the Mott gap [11].

Figure 3 shows the dynamical spin structure factor at $t = 5$ and $t = 8$. At $t = 5$, the spectral distribution is similar to that before pumping and a difference of weight appears at originally equivalent $\mathbf{q}$ points such as $(\pi, 0)$ and $(0, \pi)$ as for the case of charge dynamics. The weight at $(\pi, 0)$ is slightly larger than that of $(0, \pi)$. An interesting change occurs when t is put forward $t = 8$: the weight at $(\pi, 0)$ is now smaller than that of $(0, \pi)$. Other pairs of momenta $\mathbf{q} = (\pi/2, 0), (0, \pi/2)$ and $\mathbf{q} = (\pi, \pi/2), (\pi/2, \pi)$ also show such a behavior. This means the presence of oscillations of spectral weight with t for momenta originally equivalent but becoming inequivalent due to the applied electric field. A detailed investigation of the oscillation has lead to an antiphase type of oscillation between the originally equivalent momenta such as $(\pi, 0)$ and $(0, \pi)$, and the origin of the oscillation has been assigned as a Rabi-type one between the ground state and the Raman active B_{1g} state [11].

The oscillation period estimated form our calculation is $6.5/t_{\rm h}$, which is roughly 40fs. Recent trRIXS has reported the time resolution of 400fs. Therefore, the resolution should be improved one order of magnitude to detect the antiphase oscillation predicted by our calculation. We believe that recent rapid development of trRIXS will enable this improvement in the near future.

5. Summary

We have investigated two topics related to photoexcitation for a 2D Mott insulator described by a half-filled Hubbard model on the square lattice. One is the the spectral shape of the optical conductivity of the 2D Mott insulator. Using tDMRG implemented by the Legendre polynomial, we have found that there is a shape peak resembling to an excitonic peak at the absorption edge. The origin of the peak is attributed to AF spin correlation in the spin background. The line shape of the calculated optical conductivity is expected to be similar to experimental data for cuprate Mott insulator such as Nd_2CuO_4 once we introduces a proper broadening into the calculated spectrum [6]. Another topic is the momentum-dependent transient charge and spin dynamics in the 2D Mott insulator after photoirradiation. Using the Lanczos-type exact diagonalization technique, we have found that after turning off the pump pulse, the intensity of the dynamical spin structure factor temporally oscillates. The oscillation comes from a Rabi-type one between the ground state and the Raman-active B_{1g} state [11]. We have also found that an antiphase behavior in the oscillations between originally equivalent two momenta, for example, two orthogonal momentum directions parallel and perpendicular to the electric field of a pump pulse. We hope that trRIXS, which can observe the dynamical spin structure factor, will be able to detect these predicted oscillations in the near future.

Acknowledgments

This work was supported by CREST (Grant No. JPMJCR1661), the Japan Science and Technology Agency, by the Japan Society for the Promotion of Science, KAKENHI (Grants No. 19H01829, No. JP19H05825, No. 17K14148, No. 21H03455) from Ministry of Education, Culture, Sports, Science, and Technology (MEXT), Japan, by JST PRESTO (Grant No. JPMJPR2013), by MEXT HPCI Strategic Programs for Innovative Research (SPIRE; hp200071), and by QST President's Strategic Grant (QST Advanced Study Laboratory). Part of the numerical calculation was carried out using HOKUSAI at RIKEN, the supercomputer system at the information initiative center, Hokkaido University, the facilities of the Supercomputer Center at Institute for Solid State Physics, University of Tokyo, and the supercomputing facilities in QST.

References

[1] Tohyama T, Inoue Y, Tsutsui K and Maekawa S 2005 *Phys. Rev.* B **72** 045113
[2] Han X J, Liu Y, Liu Z Y, Li X, Chen J, Liao H J, Xie Z Y, Normand B and Xiang T 2016 *New J. Phys.* **18** 103004
[3] Dean M P M *et al.* 2016 *Nat. Mater.* **15** 601
[4] Shao C, Tohyama T, Luo H G and Lu H 2016 *Phys. Rev.* B **93** 195144
[5] White S R and Feiguin A E 2004 *Phys. Rev. Lett.* **93** 076401
[6] Shinjo K, Tamaki Y, Sota S and Tohyama T 2021 in preparation.
[7] Lu H, Sota S, Matsueda H, Bonča J and Tohyama T 2012 *Phys. Rev. Lett.* **109** 197401
[8] Shirakawa T, Miyakoshi S and Yunoki S 2020 *Phys. Rev.* B **101** 174307
[9] Wang Y, Claassen M, Moritz B and Devereaux T P 2017 *Phys. Rev.* B **96** 235142
[10] Wang Y, Chen Y, Jia C, Moritz B and Devereaux T P 2020 *Phys. Rev.* B **101** 165126
[11] Tsutsui K, Shinjo K and Tohyama T 2021 *Phys. Rev. Lett.* **126** 127404

Higher Order Auxiliary Field Quantum Monte Carlo Methods

Florian Goth
Universität Würzburg, Am Hubland, 97070 Würzburg, Germany
E-mail: `fgoth@physik.uni-wuerzburg.de`

Abstract. The auxiliary field quantum Monte Carlo (AFQMC) method has been a workhorse in the field of strongly correlated electrons for a long time and has found its most recent implementation in the ALF package (`alf.physik.uni-wuerzburg.de`). The utilization of the Trotter decomposition to decouple the interaction from the non-interacting Hamiltonian makes this method inherently second order in terms of the imaginary time slice. We show that due to the use of the Hubbard-Stratonovich transformation (HST) a semigroup structure on the time evolution is imposed that necessitates the introduction of a new family of complex-hermitian splitting methods for the purpose of reaching higher order. We will give examples of these new methods and study their efficiency, as well as perform comparisons with other established second and higher order methods in the realm of the AFQMC method.

1. Introduction

The AFQMC method has a distinguished place among the numerical methods of solid state physics for producing large scale benchmark data. Its most recent implementation is the Algorithms for Lattice Fermions (ALF) package (`alf.physik.uni-wuerzburg.de`). The use of the Trotter product formula [1] to decouple the interaction part V of the Hamiltonian $H = T+V$ from the non-interacting kinetic energy T makes this method inherently second order in terms of the imaginary time discretization $\Delta\tau$. This approximation was initially sufficient and provides a convergent approximation of all relevant quantities. Nevertheless it is of great interest to have alternatives to the established Trotter splitting that enable a better control of the discretization error as well as a better approximation of symmetries. This motivates the search for higher order methods that are efficiently usable in the AFQMC method. The AFQMC method dates back to Ref. [2] as a method for the simulation of lattice models subject to on-site interactions. The algorithm has matured and is available in two open source projects [3, 4] that allow the reliable simulation of quantum lattice Hamiltonians as well as the extraction of relevant observables. The AFQMC method can now be used for exploring the exponentially growing many-particle configuration space in polynomial time while being numerically exact: investing more CPU time will decrease the error. The notion of operator splittings was shown with the Trotter product formula [1] and work on difference schemes for partial differential equations lead to the symmetric second order splitting by Strang [5]. For a detailed account and information on the development of splitting methods we refer to Ref. [6]. Although Suzuki has explicitly expressed the hope in [7] that his methods will be of use in Monte Carlo simulations, in solid-state physics and the AFQMC method in particular there has not been a widespread use of higher order approximations to the partition function Z. All simulations in this paper have been done with a fork of the ALF package version 1.0 [3] called splitALF [1].

2. Derivation

2.1. Splitting methods

Given a vector ψ from a Hilbert space $\mathcal{H}$ and the autonomous differential equation $\dot{\psi}(\tau) = -H\psi(\tau)$ with the Hamilton operator H we can write its solution as $\psi(\tau) = U_H(\tau, 0)\psi(0)$ with the time evolution operator $U_H(\tau) = \exp(-\tau H)$. If now H has an additive substructure, where we will restrict ourselves for the purpose of this paper to the case of two non-commuting constituents, T and V, $H = T + V$, a first order approximation in $\Delta\tau$ is given by the expression

$$\begin{aligned} U_H(\tau) = e^{-\tau(T+V)} &= \exp(-\tau T)\exp(-\tau V) + \mathcal{O}(\tau^2) \\ &= U_T(\tau)U_V(\tau) + \mathcal{O}(\tau^2) \end{aligned} \tag{1}$$

where it is now hoped that the individual flows of T and V, U_T and U_V are in a to be precised way easier to calculate than the full problem $U_H(\tau)$. Of course we do not have to stop at a single application of T and V but we can study splittings that consist of multiple of these stages, and we obtain the splitting method X_sp defined by the tuple of parameters $\mathrm{X}_sp = (\vec{t}, \vec{v})$

$$U_H^{X_sp}(\tau) = \prod_{k=1}^{s+1} U_T(t_k\tau)U_V(v_k\tau) + \mathcal{O}(\tau^{p+1}) \tag{2}$$

where $s+1$ counts the number of stages, i.e. the number of evaluations of T and V. p defines the order of the method. For the method to be at least first order accurate, the consistency equations, $\sum_k^{s+1} t_k = \sum_k^{s+1} v_k = 1$, have to hold.

2.2. The Hubbard model

The physical setting is determined by the Hamilton Operator H that is acting on the fermionic Hilbert space and is in our case given by the Hubbard model, a model often studied with AFQMC (see [3, 8] and references therein). The Hubbard Hamiltonian $H = T + V$, is a very simple model that considers electrons on a lattice with kinetic energy T and interaction V. From this we have the representation which is used in ALF,

$$H = T - U\sum_{i=1}^{L_N} V_i^2 = \sum_{i,j=1}^{L_N}\sum_{\sigma} c_{i\sigma}^{\dagger} T_{ij} c_{j\sigma} - \frac{U}{2}\sum_{i=1}^{L_N}\left(\sum_{\sigma}\sigma n_{i\sigma}\right)^2 \tag{3}$$

with $U > 0$, $[V_i, V_j] = 0$, $n_{i\sigma} = c_{i\sigma}^{\dagger}c_{i\sigma}$, and $c_{i\sigma}$ ($c_{i\sigma}^{\dagger}$) denoting the annihilation (creation) operators that annihilate (create) an electron in state i with spin σ. σ enumerates the two spin directions $\uparrow, \downarrow$. We study this model on a two-dimensional regular square lattice of linear length L_N with periodic boundary conditions, hence $T_{ij} = \delta_{i,j+1} + \delta_{i+1,j}$.

2.3. The AFQMC method

For our purposes the partition function Z of a physical system determined by the Hamilton operator H is defined as the trace taken in the many particle Fock space over the fermionic degrees of freedom of the time evolution operator U_H, $Z = \mathrm{Tr}\,(U_H(\beta)) = \mathrm{Tr}\left(e^{-\beta H}\right)$ with inverse temperature β. We split up the propagation into L_τ time slices

$$Z = \mathrm{Tr}(U_H(\beta)) = \mathrm{Tr}\left((U_H(\Delta\tau))^{L_\tau}\right). \tag{4}$$

[1] splitALF can be obtained from `https://git.physik.uni-wuerzburg.de/fgoth/splitALF`.

Since we only have two operators we can approximate the exact U_H with the approximate flow of a method $\mathrm{X}_s p$ from (2) to obtain

$$\begin{aligned} Z &= \mathrm{Tr}\left(\prod_{\tau=1}^{L_\tau}\prod_{k=1}^{s+1} U_T(t_k\Delta\tau)U_V(v_k\Delta\tau)\right) + \mathcal{O}(\Delta\tau^p) \\ &= \mathrm{Tr}\left(\prod_{\tau=1}^{L_\tau}\prod_{k=1}^{s+1} e^{-\Delta\tau t_k T} e^{\Delta\tau v_k U \sum_{i=1}^{L_N} V_i^2}\right) + \mathcal{O}(\Delta\tau^p) \\ &= Z^{\mathrm{X}_s p} + \mathcal{O}(\Delta\tau^p), \end{aligned} \tag{5}$$

creating the splitting method dependent approximation $Z^{\mathrm{X}_s p}$. To reduce the interaction which is quartic in the fermion operators to a quadratic form we use on every lattice site, imaginary time slice and for every stage a discrete approximation to the HST of order q, and obtain

$$\exp(v_k\Delta\tau U V_i^2) = \sum_{l_{i,\tau,k}=1}^{N(q)} w(l_{i,\tau,k}) e^{\sqrt{v_k\Delta\tau U}x(l_{i,\tau,k})V_i} + \mathcal{O}\left(\Delta\tau^q\right). \tag{6}$$

$w(l_{i,\tau,k})$ denotes the weights and $x(l_{i,\tau,k})$ the abscissas of a suitable chosen quadrature rule of order $q = p + 1$ and hence $N(q)$ integration nodes. We arrive for the partition function at

$$Z = \sum_{C\in\mathcal{C}}\left[\prod_{i,\tau,k=1} w(l_{i,\tau,k})\right] \mathrm{Tr}\left(\prod_{\tau}^{L_\tau}\prod_{k=1}^{s+1} e^{-\Delta\tau t_k T}\prod_{i=1}^{L_N} e^{\sqrt{\Delta\tau v_k U}x(l_{i,\tau,k})V_i}\right) + \mathcal{O}\left(\Delta\tau^p\right). \tag{7}$$

$\sum_{C\in\mathcal{C}}$ denotes a summation over the entire space $\mathcal{C}$ of configurations C that we will later on sample stochastically. $\mathcal{C}$ is finite dimensional and discrete:

$$\mathcal{C} = \{l_{i,\tau,k}\} \quad : \quad 1 \le i \le L_N, 1 \le \tau \le L_\tau, 1 \le k \le s+1 \tag{8}$$

and each integer $l_{i,\tau,k}$ has values given by the order q of the Gauss-Hermite quadrature. The time evolution of the interaction operator is now determined by $\sqrt{v_k\Delta\tau U}$. For a well-defined Monte Carlo sampling over positive probability densities we want the quantities under the square root to be real. Using the identity proven in [9] valid for matrices $A^{(k)}$ with entries $A_{ij}^{(k)}$

$$\mathrm{Tr}\left(\prod_k \exp\left(\sum_{ij} c_i^\dagger A_{ij}^{(k)} c_j\right)\right) = \det\left(1 + \prod_k e^{A^{(k)}}\right) \tag{9}$$

that enables us to map the trace over the fermionic Fock space on the left hand side to a determinant of a product of exponentiated matrices on the right hand side, we arrive at $Z = \sum_{C\in\mathcal{C}} P(C) + \mathcal{O}\left(\Delta\tau^p\right)$ with the probability density $P(C)$ and

$$P(C) = \left[\prod_{i,\tau,k} w(l_{i,\tau,k})\right] \det\left(1 + \coprod_{\tau=1}^{L_\tau} B_\tau^{\mathrm{X}_s}\right) \quad \text{and} \tag{10}$$

$$B_\tau^{\mathrm{X}_s} = \prod_{k=1}^{s+1} e^{-\Delta\tau t_k T}\prod_{i=1}^{L_N} e^{\sqrt{\Delta\tau v_k U}x(l_{i,\tau,k})V_i}. \tag{11}$$

2.4. Metropolis Monte Carlo and observables

We will very briefly mention in this section the relevant steps for the Metropolis update scheme and follow closely Ref. [9]. We construct a Markov chain Monte Carlo method to generate samples to evaluate Z. Starting from a configuration C (initially randomly chosen integration nodes), a move consists of changing the value of one integration node $x_{i,\tau,k} = x_n$ to another $x_{i,\tau,k} = x'_n$ to obtain a new configuration C' and accept according to the Metropolis-Hastings rule $r_{MH} = \min\left(P(C')/P(C), 1\right)$. This gives an ergodic Markov chain Monte Carlo method and we measure observables according to

$$\langle O \rangle = \frac{\mathrm{Tr}\left(U_H(\beta, 0)O\right)}{\mathrm{Tr}\left(U_H(\beta, 0)\right)} = \sum_C P(C)O(C) \tag{12}$$

where $O(C)$ is the numerical value of an observable O evaluated on the configuration C. Since each configuration C is equivalent to a non-interacting system, Wick's theorem can be used to find expressions for arbitrary physical observables [10, 3, 9]. Nevertheless, for later reference in this paper we have to mention the sign, or more generally, phase problem. To mitigate this issue one employs a reweighting procedure. This requires the notion of the sign σ of a configuration $\sigma(C) = P(C)/|P(C)|$. With that we obtain for an observable [3] the modified equation

$$\langle O \rangle = \frac{\sum\limits_C |P(C)|\sigma(C)O(C)}{\sum\limits_C |P(C)|\sigma(C)} \tag{13}$$

with the now real and non-negative probability distribution $|P(C)|$. The denominator in (13) $\sigma = \sum\limits_C |P(C)|\sigma(C)$, gives rise to the notion of an average sign of a particular simulation.

3. Comparison of second and higher order methods in the AFQMC algorithm

In this section we will study the scaling behaviour with respect to the step size $\Delta\tau$ of an important physical quantity, the total energy $E^X(\Delta\tau) = \langle H \rangle^X(\Delta\tau)$ on a very small test case where analytical results are available. The systematic deviation with respect to $\Delta\tau$ enables us to access the leading order error term. Depending on the precision that we are able to achieve for a particular class of simulations, a double logarithmic plot will expose the power law behaviour. In the case of simulations in complex arithmetic we will suppress any occuring artificial imaginary parts of the energy and the average sign, since they are small for the considered parameters. The intrinsic method dependent parameters of ALF that control its numerical stability have been chosen such that numerically stable simulations were performed with all methods.

3.1. Second order methods

In addition to the well-known Strang splitting $\mathrm{S}_1 2$ [5], we have the minimum-norm Integrator $\mathrm{S}_2 2$ by McLachlan [11]. Additionally, we have the family of second order methods given by Blanes *et al* in [12]. They are designed to minimize the energy error in HMC simulations and thereby optimize the acceptance rate. The associated methods $\mathrm{SE}_n 2$ are symmetric, of second order and have $n = 2, 3, 4$ evaluations of U_V.

Figure 1 gives an overview of the scaling of the second order methods with respect to the step size $\Delta\tau$. Since all involved splitting coefficients are positive the average sign is one. We see that the non-symmetric Euler approximation $\mathrm{NS}_1 1$ is of second order but with a big prefactor in comparison to the symmetric Strang splitting $\mathrm{S}_1 2$, an observation that already occured in [13]. The improved methods $\mathrm{SE}_n 2$ provide substantial benefits over the conventional Strang splitting. Combined with the fact that they can be easily expanded to splittings consisting of an arbitrary number of operators we highly recommend to use e.g. at least the $\mathrm{SE}_2 2$ method for the construction of the approximate partition function.

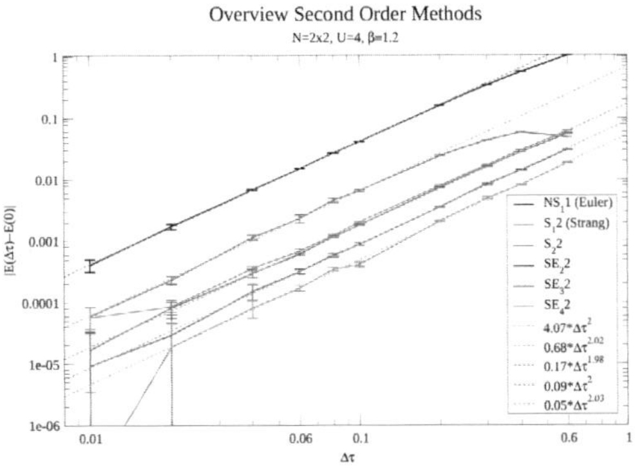

Figure 1. Scaling behaviour with respect to the step size $\Delta\tau$ of the considered symmetric second order methods studied on the systematic deviations of the total energy, $E(0) - E(\Delta\tau)$. The exact value $E(0) = -0.89200074$ is known from Exact Diagonalization calculations.

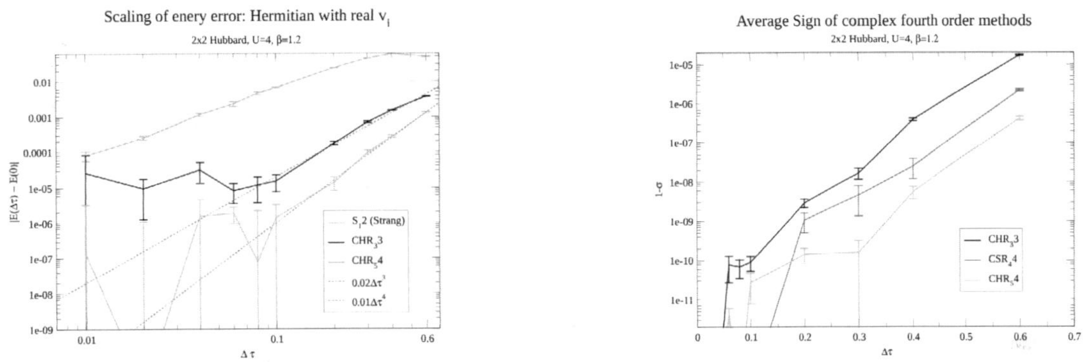

Figure 2. Comparison of the hermitian, but not time reversal symmetric methods $\mathrm{CHR}_3 3$ and $\mathrm{CHR}_5 4$ where $\vec{v}$ is real and positive. We see that within their error bars the methods approach the expected polynomial decay. We note that the resolution of the systematic $\Delta\tau$ error poses a numerical challenge for the high-order methods. This is another reason why this section only considers third and fourth order methods. Note that in the right panel we give $1 - \sigma$ instead of σ and we see that smaller $\Delta\tau$ leads to an improved average sign.

3.1.1. Particular Higher order methods In the following we report on some particular solutions that we considered. No attempt at an optimization in any particular kind has been done. We note that all given hermitian methods also fulfill the property $\Re(t_i) > 0$ and hence could be suitable for the study of evolution equations [14, 15]. $\mathrm{CHR}_3 3$ and $\mathrm{CHR}_5 4$ will be the two methods that we study deeper in this section; methods $\mathrm{CHR}_9 5$ and $\mathrm{CHR}_{15} 6$ will be considered briefly for comparison afterwards. Figure 2 gives the scaling behaviour of the error as well as the behaviour of the average sign. We see that the proposed methods are able to reach their expected scaling behaviour and introduce a slight sign problem on the toy example. Hence it is of interest to study the behaviour for larger β and bigger system sizes. Figure 3 studies the average sign with respect to a variation of the step-size $\Delta\tau$, the inverse temperature β and the system size N. We see that with an increase of β and N the average sign worsens exponentially. We observe that increases in the step size lead to a worse average sign, a property that has been noted for the AFQMC algorithm in [16]. All further details can be found in the preprint [17].

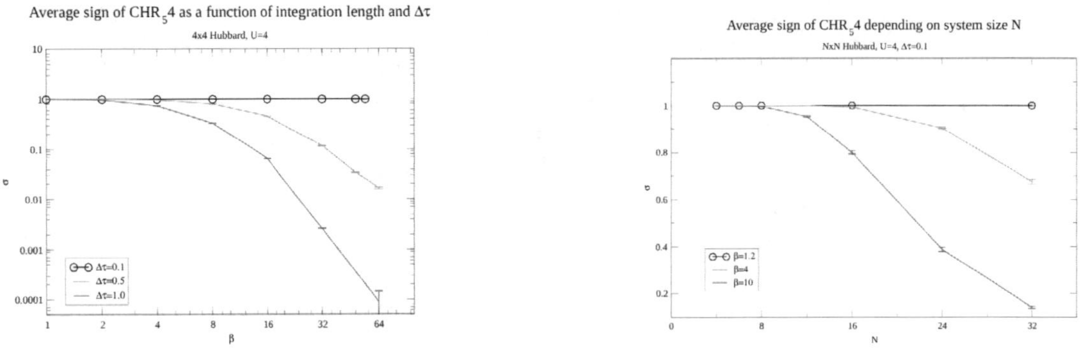

Figure 3. In the left panel we have the decay of the average sign for a 4×4 system at $U = 4$ with respect to β for three different discretizations. We see a strong dependence of the average sign on $\Delta\tau$ and obtain an exponential decay with respect to β for large $\Delta\tau$. The right panel shows the decay of the average sign with respect to the linear system size N for three values of β. We observe again a behaviour reminiscent of an exponential decay.

Acknowledgments

I thank the DFG for funding through the SFB 1170 "Tocotronics" under the grant number Z03. The author gratefully acknowledges the Gauss Centre for Supercomputing e.V. (www.gauss-centre.eu) for funding this project by providing computing time on the GCS Supercomputer SuperMUC at Leibniz Supercomputing Centre (www.lrz.de). The author gratefully acknowledges the compute resources and support provided by the Universität Würzburg IT Center and the German Research Foundation (DFG) through grant No. INST 93/878-1 FUGG. I acknowledge useful discussions with Prof. Fakher F. Assaad.

References

[1] Trotter H F 1959 *Proceedings of the American Mathematical Society* **10** 545–551
[2] Blankenbecler R, Scalapino D J and Sugar R L 1981 *Phys. Rev. D* **24** 2278–2286
[3] Bercx M, Goth F, Hofmann J S and Assaad F F 2017 *SciPost Phys.* **3**(2) 013
[4] Lee C R, Chiesa S, Varney C N, Khatami E, Bai Z, D'Azevedo E F, Jarrell M, Maier T, Savrasov S Y, Scalettar R T and Tomko K 2010 Quest: Quantum electron simulation toolbox URL `http://quest.ucdavis.edu/`
[5] Strang G 1968 *SIAM journal on numerical analysis* **5** 506–517
[6] McLachlan R I and Quispel G R W 2002 *Acta Numerica* **11** 341–434
[7] Suzuki M 1990 *Physics Letters A* **146** 319 – 323 ISSN 0375-9601
[8] Varney C N, Lee C R, Bai Z J, Chiesa S, Jarrell M and Scalettar R T 2009 *Phys. Rev. B* **80**(7) 075116 (*Preprint* `0903.2519v2`)
[9] Evertz H and Assaad F 2007 *World line and determinantal Quantum Monte Carlo methods for spins, phonons, and electrons* (*Lecture Notes in Physics* vol 739) (Springer) pp 277–356 1st ed ISBN 978-3-540-74685-0
[10] Bai Z, Chen W, Scalettar R and Yamazaki I 2009 *Multi-scale Phenomena in Complex Fluids: Modeling, Analysis and Numerical Simulation* ed Hou T Y (World Scientific)
[11] McLachlan R I 1995 *SIAM Journal on Scientific Computing* **16** 151–168
[12] Blanes S, Casas F and Sanz-Serna J M 2014 *SIAM Journal on Scientific Computing* **36** A1556–A1580
[13] Assaad F F and Herbut I F 2013 *Phys. Rev. X* **3**(3) 031010
[14] Blanes S, Casas F and Murua A 2010 *SeMA Journal* **50** 47–60 (*Preprint* `1001.1549`)
[15] Hansen E and Ostermann A 2009 *BIT Numerical Mathematics* **49** 527–542 ISSN 0006-3835
[16] Mikelsons K, Macridin A and Jarrell M 2009 *Physical Review E* **79** 057701 (*Preprint* `0903.0559`)
[17] Goth F 2020 Higher order auxiliary field quantum monte carlo methods (*Preprint* `2009.04491`)

Journal of Physics: Conference Series **2207** (2022) 012030 doi:10.1088/1742-6596/2207/1/012030

QMC study of the chiral Heisenberg Gross-Neveu universality class

Yuichi Otsuka[1,2], Kazuhiro Seki[2], Sandro Sorella[1,3,4], and Seiji Yunoki[1,2,5,6]

[1]Computational Materials Science Research Team, RIKEN Center for Computational Science (R-CCS), Kobe, Hyogo 650-0047, Japan
[2]Quantum Computational Science Research Team, RIKEN Center for Quantum Computing (RQC), Saitama 351-0198, Japan
[3]SISSA – International School for Advanced Studies, Via Bonomea 265, 34136 Trieste, Italy
[4]Democritos Simulation Center CNR–IOM Instituto Officina dei Materiali, Via Bonomea 265, 34136 Trieste, Italy
[5]Computational Quantum Matter Research Team, RIKEN Center for Emergent Matter Science (CEMS), Wako, Saitama 351-0198, Japan
[6]Computational Condensed Matter Physics Laboratory, RIKEN, Wako, Saitama 351-0198, Japan

E-mail: otsukay@riken.jp

Abstract. We investigate a quantum criticality of an antiferromagnetic phase transition in the Hubbard model on a square lattice with a d-wave pairing field by large-scale auxiliary-field quantum Monte Carlo simulations. Since the d-wave pairing filed induces Dirac cones in the non-interacting single-particle spectrum, the quantum criticality should correspond to the chiral Heisenberg universality class in terms of the Gross-Neveu theory, which is the same as those expected in the Hubbard model on the honeycomb lattice, despite the unit cells being different (e.g., they contain one and two sites, respectively). We show that both the two phase transitions, expected to occur on the square and on the honeycomb lattices, indeed have the same quantum criticality. We also argue that details of the models, i.e., the way of counting the total number N of fermion components and the anisotropy of the Dirac cones, do not change the critical exponents.

1. Introduction

Accurate calculation of critical exponents of phase transitions to identify universality classes is one of the major issues in computational physics, especially for Monte Carlo simulations. There are many successful examples found in the studies of classical [1, 2, 3] and quantum [4, 5, 6, 7] spin systems. Recently, another family of the quantum phase transitions that involve fermionic degrees of freedom has attracted much attention in this context because sign-problem-free quantum Monte Carlo (QMC) methods as well as modern analytical techniques such as functional renormalization-group approach are applicable to investigate the fermionic quantum criticality.

The most well-studied example would be the Hubbard model on the honeycomb lattice (hereinafter referred to as the honeycomb lattice model), of which the antiferromagnetic (AF) semimetal(SM)-insulator phase transition was examined from the view point of the Mott transition [8, 9, 10, 11, 12]. Later, since the connection between the effective theory of the honeycomb lattice model and the Gross-Neveu (GN) model in high-energy physics was pointed out [13, 14], the focus has been shifted from the model-dependent quantum phase transitions to the universal nature of the quantum criticality because the existence of the universality classes for the interacting Dirac fermions had been well formulated by the GN model [15]. For the AF transition, which corresponds to the chiral-Heisenberg universality class in terms of the GN theory, reliable estimations of the critical exponents have been obtained employing the honeycomb lattice model [16, 17, 18] and the Hubbard model on the square lattice with π-flux [19, 20] (referred to as the π-flux model in the following).

In this paper, we revisit the chiral-Heisenberg universality class using another lattice model, the square-lattice Hubbard model with a d-wave pairing field [21], which we call the d-SC model. If we consider brick-wall square lattices [22, 23], it can be seen that the honeycomb lattice model and the π-flux model are smoothly connected to each other even without assuming the universality class. On the other hand, the present d-SC model is quite different from these two models as discussed in the next section, while the low-energy effective model is the same for all the three models, i.e., the GN model breaking the $SU(2)$ symmetry. Therefore, the examination of the critical exponents for the d-SC model is expected to serve as an independent and nontrivial check of the previous estimations.

2. Model and method

The Hamiltonian of the d-SC model reads as follows:

$$H = H_{\rm BCS} + H_U, \tag{1}$$

where

$$H_{\rm BCS} = \sum_{\langle i,j \rangle} \left\{ \begin{pmatrix} c_{i\uparrow}^\dagger & c_{i\downarrow} \end{pmatrix} \begin{pmatrix} -t & \Delta_{ij} \\ \Delta_{ij}^* & t \end{pmatrix} \begin{pmatrix} c_{j\uparrow} \\ c_{j\downarrow}^\dagger \end{pmatrix} + {\rm h.c} \right\} \tag{2}$$

and

$$H_U = U \sum_i n_{i\uparrow} n_{i\downarrow}. \tag{3}$$

Here, t is the transfer integral between the nearest sites, chosen as an energy unit (t=1), and Δ_{ij} denotes the d-wave pairing field with its amplitude being uniform, i.e., $|\Delta_{ij}| = \Delta$. We consider the model at half filling. The Hubbard interaction denoted by U triggers the AF transition at the strong coupling regime, where the mass gap opens. In the noninteracting limit ($U = 0$) for finite Δ, the ground-state is the SM phase having four Dirac points at the Fermi level in the momentum space. This state is different from those of the honeycomb lattice or the π-flux model which has the two Dirac cones. In addition, the unit cell of the d-SC model has one site, whereas the honeycomb lattice or the π-flux model has two sublattices. Thus, at the level of the lattice model, they are indeed different. However, in the low energy continuum limit, these models should be described by the GN model with the same number of the fermion components, $N = 8$, with the spin degrees of freedom considered. This is why we expect the same universality class for the different lattice models. Furthermore, the d-SC model has another unique feature that an anisotropy of the Dirac cone can be tuned by changing Δ. It is isotropic as in the case of the honeycomb lattice or the π-flux model only at $\Delta = 1$, and otherwise the velocity at the Dirac point depends on the direction in the momentum space. Taking advantage of this feature, we also study whether the anisotropy affects the quantum criticality.

Since the square lattice is bipartite and the particle-hole symmetry holds at half filling, we investigate the d-SC model by the auxiliary-field QMC method without facing the negative-sign

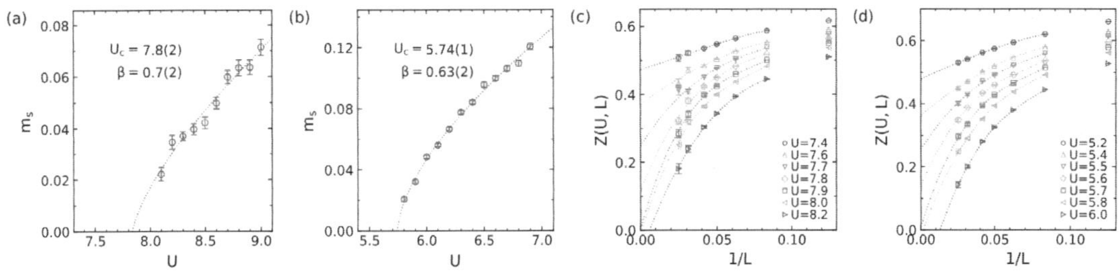

Figure 1. U-dependence of the AF order parameter for (a) $\Delta = 1.0$ and (b) $\Delta = 0.5$, and $1/L$-extrapolation of the quasiparticle weight for (c) $\Delta = 1.0$ and (d) $\Delta = 0.5$.

problem [24, 25, 26]. The ground-state expectation value of a physical observable O is evaluated by projection from a left (right) trial wave function $\langle\psi_{\rm L}|$ ($|\psi_{\rm R}\rangle$) as

$$\langle O\rangle = \frac{\langle\psi_{\rm L}|e^{-\frac{\tau}{2}H}Oe^{-\frac{\tau}{2}H}|\psi_{\rm R}\rangle}{\langle\psi_{\rm L}|e^{-\tau H}|\psi_{\rm R}\rangle}, \tag{4}$$

where τ is projection time and is divided by the Suzuki-Trotter decomposition into $\tau/M = \Delta\tau$ with M being integer. We set τ to be proportional to linear dimension of the square lattice L and chose $\Delta\tau = 0.1$ to reduce the systematic errors compared to the stochastic errors. The simulations are performed on finite-size clusters of L=8, 12, 16, 20, 24, 32, 40 with periodic boundary conditions for the isotropic ($\Delta = 1.0$) and anisotropic ($\Delta = 0.5$) cases. Therefore, the Dirac points at $(\frac{\pi}{2}, \frac{\pi}{2})$ and other symmetry equivalent momenta are allowed in all these clusters.

3. Results

We calculate the spin structure factor $S(\boldsymbol{k}) = L^{-2}\sum_{i,j} e^{i\boldsymbol{k}(\boldsymbol{r}_i-\boldsymbol{r}_j)}\langle \boldsymbol{S}_i \cdot \boldsymbol{S}_j\rangle$ in the standard notation and the quasiparticle weight $Z(U,L) = D_\sigma(U,L)/D_\sigma(0,L)$ estimated from the equal-time Green's function $D_\sigma(U,L) = L^{-2}\sum_i\langle c^\dagger_{j\sigma}c_{i\sigma}\rangle$ at the maximum distance [27]. Then, the obtained data are analyzed; by a conventional method to first extrapolate the results to the thermodynamics limit ($1/L \to 0$); by a more sophisticated method called the crossing-point analysis [6]; and by a rather involved method of data collapse.

The critical point $U_{\rm c}$ dividing the SM and the AF insulator is in principle obtained as the value of U at which the order parameter sets in. In Figs. 1(a) and 1(b), we plot the staggered magnetization, that is, the AF order parameter, calculated as $m_s = \lim_{1/L\to 0}\sqrt{S(\pi,\pi)/L}$ as a function of U. From these plots, the critical points and the critical exponent β are estimated by fitting with the critical behavior $m_s \sim (U/U_{\rm c}-1)^\beta$ as $U_{\rm c}$=7.8(2) [5.74(1)] and β=0.7(2) [0.63(2)] for Δ=1.0 (0.5). Since the statistical errors are larger for the strong-coupling region, the error bars for Δ=1.0 are large, which makes it difficult to safely conclude that the exponents are consistent between Δ=1.0 and 0.5 or between the d-SC model and the honeycomb lattice or the π-flux model. We also find large error bars for the quasiparticle weight as shown in Figs. 1(c) and 1(d). In addition, it is not obvious which fitting function is suited both for the SM and the insulating phase to extrapolate $Z(U,L)$ to the thermodynamics limit $1/L = 0$ at each U. We are thus not able to plot the quasiparticle weight as the function of U, although the rough estimates of $U_{\rm c}$ from Figs. 1(c) and 1(d) seem consistent with those obtained from m_s.

Based on the previous studies of the honeycomb lattice model [10, 11, 12], it is anticipated that the conventional method to deal directly with the order parameters discussed above tends to overestimate the critical points. Furthermore, it is not trivial to take into account a possible contribution from correction terms to the simplest scaling ansatz. Thus, here

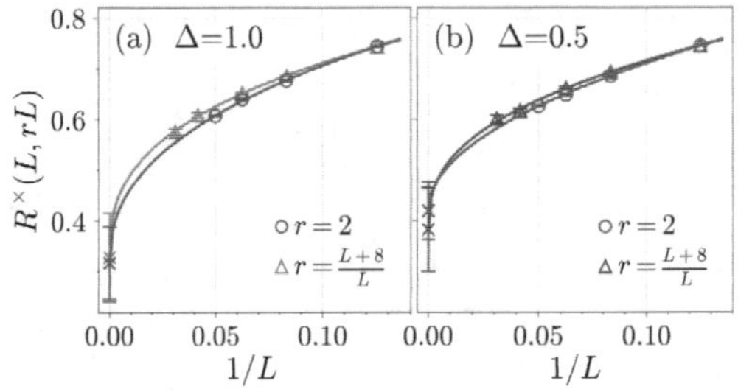

Figure 2. $1/L$-extrapolation of crossing points of $R_{m^2}(U, L)$ for (a) $\Delta = 1.0$ and (b) $\Delta = 0.5$.

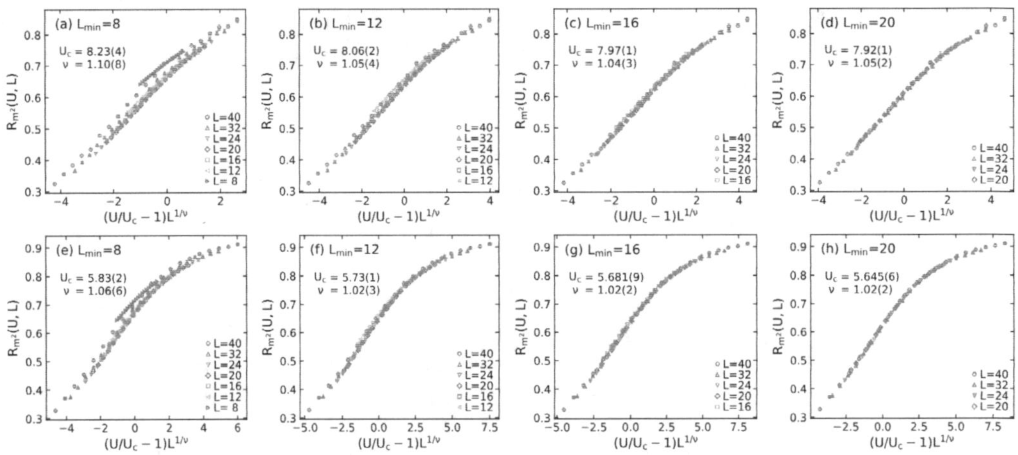

Figure 3. Data-collapse fits of $R_{m^2}(U, L)$ for various $L_{\min}$; (a)-(d) $\Delta = 1.0$ and (e)-(f) $\Delta = 0.5$.

we take a more sophisticated approach. First, we calculate the correlation ratio defined as $R_{m^2}(U, L) = 1 - \frac{S(\boldsymbol{K}+\boldsymbol{b}/L)}{S(\boldsymbol{K})}$, where $\boldsymbol{K} = (\pi, \pi)$ is the AF ordering momentum, and $\boldsymbol{b}$ is the smallest reciprocal-lattice vector [28]. Like the Binder ratio, this quantity has the advantage of being size-independent at the critical point when the correction terms are negligible. Conversely, we can know that the effects of the correction are non-negligible if we observe that curves of $R_{m^2}(U, L)$ as the function of U cross at different points of U for various L. Since such a drift is indeed noticed in our results, we employ the crossing-point analysis; we determine the crossing points $U^{\times}(L, rL)$ at which the curves of L and rL cross and extrapolate them to $1/L = 0$ assuming the critical behavior of $U^{\times}(L, rL) = U_{\rm c} + cL^{-(\omega+1/\nu)}$, where c is a constant, ω is an effective correction exponent, and ν is the correlation-length exponent [6]. It turns out that this analysis yields reasonable estimates of the critical points and the exponents to confirm the expected universal nature of the quantum criticality [21]. Additionally, in Fig. 2, we show that $R^{\times}(L, rL)$, the values of $R_{m^2}(U, L)$ at the crossing points, also follow a similar universal behavior of $R^{\times}(L, rL) = R^{\times}_{\rm c} + dL^{-\omega}$ with $R^{\times}_{\rm c}$=0.33(10) [0.40(7)] and ω =0.43(13) [0.46(11)] for Δ=1.0 (0.5).

Finally, we show details of the data-collapse fits of $R_{m^2}(U, L)$ in Fig. 3, where we collapse the data of $L_{\min} \leq L \leq L_{\max}(= 40)$ changing $L_{\min}$. Since we utilize the simplest form of the finite-size-scaling ansatz without the correction terms, $R_{m^2}(U, L) = f_R(uL^{1/\nu})$, where $f_R(\cdot)$ denotes the scaling function, and $u = (U - U_{\rm c})/U$, the non-negligible contribution from the correction terms are evident for smaller $L_{\min}$. The values of $U_{\rm c}$ and ν estimated from the collapse fits for each $L_{\min}$ can be extrapolated to the thermodynamics limit by plotting the data as a function of $1/L_{\min}$, and the results are consistent with those obtained by the crossing-point analysis within the error bars. Readers interested in more details of the critical exponents and comparison with analytical results are referred to Ref. [21].

4. Summary

We have investigated the square-lattice Hubbard model with a d-wave pairing field by large-scale auxiliary-field quantum Monte Carlo simulations with the aim to revisit the chiral-Heisenberg universality class. The critical exponents estimated by several ways are overall consistent with each other and also with those obtained in the previous studies of the Hubbard model on the honeycomb lattice and on the square lattice with π-flux, suggesting that all these models belong to the same chiral-Heisenberg universality class described by the Gross-Neveu model with the same number of the fermion components $N = 8$. We also confirm that the anisotropy of the Dirac cones does not affect the quantum criticality.

Acknowledgments

The authors thank F. F. Assaad, T. Sato, F. Parisen Toldin, and Z. Wang for valuable comments. This work has been supported by Grant-in-Aid for Scientific Research from MEXT Japan (under Grant Nos. JP18K03475, JP18H01183, JP19K23433, JP21H04446, and JP21K03395) and by PRIN2017 MIUR prot.2017BZPKSZ. The numerical simulations have been performed on K computer provided by the RIKEN Center for Computational Science (R-CCS) through the HPCI System Research project (Project IDs: hp170162 and hp170328), and the HOKUSAI supercomputer at RIKEN (Project IDs: G20006 and Q21525).

References

[1] Peczak P, Ferrenberg A M and Landau D P 1991 *Phys. Rev. B* **43** 6087
[2] Holm C and Janke W 1993 *Phys. Rev. B* **48** 936
[3] Campostrini M, Hasenbusch M, Pelissetto A, Rossi P and Vicari E 2002 *Phys. Rev. B* **65** 144520
[4] Wang L, Beach K S D and Sandvik A W 2005 *Phys. Rev. B* **73** 014431
[5] Wenzel S and Janke W 2009 *Phys. Rev. B* **79** 014410
[6] Shao H, Guo W and Sandvik A W 2016 *Science* **352** 213
[7] Ma N, Weinberg P, Shao H, Guo W, Yao D X and Sandvik A W 2018 *Phys. Rev. Lett.* **121** 117202
[8] Sorella S and Tosatti E 1992 *Europhys. Lett.* **19** 699
[9] Paiva T, Scalettar R T, Zheng W, Singh R R P and Oitmaa J 2005 *Phys. Rev. B* **72** 085123
[10] Meng Z Y, Lang T C, Wessel S, Assaad F F and Muramatsu A 2010 *Nature* **464** 847
[11] Sorella S, Otsuka Y and Yunoki S 2012 *Sci. Rep.* **2** 992
[12] Otsuka Y, Yunoki S and Sorella S 2013 *J. Phys. Conf. Ser.* **454** 012045
[13] Herbut I F 2006 *Phys. Rev. Lett.* **97** 146401
[14] Assaad F F and Herbut I F 2013 *Phys. Rev. X* **3** 031010
[15] Rosenstein B, Kovner A, Hoi-Lai Yu and Kovner A 1993 *Phys. Lett. B* **314** 381
[16] Parisen Toldin F, Hohenadler M, Assaad F F and Herbut I F 2015 *Phys. Rev. B* **91** 165108
[17] Otsuka Y, Yunoki S and Sorella S 2016 *Phys. Rev. X* **6** 011029
[18] Ostmeyer J, Berkowitz E, Krieg S, Lähde T A, Luu T and Urbach C 2021 *Phys. Rev. B* **104** 155142
[19] Otsuka Y and Hatsugai Y 2002 *Phys. Rev. B* **65** 073101
[20] Otsuka Y, Yunoki S and Sorella S 2014 *JPS Conf. Proc.* **3** 013021
[21] Otsuka Y, Seki K, Sorella S and Yunoki S 2020 *Phys. Rev. B* **102** 235105
[22] Ixert D, Assaad F F and Schmidt K P 2014 *Phys. Rev. B* **90** 195133
[23] Hatsugai Y, Fukui T and Aoki H 2006 *Phys. Rev. B* **74** 205414
[24] Blankenbecler R, Scalapino D J and Sugar R L 1981 *Phys. Rev. D* **24** 2278
[25] Hirsch J E 1985 *Phys. Rev. B* **31** 4403
[26] White S R, Scalapino D J, Sugar R L, Loh E Y, Gubernatis J E and Scalettar R T 1989 *Phys. Rev. B* **40** 506
[27] Seki K, Otsuka Y, Yunoki S and Sorella S 2019 *Phys. Rev. B* **99** 125145
[28] Kaul R K 2015 *Phys. Rev. Lett.* **115** 157202

Journal of Physics: Conference Series **2207** (2022) 012031 doi:10.1088/1742-6596/2207/1/012031

Thermal pure quantum matrix product states: a simple numerical protocol for finite temperature

Atsushi Iwaki and Chisa Hotta

Department of Basic Science, University of Tokyo, Meguro-ku, Tokyo 153-8902, Japan

E-mail: iwaki-atsushi413@g.ecc.u-tokyo.ac.jp

Abstract. We present several features of a thermal pure quantum matrix product state (TPQ-MPS). TPQ-MPS can simulate finite temperature quantum many-body systems with a numerical cost comparable to the matrix product algorithm for the ground state in one-dimensional systems. Starting from a random matrix product state with auxiliary sites attached to the edges of the system, we successively operate the Hamiltonian with a shift-constant and obtain a series of TPQ-MPS that represents the microcanonical pure state at the highest temperature to the one at the lowest available temperature. How the parameters of the TPQ-MPS protocol influence the effective bond dimension χ of the matrix and the temperature distribution of these microcanonical states is examined.

1. Introduction

Although thermodynamic properties contain rich information of the phases of matter in laboratories, a meaningful comparison with theories is limited. This is basically because a thermal equilibrium "Gibbs state" is a mixed state in the standard ensemble framework whose exponentially large numbers of summation cannot be straightforwardly treated in sufficiently large systems. The stochastic ways to construct such ensemble like quantum Monte Carlo methods are available only in particular systems. A key idea for avoiding the mixture of exponentially large numbers of states had been to replace the Gibbs state by some single pure states in two different contexts; one is the thermofield dynamics that doubles the system by attaching the ancilla (bath) and purifies them altogether [1]. It would then efficiently give the proper mixed state of the subsystem through the process of tracing out the ancilla. A series of purification methods that rely on the ancilla construction has been developed [2, 3, 4, 5]. The other relies on the concept of typicality; in an isolated system, there is a single pure quantum state that can be considered as a replacement of a Gibbs state at that temperature in the sense that they give equivalent local physical quantities [6, 7, 8, 9, 10]. There are several variants of such states [11, 12, 13, 14, 15, 16, 17, 18, 19] which we overall regard as thermal pure quantum (TPQ) states. The standard way of constructing the TPQ state for temperature $k_BT = \beta^{-1}$ is to operate $e^{-\beta\hat{H}/2}$ for a given Hamiltonian $\hat{H}$ to the infinite temperature random state. The "exact" description of TPQ state requires a full Hibert space dimension which increases exponentially with system size N. Indeed, for a finite temperature exact diagonalization (ED) or its variants and microcanonical-TPQ (mTPQ) method, the calculation is possible for a system size of $N \lesssim 30$.

We recently developed a thermal pure quantum matrix product state (TPQ-MPS) which combines the mTPQ method and the matric product state (MPS) [20]. Previously, the MPS description of the TPQ state was considered to be inappropriate, since the MPS has an entanglement that follows an area law; the entanglement entropy $S_n = -\mathrm{Tr}(\rho_n \ln \rho_n)$ based on the reduced density matrix operator ρ_n of a subsystem size-n obtained by cutting the system into two parts is known to follow $S_n \le \ln \chi$, where χ is the bond dimension of the MPS which is a finite constant. This constraint does not deteriorate the description of the ground state in 1D that has the Schmidt coefficients of ρ_n distributing in an exponentially decaying manner. Whereas, a finite temperature TPQ state has $S_n \propto n$ by construction, which does not match the nature of MPS by definition. In fact, the major description of finite temperature state in MPS in the past years have focused on how to efficiently construct a mixture of pure states; for example, minimally entangled typically thermal state (METTS) built on a product state generated by the Markov chain process [21, 22]. Several developments of METTS, e.g. an idea to improve the efficiency of samplings in METTS were also proposed. These devices were required since the simplest use of MPS without importance sampling often requires a huge number of samples [23, 24, 25, 26]. However, our construction of TPQ-MPS is different from the standard MPS; it has auxiliary sites to both ends of the system, which work as entanglement baths (see the inset of Fig.1). We demonstrated that the relationship $S_n \propto n$ is kept up to $n \sim N$ when we prepare the subsystem-n at the center of the system. In fact, although there is still an upper bound $S_n \le 2\ln\chi$, the standard values of $\chi \sim 40 - 100$ used in the MPS calculation is sufficient to accommodate the required entanglement of the system at temperatures lower than the typical energy scale of the model such as exchange coupling of the magnetic moments.

The TPQ-MPS is an approximate description of the TPQ state (which we call full-TPQ state) in that we use a finite bond dimension χ and make use of the auxiliary degrees of freedom. We already notice that the quality of the numerical results of TPQ-MPS depends not much on χ but on the shift-constant l. This constant l is introduced in the full-TPQ calculation and plays an important role to reduce the numerical cost. Since our TPQ-MPS is an approximation, the clarification of the relationships between l and χ is needed, which will provide a clue to improve the quality of the results.

2. TPQ-MPS method

The TPQ-MPS has the following form with auxiliaries attached to both sides of the system,

$$|\psi\rangle = \sum_{\alpha_L\{i_n\}\alpha_R} \langle \mathrm{aux}^L_{\alpha_L}| A^{[1]i_1}\cdots A^{[N]i_N} |\mathrm{aux}^R_{\alpha_R}\rangle\, |\alpha_L, i_1\cdots i_N, \alpha_R\rangle, \tag{1}$$

where $A^{[m]i_m}$ is the $(\chi\times\chi)$ matrix on the m-th site with local degrees of freedom $i_m = 1,\cdots,d$, and $|\mathrm{aux}^{L/R}_{\alpha_{L/R}}\rangle$ is the right/left auxiliary state with $\alpha_{L/R} =$ 1-χ. The auxiliary states have no physical interactions with the main part of the system.

We first prepare an initial state $|\psi_0\rangle$ which is the random matrix product state (RMPS) that represents the infinite temperature state in principle [27, 28]. Its explicit form can be constructed by preparing a tensor $c_{\boldsymbol{i}j\boldsymbol{k}} \in \mathbb{C}$ that follows the complex Gaussian distribution and then by obtaining the canonical form via successive Schmidt-decomposition as [29]

$$|\psi_0\rangle = \sum_{\boldsymbol{i}j\boldsymbol{k}} c_{\boldsymbol{i}j\boldsymbol{k}} |\boldsymbol{i}\rangle |j\rangle |\boldsymbol{k}\rangle = \sum_{\boldsymbol{i}j\boldsymbol{k}} \sum_{\alpha\beta} \Gamma^{[L]\boldsymbol{i}}_{\alpha} \lambda^{[L]}_{\alpha} \Gamma^{[C]j}_{\alpha\beta} \lambda^{[R]}_{\beta} \Gamma^{[R]\boldsymbol{k}}_{\beta} |\boldsymbol{i}\rangle |j\rangle |\boldsymbol{k}\rangle \tag{2}$$

where j represents one degree of freedom and $\boldsymbol{i}$ and $\boldsymbol{k}$ represent sets of degrees of freedom on the left and right-hand side, respectively.

The following protocol is the same as the full-TPQ method [17, 18] except that the wave function is described in the form Eq.(1) and has a truncation process of keeping the bond dimension finite. Using Eq.(2) as an initial state ($k = 0$), we successively generate a series of unnormalized microcanonical TPQ-MPS state $k = 1, 2, \cdots$ as

$$|k\rangle = (l - \hat{h})^k |\psi_0\rangle. \tag{3}$$

where $\hat{h}$ is the Hamiltonian divided by N and l is a real constant larger than the maximal eigenvalue of $\hat{h}$, which is given by hand. Here, the k-th TPQ-MPS state has a micricaninical temperature approximated as $k_B T^{(k)} = N(l - u_k^*)/2k$ (with higher order correction given in Refs.[17, 30]). Its microcanonical energy is given as $u_k^* = \langle k|\hat{h}|k\rangle / \langle k|k\rangle$. The matrix product operator (MPO) $(l - \hat{h})$ has a bond-dimension D that depend on $\hat{h}$. Therefore, applying this MPO at the k-th step of iteration increases the matrix dimension from χ to $D\chi$. The matrix $(l - \hat{h})|k - 1\rangle$ is transformed to the canonical form, and then the dimension of the matrix is truncated to obtain $|k\rangle$. Previously in Ref.[20], we truncated this bond dimension to a fixed value $\chi \le \chi_{\rm ini}$, and kept this maximum value throughout the calculation. However, we found that such construction makes the numerical result depend quite much on the value of constant l in Eq.(3). Therefore, in this paper, we set the upper bound of χ to be free and reduce the bond dimension by keeping the following condition for the truncation error;

$$\sum_{\alpha > \chi} \lambda_\alpha^2. \tag{4}$$

The process of (i) operating $(l - \hat{h})$, (ii) taking a canonical form, and (iii) truncating the matrices is repeated until $|k\rangle$ reaches low enough temperature $\beta_{\rm max}^{-1}$. The advantage of this method over other MPS-based finite temperature methods is that the physical quantities $\langle \hat{A} \rangle$ for all temperatures between $\beta_{\rm min}^{-1} < \beta < \beta_{\rm max}^{-1}$ can be obtained immediately using the sets of $\langle k|\hat{A}|k\rangle$ and $\langle k|\hat{A}|k+1\rangle$ by the relationship [18],

$$\langle \beta|\hat{A}|\beta\rangle = e^{-N\beta l} \sum_{k=0} \left\{ \frac{(N\beta)^{2k}}{(2k)!} \langle k|\hat{A}|k\rangle + \frac{(N\beta)^{2k+1}}{(2k+1)!} \langle k|\hat{A}|k+1\rangle \right\} \tag{5}$$

with $|\beta\rangle = e^{-\beta \hat{H}/2} |\psi_0\rangle$. Here, $\beta_{\rm min}$ corresponds to the temperature at the first step $k_B T^{(1)}$. In fact, the usual canonical thermal pure state or related methods need to operate $e^{-\beta \hat{H}/2}$, which has a comparable cost with evaluating $\{|k\rangle\}$, for each temperature β^{-1} independently. There is a finite fluctuation of $\langle \hat{A} \rangle$ that depends on the initial choice of $|\psi_0\rangle$. In a full-TPQ state, this fluctuation of local operators is suppressed by $e^{-\Theta(N)}$ and is negligible for $N \gtrsim 30$, while for small-N, we need to take $N_{\rm samp}$ averages. In TPQ-MPS, the fluctuation depends on the initial bond dimension $\chi_{\rm ini}$, not on the system size. For $\chi_{\rm ini} \gtrsim 10$, the fluctuation is usually small enough for most of the local operators $\hat{A}$. We typically take $N_{\rm samp} = 10$ averages for $N = 64$.

3. Numerical examinations

Here, we show several details of the numerical results to understand the features of TPQ-MPS. For this purpose, we use the transverse Ising model, $\hat{H} = -J \sum_{i=1}^{N-1} \hat{\sigma}_i^z \hat{\sigma}_{i+1}^z - g \sum_{i=1}^{N} \hat{\sigma}_i^x$, with $\sigma^z = \pm 1$, whose MPO has bond dimension $D = 3$. We choose the parameter values as $J = 1$ and $g = 0.5$ and present the results for $N = 64$.

Figure 1(a) shows the microcanoincal temperature $k_B T^{(k)}$ of $|k\rangle$ obtained by using $\chi_{\rm ini} = 40$ and $l = 1, 5, 10$. For larger l the deviation of state from $|k\rangle$ to $|k+1\rangle$ is smaller, which means that the system should have smaller truncation errors at each step. However, since the state evolves slowly, a larger $k_{\rm max}$ is required to reach the same $\beta_{\rm max}^{-1}$ for larger l. The coefficient of $\langle k|\hat{A}|k\rangle$ in Eq.(5) is plotted in the inset of Fig.1(a); it is the distribution of $|k\rangle$ that contributes to the wave

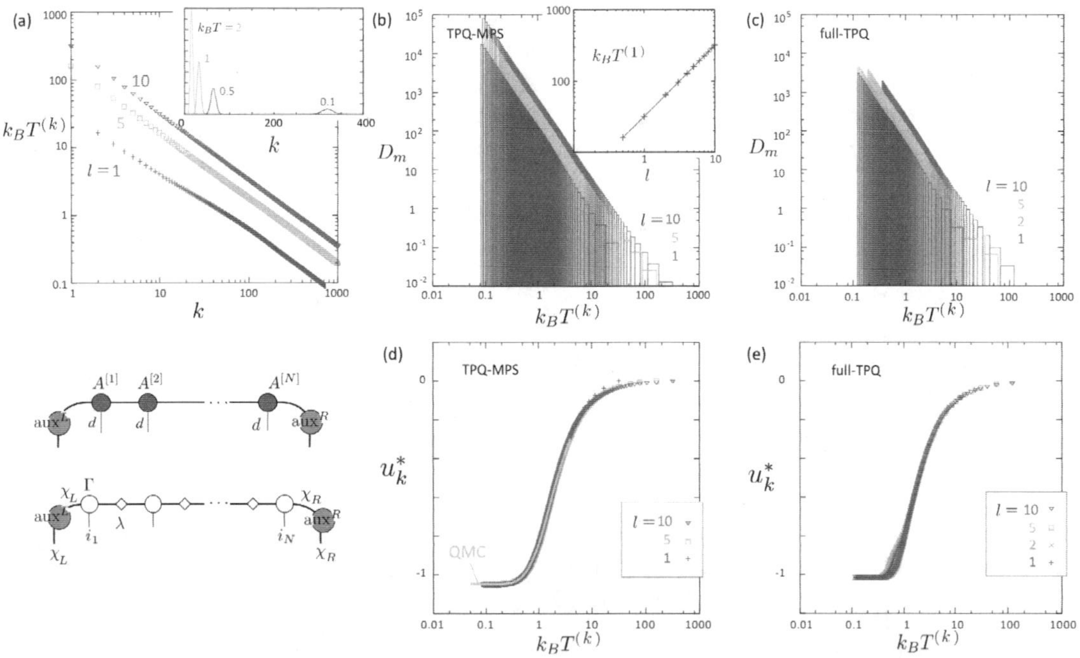

Figure 1. (a) Microcanonical temperature $k_BT^{(k)}$ of the k-th TPQ-MPS state for several choices of l. The inset is the coeficient of $\langle k|\hat{A}|k\rangle$ of Eq.(5). (b) Discritized density of k-states $D_m(k_BT^{(k)})$ for several choices of l. The inset shows the starting temperature $k_BT^{(1)}$ of each calculation for different l. (d) Microcanonical energy u_k^* corresponding to the calculation in panel (b), compared with the QMC results for $N = 64$. (c,e) The same quantity as (b,d) for full-TPQ method with $N = 24$.

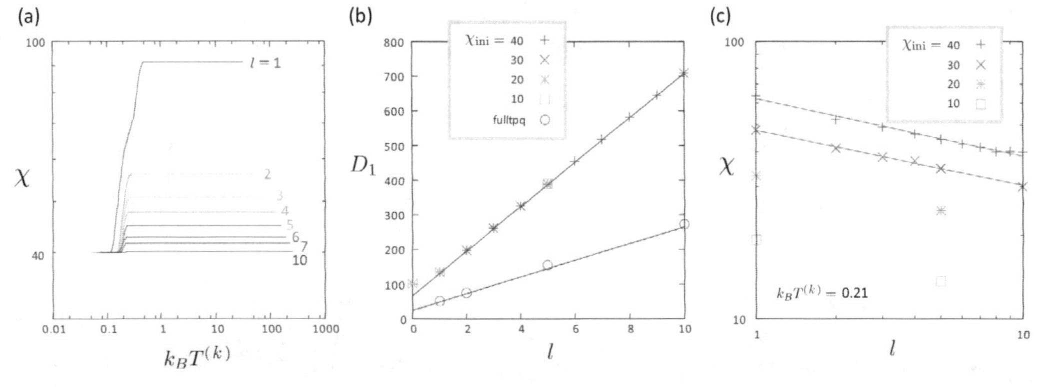

Figure 2. (a) Bond dimension χ at the major part of the system as a function of $k_BT^{(k)}$ for various l. We took $N = 64$ TPQ-MPS with $\chi_{\rm ini} = 40$. (b) D_1 from Fig.1(b) and (c) χ at $k_BT^{(k)} = 0.21$ from panel (a) plotted as functions of l for $\chi_{\rm ini} = 10, 20, 30, 40$. Solid lines represent $D_1 = N(l - u_{\rm gs})$ and $\chi \propto l^{-0.2}$. D_1 obtained from full-TPQ calculation with $N = 24$ is also shown for comparison.

function at temperatures $k_BT = 0.1, 0.5, 2$, and describes the shape of the energy shell of the present extended microcanonical ensemble. We examine the density of $\{|k\rangle\}$-states as a function $k_BT^{(k)}$, $D_m(k_BT) = \sum_{k=1}^{k_{\max}} \delta(k_BT - k_BT^{(k)})$ which is obtained as a discritized histgram. One finds that it increases in power as $D_m \sim D_1(l)(k_BT)^{-2}$ with decreasing the temperature as shown in Fig.1(b). For larger l, the larger $D_1(l)$ (correponding to D_m at $k_BT = 1$) is observed. Here, we present a simple analytical explanation. For this purpose, we introduce a continuous variable $\kappa = k/N$ and consider the continuous limit of $D_m(k_BT)$.

$$D_m(k_BT) \simeq N \int_0^\infty d\kappa \delta \left(k_BT - \frac{l - u_\kappa^*}{\kappa} \right) \tag{6}$$

At low enough temperature, u_κ^* is replaced by the energy density of the ground state $u_{\rm gs}$. Then, we can perform the integral as

$$D_m(k_BT) \simeq N \int_0^\infty d\kappa \delta \left(k_BT - \frac{l - u_{\rm gs}}{\kappa} \right) = \frac{N(l - u_{\rm gs})}{(k_BT)^2}, \tag{7}$$

and we find that $D_m(k_BT)$ scalles as $(k_BT)^{-2}$ and $D_1(l) \simeq N(l - u_{\rm gs})$. The temperature of the starting point of the calculation, $k_BT^{(1)}$, just after operating MPO once, is higher for larger l (see the inset of Fig.1(b)). Here, a sufficient $D_m(k_BT)$ from the energy shell corresponding to $k_BT^{(k)} \sim \beta^{-1}$ is required to accurately obtain the physical quantities at temperature β^{-1} in Eq.(5).

For comparison, we also plot $D_m(k_BT)$ for a full TPQ method with $N = 24$ obtained by a single set of $\{|k\rangle\}$. The constant l is originally suggested to be taken closer to the upper bound of the energy eigenvalues of $\hat{h}$ [17], while the dependence of the quality of the numerical results of l in full-TPQ method was not examined so far. As shown in Fig.1(c) $D_m(k_BT)$ shows a similar l-dependence as those of TPQ-MPS. We also compare the accuracy of u_k^* of TPQ-MPS in Fig.1(d) and full-TPQ in Fig.1(e); one finds that the TPQ-MPS compared with the energy of the QMC result of the same size is in good accuracy. Whereas, the ones for full-TPQ is not necessarily giving enough accurate results, although by taking sample averages the accuracy shall be improved. However, the above comparison confirms that the TPQ-MPS protocol works comparably well or better compared to full-TPQ.

In the present TPQ-MPS calculation, we chose the initial bond dimension $\chi_{\rm ini} = 40$ for all matrices and during the calculations, the values of χ are determined independently for each site to keep the truncation error smaller than 10^{-7}. The values of χ in the major part of the system is equal, whereas χ_L and χ_R at the left and right auxiliaries becomes smaller than those at the center when k becomes larger and the temperature becomes lower. In the low temperature limit, we find $\chi_L = \chi_R \to 1$. We show in Fig.2(a) the evolution of χ of the major part of the system as a function of $k_BT^{(k)}$ for different choices of l. One finds that except at the very low temperature χ is independent of temperature, and can often exceed the initial value $\chi > \chi_{\rm ini}$ for small l, while the larger l requires a smaller χ. This is because the evolution of the TPQ-MPS state by operating $(l - \hat{h})$ is smaller for larger l.

In Fig.2(b) we plot D_1 and χ at $k_BT^{(k)} \sim 0.21$ as functions of l for various choices of $\chi_{\rm ini}$. One finds that $D_1(l) \simeq N(l - u_{\rm gs})$ holds well for all different choices of $\chi_{\rm ini}$. Basically, regardless of $\chi_{\rm ini}$ the values of χ to keep the distribution of $|k\rangle$ decreases as $\chi = c\, l^{-0.2}$, while the constant c change depending on the choice of $\chi_{\rm ini}$.

These tendencies show that the amount of information χ^2 stored in TPQ-MPS per site can be a measure of keeping the quality of the TPQ-MPS state from the loss of information by the truncation, which becomes more serious for smaller l. While taking larger l will be advantageous against the truncation, it requires larger $k_{\max}$ to reach low enough temperatures. Therefore, there should be a trade-off between the two effects; a larger l that provides larger D_m and higher $k_BT^{(1)}$ tends to give a more accurate evaluation of physical quantities at relatively high temperatures. However, a rapid increase of $k_{\max}$ with l requires the truncation to take place as many times at lower temperatures. Therefore, it shall be enough to keep l to those giving $k_BT^{(1)} \gtrsim 100$.

Although the overall behavior of D_m and χ does not depend much on $\chi_{\rm ini}$, the larger $\chi_{\rm ini}$ gives a better evaluation of the physical quantities, although its dependence is not distinct when $\chi_{\rm ini} \gtrsim 30$. The non-local properties may depend more sensitively on $\chi_{\rm ini}$; entanglement entropy and the thermodynamic entropy density evaluated from the Schmidt-coefficients apparently extrapolate to the exact values for a given N with increasing $\chi_{\rm ini}$ [20].

4. Summary

We showed how the microcanonical TPQ-MPS state evolves with k by examining the distribution of states and the density D_m as a function of microcanonical temperature $k_BT^{(k)}$. We showed that the starting temperature $k_BT^{(1)}$ becomes higher and the evolution of states is smaller for larger l, which requires smaller χ to keep the truncation error small enough. However, larger l requires a larger number of states $k_{\max}$ to reach the low enough temperature. The optimal choice of l depends on the model while taking l enough large to have $k_BT^{(1)} \sim 100$. The bond dimension χ should be set free while still, it depends on the initial value, $\chi_{\rm ini}$, and although it does not influence much the local physical quantities, evaluating rather non-local properties like entanglement or entropy requires an extrapolation about $\chi_{\rm ini}$. The present analysis would give a reference for obtaining standard and reliable results within our framework.

References

[1] Umezawa H, Matsumoto H and Tachiki M 1982 *Thermo Field Dynamics and Condensed States* (Amsterdam: North Holland)
[2] Verstraete F, García-Ripoll J J and Cirac J I 2004 *Phys. Rev. Lett.* **93**(20) 207204
[3] Zwolak M and Vidal G 2004 *Phys. Rev. Lett.* **93**(20) 207205
[4] Feiguin A E and White S R 2005 *Phys. Rev. B* **72**(22) 220401
[5] Nomura Y, Yoshioka N and Nori F 2021 *Phys. Rev. Lett.* **127**(6) 060601
[6] von Neumann J 1929 *Z. Phys.* **57** 30–70
[7] Sugita A 2006 *RIMS kokyuroku* **1507** 147–159
[8] Popescu S, J Short A and Winter A 2006 *Nature Phys.* **2** 754–758
[9] Goldstein S, Lebowitz J L, Tumulka R and Zanghì N 2006 *Phys. Rev. Lett.* **96**(5) 050403
[10] Reimann P 2007 *Phys. Rev. Lett.* **99**(16) 160404
[11] Imada M and Takahashi M 1986 *Journal of the Physical Society of Japan* **55** 3354–3361
[12] Hams A and De Raedt H 2000 *Phys. Rev. E* **62**(3) 4365–4377
[13] Iitaka T and Ebisuzaki T 2003 *Phys. Rev. Lett.* **90**(4) 047203
[14] Iitaka T and Ebisuzaki T 2004 *Phys. Rev. E* **69**(5) 057701
[15] Machida M, Iitaka T and Miyashita S 2005 *Journal of the Physical Society of Japan* **74** 107–110
[16] Machida M, Iitaka T and Miyashita S 2012 *Phys. Rev. B* **86**(22) 224412
[17] Sugiura S and Shimizu A 2012 *Phys. Rev. Lett.* **108**(24) 240401
[18] Sugiura S and Shimizu A 2013 *Phys. Rev. Lett.* **111**(1) 010401
[19] Hyuga M, Sugiura S, Sakai K and Shimizu A 2014 *Phys. Rev. B* **90**(12) 121110
[20] Iwaki A, Shimizu A and Hotta C 2021 *Phys. Rev. Research* **3**(2) L022015
[21] White S R 2009 *Phys. Rev. Lett.* **102**(19) 190601
[22] Stoudenmire E M and White S R 2010 *New Journal of Physics* **12** 055026
[23] Garnerone S and de Oliveira T R 2013 *Phys. Rev. B* **87**(21) 214426
[24] Garnerone S 2013 *Phys. Rev. B* **88**(16) 165140
[25] Iitaka T 2020 *arXiv:2006.14459*
[26] Goto S, Kaneko R and Danshita I 2021 *Phys. Rev. B* **104**(4) 045133
[27] Garnerone S, de Oliveira T R and Zanardi P 2010 *Phys. Rev. A* **81**(3) 032336
[28] Garnerone S, de Oliveira T R, Haas S and Zanardi P 2010 *Phys. Rev. A* **82**(5) 052312
[29] Vidal G 2003 *Phys. Rev. Lett.* **91**(14) 147902
[30] Yoneta Y and Shimizu A 2019 *Phys. Rev. B* **99**(14) 144105

Journal of Physics: Conference Series **2207** (2022) 012032 doi:10.1088/1742-6596/2207/1/012032

Quantum Monte Carlo simulations of highly frustrated magnets in a cluster basis: The two-dimensional Shastry-Sutherland model

Andreas Honecker[1], Lukas Weber[2], Philippe Corboz[3], Frédéric Mila[4], Stefan Wessel[2]

[1] Laboratoire de Physique Théorique et Modélisation, CNRS UMR 8089, CY Cergy Paris Université, Cergy-Pontoise, France
[2] Institute for Theoretical Solid State Physics, RWTH Aachen University, Germany
[3] Institute for Theoretical Physics, University of Amsterdam, The Netherlands
[4] Institute of Physics, Ecole Polytechnique Fédérale de Lausanne (EPFL), Switzerland

E-mail: andreas.honecker@cyu.fr

Abstract. Quantum Monte Carlo (QMC) simulations constitute nowadays one of the most powerful methods to study strongly correlated quantum systems, provided that no "sign problem" arises. However, many systems of interest, including highly frustrated magnets, suffer from an average sign that is close to zero in standard QMC simulations. Nevertheless, a possible sign problem depends on the simulation basis, and here we demonstrate how a suitable choice of cluster basis can be used to eliminate or at least reduce the sign problem in highly frustrated magnets that were so far inaccessible to efficient QMC simulations. We focus in particular on the application of a two-spin (dimer)-based QMC method to the thermodynamics of the spin-1/2 Shastry-Sutherland model for $SrCu_2(BO_3)_2$.

1. Introduction

Quantum many-body systems constitute a challenge for a quantitatively accurate treatment. In this context, Quantum Monte Carlo (QMC) methods have become one of the most powerful tools [1–3]. However, in some cases, including geometrically frustrated magnets, their use is limited by a phenomenon known as the "sign problem", namely a cancellation of positive and negative weights of configurations [4–6]. In the present contribution, we illustrate how these problems can be overcome or at least alleviated by working in a suitable cluster basis [7–10]. For the purpose of simplicity, we will focus on the application of a dimer basis to a highly frustrated two-dimensional spin-1/2 model known as the "Shastry-Sutherland model" [11] that is realised in the compound $SrCu_2(BO_3)_2$ [12, 13].

2. Method and the Shastry-Sutherland model

One flexible QMC approach is based on the representation of the partition function Z as a high-temperature series [14, 15], leading to the "stochastic series expansion" (SSE). The SSE is based on a splitting of the Hamiltonian H into bond terms H_b. These terms are split in turn into different types, such that the action of each bond term $H_{(b,t)}$ on a given basis state $|\alpha\rangle$ either vanishes or yields exactly one other basis state $|\alpha'\rangle$. Note that one of the types t is diagonal while all others are off-diagonal. Using $H = \sum_{(b,t)} H_{(b,t)}$, one finds in the given basis

$$Z = \mathrm{Tr}\, \mathrm{e}^{-\beta H} = \sum_{n=0}^{\infty} \frac{\beta^n}{n!} \sum_{|\alpha_n\rangle = |\alpha_1\rangle} \sum_{(b_1,t_1),\ldots,(b_n,t_n)} \prod_{i=1}^{n} \langle \alpha_{i+1}| \left(-H_{(b_i,t_i)}\right) |\alpha_i\rangle = \sum_c W_c\,, \tag{1}$$

where $\beta = 1/T$ is the inverse temperature (we set the Boltzmann constant $k_B = 1$). Important developments concern the representation and efficient Monte Carlo sampling of the sum in Eq. (1), as well as measurement of physical quantities, and we refer to Refs. [10, 16–18] for detailed discussions of these issues.

Here, our main concern will be a different one: the weights W_c of a configuration c in Eq. (1) are given by products of matrix elements $\langle \alpha_{i+1}| \left(-H_{(b_i,t_i)}\right) |\alpha_i\rangle$ and there is in general no guarantee that they are non-negative, *i.e.*, that the expression Eq. (1) does indeed have a probabilistic interpretation. For diagonal terms of the Hamiltonian, one can add a constant shift ϵ to render $-H_{(b_i,t_i)} + \epsilon$ non-negative. When off-diagonal terms are negative, one uses the absolute value of the weights $|W_c|$ for sampling such that the average of an observable A is given as the ratio of the averages of the observable and of the sign [1, 4],

$$\langle A\rangle = \frac{\sum_c W_c A_c}{\sum_c W_c} = \frac{\sum_c \mathrm{sign}(W_c)\ |W_c|\ A_c}{\sum_c \mathrm{sign}(W_c)\ |W_c|} = \frac{\langle \mathrm{sign}\, A\rangle_{|\cdot|}}{\langle \mathrm{sign}\rangle_{|\cdot|}}\,. \tag{2}$$

This allows one to run a QMC simulation for any given problem. However, if indeed negative contributions appear in the sum in Eq. (1), there is a cancellation of negative and positive contributions to $\langle \mathrm{sign}\rangle_{|\cdot|}$, and if these are comparable in magnitude one may not be able to distinguish the normalisation $\langle \mathrm{sign}\rangle_{|\cdot|}$ from zero within the error bars of the calculation, thus rendering the results meaningless. This problem is known as the "sign problem", and we will provide some examples in section 3. Nevertheless, because the sign of the terms in Eq. (1) depends on the choice of basis, one may be able to eliminate or at least alleviate it by a simulation basis that is suitably chosen for a given problem, and we will illustrate one such approach in the following.

In order to be specific, we will now focus on the Shastry-Sutherland model, which is represented schematically in Fig. 1. The Hamiltonian in the single-site basis is

$$H = J_D \sum_{\langle i,j\rangle} \vec{S}_i \cdot \vec{S}_j + J \sum_{\langle\langle i,j\rangle\rangle} \vec{S}_i \cdot \vec{S}_j\,, \tag{3}$$

where $\vec{S}_i$ are spin-1/2 operators for site i. J_D is the intra-dimer coupling (denoted by $\langle i,j\rangle$) and the inter-dimer coupling ($\langle\langle i,j\rangle\rangle$) J corresponds to a square lattice, as shown in Fig. 1. This model was constructed [11] to ensure that the ground state is an exact product of singlets formed on the dimer bonds for small and intermediate coupling ratios J/J_D. More recently, this model was also found to be realised in the quantum spin system $SrCu_2(BO_3)_2$ [12, 13], which increased the interest in calculating its thermodynamic properties.

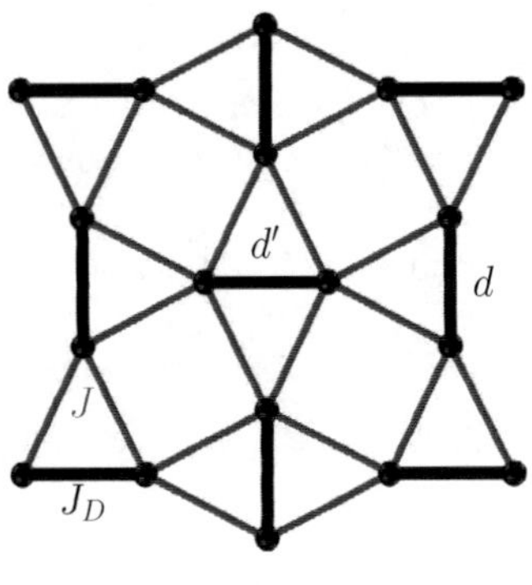

Figure 1. Schematic representation of the Shastry-Sutherland model. The black dots correspond to sites carrying spins 1/2. Dimer couplings J_D are shown by thick black lines and inter-dimer couplings J by thinner blue lines. Two neighbouring dimers are designated by d and d'.

As a consequence of the triangular arrangement of the J and J_D interactions, this problem is geometrically

Journal of Physics: Conference Series **2207** (2022) 012032 doi:10.1088/1742-6596/2207/1/012032

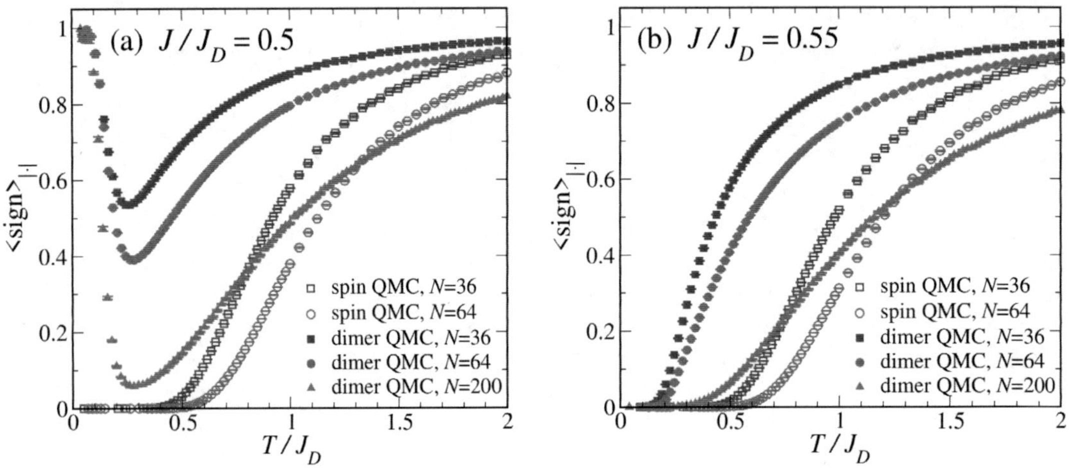

Figure 2. Average sign $\langle \text{sign} \rangle_{|\cdot|}$ computed as a function of temperature for the Shastry-Sutherland model with (a) $J/J_D = 0.5$ and (b) $J/J_D = 0.55$. Open symbols are for the single-spin basis, filled symbols for the dimer basis. Data for $N = 200$ spins are reproduced from Ref. [19].

frustrated. Consequently, QMC simulations in the single-spin basis (3) suffer from a severe sign problem, as we will show in section 3. However, one may pass to the total-spin and spin-difference operators, respectively $\vec{T}_d = \vec{S}_{d,1} + \vec{S}_{d,2}$ and $\vec{D}_d = \vec{S}_{d,1} - \vec{S}_{d,2}$, for the two spins $\vec{S}_{d,1}$ and $\vec{S}_{d,2}$ on a given dimer d [19]. The coupling of the two spins in dimer d can then be expressed through $\vec{T}_d^2$ and the inter-dimer coupling for the two dimers d and d' indicated in Fig. 1 takes the form

$$H_{dd'} = \tfrac{1}{2}\, J\, \vec{T}_d \cdot \vec{T}_{d'} - \tfrac{1}{2}\, J\, \vec{T}_d \cdot \vec{D}_{d'} \,. \tag{4}$$

The first term corresponds just to a composite-spin model on a square lattice, which is unfrustrated and thus poses no QMC sign problem. The sign of the coefficient of the second term in Eq. (4) depends on the assignment of the labels 1 and 2 in dimer d' and varies for the different inter-dimer coupling terms throughout the lattice. There are also signs involved in the construction of the operator $\vec{D}_d$ such that it is not immediately clear to which extent this second term may still give rise to a sign problem. Nevertheless, the operator $\vec{D}_d$ yields dimer triplet states when applied to the singlet state of the dimer d, $|S\rangle_d = \frac{1}{\sqrt{2}}\,(|\uparrow\downarrow\rangle_d - |\downarrow\uparrow\rangle_d)$. Because the ground state is an exact product of such dimer singlets for $J \ll J_D$, one may expect the difference operators $\vec{D}_d$ not to contribute to the leading orders at low temperatures, such that a potential sign problem remains manageable. We will show in the next section that indeed this turns out to be true in the regime where J is not too large.

3. Results

We now present some QMC results for the Shastry-Sutherland model where we will focus on the two representative cases $J/J_D = 0.5$ and 0.55.

Figure 2 presents the average sign $\langle \text{sign} \rangle_{|\cdot|}$. In the single-spin basis (open symbols), the geometric frustration leads to the expected sign problem (see section 2), which manifests itself as the rapid drop of the average sign at low temperatures. If a sign problem arises, the average sign generically becomes exponentially small in N/T [1, 4]. This is verified by the single-site results, where the average sign not only drops at low temperatures but also decreases with increasing system size N. Working in the dimer basis improves the sign problem very significantly, as is illustrated by the full symbols in Fig. 2. There is still a sign problem, as evidenced by the regime

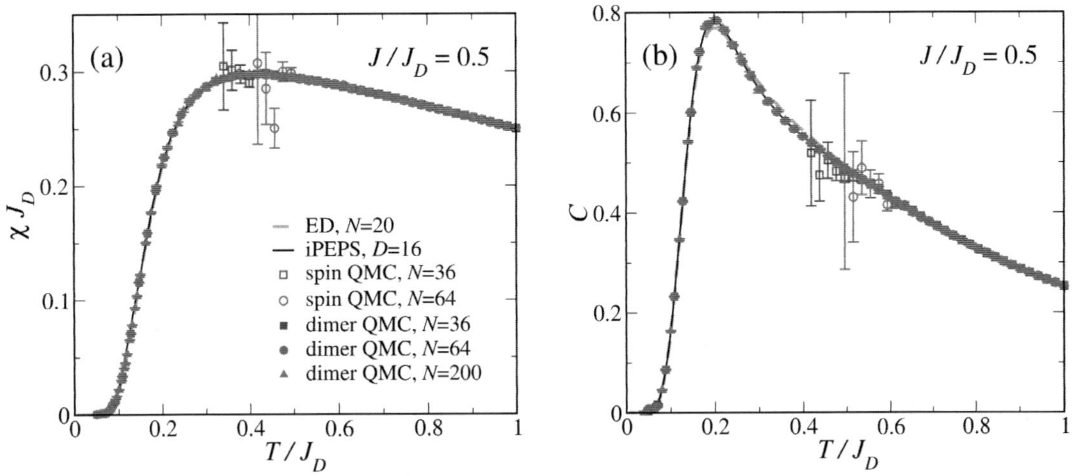

Figure 3. (a) Magnetic susceptibility χ and (b) specific heat C per dimer of the Shastry-Sutherland model at $J/J_D = 0.5$. Exact diagonalisation (ED) results for $N = 20$ and QMC data for $N = 200$ spins are reproduced from Ref. [19], $D = 16$ iPEPS data from Ref. [20].

in which $\langle \text{sign} \rangle_{|\cdot|} < 1$. However, for $N = 32$ and 64 spins, the average sign is always bigger in the dimer basis than in the single-site basis. Most remarkably, at least in the case $J/J_D = 0.5$ (Fig. 2(a)) the average sign does not vanish as temperature goes to zero, but instead recovers after going through a minimum around $T/J_D \approx 0.25$. This minimum still drops with increasing N, but one can go to systems as large as $N = 200$ and still ensure that $\langle \text{sign} \rangle_{|\cdot|} \gtrsim 0.06$, such that the resulting loss of accuracy can be compensated by improving the statistics.

At $J/J_D = 0.55$ (Fig. 2(b)), the average sign no longer recovers for $T \to 0$, even in the dimer basis. This difference between the cases $J/J_D = 0.5$ and 0.55 can be traced to an algorithmic phase transition at $J/J_D = 0.526(1)$ [19], where the effective model defined by the absolute value of the weights in Eq. (2) undergoes a transition out of the dimer phase. Nevertheless, $\langle \text{sign} \rangle_{|\cdot|}$ in the dimer basis for $N = 200$ spins retains a larger value than that in the single-site basis for the much smaller system $N = 32$ at temperatures $T/J_D \lesssim 0.7$, as shown in Fig. 2(b).

Next we present results for thermodynamic observables at $J/J_D = 0.5$ and 0.55 which are shown respectively in Figs. 3 and 4. As before, QMC results in the single-spin basis are shown by open symbols and results obtained in the dimer basis by filled symbols. For reference, we have included exact diagonalisation (ED) results for $N = 20$ spins [19] and infinite projected entangled-pair states (iPEPS) data for a bond dimension $D = 16$ [20]. As thermodynamic observables we have selected the magnetic susceptibility χ and specific heat C; the temperature window $0 \leq T \leq J_D$ is chosen to include the maximum of these two quantities.

The single-site basis is barely able to resolve the maximum of χ at $J/J_D = 0.5$ (Fig. 3(a)) or get close to it for $J/J_D = 0.55$ (Fig. 4(a)), provided that one restricts a system size as small as $N = 36$. However, single-site QMC generally loses accuracy already at significantly higher temperatures, such that the maxima are out of reach and only the high-temperature regime is accessible, as was to be expected according to the average sign shown in Fig. 2. On the other hand, for $J/J_D = 0.5$, the dimer basis allows one to obtain accurate QMC data for systems with up to $N = 200$ spins and at all temperatures, as shown in Fig. 3. For $J/J_D = 0.55$, the sign problem of Fig. 2(b) prevents one from reaching the zero-temperature limit, but the maximum of the magnetic susceptibility can still be well resolved in the dimer basis for $N \leq 128$ spins (Fig. 4(a)). In order to resolve the maximum of C (Fig. 4(b)) by QMC in the dimer basis, one needs to compromise on system size, *i.e.*, restrict to $N = 36$, and on accuracy, but even $N = 128$

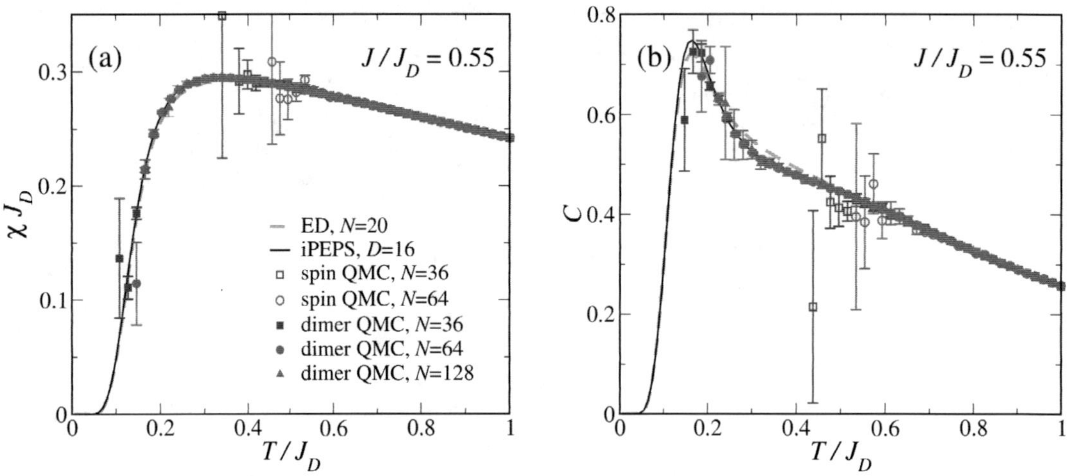

Figure 4. Same as Fig. 3, but for $J/J_D = 0.55$. Exact diagonalisation (ED) results for $N = 20$ and QMC data for $N = 128$ spins are reproduced from Ref. [19].

still works significantly better in the dimer basis than $N = 36$ in the single-spin basis, again as expected from the results for the average sign shown in Fig. 2.

The iPEPS data for $D = 16$ can be considered as representative of the thermodynamic limit in the present parameter regime and at the level of resolution in Figs. 3 and 4 [20]. Indeed, all of our QMC results agree with iPEPS within the corresponding statistical error bars. On the other hand, small (albeit distinct) finite-size effects can be seen in the $N = 20$ ED data, in particular around the maximum of C and at temperatures above it. These finite-size effects become increasingly pronounced at the larger value $J/J_D = 0.55$ (Fig. 4(b)). This highlights the importance of having access to larger systems in order to ensure quantitatively accurate results.

Concerning the physics of these thermodynamic quantities, we observe the emergence of low-temperature maxima in the magnetic susceptibility and the specific heat that move to lower temperature and sharpen as J/J_D increases [19, 20]. In the present results, this is seen most clearly for the specific heat, shown in Figs. 3(b) and 4(b).

4. Conclusion

We have shown how a dimer basis can be used to perform accurate QMC simulations for the thermodynamic properties of a highly frustrated two-dimensional spin-1/2 model, the Shastry-Sutherland model, that would be out of reach of conventional approaches. The compound $SrCu_2(BO_3)_2$ is believed to correspond to the parameter ratio $J/J_D \approx 0.63$ [13, 20]. This ratio is still in the dimer phase of the Shastry-Sutherland model, but rather far beyond the algorithmic phase transition at $J/J_D = 0.526(1)$ [19]. Thus the parameter regime relevant to $SrCu_2(BO_3)_2$ remains unfortunately inaccessible to accurate QMC simulations even in the dimer basis, and one needs to resort to other methods such as iPEPS [20, 21].

Still, there are other cases to which the ideas illustrated here can be applied. For example, in the case of a frustrated ladder, only boundary terms contribute to the sign problem, such that it remains sufficiently mild to allow for a detailed study of the entire phase diagram [8, 22]. Another case is a fully frustrated bilayer model, where the counterpart of the $\vec{T}_d \cdot \vec{D}_{d'}$ term in Eq. (4) is absent and thus the dimer basis eliminates the sign problem completely [9, 23, 24]. A detailed QMC study of the phase diagram then becomes possible, allowing one to trace a first-order phase transition from zero temperature to finite temperature until it terminates at an Ising critical point [21, 24].

The cluster-basis approach is is not restricted to dimers and we have demonstrated recently that it also works with a trimer (three-spin) basis [10]. This permits one to generalise the investigation of the fully frustrated bilayer model to a trilayer analogue, and study for example the influence of the ground-state entropy that arises in the latter model on the slope of the first-order line at finite temperatures [10].

Acknowledgments
We would like to thank Bruce Normand for numerous collaborations on related problems and for a critical reading of the manuscript.

References

[1] Evertz H G 2003 *Adv. Phys.* **52** 1–66 URL https://doi.org/10.1080/0001873021000049195
[2] Troyer M, Alet F, Trebst S and Wessel S 2003 *AIP Conf. Proc.* **690** 156–169 URL https://doi.org/10.1063/1.1632126
[3] Sandvik A W 2010 *AIP Conf. Proc.* **1297** 135–338 URL https://doi.org/10.1063/1.3518900
[4] Troyer M and Wiese U J 2005 *Phys. Rev. Lett.* **94** 170201 URL https://doi.org/10.1103/PhysRevLett.94.170201
[5] Hangleiter D, Roth I, Nagaj D and Eisert J 2020 *Sci. Adv.* **6** eabb8341 URL https://doi.org/10.1126/sciadv.abb8341
[6] Hen I 2021 *Phys. Rev. Research* **3** 023080 URL https://doi.org/10.1103/PhysRevResearch.3.023080
[7] Nakamura T 1998 *Phys. Rev. B* **57** R3197–R3200 URL https://doi.org/10.1103/PhysRevB.57.R3197
[8] Honecker A, Wessel S, Kerkdyk R, Pruschke T, Mila F and Normand B 2016 *Phys. Rev. B* **93** 054408 URL https://doi.org/10.1103/PhysRevB.93.054408
[9] Alet F, Damle K and Pujari S 2016 *Phys. Rev. Lett.* **117** 197203 URL https://doi.org/10.1103/PhysRevLett.117.197203
[10] Weber L, Honecker A, Normand B, Corboz P, Mila F and Wessel S 2022 to appear in *SciPost Phys.* (*Preprint* 2105.05271v4) URL https://arxiv.org/abs/2105.05271v4
[11] Shastry B S and Sutherland B 1981 *Physica B+C* **108** 1069–1070 URL https://doi.org/10.1016/0378-4363(81)90838-X
[12] Kageyama H, Yoshimura K, Stern R, Mushnikov N V, Onizuka K, Kato M, Kosuge K, Slichter C P, Goto T and Ueda Y 1999 *Phys. Rev. Lett.* **82** 3168–3171 URL https://doi.org/10.1103/PhysRevLett.82.3168
[13] Miyahara S and Ueda K 2003 *J. Phys.: Condens. Matter* **15** R327–R366 URL https://doi.org/10.1088/0953-8984/15/9/201
[14] Handscomb D C 1964 *Math. Proc. Cambridge Phil. Soc.* **60** 115–122 URL https://doi.org/10.1017/S030500410003752X
[15] Sandvik A W 1992 *J. Phys. A: Math. Gen.* **25** 3667–3682 URL https://doi.org/10.1088/0305-4470/25/13/017
[16] Syljuåsen O F and Sandvik A W 2002 *Phys. Rev. E* **66** 046701 URL https://doi.org/10.1103/PhysRevE.66.046701
[17] Syljuåsen O F 2003 *Phys. Rev. E* **67** 046701 URL https://doi.org/10.1103/PhysRevE.67.046701
[18] Alet F, Wessel S and Troyer M 2005 *Phys. Rev. E* **71** 036706 URL https://doi.org/10.1103/PhysRevE.71.036706
[19] Wessel S, Niesen I, Stapmanns J, Normand B, Mila F, Corboz P and Honecker A 2018 *Phys. Rev. B* **98** 174432 URL https://doi.org/10.1103/PhysRevB.98.174432
[20] Wietek A, Corboz P, Wessel S, Normand B, Mila F and Honecker A 2019 *Phys. Rev. Research* **1** 033038 URL https://doi.org/10.1103/PhysRevResearch.1.033038
[21] Larrea Jiménez J, Crone S P G, Fogh E, Zayed M E, Lortz R, Pomjakushina E, Conder K, Läuchli A M, Weber L, Wessel S, Honecker A, Normand B, Rüegg C, Corboz P, Rønnow H M and Mila F 2021 *Nature* **592** 370–375 URL https://doi.org/10.1038/s41586-021-03411-8
[22] Wessel S, Normand B, Mila F and Honecker A 2017 *SciPost Phys.* **3** 005 URL https://doi.org/10.21468/SciPostPhys.3.1.005
[23] Ng K K and Yang M F 2017 *Phys. Rev. B* **95** 064431 URL https://doi.org/10.1103/PhysRevB.95.064431
[24] Stapmanns J, Corboz P, Mila F, Honecker A, Normand B and Wessel S 2018 *Phys. Rev. Lett.* **121** 127201 URL https://doi.org/10.1103/PhysRevLett.121.127201

Compressed sensing based on QUBO formulation

Kazue Kudo[1,2]

[1] Department of Computer Science, Ochanomizu University, Tokyo 112-8610, Japan
[2] Graduate School of Information Sciences, Tohoku University, Sendai 980-8579, Japan

E-mail: kudo@is.ocha.ac.jp

Abstract. Ising machines efficiently solve the combinatorial optimization problems described by the Ising model or the quadratic unconstrained binary optimization (QUBO) formulation. A hybrid method based on the QUBO formulation for compressed sensing is proposed. The proposed method comprises alternative steps of discrete and continuous optimization. In the discrete optimization step, the objective function is described by the QUBO formulation. Successful examples obtained via the proposed method are demonstrated. The performance of the proposed method depends highly on the initial conditions.

1. Introduction

Compressed sensing enables the reconstruction of a sparse signal. When a high-dimensional signal is "sparse," the signal is represented by a linear combination of a small number of vectors on an appropriate basis. The success of compressed sensing [1–3] has been demonstrated in magnetic resonance imaging [4, 5], scanning tunneling microscopy [6], and X-ray absorption fine structure [7], to name a few. The least absolute shrinkage and selection operator (LASSO) is a successful method for compressed sensing [8]. LASSO can efficiently solve a problem of L_1-regularization compressed sensing, which is a convex optimization problem. In contrast, compressed sensing based on L_0 regularization is a combinatorial problem and requires an exhaustive search for an appropriate combination of basis vectors. For L_0-regularization compressed sensing, simulated annealing (SA) works well in certain situations [9, 10].

In recent years, quantum annealers [11] and other Ising machines [12–17] have attracted attention as new types of computers for solving combinatorial optimization problems. These Ising machines accept a problem described by the Ising model or the quadratic unconstrained binary optimization (QUBO) formulation and return binary variables that approximately minimize the objective function corresponding to the problem. Methods that employ Ising machines are promising for compressed sensing based on L_0 regularization as they can efficiently solve combinatorial optimization problems. A hybrid system consisting of a coherent Ising machine and a digital processor performs well for L_0-regularization compressed sensing [18].

This paper proposes a hybrid method based on the QUBO formulation for L_0-regularization compressed sensing. The proposed method comprises alternative steps of discrete and continuous optimization. In the discrete optimization step, the objective function described in the QUBO formulation can be minimized using an Ising machine. The proposed method functions also when SA is employed instead of an Ising machine. Numerical results demonstrated in this study use SA for the discrete optimization step. The proposed method works well in certain cases, although the results are highly dependent on the initial conditions.

2. Methods

We here consider the problem of reconstructing an N-dimensional vector $\boldsymbol{x}$ from the linear equation given by:

$$\boldsymbol{y} = A\boldsymbol{x} + \boldsymbol{\xi}, \tag{1}$$

where $\boldsymbol{y}$ and $\boldsymbol{\xi}$ are M-dimensional vectors, and A is an $M \times N$ matrix with $M < N$. This is an underdetermined problem as $M < N$. To solve this, the number of nonzero components of $\boldsymbol{x}$ are assumed to be sufficiently smaller than M. In other words, $\boldsymbol{x}$ is sparse.

Under the assumption that the components of $\boldsymbol{\xi}$ are Gaussian random numbers with zero mean, the maximum likelihood estimation for $\boldsymbol{x}$ corresponds to minimizing the function $F(\boldsymbol{x}) = \frac{1}{2}\|\boldsymbol{y} - A\boldsymbol{x}\|_2^2$, where $\|\cdot\|_k$ denotes the L_k norm. The sparse condition of $\boldsymbol{x}$ requires an additional regularization term in the function. For L_0 regularization, the objective function to be minimized is given as follows:

$$F(\boldsymbol{x}, \boldsymbol{c}) = \frac{1}{2}\|\boldsymbol{y} - A(\boldsymbol{x} \circ \boldsymbol{c})\|_2^2 + \lambda\|\boldsymbol{c}\|_0, \tag{2}$$

where $\boldsymbol{c} \in \{0,1\}^N$, $\lambda > 0$, and $\circ$ denotes the Hadamard product, that is, $(\boldsymbol{x} \circ \boldsymbol{c})_i = x_i c_i$.

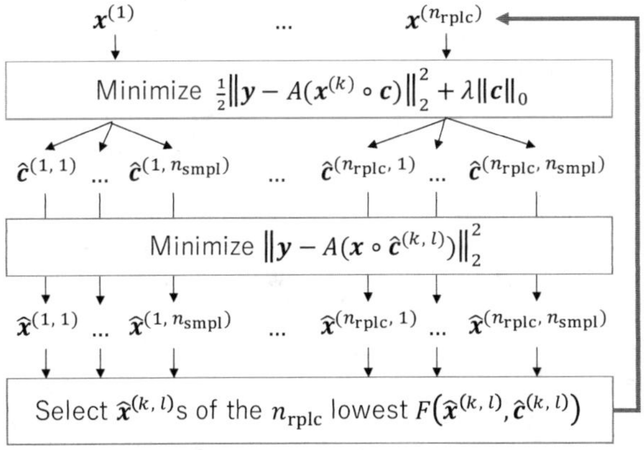

Figure 1. Schematic of the hybrid algorithm. The minimization step in the orange box is assumed to be executed by an Ising machine.

The proposed hybrid algorithm consists of alternative optimization of the following: (i) Minimization with respect to $\boldsymbol{c}$ using an Ising machine for a fixed $\boldsymbol{x}$, and (ii) minimization with respect to $\boldsymbol{x}$ for a fixed $\boldsymbol{c}$. Figure 1 illustrates the schematic of the algorithm. First, the $n_{\rm rplc}$ candidates of $\boldsymbol{x}$ are prepared, and the same number of candidates after each set of alternative optimizations are stored. The alternative optimization is repeated, with λ increased gradually. The detailed algorithm is as follows:

1. Give initial $\boldsymbol{x}^{(k)}$ $(k = 1, \ldots, n_{\rm rplc})$. Set λ as a small positive value.
2. Perform the minimization of $F(\boldsymbol{c}) = \frac{1}{2}\|\boldsymbol{y} - A(\boldsymbol{x}^{(k)} \circ \boldsymbol{c})\|_2^2 + \lambda\|\boldsymbol{c}\|_0$ with a fixed $\boldsymbol{x}^{(k)}$, $n_{\rm smpl}$ times, to obtain approximate solutions $\hat{\boldsymbol{c}}^{(k,l)}$ $(l = 1, \ldots, n_{\rm smpl})$.
3. Minimize $F(\boldsymbol{x}) = \|\boldsymbol{y} - A(\boldsymbol{x} \circ \hat{\boldsymbol{c}}^{(k,l)})\|_2^2$ with a fixed $\hat{\boldsymbol{c}}^{(k,l)}$.
4. Select $n_{\rm rplc}$ samples of $\hat{\boldsymbol{x}}^{(k,l)}$ in ascending order of $F(\hat{\boldsymbol{x}}^{(k,l)}, \hat{\boldsymbol{c}}^{(k,l)})$ and rename them as $\boldsymbol{x}^{(k)}$.
5. If $\lambda > \lambda_{\rm max}$, stop. Otherwise, return to step 2 with $\lambda \leftarrow \lambda + d\lambda$.

Step 2 is supposed to be executed by an Ising machine. The objective function $F(\boldsymbol{c})$ is rewritten as

$$F(\boldsymbol{c}) = \lambda\sum_{j=1}^{N} c_j + \frac{1}{2}\left[\sum_{i=1}^{M} y_i^2 - 2\sum_{j=1}^{N}\left(\sum_{i=1}^{M} y_i B_{ij}\right)c_j + \sum_{j,k=1}^{N}\left(\sum_{i=1}^{M} B_{ij}B_{ik}\right)c_j c_k\right], \tag{3}$$

where $B_{ij} = A_{ij}x_j^{(k)}$. Disregarding the constant terms, the Hamiltonian in a QUBO formulation is obtained.

$$H = \sum_{j=1}^{N} \left(\lambda - (B^\top \boldsymbol{y})_j\right) c_j + \frac{1}{2} \sum_{j,k=1}^{N} (B^\top B)_{jk} c_j c_k. \tag{4}$$

The calculation in step 3 is performed using a classical computer. Let a submatrix A_{sub} be a matrix that extracts only the columns of A that correspond to nonzero components of $\hat{\boldsymbol{c}}^{(k,l)}$. Then, the subvector of $\hat{\boldsymbol{x}}$ corresponding to the nonzero components of $\hat{\boldsymbol{c}}^{(k,l)}$ is as follows:

$$\hat{\boldsymbol{x}}_{\text{sub}} = (A_{\text{sub}}^\top A_{\text{sub}})^{-1} A_{\text{sub}}^\top \boldsymbol{y}. \tag{5}$$

However, if $\det(A_{\text{sub}}^\top A_{\text{sub}})$ is too small, the gradient descent is used instead of Equation (5). In other words, the following is repeated.

$$\boldsymbol{x}_{\text{sub}} \leftarrow \boldsymbol{x}_{\text{sub}} - \eta \boldsymbol{\nabla} f(\boldsymbol{x}_{\text{sub}}), \tag{6}$$

where η is a small positive quantity, and

$$\boldsymbol{\nabla} f(\boldsymbol{x}_{\text{sub}}) = -A_{\text{sub}}^\top \boldsymbol{y} + (A_{\text{sub}}^\top A_{\text{sub}})\boldsymbol{x}_{\text{sub}}, \tag{7}$$

to obtain $\hat{\boldsymbol{x}}_{\text{sub}}$.

3. Numerical experiments

First, a matrix A and vectors $\boldsymbol{x}_{\text{true}}$ and $\boldsymbol{\xi}$ are provided, then a vector $\boldsymbol{y} = A\boldsymbol{x}_{\text{true}} + \boldsymbol{\xi}$ is determined. Here, A is an $M \times N$ matrix whose components are Gaussian random numbers with mean 0 and variance $1/N$, and $\boldsymbol{x}_{\text{true}}$ is an N-dimensional vector with K nonzero components that are Gaussian random numbers with mean 0 and variance N/K. The components of the N-dimensional vector $\boldsymbol{\xi}$ are Gaussian random numbers with mean 0 and variance σ_ξ^2. Here, $N = 100$, $M = 25$, $K = 5$, and $\lambda_{\text{max}} = 0.1$ are considered.

Figure 2 demonstrates a few successful examples obtained using the proposed hybrid algorithm. In step 1, the initial samples of $\boldsymbol{x}$ are provided randomly. In step 2, SA is used instead of an Ising machine for the simulations of this study. In the noiseless case, as shown in figures 2(a) and 2(b), the obtained solution coincides with the true solution; thus, $A\boldsymbol{x}$ also coincides with $\boldsymbol{y}$. The obtained solution is remarkably close to the true solution in the noisy case, as shown in figure 2(c). Thus, $\boldsymbol{y}$ and $A\boldsymbol{x}$ are similar in figure 2(d). Incidentally, there would be a difference between $\boldsymbol{y}$ and $A\boldsymbol{x}$ even if $\boldsymbol{x}$ coincides with $\boldsymbol{x}_{\text{true}}$ owing to noise.

Table 1. Correct answer rates in noiseless ($\sigma_\xi = 0.0$) and noisy ($\sigma_\xi = 0.1$) cases.

		$\sigma_\xi = 0.0$, $\epsilon_x < 10^{-7}$			$\sigma_\xi = 0.1$, $\epsilon_x < 0.1$		
n_{rplc}	n_{smpl}	$d\lambda$			$d\lambda$		
		0.001	0.005	0.01	0.001	0.005	0.01
40	2	0.20	0.31	0.13	0.26	0.22	0.23
40	10	0.15	0.24	0.25	0.15	0.21	0.24
80	2	0.39	0.34	0.28	0.32	0.37	0.50
80	10	0.29	0.38	0.44	0.20	0.36	0.40

As the proposed method is a heuristic algorithm, it does not always provide correct answers. Table 1 shows the proportion of correct answers obtained from 100 simulations.

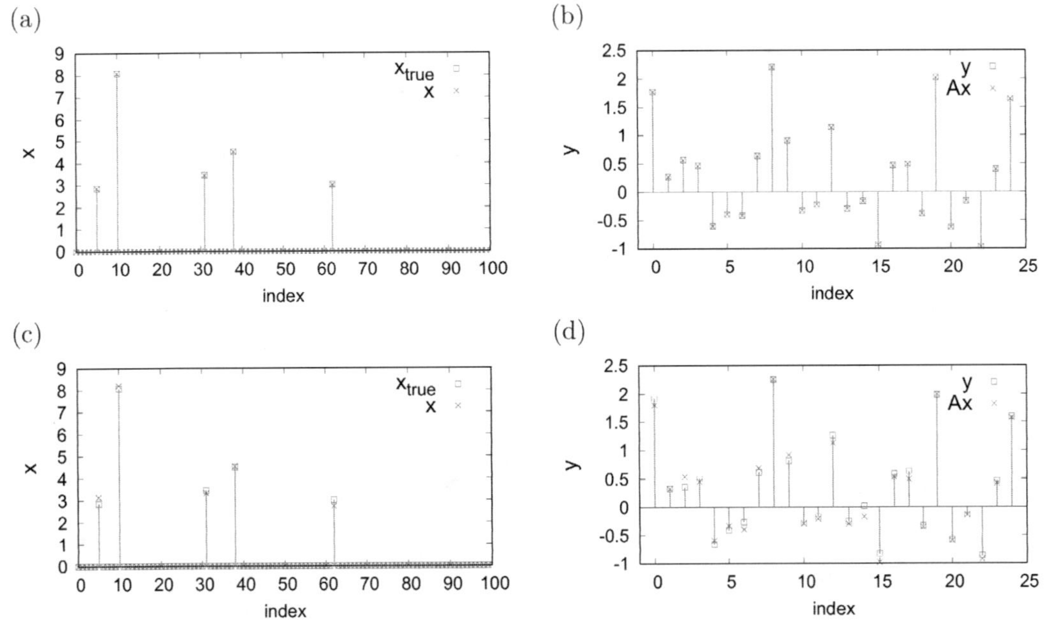

Figure 2. Example plots in (a, b) noiseless ($\sigma_\xi = 0.0$) and (c, d) noisy ($\sigma_\xi = 0.1$) cases. (a, c) True solution $\boldsymbol{x}_{\text{true}}$ and obtained ones $\boldsymbol{x}$. (b, d) Input vector $\boldsymbol{y}$ and vectors $A\boldsymbol{x}$ calculated from the obtained solutions.

Here, the correct answer is defined as a solution with a small root mean squared error (RMSE) $\epsilon_x = \sqrt{\|\boldsymbol{x}_{\text{true}} - \boldsymbol{x}\|_2^2/N}$, where $\epsilon_x < 10^{-7}$ in the noiseless case ($\sigma_\xi = 0.0$), and $\epsilon_x < 0.1$ in the noisy case ($\sigma_\xi = 0.1$). The dependence on $d\lambda$ is ambiguous, and a slow increase in λ does not always provide good results. Compared with n_{smpl} dependence, n_{rplc} dependence is more prominent. As n_{rplc} determines the number of initial samples, the n_{rplc} dependence implies that the variation in the initial samples affects the final solutions. Therefore, initial samples of $\boldsymbol{x}$ are critical.

The hybrid algorithm can provide satisfactory results when the initial samples include a good approximation of the true solution. The L_1-regularization problem with the objective function $F(\boldsymbol{x}) = \frac{1}{2}\|\boldsymbol{y} - A\boldsymbol{x}\|_2^2 + \lambda_1\|\boldsymbol{x}\|_1$ can be solved efficiently using LASSO. As LASSO provides a good approximation of the original problem, the LASSO solution is used as the initial $\boldsymbol{x}$. Figures 3(a) and 3(b) show the LASSO solution in the noisy case ($\sigma_\xi = 0.1$), where hyperparameter λ_1 is tuned by cross-validation. Although several nonzero components appear in $\boldsymbol{x}$, the solution is a good approximation. Using the LASSO solution as the initial sample, the hybrid algorithm was performed with $n_{\text{rplc}} = 1$ and $n_{\text{smpl}} = 2$, and the results obtained are depicted in figures 3(c) and 3(d). The obtained solution is remarkably close to the true solution, and $\boldsymbol{y}$ and $A\boldsymbol{x}$ are similar. Compared with the time consumed by the hybrid algorithm, LASSO consumes considerably less time. Moreover, $n_{\text{rplc}} = 1$ is sufficient when the initial $\boldsymbol{x}$ is a good approximation. Therefore, the computation time can be reduced greatly by using the LASSO solution as the initial $\boldsymbol{x}$.

The simulation shown here is limited to the case where $N = 100$, $M = 25$, and $K = 5$. In general, it is more difficult to obtain correct answers is for smaller-M/N and larger-K/N cases, as theoretically predected [9, 10]. On the other hand, the performance of the proposed algorithm depends on not only the initial condition but also the components of A and $\boldsymbol{y}$. More comprehensive works will reveal the properties of this method in the future.

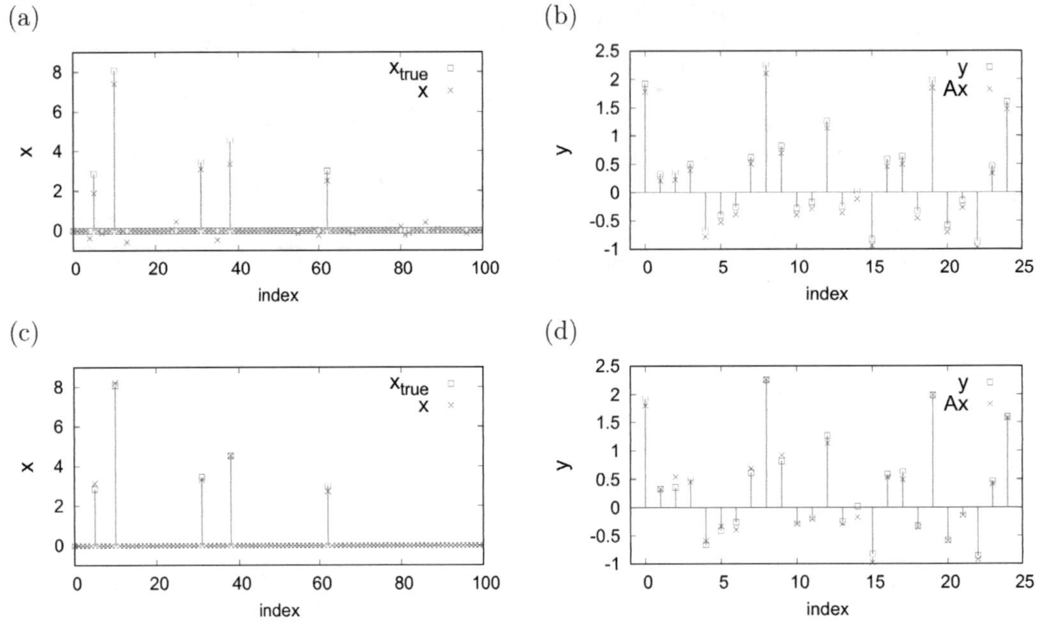

Figure 3. Solutions in the noisy case ($\sigma_\xi = 0.1$) obtained by (a, b) LASSO, and (c, d) the hybrid algorithm in which the LASSO solution is used as the initial $\boldsymbol{x}$. (a, c) True solution $\boldsymbol{x}_{\text{true}}$ and obtained ones $\boldsymbol{x}$. (b, d) Input vector $\boldsymbol{y}$ and vectors $A\boldsymbol{x}$ calculated from the obtained solutions.

4. Conclusion

A hybrid algorithm based on the QUBO formulation for L_0-regularization compressed sensing was proposed. The hybrid algorithm using an Ising machine is expected to work efficiently. Successful cases in which the algorithm works well were demonstrated. The performance of the hybrid algorithm depends highly on the initial samples. If the initial sample is a good approximation of the answer to the given problem, the hybrid algorithm provides a satisfactory result.

Acknowledgments

This work was partially supported by the JSPS KAKENHI (Grant Number JP18K11333).

References

[1] Donoho D 2006 *IEEE Trans. Inf. Theory* **52** 1289–1306
[2] Candès E, Romberg J and Tao T 2006 *IEEE Transactions on Information Theory* **52** 489–509
[3] Candès E J, Romberg J K and Tao T 2006 *Comm. Pure Appl. Math.* **59** 1207–1223
[4] Lustig M, Donoho D and Pauly J M 2007 *Magn. Resonance Med.* **58** 1182–1195
[5] Lustig M, Donoho D L, Santos J M and Pauly J M 2008 *IEEE Signal Process. Mag.* **25** 72–82
[6] Nakanishi-Ohno Y, Haze M, Yoshida Y, Hukushima K, Hasegawa Y and Okada M 2016 *J. Phys. Soc. Jpn.* **85** 093702
[7] Akai I, Iwamitsu K, Igarashi Y, Okada M, Setoyama H, Okajima T and Hirai Y 2018 *J. Phys. Soc. Jpn.* **87** 074003
[8] Tibshirani R 1996 *J. R. Stat. Soc.: Ser. B* **58** 267–288
[9] Obuchi T and Kabashima Y 2016 *J. Phys.: Conf. Ser.* **699** 012017
[10] Obuchi T, Nakanishi-Ohno Y, Okada M and Kabashima Y 2018 *J. Stat. Mech.* **2018** 103401
[11] Johnson M W, Amin M H S, Gildert S, Lanting T, Hamze F, Dickson N, Harris R, Berkley A J, Johansson J, Bunyk P, Chapple E M, Enderud C, Hilton J P, Karimi K, Ladizinsky E, Ladizinsky N, Oh T, Perminov

I, Rich C, Thom M C, Tolkacheva E, Truncik C J S, Uchaikin S, Wang J, Wilson B and Rose G 2011 *Nature* **473** 194–198
[12] McMahon P L, Marandi A, Haribara Y, Hamerly R, Langrock C, Tamate S, Inagaki T, Takesue H, Utsunomiya S, Aihara K, Byer R L, Fejer M M, Mabuchi H and Yamamoto Y 2016 *Science* **354** 614–617
[13] Inagaki T, Haribara Y, Igarashi K, Sonobe T, Tamate S, Honjo T, Marandi A, McMahon P L, Umeki T, Enbutsu K, Tadanaga O, Takenouchi H, Aihara K, Kawarabayashi K, Inoue K, Utsunomiya S and Takesue H 2016 *Science* **354** 603–606
[14] Aramon M, Rosenberg G, Valiante E, Miyazawa T, Tamura H and Katzgraber H G 2019 *Front. Phys.* **7** 48
[15] Hamerly R, Inagaki T, McMahon P L, Venturelli D, Marandi A, Onodera T, Ng E, Langrock C, Inaba K, Honjo T, Enbutsu K, Umeki T, Kasahara R, Utsunomiya S, Kako S, Kawarabayashi K i, Byer R L, Fejer M M, Mabuchi H, Englund D, Rieffel E, Takesue H and Yamamoto Y 2019 *Sci. Adv.* **5** eaau0823
[16] Goto H, Tatsumura K and Dixon A R 2019 *Sci. Adv.* **5** eaav2372
[17] Goto H, Endo K, Suzuki M, Sakai Y, Kanao T, Hamakawa Y, Hidaka R, Yamasaki M and Tatsumura K 2021 *Sci. Adv.* **7** eabe7953
[18] Aonishi T, Mimura K, Okada M and Yamamoto Y 2021 (*Preprint* `arXiv:2102.11412`)

Quantum Game Theory approach for data network routing: a solution for the congestion problem

Silva Agustin
Instituto de Investigaciones Científicas y Tecnológicas, 7600 Mar del Plata, Argentina
E-mail: agustinsilva447@gmail.com

Zabaleta Omar G.
Instituto de Investigaciones Científicas y Tecnológicas, 7600 Mar del Plata, Argentina
E-mail: gzabaleta@fi.mdp.edu.ar

Arizmendi Constancio M.
Instituto de Investigaciones Científicas y Tecnológicas, 7600 Mar del Plata, Argentina
E-mail: miguelarizmendi2001@yahoo.com.ar

Abstract. Because of the sustained growth of information and mobile users transmitting a great amount of data packets, modern network performances are being seriously affected by congestion problems. In fact, congestion management is a challenging task that can be roughly summarized as a trade off between transmission latency and cost. In this context, a novel framework is proposed for solving the congestion problem on communication networks using a quantum game theory model, where network packets compete selfishly for the shortest route. Simulations show that final network routing and traveling times achieved with the quantum game version outperform those obtained classically. This leads to the opportunity of developing full-stack protocols that may be capable of taking advantage of the quantum properties for optimizing communication systems. Due to its generality, the game theoretic approach can be applied both in local networks and in more complex networks such as 5G or LTE-A.

1. Introduction

Congestion is a significant issue in the more diverse environments, from supermarket queuing, urban traffic, transport, local network to 5G and LTE-A networks.

Congestion problems arise when users compete over a set of limited resources causing an increase in the latency to all contenders. Latency, for its part, depends on the number of agents that use a resource. A common example of congestion is at rush hour when the city's vehicle flow is saturated due to excess demand for roads by drivers. When the number of vehicles surpass the capacity of the roads, there is a decrease in velocities that cause waste of time and excessive fuel consumption. Likewise, delays, loss of efficiency and increase of transmission time occur when networks are too demanded.

At present, the constant increase in the number of packets sent over networks evidences the congestion problems and suggests that more research is needed when designing efficient communication systems. By their nature, the congestion problems become appropriate for being modeled by Game Theory [1, 2], where selfish decisions of agents (packets on the network or vehicles in a city) cause the whole system performance to be adversely affected.

Furthermore, today we know that if game theory models harness the capabilities of quantum computing [3, 4] better outcomes can emerge. Moreover, the authors of [5] present an interesting work that address the Nash equilibria and correlated equilibria of classical and quantum games in the context of their Pareto efficiency. They focus their study in three classical games: the prisoner's dilemma, battle of the sexes and the game of chicken and show that the Nash equilibria of these games in quantum mixed Pauli strategies are closer to Pareto optimal results than their classical counter-parts. In addition, in [6] a quantum game is applied to diminish spectrum allocation times and power consumption.

In this work we propose a routing protocol designed from the Quantum Game Theory formalism that significantly outperforms its classical equivalent. We demonstrate how to avoid congestion in a network and decrease the traveling time per packet in a system with a high number of packets that autonomously decide which decision is the best for everyone of them.

The work is organized as follows. In section 2, the problem and the system are presented in detail. In section 3, we explain the two possible strategies for modeling the network and develop in depth the quantum model. In section 4 the results of each protocol are compared and their performance is graphically analyzed. Finally, the work is concluded in section 5 with a debate on its consequences.

2. Traffic Congestion Model

The communication network model has $n1$ nodes and $n2$ packets to be sent between nodes. The net is generated using the Erdős–Rényi–Gilbert model [7] with $G(nodes = n1, probability = 0.5)$ (where $nodes$ is the number of nodes and $probability$ the probability that two nodes are connected to each others) with the condition that every node has at least one connection to other node. An example of a $n1 = 10$ nodes network is shown in Fig. 1 where the packets traveling through are represented with different colors.

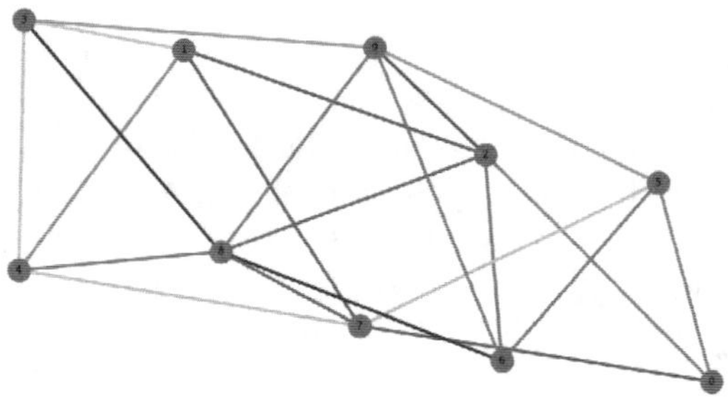

Figure 1. Network with 10 nodes and packets traveling through the edges represented in different colors.

The traveling time of the packets in a channel depends on the number of packets traveling at that particular channel. In a congested channel it will take much longer traveling time for a given packet than in a clear one. The maximum capacity assumed of every channel is one packet, with linearly declining performance from there on.

In other words, as the number of packets traveling through the same channel increases, the traveling time for each packet increases. This situation is analogous to city transport where the time a vehicle takes to travel a route increases as the number of vehicles on that route increases. If this situation is not handled well, this may result in heavy traffic congestion [8].

Journal of Physics: Conference Series **2207** (2022) 012034 doi:10.1088/1742-6596/2207/1/012034

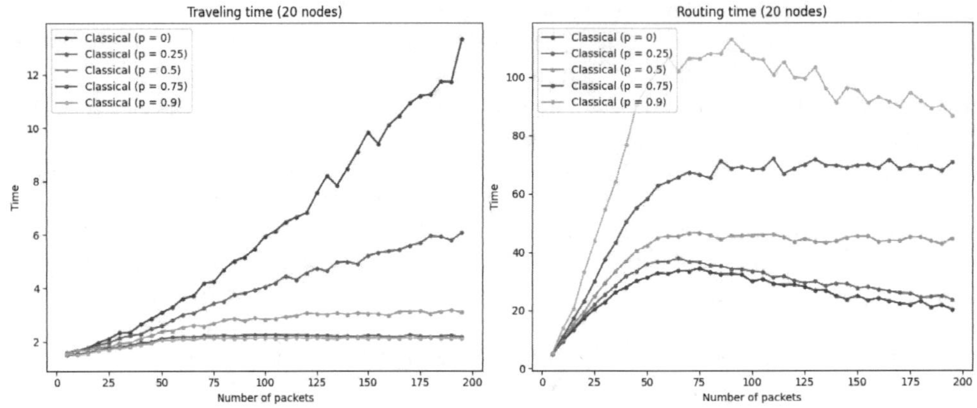

Figure 2. Graphs for different probabilities p of: (a) Traveling time as a function of the number of packets. (b) Routing time as a function of the number of packets.

Every packet is thought as a player in a game that prefers the shortest path channel. The traveling channel choice is made by playing a game. Whenever many players choose the same channel, the traveling time for that channel increases. Packet congestion makes traveling time much longer. Finally, in the game played on every congested channel, each player has two options: to travel through the preferred channel or to search for another path.

If every player decides to choose his shortest path, some network channels will be overloaded with all the packets that are disputing for it, slowing down the network. Therefore, if players behave in a selfish way, always choosing their shortest path, the whole communication network will be harmed. This simple model reflects the massive problem faced in modern communication technologies which poses a severe challenge to the existing routing strategies.

3. Classical and Quantum Strategies

The two strategies: choosing the preferred shortest channel and the search for another one, will be called options 1 and 0 respectively. The playing games may be of different kind, that is classical or quantum. We consider first classical games. The player's mixed classical strategies are probabilistic, that is, option 0 is chosen with probability p and option 1 with probability $1 - p$. Then, probability p near *zero* corresponds to greedy players, since players will tend to always take the shortest path even when many players are competing for that channel. A value of p closer to *one* will create players more patiently looking for another clearer route.

In order to analyze the performance of the system two variables are going to be measured: routing and traveling time. The traveling time will be how long it took for each packet to go from origin to final destination. On the other hand, the routing time is associated with the number of tries needed to find the final route. It is important to emphasize the fact that whenever a channel is congested, packets are going to play for it using classic or quantum strategies depending on the nature of the protocol. The more games a packet plays, the longer its routing time.

Simulations of the classical game are shown in figure 2 where all players have the same probability p. Fig 2(a) displays the average traveling time per packet increase with the increasing number of packets for different probabilities p. The variation of the average routing time with the number of packets is shown alongside in Fig. 2(b).

The simulations were done on networks of $n1 = 20$ nodes and $n2 = 200$ packets and averaging the performance of 50 different cases. The qualitative behavior of the network, that is, the dynamics reflected in the graphs, is independent of the number of nodes. The only difference is

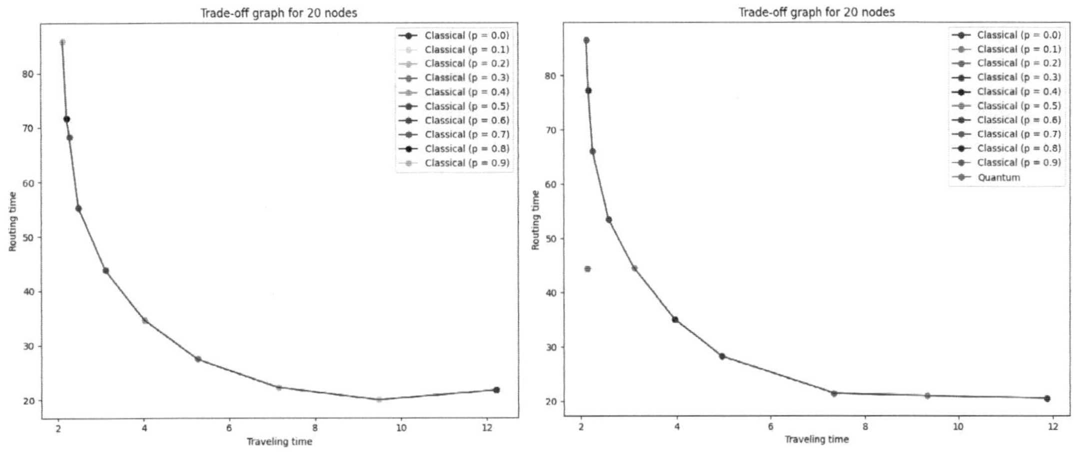

Figure 3. Trade-Off between traveling and routing time: (a) Only classical for different values of p. (b) Both classical and quantum.

that the network stabilizes for a larger number of packets when the number of nodes increases. Analyzing the graphs it is possible to conclude that there is a kind of trade-off. Given a fixed number of packets and a low p, the traveling time per packet increases but the routing time decreases. On the other hand, when p increases, the traveling time per packet decreases but the routing time increases.

This effect can be seen in Fig. 3 (a) where traveling and routing time for different probabilities p are shown. These results are obtained when the system reaches steady state with a high number of packets ($n2 = 200$). As far as the classic game with mixed strategies is considered this is the result of changing the channel choice option probability p.

To study the quantum game for channel choice we follow the EWL protocol for 2 players [4]. The circuit for this protocol can be seen in Figure 4, where the first unitary gate correspond to $J = \frac{1}{\sqrt{2}}(I^{\otimes N} + i\sigma_x^{\otimes N})$ with $N = 2$ and the second to $J^\dagger$ (J conjugate transpose).

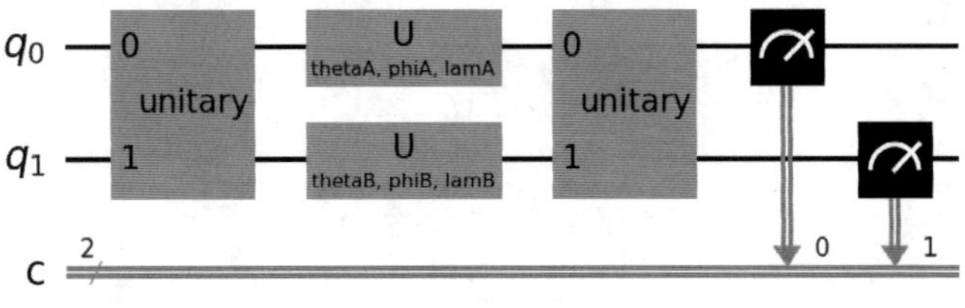

Figure 4. Circuit corresponding to the EWL protocol for 2 players.

The U gates are the different strategies that every players choose individually, where its more general representation is the unitary matrix: $U(\theta, \phi, \lambda) = \begin{pmatrix} \cos(\frac{\theta}{2}) & -e^{i\lambda}\sin(\frac{\theta}{2}) \\ e^{i\phi}\sin(\frac{\theta}{2}) & e^{i(\phi+\lambda)}\cos(\frac{\theta}{2}) \end{pmatrix}$ [9]. We can already highlight the fact that while classical players could modify only one parameter to choose his mixed strategy: p, quantum players have three parameters to choose in order to select his own pure strategy: θ, ϕ and λ. The game finishes when measuring the state of the qubits to know the output of the circuit and, therefore, the final action of each player.

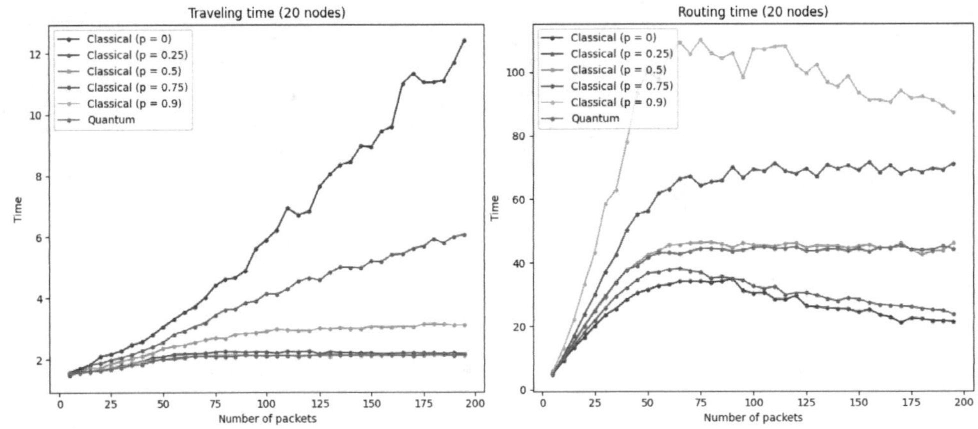

Figure 5. Graphs for different probabilities p and the quantum case: (a) Traveling time as a function of the number of packets. (b) Routing time as a function of the number of packets.

4. Results

In this section we compare the results of the protocol with quantum players against those shown previously with classical players. Every quantum player applies rotations to his qubit in the three axes: X, Y and Z. The matrices for the rotations are:

$$R_X(\varphi) = \begin{pmatrix} \cos(\frac{\varphi}{2}) & -i\sin(\frac{\varphi}{2}) \\ -i\sin(\frac{\varphi}{2}) & \cos(\frac{\varphi}{2}) \end{pmatrix}; R_Y(\varphi) = \begin{pmatrix} \cos(\frac{\varphi}{2}) & -\sin(\frac{\varphi}{2}) \\ \sin(\frac{\varphi}{2}) & \cos(\frac{\varphi}{2}) \end{pmatrix}; R_Z(\varphi) = \begin{pmatrix} e^{-i\frac{\varphi}{2}} & 0 \\ 0 & e^{i\frac{\varphi}{2}} \end{pmatrix}$$

Therefore, every player must choose 3 angles that we will call φ_X, φ_Y and φ_Z . In our design, in two-player games, we propose the strategy $\varphi_X = \frac{\pi}{2}$, $\varphi_Y = \frac{\pi}{4}$ and $\varphi_Z = 0$ for every player as shown in figure 6.

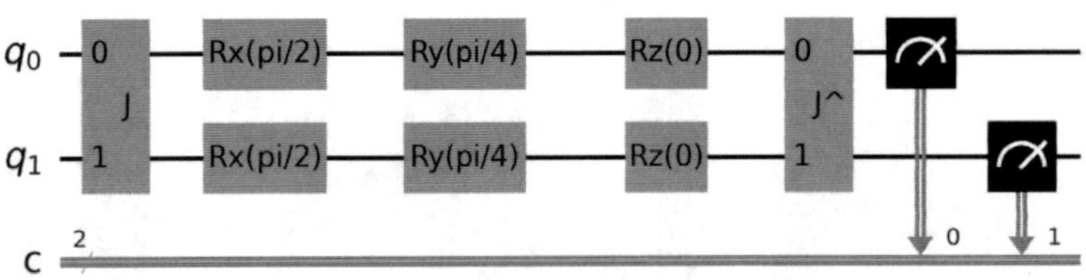

Figure 6. Circuit corresponding to the routing protocol for 2 players.

In Figs 5(a) and 5(b) the quantum protocol performance is shown with the classical performance. The quantum game shows a lowest traveling time and medium routing time compared to the classical one. With this in mind we can recalculate graph 3(a) and add the quantum case. By doing this we obtain figure 3(b) where we can observe how the classical barrier is broken if we work with a quantum protocol instead of the classical one. Thus we will obtain a performance that enhance any classical performance by simultaneously reaching less routing time and less traveling time.

This advantage may be understood considering the quantum state of the circuit for two players (figure 6) just before measuring, is: $|\psi_{out}\rangle = J^\dagger(R_z(0)\otimes R_z(0))(R_y(\frac{\pi}{4})\otimes R_y(\frac{\pi}{4}))(R_x(\frac{\pi}{2})\otimes R_x(\frac{\pi}{2}))J|00\rangle = -j\frac{|01\rangle+|10\rangle}{\sqrt{2}}$.

That is, state $|01\rangle$ will be measured with a 50% probability and state $|10\rangle$ with a 50% probability. This strategy turns out to be Pareto optimal since no player can improve his performance without worsening someone else's. This $|\psi_{out}\rangle = -j\frac{|01\rangle+|10\rangle}{\sqrt{2}}$ means that one of the two players will always take the channel and the other will not. By avoiding state $|11\rangle$ we are avoiding the case where the two players take over the channel, therefore, avoiding network congestion.

5. Conclusions

In the present work, the congestion problem was modeled using the techniques of network design and game theory. It was verified that when classical players behave selfishly and try to maximize their own benefit, both themselves and the network as a whole suffer a significant drop in performance.

The model was then extended by adding quantum strategies as proposed in the EWL protocol in the framework of quantum game theory. Quantum strategies were represented by a three-parameter one-qubit quantum gate model. Then, it was verified that the possibilities of the players were expanded when they went from being classical to quantum.

The final routing and traveling times achieved in the network with the quantum game model outperform those obtained classically. Although these values correspond to a Pareto-optimum value of the system, it will remain for future works to find the Nash equilibrium and whether this also corresponds to a Pareto-optimum or not. Another interesting scenario to consider in future works would be to calculate the shortest path using quantum algorithms to take full advantage of the properties of quantum computing to design routing protocols.

We have shown that a quantum game formalism applied to the communication network enhances its efficiency when dealing with congestion problems. Consequently, a new world of opportunities could emerge in these types of complex systems when taking advantage of the possibilities offered by quantum computing. Moreover, they might bring with them solutions that create remarkably more efficient systems.

References

[1] Oskar Morgenstern and John Von Neumann. *Theory of games and economic behavior.* Princeton university press, 1953.

[2] John F Nash et al. "Equilibrium points in n-person games". In: *Proceedings of the national academy of sciences* 36.1 (1950), pp. 48–49.

[3] David A Meyer. "Quantum strategies". In: *Physical Review Letters* 82.5 (1999), p. 1052.

[4] Jens Eisert, Martin Wilkens, and Maciej Lewenstein. "Quantum games and quantum strategies". In: *Physical Review Letters* 83.15 (1999), p. 3077.

[5] Marek Szopa. "Efficiency of classical and quantum games equilibria". In: *Entropy* 23.5 (2021), p. 506.

[6] Omar Gustavo Zabaleta, Juan Pablo Barrangú, and Constancio M Arizmendi. "Quantum game application to spectrum scarcity problems". In: *Physica A: Statistical Mechanics and its Applications* 466 (2017), pp. 455–461.

[7] Edgar N Gilbert. "Random graphs". In: *The Annals of Mathematical Statistics* 30.4 (1959), pp. 1141–1144.

[8] Mohammad Reza Jabbarpour et al. "Applications of computational intelligence in vehicle traffic congestion problem: a survey". In: *Soft Computing* 22.7 (2018), pp. 2299–2320.

[9] Michael A Nielsen and Isaac Chuang. *Quantum computation and quantum information.* 2002.

Journal of Physics: Conference Series **2207** (2022) 012035 doi:10.1088/1742-6596/2207/1/012035

Four-body variational calculation of a hydrogen-like atom involving an excited muonic molecule

Takuma Yamashita[1,2], **Motoaki Niiyama**[2], **Kazuhiro Yasuda**[2], **Yasushi Kino**[2]

[1]Institute for Excellence in Higher Education, Tohoku University, Sendai, Miyagi, 980-8576, Japan
[2]Department of Chemistry, Tohoku University, Sendai, Miyagi, 980-8578, Japan

E-mail: tyamashita@tohoku.ac.jp

Abstract. We report a four-body variational calculation of a hydrogen-like atom consisting of an excited muonic molecule consisting of d, t, and μ, and a ground state electron. Due to the compact size of the muonic molecule dtμ, it behaves as a quasi-nucleus for the electron; however, the system is actually a resonance state because the de-excitation energy of dtμ is sufficient to ionize the electron. We calculate resonance energy levels of the four-body system dtμe using a Gaussian expansion method and a stabilization method. Our best calculation results in a four-body energy of $-100.159\,88$ a.u. which is in good agreement with the value estimated by the first order perturbation theory [Harston *et al.* Zeitschrift für Physik D **22**, 635 (1992)]. We indicate that implementation of the electron-induced polarization of dtμ will be indispensable for the further precise determination of the resonance energy. The present result is the first step towards a full four-body calculation of the muonic molecule under the presence of the electron, which is of importance for development of a muon catalyzed fusion kinetics model.

1. Introduction

A muonic molecule consisting of a negative muon (μ) and two nuclei of hydrogen, e.g. ppμ, pdμ, ddμ, dtμ, and ttμ, has a compact structure whose size is more than 100 times smaller than a hydrogen molecular ion, e.g. H_2^+. Since the muonic molecule is a positively charged ion, it can behave as a quasi-nucleus in atomic/molecular systems. A hydrogen-like atom consisting of the muonic molecule and an electron (e), in particular dtμe, has attracted attention in studies of muon catalyzed fusion (μCF) [1–4].

The formation of the muonic molecules in D_2/T_2 mixture has been investigated both theoretically and experimentally in the context of the μCF. The dtμ is considered to be formed in the D_2 molecule resonantly in a collision between $t\mu(1s)$ and D_2,

$$t\mu(1s) + D_2 \rightarrow (dt\mu)e - de \tag{1}$$

by transferring the binding energy of the dtμ excited state ($\Lambda = \upsilon = 1$; Λ denotes the rotational quantum number and υ the vibrational quantum number), 660 meV [5], to rovibrational excitation energy of the six-body system, (dtμ)e-de [6]. Here, (dtμ)e is a hydrogen-like atom whose nucleus is replaced by the dtμ molecule.

Since the process (1) is a resonant collision, the precise energy levels of dtμ under the presence of the electron plays an indispensable role in the prediction of the formation rate of dtμ. While the three-body properties (energy levels, structures and lifetimes) of the muonic molecules have been investigated precisely [5, 7], the four-body properties of these muonic molecules in the electron cloud are still on

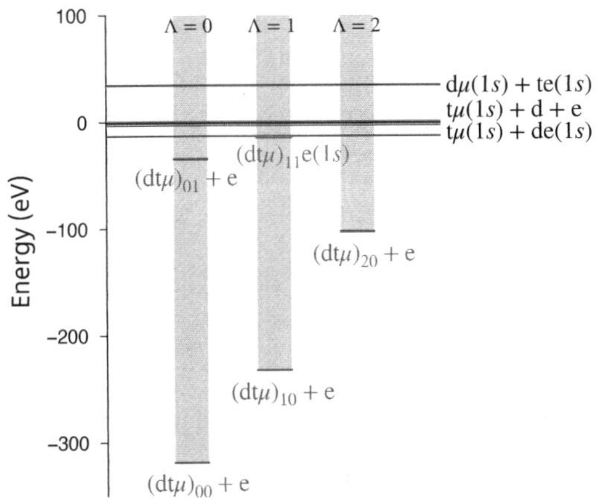

Figure 1. A resonance energy level of $(dt\mu)_{11}e(1s)$ (red line), the Auger decay threshold energies (blue lines), and neutral break-up threshold energies (black lines) are illustrated. The gray shades indicate the continuum state of the Auger electron. Three columns are drawn for each $\Lambda = 0, 1$, and 2.

discussion. In the case of $dt\mu e$, the four-body energy $E_{\mathrm{dt}\mu\mathrm{e}}$ can be expressed as

$$E_{\mathrm{dt}\mu\mathrm{e}} = \varepsilon_{\mathrm{dt}\mu} + \varepsilon_{\mathrm{e}} + \Delta E^{(\mathrm{FS})}, \tag{2}$$

where $\varepsilon_{\mathrm{dt}\mu}$ is the three-body energy of dtμ, ε_{e} is the binding energy of 1s electron to a point charge having the same mass as the dtμ. The third term $\Delta E^{(\mathrm{FS})}$ is called an energy shift by finite size effects of the quasi-nucleus dtμ. While the dtμ has a compact structure, the three-body system has a finite charge distribution, and might be deformed under the presence of the electron. Harston *et al.* adopted a perturbative approach [8, 9] to calculate $\Delta E^{(\mathrm{FS})}$; however, the first order perturbation energy $\Delta E^{(1)} = 18.253$ meV of the $\Delta E^{(\mathrm{FS})}$ is found to be comparable in opposite sign to the second order perturbation energy, $\Delta E^{(2)} = -17.752$ meV, which implies a slow convergence of the perturbative expansion of $\Delta E^{(\mathrm{FS})} = \sum_i \Delta E^{(i)}$. Thus, a full variational calculation is desirable.

Recent experimental and theoretical developments shed a light on the fundamental muonic atom processes. A high energy-resolution X-ray spectrometer utilizing superconducting transition has revealed dynamic de-excitation processes of muonic atom [10] under the electron refilling. A muon beam at the J-PARC facility has been upgraded in its intensity. A project that aims at development of ultra-slow muon beam [11], utilizing the μCF process as a muon moderator is in progress [12–14]. μCF kinetics models are also reconsidered in some theories [15, 16].

In this work, inspired by these situations, we report a variational treatment of the dtμe four-body system without using the perturbation theory. We expand the four-body wave function in terms of Gaussian basis functions and convert the Schrödinger equation to a generalized eigenvalue problem. Since the muonic molecules in the electron cloud are unstable against the Auger transitions, we treat the dtμe as a resonance state.

Atomic units (a.u.; $m_{\mathrm{e}} = \hbar = e = 1$) are used throughout this paper, except where mentioned otherwise.

2. Theory

Hereafter, for brevity, we denote the rovibrational state of dtμ as $(\mathrm{dt}\mu)_{\Lambda v}$. When an electron attaches to the $(\mathrm{dt}\mu)_{\Lambda v}$ and occupies the 1s orbital, the four-body system is denoted as $(\mathrm{dt}\mu)_{\Lambda v}\mathrm{e}(1s)$. In this work, we have a particular focus on the resonance energy level of $\mathrm{dt}\mu_{11}\mathrm{e}(1s)$.

Figure 1 displays the expected $(\mathrm{dt}\mu)_{11}\mathrm{e}(1s)$ energy levels taking the t$\mu(1s)$ + d + e three-body break-up threshold energy to be zero. The $(\mathrm{dt}\mu)_{11}\mathrm{e}(1s)$ level is located just below the t$\mu(1s)$ + de$(1s)$ threshold energy, and possesses a total orbital angular momentum $J = 1$. The $(\mathrm{dt}\mu)_{11}\mathrm{e}(1s)$ has several Auger decay channels, namely, $(\mathrm{dt}\mu)_{\Lambda' v'}$ + e where $(\Lambda', v') = (0, 0), (0, 1), (1, 0), (2, 0)$. Thus, $(\mathrm{dt}\mu)_{11}\mathrm{e}(1s)$ should be treated as a resonance state which is unstable against the Auger decay processes.

Journal of Physics: Conference Series **2207** (2022) 012035 doi:10.1088/1742-6596/2207/1/012035

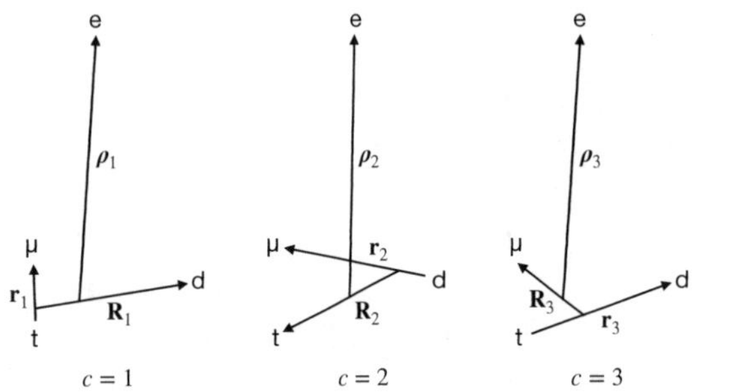

Figure 2. Jacobi coordinate systems $c = 1, 2$, and 3.

In the present calculation, we consider a Schrödinger equation $(H - E_\nu)\Psi_\nu = 0$ where the four-body Hamiltonian H contains kinetic energy operators (separated from the center-of-mass motion) and all inter-particle Coulomb potential energy operators. We adopt a Gaussian expansion method (see reviews [17–19] and our recent calculations [20–24]) for the four-body wave function Ψ_ν. The Ψ_ν is expanded in terms of Gaussian basis functions $\phi_{il}(r) = N_{il} r^l \exp(-r^2)$ (N_{il} is a normalization constant) and spherical harmonics as follows:

$$\Psi_\nu = \sum_{\Lambda,\lambda} \sum_{k}^{n} \sum_{c} \sum_{il_c} \sum_{jL_c} C^{(\nu)}_{cl_cL_c\lambda,ijk} \phi_{il_c}(b_i r_c) \phi_{jL_c}(B_j R_c) \phi_{k\lambda}(\beta_k \rho_c) \left[\left[Y_{l_c}(\hat{\mathbf{r}}_c) \otimes Y_{L_c}(\hat{\mathbf{R}}_c) \right]_\Lambda Y_\lambda(\hat{\rho}_c) \right]_{JM} \quad (3)$$

where $[\cdots]_{JM}$ denotes linear combination of the spherical harmonics which are weighted by Clebsch-Gordan coefficients as to form the eigenfunctions of the total orbital angular momentum of J and its projection M on to the quantization axis. Λ, a vector sum of $\mathbf{l}_c$ and $\mathbf{L}_c$, specifies the rotational state of dtμ and the λ denotes the orbital angular momentum of the electron with respect to the center-of-mass of the dtμ molecule. Non-linear coefficients $\{b_i\}$, $\{B_j\}$ and $\{\beta_k\}$ are chosen in accordance with geometrical progression. The linear coefficients $\{C^{(\nu)}_{cl_cL_c\lambda,ijk}\}$ are determined by the Rayleigh-Ritz variational method. Owing to the non-orthogonal basis functions of Ψ_ν, the Schrödinger equation is converted to a generalized matrix eigenvalue problem.

The c denotes an index of coordinate systems. We use three Jacobi coordinate systems $c = 1, 2$, and 3 as shown in figure 2. Since the muonic molecule dtμ has a compact structure in compared with the electron distribution, the coordinate ρ_c is chosen as a common vector from the center-of-mass of dtμ to the electron. Therefore, these three coordinate systems are introduced to facilitate the description of the d-t-μ three-body correlation in the presence of the electron.

The major components of the $(\mathrm{dt}\mu)_{11}\mathrm{e}(1s)$ state can be described by $\Lambda = 1$ and $\lambda = 0$ basis functions. Although there can be the Auger decay channels, $(\mathrm{dt}\mu)_{10}$ + e, $(\mathrm{dt}\mu)_{00}$ + e, $(\mathrm{dt}\mu)_{01}$ + e, and $(\mathrm{dt}\mu)_{20}$ + e, described by $\Lambda = 0, 2$ and compatible λ components, in the present work, we limit ourselves to include $\Lambda = 1$ and $\lambda = 0$ component. The full calculation including $0 \leq \Lambda \leq 3$ and all compatible λ is in progress and will appear in our following paper. This paper focuses on the role of $\Lambda = 1$ and $\lambda = 0$ basis functions and the validity of the four-body variational calculations.

The resonance energy levels are obtained by a stabilization method [25–29]. We introduce a stabilization parameter α and scales the all coordinates as $r_c \to \alpha^{-1} r_c$, $R_c \to \alpha^{-1} R_c$, and $\rho_c \to \alpha^{-1} \rho_c$ before the diagonalization of Hamiltonian. This procedure slightly changes the expansion ranges of the basis functions, and gives eigenvalues $\{E_\nu\}$ as a function of α. For a bound state, the eigenenergy E_ν is independent from the small change of α if the number and the range of basis functions are sufficiently large. Since the basis functions are square integrable, the generalized matrix eigenvalue problem results in a discretization of the continuum states by superimposing scattering wave functions. The present calculation provides these superimposed wave functions above the lowest threshold energy, which are so-called *pseudo*-states. Eigenvalues of the *pseudo*-states strongly depend on the choice of α, in contrast to the bound states.

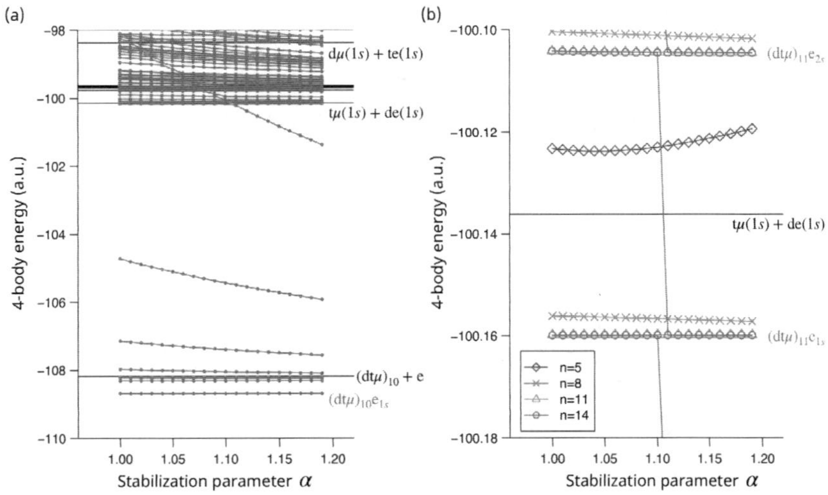

Figure 3. Four-body eigenvalues are plotted against the stabilization parameter α. (a) The eigenvalues are obtained by $n = 14$ basis functions for the electron. (b) A close-up view of panel (a) in the vicinity of the resonance energy of $(\mathrm{dt}\mu)_{11}\mathrm{e}(1s)$. The different colors correspond to the different basis functions as indicated by the legends.

Although a resonance state is basically a scattering state, it has a large amplitude in the interaction region like a bound state. Therefore, in the vicinity of the resonance energy, within the energy width of the resonance state, the eigenvalues are stable against α and can be distinguished from the non-resonant states. This procedure has been adopted to a variety of atomic/molecular systems.

3. Results and Discussion

The non-linear parameters of $\{b_i\}$ and $\{B_j\}$ are optimized so that the binding energy of the subsystem dtμ is reproduced in 6-digit accuracy. The number of basis functions that describe the dtμ quasi nucleus is $N_{\mathrm{dt}\mu} = 2309$.

Figure 3(a) shows the eigenvalues $\{E_\nu\}$ as a function of the stabilization parameter α calculated using $n = 14$ basis functions (thus, the total number of basis functions is $N_{\mathrm{dt}\mu}n = 32326$). We see three stable levels below the $(\mathrm{dt}\mu)_{10}$ + e threshold energy. These levels are attributed to the electronic $1s$, $2s$, and $3s$ states with the $(\mathrm{dt}\mu)_{10}$ quasi-nucleus. They are apparently bound states because we do not include the basis functions that describe Auger decay processes into $(\mathrm{dt}\mu)_{\Lambda\upsilon}$ + e, $\Lambda = 0, 2$.

Above the $(\mathrm{dt}\mu)_{10}$ + e threshold energy, the continuum states are discretized by the generalized matrix eigenvalue problem. These discretized continuum levels gradually decrease as the α increases because the increase of α elongates the expansion range along with ρ_c and decreases the kinetic energy of the electron.

Figure 3(b) is a close-up view of the figure 3(a) in the vicinity of the $(\mathrm{dt}\mu)_{11}\mathrm{e}_1$ energy, $E_{\mathrm{dt}\mu\mathrm{e}}$. We have examined the convergence of $E_{\mathrm{dt}\mu\mathrm{e}}$ as increasing the number of basis functions, n in eq. (3). Among the results of $n = 5, 8, 11$, and 14, $n \geq 8$ result in the resonance energy below the t$\mu(1s)$ + de$(1s)$ threshold energy. The $E_{\mathrm{dt}\mu\mathrm{e}}$ shows a good agreement between the $n = 11$ and $n = 14$.

Table 1 list the obtained $E_{\mathrm{dt}\mu\mathrm{e}}$ with the number of basis functions n and $N_{\mathrm{tot}} = N_{\mathrm{dt}\mu} \cdot n$. Here, $\Delta E^{(\mathrm{FS})}$ is calculated by the eq. (2). Our best $\Delta E^{(\mathrm{FS})}$ is found to be very close to the first order perturbation energy

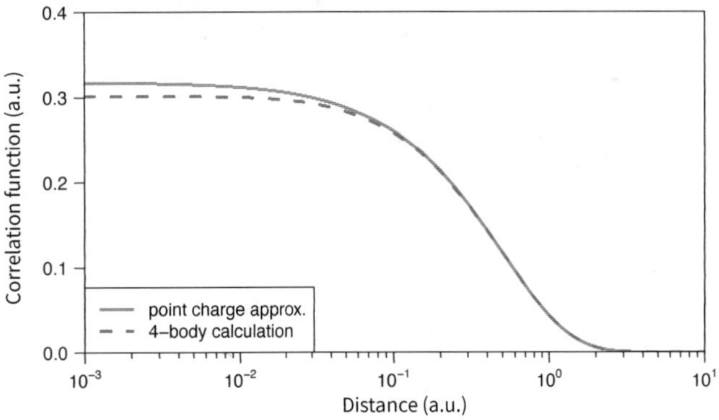

Figure 4. Four-body correlation function of $C(\rho_c)$ are compared with the correlation function of the point charge approximation.

$\Delta E^{(1)} = 18.253$ meV derived in [8]. The $\Delta E^{(1)}$ is given as

$$\Delta E^{(1)} = \langle \psi_{1s}\Phi_{\mathrm{dt}\mu}| - r_{\mathrm{de}}^{-1} - r_{\mathrm{te}}^{-1} + r_{\mu\mathrm{e}}^{-1}|\psi_{1s}\Phi_{\mathrm{dt}\mu}\rangle, \tag{4}$$

where $\Phi_{\mathrm{dt}\mu}$ is a three-body wave function of $(\mathrm{dt}\mu)_{11}$. ψ_{1s} is an electron wave function of a hydrogen atom in which the nucleus is replaced by a point charge having the same mass as the total mass of d, t, and μ. r_{xy} denotes the inter-particle distance between the particles x and y.

Harston *et al.* also reported the second order perturbation energy $\Delta E^{(2)} = -17.752$ meV, is comparable to the first order one. The fact that $\Delta E^{(\mathrm{FS})}$ of this work is similar to the $\Delta E^{(1)}$ but not to the $\Delta E^{(1)} + \Delta E^{(2)}$ can be interpreted as follows: In the present calculation, we do not include the $\Lambda = 0, 2$ components in the basis functions. This restriction forces the quasi-nucleus $(\mathrm{dt}\mu)_{11}$ not to be drastically deformed even under the presence of the electron. Note that the structure of $(\mathrm{dt}\mu)_{11}$ can be slightly changed by the interaction with the electron from the structure of the pure three-body state. The $\Lambda = 0$ components are expected to be of importance for the evaluation of the Auger processes because the $(\mathrm{dt}\mu)_{01}$ + e is the closest dissociation threshold energy below the $(\mathrm{dt}\mu)_{11}\mathrm{e}(1s)$ resonance state. Moreover, the $\Lambda = 2$ components may describe the polarization of $(\mathrm{dt}\mu)_{11}$ induced by the electron, which may deepen the resonance energy. Our preliminary calculation including $\Lambda = 2$ component, indeed, results in a deeper resonance energy level, which gives an upper limit of $\Delta E^{(\mathrm{FS})} = 9$ meV.

Finally, in figure 4, we show an electronic correlation function obtained by

$$C(\rho_c') = \langle \Psi_v|\delta(\rho_c' - \rho_c)|\Psi_v\rangle. \tag{5}$$

For comparison, we also illustrate the correlation function of a hydrogen atom with a point charge nucleus whose mass is equal to the total mass of d, t, and μ. One can see that the short range behavior of

Table 1. Four-body resonance energies of $(\mathrm{dt}\mu)_{11}\mathrm{e}(1s)$, $E_{\mathrm{dt}\mu\mathrm{e}}$, obtained by different basis sets are listed. $\Delta E^{(\mathrm{FS})}$ is calculated by eq. (2). $\Delta E^{(1)}$ and $\Delta E^{(2)}$ are the first and second order perturbation energies of $\Delta E^{(\mathrm{FS})}$ derived in the previous perturbation calculation (see text).

n	N_{tot}	$E_{\mathrm{dt}\mu\mathrm{e}}$ (a.u.)	$\Delta E^{(\mathrm{FS})}$ (meV)
5	11545	−100.123 30	1031
8	18472	−100.156 13	121
11	25399	−100.159 83	19.7
14	32326	−100.159 88	18.3
$\Delta E^{(1)}$ [8]			18.253
$\Delta E^{(2)}$ [8]			−17.752

$C(\rho_c')$ obtained by the four-body variational calculation differs from that derived by the point charge approximation. In the latter case, the electronic wave function has a cusp at $\rho_c = 0$, while in the four-body calculation the cusp is suppressed by the finite size distribution of the charges of d$t\mu$.

4. Conclusion

We have presented the four-body variational calculation of a hydrogen-like atom having an excited d$t\mu$ molecule as a quasi-nucleus. The resonance energy level of $(\mathrm{dt}\mu)_{11}\mathrm{e}(1s)$ is calculated to be $-100.159\,88$ a.u. that is shallower than the energy calculated by assuming the $(\mathrm{dt}\mu)_{11}$ a point charge. The shift by the finite size effect is found to be 18.3 meV, which is in a good agreement with the first order perturbation energy reported by Harston *et al.* [8].

While the present calculation is a fully variational approach, the basis functions are restricted. The resonance energy level will be deepened by including the $\Lambda = 0, 2$ components because they allow the *electron-induced* polarization of $(\mathrm{dt}\mu)_{11}$ quasi-nucleus. In addition, these basis functions will significantly contribute to precise evaluation of the Auger decay lifetime of this resonance state, because the Auger decay branch of $(\mathrm{dt}\mu)_{11}\mathrm{e}(1s) \to (\mathrm{dt}\mu)_{01}$ + e is expected to be more dominant than $(\mathrm{dt}\mu)_{11}\mathrm{e}(1s) \to (\mathrm{dt}\mu)_{10}$ + e due to the energy difference. A full four-body calculation of $(\mathrm{dt}\mu)_{11}\mathrm{e}(1s)$ including *all* Auger decay branches will appear in our later paper.

Acknowledgments

This work was financially supported by JSPS KAKENHI for Grant Numbers JP18H05461, JP20K14381, and JP21H00115. The computation was conducted on the computer systems ITO at Kyushu University, FLOW at Nagoya University, and HOKUSAI at RIKEN.

References

[1] Breunlich W H, Kammel P, Cohen J S and Leon M 1989 *Annu. Rev. Nucl. Part. Sci.* **39** 311
[2] Rafelski H E, Harley D, Shin G R and Rafelski J 1991 *J. Phys. B: At. Mol. Opt. Phys.* **24** 1469
[3] Froelich P 1992 *Adv. Phys.* **41** 405
[4] Nagamine K and Kamimura M 1998 *Adv. Nucl. Phys.* **24** 150
[5] Kamimura M 1988 *Phys. Rev. A* **38** 621
[6] Vesman E A 1967 *ZhETF Pisma Redaktsiiu* **5** 113
[7] Hara S and Ishihara T 1989 *Phys. Rev. A* **40** 4232
[8] Harston M R, Shimammura I and Kamimura M 1992 *Z. Phys. D* **22** 635
[9] Harston M R, Hara S, Kino Y, Shimamura I, Sato H and Kamimura M 1997 *Phys. Rev. A* **56** 2685
[10] Okumura T *et al.* 2021 *Phys. Rev. Lett.* **127** 053001
[11] Nagamine K 1989 *Proc. Jpn. Academy Ser. B* **65** 225–228
[12] Natori H 2020 *PoS* **NuFact2019** 090
[13] Okutsu K *et al.* 2021 *Fusion Eng. Des.* **170** 112712
[14] Yamashita T *et al.* 2021 *Fus. Eng. Des.* **169** 112580
[15] Mori Y 2021 *Progress of Theoretical and Experimental Physics*
[16] Yamashita T, Okutsu K, Kino Y, Okada S and Sato M submitted
[17] Hiyama E, Kino Y and Kamimura M 2003 *Prog. Part. Nucl. Phys.* **51** 223
[18] Hiyama E 2012 *Prog. Theo. Exp. Phys.* **2012** 01A204
[19] Hiyama E and Kamimura M 2018 *Front. Phys.* **13** 132106
[20] Yamashita T, Umair M and Kino Y 2017 *J. Phys. B: At. Mol. Opt. Phys.* **50** 205002
[21] Yamashita T and Kino Y 2018 *Eur. Phys. J. D* **72** 13
[22] Yamashita T and Kino Y 2019 *Phys. Rev. A* **100** 062511
[23] Yamashita T, Kino Y, Hiyama E, Jonsell S and Froelich P 2021 *New J. Phys.* **23** 012001
[24] Yamashita T, Kino Y, Hiyama E, Jonsell S and Froelich P 2021 *Few-Body Systems* **62** 81
[25] Taylor H S, Nazaroff G V and Golebiewski A 1966 *J. Chem. Phys.* **45** 2872
[26] Holøien E and Midtdal J 1966 *J. Chem. Phys.* **45** 2209
[27] Lipsky L and Russek A 1966 *Phys. Rev.* **142** 59
[28] Eliezer I, Taylor H S and Williams J K 1967 *J. Chem. Phys.* **47** 2165
[29] Hazi A U and Taylor H S 1970 *Phys. Rev. A* **1** 1109

Journal of Physics: Conference Series **2207** (2022) 012036 doi:10.1088/1742-6596/2207/1/012036

Density-matrix renormalization-group study for a two-dimensional lattice-gas on the Ih-ice prism surface

Noriko Akutsu

Faculty of Engineering, Osaka Electro-Communication University, Hatsu-cho, Neyagawa, Osaka 572-8530, Japan

E-mail: nori3@phys.osakac.ac.jp

Abstract. The phase transition temperature of a two-dimensional (2D) lattice-gas model for the Ih-ice prism surface ((10$\bar{1}$0) surface on a wurtzite crystal structure) is calculated using the density-matrix renormalization-group (DMRG) method. Since the unit cell on the prism surface contains four atoms without inversion symmetry, the asymmetric version of the DMRG is required to obtain a precise value of the phase transition temperature. The effective bond energy on the surface is obtained by comparing the phase transition temperature with the faceting transition temperature observed for a real ice crystal. Using the effective bond energy, the temperature dependence of the coverage on sub-lattice sites is calculated. The surface tensions on the (0001) surface (the basal plane) and the prism plane are also calculated. The surface tension of the basal plane for the ice/water interface agrees well with the recent calculated values obtained by large scale molecular dynamics.

1. Introduction

Ice is an important material for environmental science, medical science, and food science, and ice crystals have been studied extensively in many fields. However, the fundamental surface quantities of ice have not yet been fully determined. One reason for this is that the crystal structure of Ih-ice, which is the wurtzite structure (Fig. 1), is too complicated to apply sophisticated methods for precise calculations. Also, the surface and interface roughness makes the surface structure complex.

Usually, crystal/vapour interfaces (surfaces) are expected to be well defined. However, for the Ih-ice/vapour surface, a metastable quasi-liquid layer (QLL) [1]–[14] exists on the surface. A recent simulation study [15] showed that ice-like molecules can be distinguished from liquid-like molecules locally by using a tetrahedral order parameter [16, 17]. In addition, detailed ice/QLL interface laser confocal microscopy observations combined with differential interference contrast microscopy (LCM-DIM) [18] have revealed kinks and steps on the surface between the ice and the QLL [19]. Hence a lattice model with a sharp surface will be a good approximate model to express crystal/vapour surfaces of Ih-ice.

For the Ih-ice/water interface, measurements of a roughening transition of the (10$\bar{1}$0) plane under 165 MPa have been reported.[20, 21] The faceting transition was observed optically and the crystal growth mode was also studied in Ref. [20, 21]. The faceting transition is known to occur at the same temperature as the roughening transition (Fig. 2) [22, 23]. Generally, for

Journal of Physics: Conference Series **2207** (2022) 012036 doi:10.1088/1742-6596/2207/1/012036

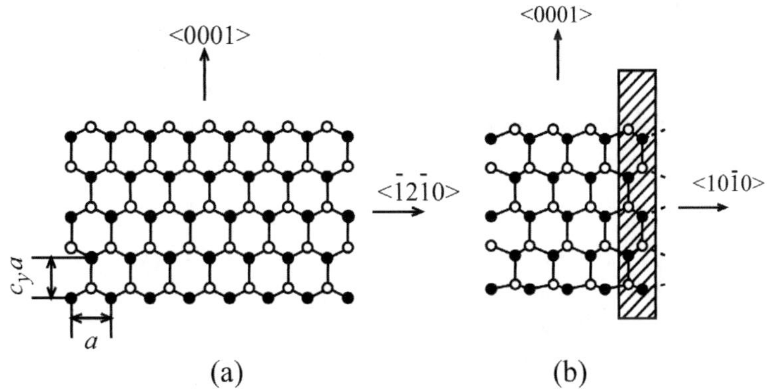

Figure 1. ($10\bar{1}0$) surface of the wurtzite structure. he open and closed circles denote A atoms and B atoms, respectively, for a two-component system, such as GaN. There are four atomic sites in the unit cell, denoted by $\mathcal{A}'$, $\mathcal{B}'$, $\mathcal{A}''$, and $\mathcal{B}''$. (a) Overhead view. (b) Side view. The shaded area indicates the lattice gas layer. Reprinted from reference [24] with permission from Nova Science Publishers, Inc.

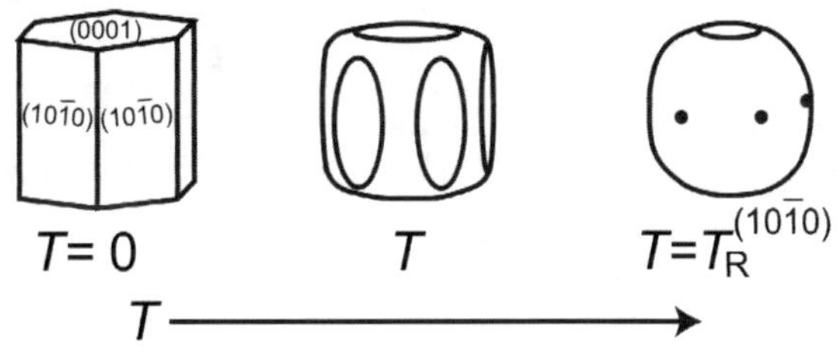

Figure 2. Schematic figure for the relationship between the roughening transition and faceting transition. $T_R{}^{(10\bar{1}0)}$ represents the roughening transition temperature on the ($10\bar{1}0$) surface.

a melt/crystal interface, the interface is considered to be "atomically rough" or diffuse. For ^{4}He, the faceted interfaces on the equilibrium shape of a crystal droplet are an example of an atomically rough but globally smooth surface [25]. Since the roughening transition is caused by global wandering of the interface [26], a globally smooth but atomically rough surface is possible. Similarly, the Ih-ice/water interface has been considered to be atomically rough. Since the faceting transition was observed, the lattice-model study will be effective as in the case of ^{4}He.

The efficacy of the tetrahedral order parameter suggests that the difference between ice and liquid is not due to the behaviour of the long-range order, but that the behaviour is that of the three dimensional (3D) Ising model. However, the 3D Ising model is also too complex to provide the exact free energy or the precise numerical values of the surface quantities. Hence, we apply a 2D lattice gas model for ice/QLL, ice/vapour, or ice/water interfaces (surfaces).

The lattice-gas model of the prism plane, which is the ($10\bar{1}0$) surface of a wurtzite crystal structure, consists of four sites in a unit cell on the surface without inversion symmetry (Fig. 1). In calculating the phase transition temperature of the lattice gas model, the mean-field or quasi-chemical approximation methods do not provide a precise phase transition temperature because they underestimate the fluctuation entropy in low dimensions [27]. Hence, we adopted the density-matrix renormalization-group (DMRG) method[28]–[31] to obtain reliable results. To overcome the lack of inversion symmetry of the crystal structure, we applied the asymmetric version of the DMRG method [24].

The aim of this paper is to obtain microscopic bond energies and surface tensions on both the basal and prism planes by comparing the calculated transition temperature to the experimentally observed faceting temperature. The transition temperatures were calculated by the product-wave-function renormalization-group (PWFRG, tensor network) method [30], which is a transfer matrix version of the DMRG method.

2. Two-dimensional lattice-gas model

For the $(10\bar{1}0)$ surface (the prism plane, Fig. 1), the surface layer consists of four rectangular sub-lattices, denoted by $\mathcal{A}'$, $\mathcal{B}'$, $\mathcal{A}''$, and $\mathcal{B}''$. We consider the surface energy by counting the number of broken bonds for a surface configuration. The lateral bond energy is denoted by ϕ. Then, the 2D lattice gas Hamiltonian for the prism plane is expressed as follows [32, 24]:

$$\begin{aligned}\mathcal{H}_{\rm LG,p} &= -\phi\sum_{(\vec{x})}[C_{\mathcal{A}'}(\vec{x})C_{\mathcal{B}'}(\vec{x}) + C_{\mathcal{A}'}(\vec{x})C_{\mathcal{B}''}(\vec{x}-\vec{c}) \\ &\quad +C_{\mathcal{A}'}(\vec{x})C_{\mathcal{B}''}(\vec{x}-\vec{a}-\vec{c}) + C_{\mathcal{A}''}(\vec{x})C_{\mathcal{B}'}(\vec{x}) \\ &\quad +C_{\mathcal{A}''}(\vec{x})C_{\mathcal{B}''}(\vec{x}) + C_{\mathcal{A}''}(\vec{x})C_{\mathcal{B}'}(\vec{x}+\vec{a})] \\ &\quad -\epsilon_{\mathcal{A}'}\sum_{\vec{x}}C_{\mathcal{A}'}(\vec{x}) - \epsilon_{\mathcal{B}'}\sum_{\vec{x}}C_{\mathcal{B}'}(\vec{x}) - \epsilon_{\mathcal{A}''}\sum_{\vec{x}}C_{\mathcal{A}''}(\vec{x}) - \epsilon_{\mathcal{B}''}\sum_{\vec{x}}C_{\mathcal{B}''}(\vec{x}),\end{aligned} \tag{1}$$

where $\epsilon_{\mathcal{A}'}$, $\epsilon_{\mathcal{A}''}$, $\epsilon_{\mathcal{B}'}$, and $\epsilon_{\mathcal{B}''}$ are the surface chemical potentials for lattice gas particles on the $\mathcal{A}'$, $\mathcal{A}''$, $\mathcal{B}'$, and $\mathcal{B}''$ sub-lattices, and $C_{\mathcal{A}'}(\vec{x}) = \{0, 1\}$ is the lattice gas variable. When $C_{\mathcal{A}'}(\vec{x}) = 1$, the sub-lattice site $\mathcal{A}'$ at the lattice point $\vec{x}$ is occupied, whereas for $C_{\mathcal{A}'}(\vec{x}) = 0$ it is empty.

From the thermodynamics at equilibrium for the prism plane, the surface chemical potentials are expressed by

$$\begin{aligned}\epsilon_{\mathcal{A}'} &= \Delta E_p - z\phi/2, \quad \epsilon_{\mathcal{B}'} = \Delta E_p - z\phi/2, \\ \epsilon_{\mathcal{A}''} &= -\Delta E_p - z\phi/2, \quad \epsilon_{\mathcal{B}''} = -\Delta E_p - z\phi/2,\end{aligned} \tag{2}$$

where $z = 3$ and ΔE_p is the energy gain for a bond inward to the crystal from a layer on the prism plane. The sub-lattice sites $\mathcal{A}''$ and $\mathcal{B}''$ protrude outward.

The partition function, surface energy, and sub-lattice coverage are calculated using the PWFRG method [30]. Details of the calculation method are given in reference [24].

3. Computational results

The calculated specific heat values are shown in Fig. 3. The inset shows the temperature dependence of the specific heat. The lines calculated by the PWFRG method show a sharp peak, whereas the values calculated by the Monte Carlo method show a low peak due to the finite size effect. The result clearly shows 2D Ising critical behaviour. From the semi-logarithmic figure of the specific heat, the transition temperature is obtained as

$$k_{\rm B}T_{\rm N}/J = 1.3012, \tag{3}$$

where $J = \phi/4$.

3.1. Ice/vapour surface

To obtain the effective bond energy, we start with the faceting transition temperature at the prism surface [4]–[8]. The faceting transition temperature on the prism plane is observed to be about 271–272 K [4]–[8]. Therefore, we consider a value of the phase transition temperature for

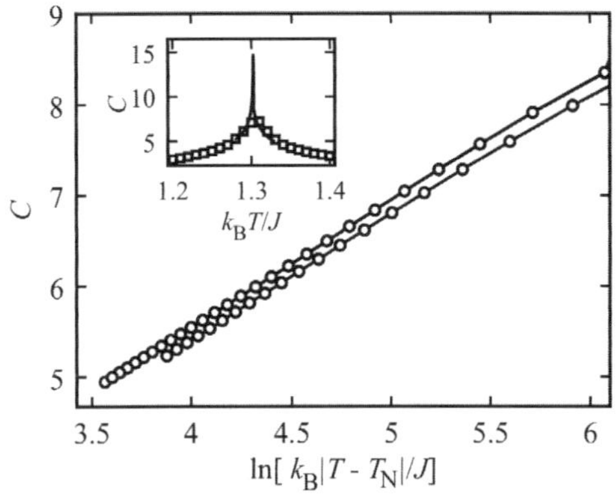

Figure 3. Logarithmic behaviour of the specific heat. ○ : PWFRG calculation. ——: connecting neighbouring ○ . The inset shows the temperature dependence of the specific heat. ——: PWFRG calculations. □: MC calculation (64 × 64). Reprinted from reference [24] with permission from Nova Science Publishers, Inc.

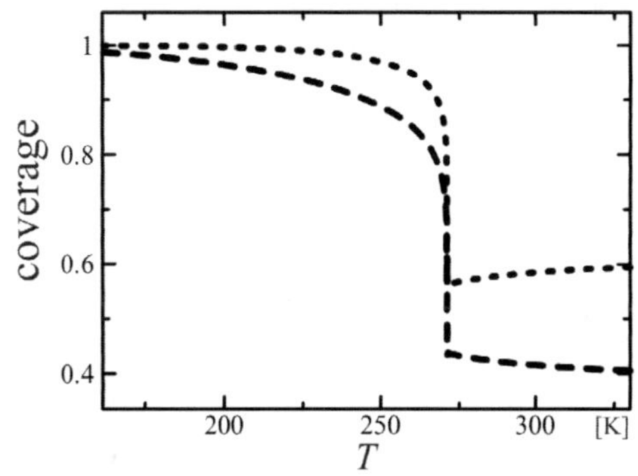

Figure 4. Temperature dependence of the coverage of the prism plane for the ice/vapour interface. ······: $\mathcal{A}'$ and $\mathcal{B}'$ sub-lattice sites. - - - -: $\mathcal{A}''$ and $\mathcal{B}''$ sub-lattice sites.

Table 1. Calculated microscopic quantities ($a = 4.52$ Å, $c = 7.36$ Å). Reprinted from reference [24] with permission from Nova Science Publishers, Inc.

Quantity	Equation	Value	
		Ice/vapour interface	ice/water interface
$T_{R,p}$ (Experiment)	$\geq T_{\rm N}^{(10\bar{1}0)}$	271 K [4]–[8]	257 K [20, 21]
$\phi_b \approx \phi_p$	$4J$, Eq. (3)	71.8 meV	68.1 meV
$T_{R,b}$	$\geq T_{\rm N}^{(0001)}$, Eq. (4)	289 K	274 K
$\gamma_{\rm surf}^{(0001)}$	$T = 0$ K	32.5 mJ/m^2	30.8 mJ/m^2
$\gamma_{\rm surf}^{(10\bar{1}0)}$	$T = 0$ K	34.6 mJ/m^2	32.8 mJ/m^2

the prism plane of -2 °C. From Eq. (3), with $T_{\rm N}^{(10\bar{1}0)} = 271$ K, $\phi_p = 4J$ is obtained to be 71.8 meV, which is a key quantity for the ice/vapour surface in the following estimations.

From a similar calculation with a lattice gas model of the basal plane (the (0001) surface), we obtain

$$k_{\rm B}T_{\rm N}^{(0001)}/J = 1.3871. \tag{4}$$

Using $\phi_p = 4J = 71.8$ meV, we obtain $T_{\rm N}^{(0001)} = 289$ K. This temperature is higher than the melting temperature of ice and is consistent with the experimental observations. The (0001) facets are observed at the melting temperature of ice. The surface tensions calculated using the value of ϕ_p are listed in Table 1.

Fig. 4 shows the temperature dependence of the coverage for respective sub-lattice sites. Since the sub-lattice sites $\mathcal{A}''$ and $\mathcal{B}''$ protrude outward, the coverages at their sites are smaller than those at the $\mathcal{A}'$ and $\mathcal{B}'$ sites.

3.2. Ice/water interface
Recently, measurements of the roughening transition for the $(10\bar{1}0)$ plane on Ih-ice in water under 165 MPa have been reported.[20, 21] The faceting transition was observed optically and the crystal growth mode was studied. It was concluded that the roughening transition temperature for the prism plane is −16 °C.

Assuming $T_{\rm N}^{(10\bar{1}0)} = -16$ °C, ϕ_p is obtained to be 68.1 meV. The calculated surface tensions for the prism and the basal planes are listed in Table 1. The surface tension for the basal plane agrees well with the result obtained by recent molecular dynamic calculations [33].

4. Conclusions
- For Ih-ice, the phase transition temperatures on the prism and basal planes for both ice/vapour and ice/water interfaces are calculated based on the 2D lattice gas model using the PWFRG method, which is a transfer matrix version of the DMRG method (Eq. (3) and Eq. (4)).
- By comparing the phase transition temperature on the prism plane to the experimentally observed values for the ice /vapour and the ice/water interfaces, the effective bond energies on the surface ϕ_p are obtained. The obtained numerical values for the bond energies, surface tensions, and the faceting temperature on the basal planes are listed in Table 1

Acknowledgement
We appreciate the assistance of Professor Y. Akutsu in providing us the general part of the computer program for the PWFRG method. This work was supported by the Collaborative Research Program of the Research Institute for Applied Mechanics, Kyushu University.

References
[1] Kuroda, T. ; Lacmann, R. Growth kinetics of ice from the vapour phase and its growth forms. 1982 *J. Cryst. Growth* **56** pp 189–205.
[2] Dosch, H. ; Lied, A; Bilgram, J. H. Glancing-angle X-ray scattering studies of the premelting of ice surfaces. 1995 *Surf. Sci.* **327** pp 145–154.
[3] Van der Veen, J. F.; Pluis, B.; van der Gon, A. W. D. Surface melting *Chemistry and physics of solid surfaces*, Vol.VII, eds. Vanselow, R.; Howe, R. F. (Springer, Berlin, 1988) chapter 16 pp 455–490.
[4] Furukawa, Y. ; Yamamoto, M.; Kuroda, T. Ellipsometric study of the transition layer on the surface of an ice crystal. 1987 *J. Cryst. Growth* **82** pp 655–677.
[5] Kouchi, A. ; Furukawa, Y.; Kuroda, T. X-ray diffraction pattern of quasi-liquid layer on ice crystal surface. 1987 *J. de Physique* **48** pp C1-675 – C1-677.
[6] Elbaum, M. Roughening transition observed on the prism facet of ice. 1991 *Phys. Rev. Lett.* **67** pp 2982–2985.
[7] Elbaum, M.; Lipson, S. G.; Dash, J. G. Optical study of surface melting on ice. 1993 *J. Cryst. Growth* **129** pp 491–505.
[8] Furukawa, Y.; Kohata, S. Temperature dependence of the growth form of negative crystal in an ice single crystal and evaporation kinetics for its surfaces. 1993 *J. Cryst. Growth* **129** pp 571–581.
[9] Nada, H.; Furukawa, Y. Anisotropy in molecular-scaled growth kinetics at ice –water interfaces. 1997 *J. Phys. Chem.* **B 101** pp 6163–6166.
[10] Furukawa, Y.; Nada, H. Anisotropic surface melting of an ice crystal and its relationship to growth forms. 1997 *J. Phys. Chem.* **B 101** pp 6167–6170.
[11] Sazaki, G.; Zepeda, S.; Nakatsubo, S.; Yokomine, M.; Furukawa, Y. Quasi-liquid layers on ice crystal surfaces are made up of two different phases. 2012 *Proc. Natl. Acad. Sci. U.S.A.* **109** pp 1052–1055.
[12] Sazaki, G.; Asakawa, H.; Nagashima, K.; Nakatsubo, S.; Furukawa, Y. How do quasi-liquid layers emerge from ice crystal surfaces? 2013 *Cryst. Growth & Design* **13** pp 1761–1766.
[13] Asakawa, H.; Sazaki, G.; Nagashima, K.; Nakatsubo, S.; Furukawa, Y. Two types of quasi-liquid layers on ice crystals are formed kinetically. 2016 *Proc. Natl. Acad. Sci. U.S.A.* **113** pp 1749–1753.
[14] Murata, K.; Asakawa, H.; Nagashima, K.; Furukawa, Y; Sazaki, G. Thermodynamic origin of surface melting on ice crystals. 2016 *Proc. Natl. Acad. Sci. U.S.A.* **113** pp E6741–E6748.

[15] Benet, J. ; Llombart, P.; Sanz, E.; MacDowell, L. G. Premelting-induced smoothening of the ice-vapor interface. 2016 *Phys. Rev. Lett.* **117** 096101.

[16] Errington, J. R.; Debenedetti, P. G. Relationship between structural order and the anomalies of liquid water. 2001 *Nature (London)* **409** pp 318–321.

[17] Conde, M. M.; Vega, C.; Ptrykiejew, A. The thickness of a liquid layer on the free surface of ice as obtained from computer simulation. 2008 *J. Chem. Phys.* **129** 014702.

[18] Sazaki, G.; Matsui, C.; Tsukamoto, K.; Usami, N.; Ujihara, T.; Fujiwara, K.; Nakajima, K. In situ observation of elementary growth steps on the surface of protein crystals by laser confocal microscopy. 2004 *J. Cryst. Growth,* **262** pp 536–542.

[19] Murata, K.; Nagashima, K.; Sazaki, G. How do ice crystals grow inside quasiliquid layers 2019 *Phys. Rev. Lett.* **122** 026102.

[20] Maruyama, M.; Kishimoto, Y.; Sawada, T. Optical study of roughening transition on ice Ih ($10\bar{1}0$) planes under pressure. 1997 *J. Cryst. Growth* **172** pp 521–527.

[21] Maruyama, M. Roughening transition of prism faces of ice crystals grown from melt under pressure. 2005 *J. Cryst. Growth* **275** pp 598–605.

[22] Jayaprakash, C.; Saam, W.F.; Teitel, S. Roughening and facet formation in crystals. 1983 *Phys. Rev. Lett.* **50** pp 2017–2020.

[23] Rottman, C.; Wortis, M. Statistical mechanics of equilibrium crystal shapes: Interfacial phase diagrams and phase transitions. 1984 *Phys. Rep.* **103** pp 59–79.

[24] Akutsu, N.;Akutsu, Y. Density-matrix renormalization-group study of lattice gas on the surface of a wurtzite crystal structure *Understanding density matrices*, Ed. Danielsen, Nadia V. (Nova Science Publishers, Inc., New York, 2019) chapter 6 pp 197–227.

[25] Keshishev, K.O.; Parshin, A.Y.; Babkin, A.V. Crystallization waves in ^{4}He. 1981 *Sov. Phys. JETP* **53** pp 362–369.

[26] Huse, D.A.; van Saarloos, W.; Weeks, J.D. Interface Hamiltonians and bulk critical behavior. 1985 *Phys. Rev. B* **32** pp 233–246.

[27] Mermin, N.D.; Wagner, H. Absence of ferromagnetism or antiferromagnetism in one-or two-dimensional isotropic Heisenberg models. 1966 *Phys. Rev. Lett.* **17** pp 1133–1136. "Erratum: Absence of ferromagnetism or antiferromagnetism in one-or two-dimensional isotropic Heisenberg models. Mermin, N. D.; Wagner, H. [Phys. Rev. Letters 1966, 17, 1133]." 1966 *Phys. Rev. Lett.* **17** pp 1307–1307.

[28] White, S. R. Density matrix formulation for quantum renormalization groups. 1992 *Phys. Rev. Lett.* **69** pp 2863–2866.

[29] Schollwöck, U. The density-matrix renormalization group. 2005 *Rev. Mod. Phys.* **77** pp 259–315.

[30] Nishino, T.; Okunishi, K. Product wave function renormalization group. 1995 *J. Phys. Soc. Jpn.* **64** pp 4084–4087.

[31] Hieida, Y.; Okunishi, K.;Akutsu, Y. Magnetization process of a one-dimensional quantum antiferromagnet: The product-wave-function renormalization group approach. 1997 *Phys. Lett. A,* **233** pp 464–470.

[32] Akutsu, N.; Akutsu, Y. Ambiguity of anisotropic interface tension for complex crystals. 1995 *J. Phys. Soc. Jpn.* **64** pp 736–756.

[33] Tanaka, K. K.; Kimura, Y. Theoretical analysis of crystallization by homogeneous nucleation of water droplets. 2019 *J. Phys. Chem. Chem. Phys.* **21** pp 2410–2418.

Journal of Physics: Conference Series **2207** (2022) 012037 doi:10.1088/1742-6596/2207/1/012037

Unified Time Evolution Approach for the Electronic Structure Calculation

Q Wang and Y Zempo

Computer and Information Sciences, Hosei University, 3-7-2 Kajino Koganei 184-8584 Japan

E-mail: qinlin.wang.3p@stu.hosei.ac.jp

Abstract. In imaginary-time evolution, the same technique, Crank–Nicolson's method, as for the analysis of real-time evolution is employed. For the verification using the model oscillators such as quantum well, Morse, and harmonic potentials, practical density functional theory (DFT) calculations are performed for hydrogen atoms and molecules. The convergence is confirmed to be faster under appropriate initial and boundary conditions. The results are consistent with the analytical solutions and those of other DFT calculations. Imaginary-time evolution is quite convenient to subsequent real-time real-space time-dependent DFT calculation. Although the operators are different, similar evolution is adopted. This technique is expected to unify into practical electronic state calculations for the self-consistent electronic structure and system's response due to perturbation because calculations are in the same framework as the time-dependent equation.

1. Introduction

According to the development of density functional theory (DFT), time-dependent DFT (TDDFT) has become a prominent method for calculating excited states of medium-sized to large molecules. It is recognized as a powerful tool, especially for optoelectronic behavior. One of its significant developments is real-time real-space TDDFT, in which the dynamic properties of electrons in occupied states are observed as time-series data [1]. The wave function is propagated by giving the external cue with the initial one as a perturbation to the steady-state and following its response in real time. This method is also stable in numerical calculations. This electronic state calculation is computationally intensive and requires large-scale parallel computing [2]. Therefore, reducing the computation time by improving the calculation methods and algorithms is essential for practical usage.

To perform dynamic the electronic state calculation using this real-time real-space TDDFT, the self-consistent ground state is required as an initial state, that is obtained from the time-independent DFT. Due to the DFT, various interesting properties are effectively calculated as solutions of the Kohn–Sham equation (KS), which considerably affects materials science. A commonly used approach starts with solving the time-independent KS using the Hamiltonian operator and its eigenvalues. On the other hand, in the calculation of atomic systems, there is an alternative way to obtain the self-consistent ground state. That is the imaginary-time evolution to solve the time-dependent KS, in which the real time t is replaced by the imaginary time τ, *i.e.*, $t \rightarrow -i\tau$. This technique has been well studied and intensively used in atomic physics, nuclear physics, particle physics, astrophysics, etc. because it provides stable and fast convergence of the calculation.

Using this technique with theoretical extensions, some information of excited states is taken into account [3]. However, we propose an imaginary-time evolution method to obtain only the initial self-consistent ground state, which is quite convenient to prepare the initial states for the real-time real-space TDDFT calculation to be carried out in the subsequent process. Although the operators are different, similar time evolution is adopted. The study presents the effectiveness of this technique for practical electronic structure calculations.

The remainder of the paper is organized as follows: In Section 2, we briefly describe the imaginary-time evolution calculation procedure. Section 3 presents the proposed technique and results. Finally, Section 4 presents the effectiveness and characteristics of the proposed technique, along with the conclusion.

2. Method

Using the same time-dependent process, imaginary-time and real-time evolution calculations are in the same framework, which is excellent. In this section, we explain the process in detail.

2.1. Real-time evolution calculation

The ground state can be derived using the KS. Nevertheless, DFT is less successful in describing optical responses and absorption spectra in which electronically excited states are involved. This difficulty is, in principle, solved by the extension of DFT to its time-dependent version, TDDFT, established by Runge and Gross [4]. Practically, there are two ways; the linear-response theory [5] or the explicit time-propagation of the system after a perturbation. We have studied the latter intuitive method. The wave function is evolved by the time evolution operator, $\psi(t) = \exp[-iHt]\psi(0)$, with the initial wave function at $t = 0$. The self-consistent electronic structure calculation for $\psi(0)$ and the time evolution procedure for $\psi(t)$ are obtained in the different techniques. In the former, a traditional technique such as Krylov's subspace method is applied to the time-independent KS. On the other hand, a technique different from the former is used in the latter such as the Crank–Nicolson method, solving the time-dependent KS as follows:

$$i\frac{\partial}{\partial t}\,\psi(\mathbf{r},t) = H\,\psi(\mathbf{r},t) \tag{1}$$

$$H = -\frac{1}{2}\nabla^2 + V_{ion}^{ps}(\mathbf{r}) + V_H(\mathbf{r},t) + V_{XC}[\rho(\mathbf{r},t)] + V_{ext}(\mathbf{r},t),$$

where V_{ps}^{ion} is an ionic pseudopotential, V_H is the Hartree potential, V_{XC} is the exchange-correlation potential, and $V_{ext}(\mathbf{r},t)$ is a perturbation of the self-consistent system at $t = 0$. The time-dependent calculation detail is explained in the literature [2]. Our method, throughout the calculation, solves Eq. (1) as our basic equation to obtain both electronic structure calculations as mentioned above. In the real-time real-space calculation, the self-consistent ground state is required. To obtain this initial state, we use the imaginary-time evolution, which is quite convenient to subsequent real-time real-space TDDFT calculation. As similar time evolution is adopted, the calculation can be performed within the usual time evolution framework. In the next section, we give a detailed explanation of this process.

2.2. Imaginary-time evolution

To investigate the electronic state of a many-electron system, we replace real time with imaginary time in the time-dependent KS equation ($\tau = -it$) and obtain the imaginary-time evolution equation. This equation is in the form of a diffusion equation, as shown in Eq. (2).

$$-\frac{\partial}{\partial \tau}\,\psi\,(\mathbf{r},\tau) = H\,\psi\,(\mathbf{r},\tau), \tag{2}$$

which is known to be numerically stable. The solution can be expressed by Eq. (3).

$$\psi(\mathbf{r},t) = \textstyle\sum_{k=1}^{\infty} C_k\phi_k(\mathbf{r})\exp(-\varepsilon_k\tau), \tag{3}$$

where ϕ_k and ε_k denote the k-th eigenstate and eigenenergy, respectively. From the completeness of the wave function, *i.e.*, $\mathbf{1} = \sum_n |\phi_n\rangle\langle\phi_n|$, the ground state with the lowest energy can be obtained using the evolution of imaginary time τ

$$\begin{aligned} \mathrm{e}^{-H\tau}\psi(\mathbf{r},\tau) &= \textstyle\sum_n e^{-E_n\tau}|\phi_n\rangle\langle\phi_n|\psi\rangle \\ &= \mathrm{e}^{-E_0\tau}\textstyle\sum_n e^{(E_0-E_n)\tau}|\phi_n\rangle\langle\phi_n|\psi\rangle \\ &\to \mathrm{e}^{-E_0\tau}\,|\phi_0\rangle\ \ (\tau\to\infty). \end{aligned} \tag{4}$$

Assuming $E_0 < E_1 < \cdots < E_n$, the wave function $\psi(\boldsymbol{r},\tau)$ will be dominated by the lowest energy state, *i.e.*, the ground state,

$$\lim_{\tau\to\infty}\psi(\mathbf{r},\tau) \approx C_0\psi_0(\boldsymbol{r})\,\mathrm{e}^{-E_0\tau}. \tag{5}$$

Thus, we can obtain the ground state with an appropriate normalization.

In the imaginary-time evolution with the time step $\Delta\tau$,

$$\psi(\boldsymbol{r},\tau+\Delta\tau) = \mathrm{e}^{-H\Delta\tau}\psi(\boldsymbol{r},\tau), \tag{6}$$

we employ the Crank–Nicolson method. There is time-reversal symmetry; backward propagation by $\Delta\tau/2$ starting from $\psi(\boldsymbol{r},\tau+\Delta\tau)$ should be to the same result as forward propagation by $\Delta\tau/2$ starting from $\psi(\boldsymbol{r},\tau)$. In this simplest approximation to the evolution operator, the condition is

$$\mathrm{e}^{\frac{\Delta\tau}{2}H}\psi(\mathbf{r},\tau+\Delta\tau) = \mathrm{e}^{\frac{-\Delta\tau}{2}H}\psi(\mathbf{r},\tau). \tag{7}$$

This method is commonly used in real-time evolution because it ensures unitarity of the imaginary-time evolution. Although there is no relevance to the imaginary-time evolution, we apply this method in our imaginary-time evolution system as well, following the same calculation procedure. In the practical use of this method, we expand the propagation operator in Eq. (7) up to the first order Taylor's expansion, which becomes a simultaneous equation. This equation is solved with the asymptotic form in the distance. As we adopt the 9-point finite difference method in real space in our calculation, the wave function is determined to satisfy the Slater-type orbital as a boundary condition of 4 points outside the calculation region. This method is an implicit technique, and this operation is equivalent to the first order Padé approximation.

As mentioned above, only the electronic state's ground state can be obtained. To obtain other eigenstates, we employed the Gram–Schmidt orthogonalization process. For a smaller number of states, the method is generally accurate. However, an increase in the number of states tends to introduce numerical inaccuracy. If up to the (n-1)-th states are known, the n-th state can be sequentially obtained as follows:

$$\begin{aligned} \tilde{\psi}_n &= \psi_n - \sum_{i=1}^{n-1}\frac{1}{|\psi_i|^2} < \psi_i|\psi_n > \psi_i \\ \psi_n &= \mathcal{N}(\tilde{\psi}_n), \end{aligned} \tag{8}$$

where $\mathcal{N}$ denotes the normalization. By substituting the ground state into ψ_0, other states can be obtained as necessary.

3. Results and discussions

To verify the imaginary-time evolution program, this technique was applied to simple systems such as quantum well potential and Morse potential. The same values as the analytical solutions were obtained in all calculations. As an application to the practical systems, we calculate simple systems using analytical solutions such as a hydrogen atom and molecule.

3.1. Model potentials

To verify of our imaginary-time evolution, two basic potentials were examined, the quantum well potential and the Morse potential, which are simplified as model oscillators. These two potentials are typical model potentials, and analytical solutions are well known. Figure 1 shows the results of the examinations in a 1D quantum well and a Morse potential, and each dotted line shows the energy level of the analytical solution. The quantum well has the size of 1.0 a.u. high and 1.0 a.u. wide. The Morse potential has the following functional form: $V(r) = 24\left(\mathrm{e}^{-4(r-3)} - 2e^{-2(r-3)}\right)$ and $E_n = -\left[5 - \sqrt{2}(n + 1/2)\right]^2$, where $n = 0, 1, 2, 3$, *i.e.*, just four eigenstates exist [6]. The calculations are performed in a size of 40 a.u. in both cases. From the comparison, the results agree well with those of analytical solutions within error 10^{-3}~10^{-4} a.u. [6, 7].

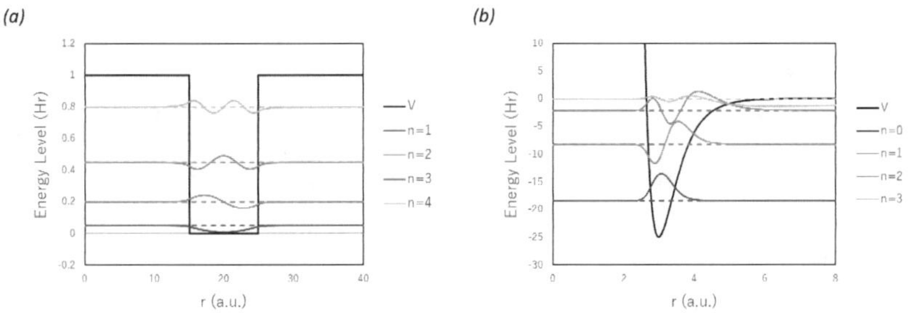

Figure 1. Eigenstates of simple model potentials: (*a*) Quantum well and (*b*) Morse potential. Each wave function is plotted at the level of each eigenvalue.

As the other specimen results showing the validity of our method, there is the familiar 3D spherical quantum harmonic oscillator with the harmonic potential $V(r) = r^2/2$, which is an exactly solvable system. Usually, the wave function is divided into radial and angular wave functions $\psi(r) = R(r)Y(\theta, \phi)$; We follow this traditional procedure. The radial wave function $R(r)$ is converted using $rR(r) = F(r)$, and $F(r)$ is calculated to avoid divergence at the origin. For simplicity, the three lowest states correspond to the rotational quantum number $l = 0$. The calculations are performed in a radial sphere size of 20 a.u. The simple exponential functions of e^{-r} and $r\mathrm{e}^{-r}$are selected as the initial trial functions and the asymptotic form in the distance for even and odd n states, respectively. In figure 2,

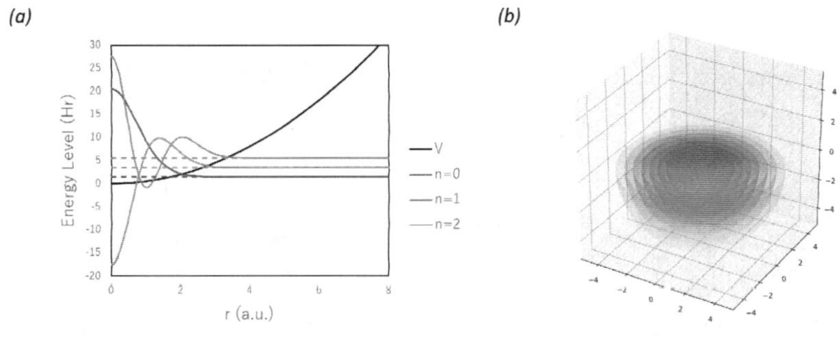

Figure 2. Eigenstates of harmonic potential: (*a*) Radial wave functions obtained by the spherical symmetry plotted with the wave function placed at the level of each eigenvalue, and (*b*) Half-plotted image of 3D harmonic potential.

each dotted line shows the energy level of the analytical solution. The results agree well with the analytical solution within error 10^{-3}~10^{-4} a.u. [6, 7].

3.2. Hydrogen and hydrogen molecule

As a practical system, we calculated the hydrogen atom and the hydrogen molecule. The hydrogen atom is an exactly solvable system; thus, it is a good test for validity. For simplicity, the two lowest states corresponding to the rotational quantum number $l = 0$ are shown in Fig. 3. Without the spherical symmetry, the wave functions $\psi(\mathbf{r}, t)$ are performed using finite difference method on the rectangular grid of size 10^3 a.u. ($\Delta x, \Delta y, \Delta z = 0.2$ a.u.). The initial and asymptotic form in the distance is the same as that in the previous 3D harmonic oscillator.

Our calculation in this section follows the definition of the Hamiltonian in Eq. (1), *i.e.*, the DFT calculation. The pseudopotential V_{ps}^{ion} is, for use in various material systems, adopted in the separable form [8]. The Hartree and exchange-correlation potentials can be determined from the electronic charge density, $\rho(\mathbf{r}, t) = \sum_i |\psi_i(\mathbf{r}, t)|^2$. The summation is over all occupied states i. The Hartree potential is determined by $\nabla^2 V_H = -4\pi\rho$, and the usual local density approximation is used as the xc potential, V_{XC}, in our study.

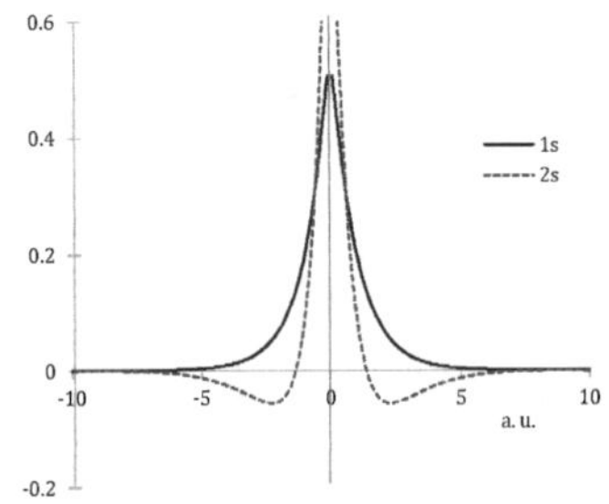

Figure 3. Two s orbitals of hydrogen atom with $l = 0$ plotted on a straight line with the atomic position at the origin

Figure 4 shows the bonding and anti-bonding states of molecular hydrogen. The ground state, the bound state, was obtained using the imaginary-time evolution method, and the excited state, the anti-bonded state, was obtained through Gram–Schmidt orthogonalization to the well-converged ground state. The results agree well with those of other DFT methods. As mentioned above, the initial wave function and the asymptotic form in the distance are required for a good convergence. We adopt the superposition of the Slater-type orbital as the ground state. We also experienced this in the real-time real-space calculation.

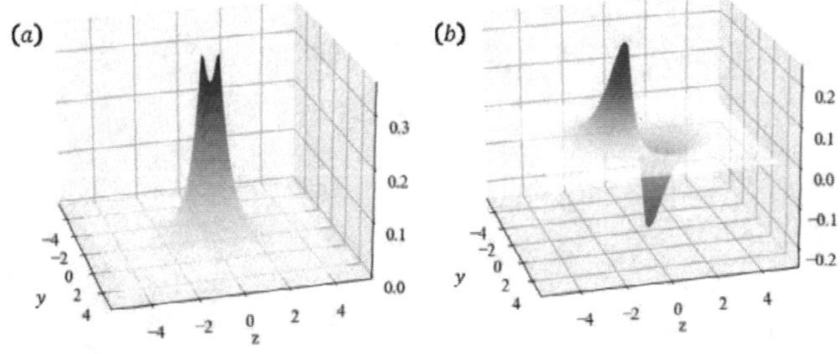

Figure 4. Orbitals of hydrogen molecule of (a) bonding and (b) anti-bonding states

The same technique used in real-time evolution is adopted in our imaginary-time evolution. The main thesis of this study is that for the self-consistent electronic structure and the response of the system due to perturbation, the calculations are in the same framework according to the time-dependent KS in Eq. (1). Using the same mesh and boundary conditions throughout the calculations is expected to be effective for real-time real-space TDDFT. However, because this technique was verified in small molecules, investigating the restrictions and convergence will be the goal of our future research.

4. Conclusion

The imaginary-time evolution method was used to analyze the static electronic states. This method is confirmed to have an advantage of faster convergence under appropriate initial and boundary conditions. Moreover, because a similar process as the analysis of real-time evolution calculation was used, the calculation procedure is expected to be simple, thus reducing the total computational complexity. In the calculations for model oscillators such as quantum well, Morse and harmonic potentials, and for hydrogen atoms and molecules, Eq. (2) is solved using the Crank–Nicolson method, which is also used for real-time evolution. The results agree well with the analytical solutions and those of other DFT techniques. The imaginary-time evolution technique provides faster convergence to obtain the initial state. It is also quite convenient to the subsequent real-time real-space TDDFT calculation, because the similar time evolution is adopted. It is expected that the method can be unified into practical electronic state calculations. However, we observed that the initial and boundary conditions affect the convergence. Moreover, this technique was verified in small molecules; thus, the restrictions and convergence should be examined in future work.

Acknowledgment

The authors would like to express their acknowledgments to Professors Satoru S. Kano and Nobuhiko Akino for useful discussions and encouragements. This work was partially supported by JSPS Grants-in-Aid for Scientific Research (C) Grant number 16K05047, and Sumitomo Chemical, Co. Ltd.

References

[1] Yabana K and Bertsch G F 1996 *Phys. Rev.* B **54** 4484
[2] Zempo Y, Akino N, Ishida M, Ishitobi M and Kurita Y 2008 *J. Phys. Condens. Matter.* **20** 064231
[3] Cedric F, Grigory K, Efstratios M and Efthimios K 2019 *J. Chem. Theory Comput.* **15** 6036
[4] Casida M E, 1995 D.P. Chong (Ed.) Recent Advances in Density Functional Methods Part I, World Scientific, Singapore, p155.
[5] Runge E and Gross E K U 1984 *Phys. Rev. Lett.* **52** 997
[6] Shore B W 1973 *J. Chem. Phys.* **58** 3855
[7] Roy A K 2014 *J. Math. Chem.* **52** 2645
[8] Troulier N and Martins J L 1991 *Phys. Rev.* B **43** 1993

Journal of Physics: Conference Series **2207** (2022) 012038 doi:10.1088/1742-6596/2207/1/012038

Numerical analysis of electronic state of CNT/BNNT heterojunction

Mitsuyoshi Tomiya, Shoichi Sakamoto, Toshihiro Sagara, Masaki Yamamoto, Masahiro Ishii

Seikei University, Faculty of Science and Technology

tomiya@st.seikei.ac.jp

Abstract. The *I-V* characteristics of CNT/BNNT heterojunctions, which comprise quantum double barrier systems (DBS), were numerically investigated using density functional theory (DFT) calculations and the non-equilibrium Green's function (NEGF) method. First, we performed a structural optimisation of the DBS of CNT/BNNT. Next, the *I-V* characteristics of the heterojunctions were calculated using the SIESTA package. The *I-V* characteristics remain flat up to a certain bias voltage, and then the current increases with increasing voltage. Our CNT/BNNT double-barrier heterojunctions have resonant states inside the barriers as semiconductor-based double-barrier diodes; these states determine the transmission function. An energy gap was centred at the Fermi energy and sharp peaks in the gap were observed. It was also found that the energy interval of the peaks near the Fermi energy was dependent on the separation length of the two BNNT barriers. Using projected density of states (PDOS) analysis, we also investigated the local behaviour of electrons in the devices.

1. Introduction: CNT, BNNT, DBS of CNT/BNNT

Carbon nanotubes (CNTs) are considered promising for use as semiconductors. They are relatively similar to graphene and are already being mass-produced. The hexagonal boron-nitride nanotube (h-BNNT) of any chirality is known to be a semiconductor with a wide bandgap. The heterojunctions CNT (9,0):zigzag-type CNT(ZCNT) and h-BNNT (9,0), CNT (5.5):armchair type (ACNT) and h-BNNT (5,5) were fabricated, and the optimised structures were numerically obtained. Upon embedding the two rings of h-BNNT into CNTs, quantum tunnelling through the double barrier system (DBS) is expected to occur. These conditions are quite similar to those of carbon nanoribbons with two embedded h-BN strips [1].

The structures of the models were optimised using the SIESTA package [2,3,4]. The wave functions of the electrons were also enumerated using SIESTA. The transmission functions and currents passing through the models due to the application of voltage were also obtained by TranSIESTA, which is based on the density functional theory (DFT) calculations and the non-equilibrium Green's function (NEGF) method [1]. The cut-off energy parameter, which expresses the quality of the plane wave bases, was set at 350 Ry. The number of wavevector samplings in the first Brillouin zone of the wavenumber space was $64 = 4 \times 4 \times 4$. We adopted the Kleinman-Bylander potential as the pseudo potential and PBE functional [5] of GGA [6].

After the optimisation of the structures of the tube materials by SIESTA, we set the C-C, B-N, C-N, and C-B bond lengths at 0.144 nm, 0.145 nm, 0.142 nm, and 0.151 nm, respectively. The B-N bond was slightly longer than the C-C bond. Therefore, h-BNNT was slightly larger than the CNT with hexagonal structure, and the numerical optimisation often required a heavy calculation task.

ACNT/h-BNNT (5,5) was constructed by embedding two BNNT (5,5) rings in the armchair type of CNT (5,5) (Figure 1), and ZCNT/h-BNNT (9,0) was numerically constructed by embedding two BNNT (9,0) rings in the zigzag type of CNT (9,0). In Figure 1, ACNT/h-BNNT (5,5) is illustrated, and ZCNT/BNNT (9,0) is fabricated in a similar manner. Of course, the chirality of the embedded BNNT ring must be the same as that of the CNT (9,0).

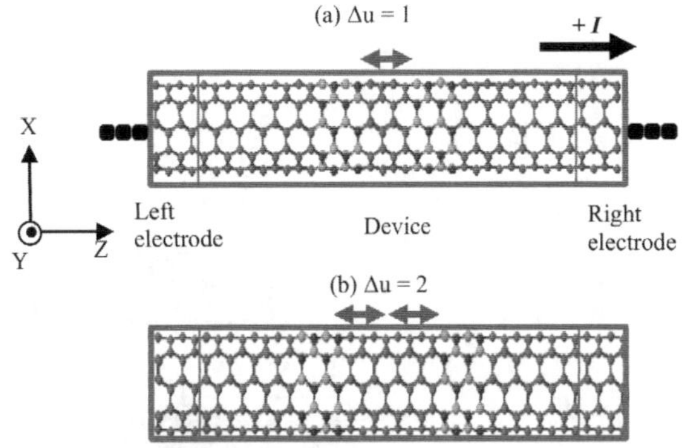

Figure 1. The ball-and-stick models of ACNT/BNNT (5,5) are presented. Between two h-BN rings, the short CNTs ((a) $\Delta u = 1$, (b) $\Delta u = 2$) are placed. The grey balls are carbon atoms, the blue ones are nitrogen, and the pink ones are boron. At the middle of our model is the device being the left and right electrodes are attached therewith. The electrodes are made of the same ACNT (5,5) that is conductive. The length of the device is 4.85 nm and its diameter is 0.69 nm. The red arrow explains how to count the number of units in the distance between two barriers.

The NEGF method can be used to calculate an open system along the direction of the electric current [1]. To enumerate the transmission function and the projected density of states (PDOS, which will also be noted in Section 4) by TranSIESTA, the left and right electrodes are added to the CNT/BNNT. We chose to involve the magnetic degree of freedom of spin in our calculations; however, every CNT/BNNT heterojunction that we have chosen in this study showed no difference. For example, no split of the band structure due to electron spin was found.

The chiral vector (n, m) indicates the type of CNT. $(n, 0)$ and (n, n) represent the CNT as a zigzag type, and an armchair type, respectively. In this study, the zigzag type is chosen to be CNT (9,0): ZCNT, because we selected $n = 3i(i \in \mathbb{Z}^+)$ and $i = 3$ to make the CNT conductive, which otherwise, would become an insulator or a wide-bandgap semiconductor. The armchair type was chosen to be CNT (5,5): ACNT, such that the size was almost the same as ZCNT, and because ACNT is always conductive.

On other hand, the bandgaps of h-BNNTs were about 5.0 eV [7]. For instance, our numerical optimization indicated that the band gaps of ZCNT (9,0) and ACNT (5,5) were 3.71 eV and 4.46 eV.

The distance between two BNNT barriers, Δu, is indicated by the number of ring units of the carbon hexagonal lattices (Figure 1), even though the boundaries of CNT and BNNT are not straight lines. The unit size was estimated to be 0.432 nm (ZCNT) and 0.333 nm (ACNT) from the structural optimisation, according to our calculation from first principle. The size was measured in the direction of the current.

SIESTA can also be used to calculate the wave functions of electrons in real space. This function can visualise the electron distribution in the devices. It can then be used to investigate how the electric charge distributions affect the electrical properties.

2. DBS of zigzag type CNT/BNNT (9,0): ZCNT/BNNT (9,0)

Two h-BN rings act as DBS, which were originally introduced to the semiconductor by Esaki [8]. The quasi-bound states originating inside the double barrier system play an important role in our model, and they can be numerically found inside the barriers (Figure 2).

Most of the sharp peaks near the Fermi energy in $T(E)$ of ZCNT/BNNT (9,0) correspond to the quasi-bound states inside the DBS (Figure 2). The energy interval, ΔE, can be estimated from the energy gap of the two nearest peaks in the transmission function, as shown in Figure 2. The separation distance between the two barriers, Δu, is inversely proportional to ΔE: $\Delta E \times \Delta u \cong 3.31$. (The approximation straight line for the plotted data set$(\Delta u, 1/\Delta E)$ is nicely fitted as $1/\Delta E \cong 0.3019\Delta u$.) This is another example that can be applied to the 1D Dirac equation model [1].

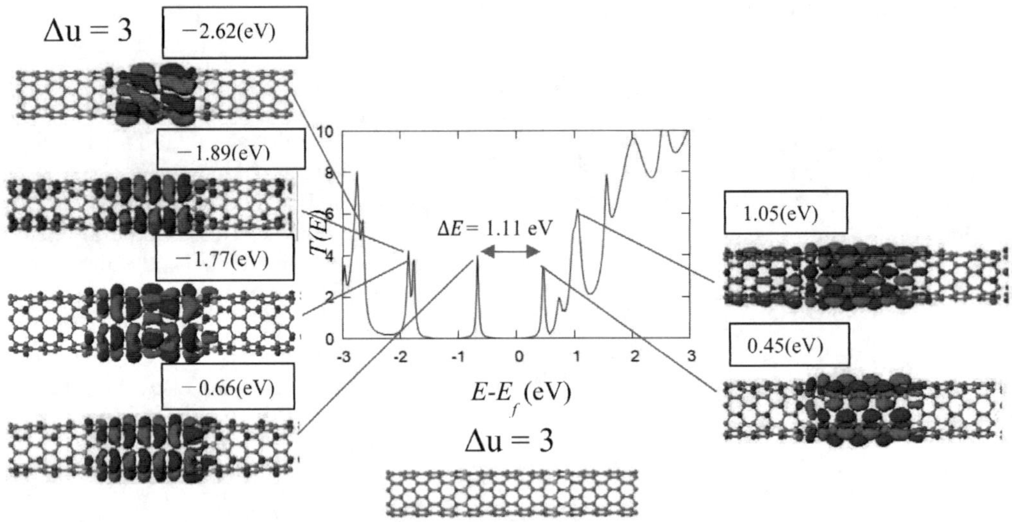

Figure 2. The transmission function of ZCNT/BNNT (9,0):$\Delta u = 3$ and the wave functions that correspond to the sharp peaks in $T(E)$ are presented. The two-way arrow indicates ΔE.

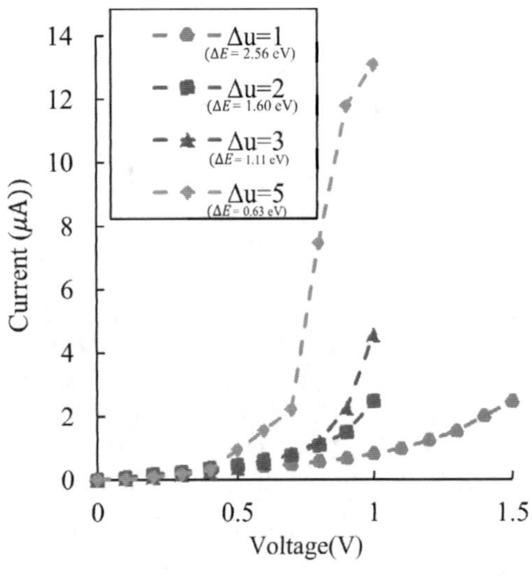

Figure 3. *I-V* characteristics of CNT/BNNT (9,0) is illustrated. When the bias voltage is under $\Delta E/e$, the current hardly flows. The voltage gets higher over the energy interval, then current jumps up.

Figure 2 shows $T(E)$ and the corresponding wave functions at sharp peaks in the neighbourhood of the Fermi energy. The DBS is introduced in ZCNT and transmission function is considerably smaller than that of pure ZCNT (9,0) especially nearby the Fermi energy. However, it can create a current large enough to construct electric devices when the bias voltage is over $\Delta E/e$ (Figure 3). The resonant tunnelling of DBS, which produces the sharp peaks at the quasi-stable states inside DBS, makes the shape of $T(E)$. Without the quantum effect, $T(E)$ would be just zero near the Fermi energy up to the energy range whose energy gap is about the same as the energy gap of h-BNNT barriers: $3 \sim 5eV$.

The value of $T(E)$ is zero at the Fermi energy and with $\Delta E = 1.11\text{eV}$ ($\Delta u = 3$). Thus, our ZCNT/BNNT (9,0) is a semiconductor. The two h-BNNT rings block the transportation of electrons through the device, except for quantum resonant tunnelling at the energies of the quasi-stable states inside the DBS. The occurrence of the states inside the DBS was confirmed by observing the wave functions in the devices (Figure 2). These results were also confirmed in the device models for other Δu distances.

The *I-V* characteristics of CNT/BNNT (Figure 3) can be calculated using TranSIESTA, which includes the Landauer formula and NEGF method. The electrical current is derived using the following formula:

$$I(V) \approx \frac{2e}{h}\int_{E_f-\frac{V}{2}}^{E_f+\frac{V}{2}} T(E)dE \text{ , (1)}$$

where V is the bias voltage. It has a certain range of nearly zero flat current behaviour and suddenly jumps up at a bias voltage of approximately $\Delta E/e$. This means that the devices are semi-conductors.

3. DBS of armchair type CNT/BNNT (5,5): ACNT/BNNT (5,5)

In $T(E)$ of ACNT/BNNT (5,5), most of the sharp peaks, which are found near the Fermi energy, correspond to the quasi-bound states inside DBS (Figure 4) as ZCNT/BNNT (9,0); however, ACNT/BNNT (5,5) does not have a clear relation between ΔE and Δu; for example $\Delta E \times \Delta u = 2.92$ for $\Delta u = 4$, $\Delta E \times \Delta u = 1.04$ for $\Delta u = 2$, and $\Delta E \times \Delta u = 1.65$ for $\Delta u = 1$. However, its *I-V* characteristics show a moderate negative differential resistance ($\Delta u > 1$), which must be effective (Figure 5).

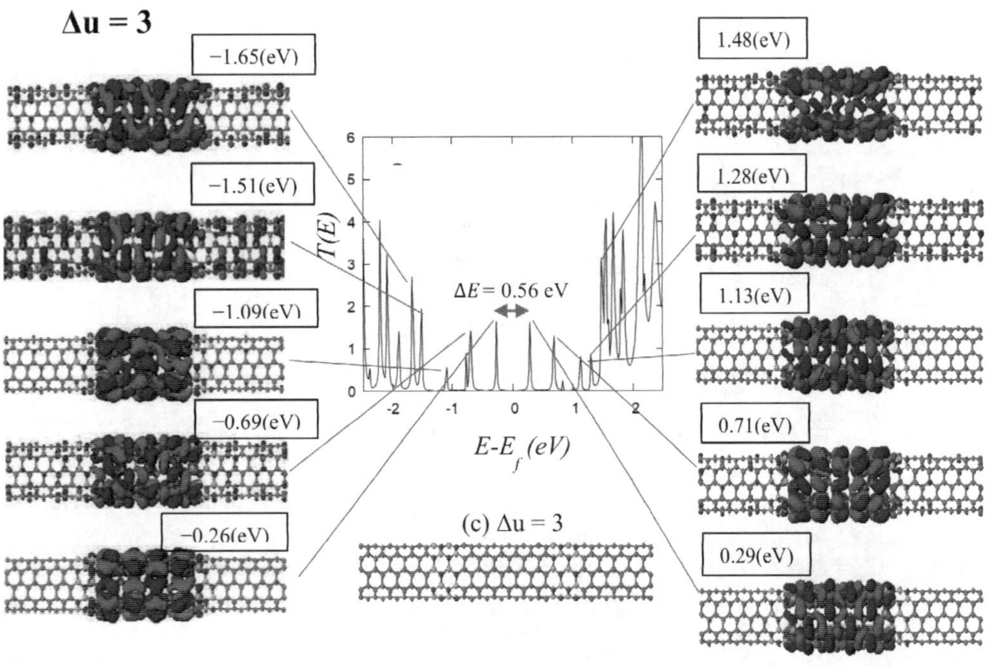

Figure 4. The transmission function of ACNT/BNNT (5,5): $\Delta u = 3$ and the wave functions that correspond to the sharp peaks in $T(E)$ are presented. The two-way arrow indicates ΔE.

Journal of Physics: Conference Series **2207** (2022) 012038 doi:10.1088/1742-6596/2207/1/012038

The value of $T(E)$ is also zero at the Fermi energy, and with $\Delta E = 0.56(\Delta u = 3)$, our ACNT/BNNT (5,5) are also semiconductors. Quantum resonant tunnelling plays a very important role in ZCNT/BNNT (9,0).

The *I-V* characteristics of ACNT/BNNT (5,5) calculated by Tran-SIESTA are also essentially the same as those of ZCNT/BNNT (9,0) below ΔE. After the bias exceeded $\Delta E/e$, the currents approached the plateau. Most of them show delicate behaviours and moderate negative differential resistances; however, the case of $\Delta u = 1$ is not the same as it has a flat plateau, similar to ZCNT/BNNT (9,0).

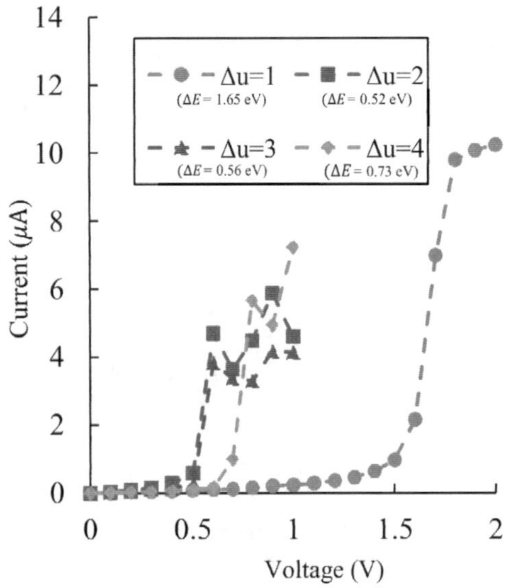

Figure 5. *I-V* characteristics of ACNT/BNNT (5,5) is illustrated. The bias voltage is under $\Delta E/e$, the current hardly flows. The voltage gets higher over the energy interval, then current jumps up as CNT/BNNT (9,0).

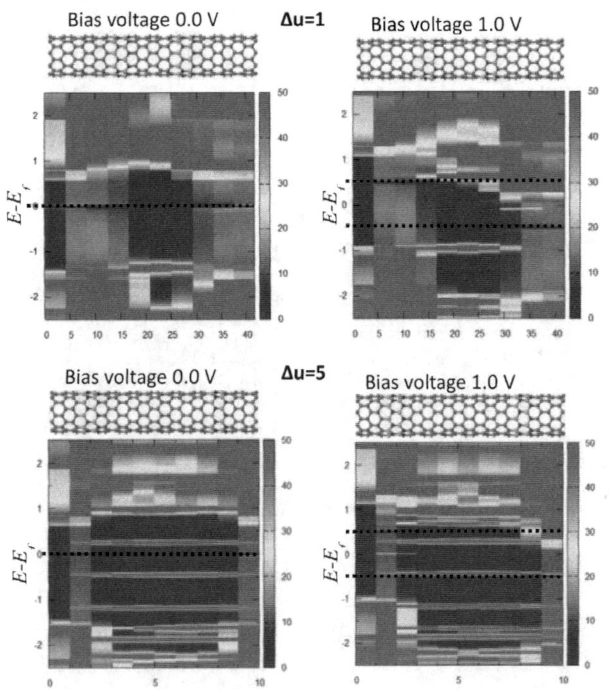

Figure 6. PDOS colour maps of ZCNT/h-BNNTs are presented. PDOS values are represented by colour bars: red represents the maximum value, blue is zero, and so on. The dotted lines represent the bias window. With zero bias voltage, the dotted lines just indicate the position of the Fermi energy.

4. PDOS Analysis

PDOS represents the projected density of states, which is DOS calculated partially in the direction of the current. Here, we separate the CNT/BNNT by hexagonal ring units and present PDOS which is calculated for each part. A higher value of PDOS means that electrons can exist at the energy with a higher probability in that part. For example, the nanodevice, which has peaks of PDOS continues from the left part to the right part around the Fermi energy and makes the band, is expected to be conductive. On the other hand, if the band is disconnected in some part, the current cannot flow. Even if the DOS of whole device has peaks around the Fermi energy. PDOS colour maps are drawn as shown in Figure 5, taking the vertical axis as the energy of the electron, and the horizontal axis as the current direction, i.e., the length direction of our devices.

ZCNT/BNNT (9,0) has almost evenly spaced PDOS peaks inside the double barriers (Figure 6: red horizontal thick lines appear in the middle of the DBS colour maps), and the energy spacings are almost the same as the energy intervals of $T(E)$. This can also explain the relation $\Delta E \times \Delta u = 3.3$ by our 1D Dirac model [1].

5. Conclusions

The transmission functions, wave functions, and quantum transport of electric currents in the DBSs of ZCNT/BNNT (9,0) and ACNT/BNNT (5,5) were numerically evaluated. In particular, quantum resonant tunnelling occurs in the models.

The transmission functions have energy gaps that originate from the embedded BNNT rings, and several sharp peaks in the energy gap. The interval of the peaks, ΔE are crucially dependent on the separation distance of two BNNT barriers, Δu . Furthermore, ΔE and Δu have good inverse proportionality for ZCNT/BNNT (9,0). On the other hand, we could not find the same kind of clear relation for ACNT/BNNT (5,5).

In the *I*-*V* characteristics, our calculations show a nearly zero flat behaviour of current at a bias voltage smaller than $\Delta E/e$, and the current suddenly and smoothly increases around the energy interval. Then, it approaches a plateau. This behaviour is common to ZCNT/BNNT (9,0) and ACNT/BNNT (5,5) with $\Delta u = 1$. The situation is quite different for ACNT/BNNT (5,5) with $\Delta u > 1$. As the bias voltage increases, the current exhibits a more complicated and delicate behaviour in such device models. They showed moderate negative differential resistance.

These properties were also confirmed by the wave functions inside the DBS and PDOS. PDOS is a powerful tool for investigating what happens inside nanodevices.

References

[1] Wakai T, Sakamoto S and Tomiya M 2018 *I-V* characteristics of graphene nanoribbon/h-BN heterojunctions and resonant tunneling *J. Phys. Condens. Matter* **30** 265302
[2] Soler J M, Artacho E, Gale J D, Garcıa A, Junquera J, Ordejón P and Sanchez-Portal D 2002 The Siesta method for ab initio order-N materials simulation *J. Phys. Condens. Matter* **14** 2745
[3] Sánchez-Portal D, Ordejón P and Canadell E 2004 Computing the properties of materials from first principles with SIESTA *Struct. Bond.* **113** 103
[4] Sánchez-Portal D, Ordejon P, Artacho E and Soler J M 1997 Density-functional method for very large systems with LCAO basis sets *Int. J. Quant. Chem.* **65** 453
[5] Haas P, Tran F, Blaha P and Schwarz K 2011 Construction of an optimal GGA functional for molecules and solids *Phys. Rev. B* **83** 205117
[6] Perdew J P, Burke K and Ernzerhof M 1996 Generalized Gradient Approximation Made Simple *Phys. Rev. Lett.* 77 3865
[7] Golberg D, Bando Y, Huang Y, Terao T, Mitome M, Tang C and Zhi C 2010 Boron nitride nanotubes and nanosheets *ACS Nano* **4** 2979
[8] Tsu R and Esaki L 1973 Tunneling in a finite superlattice *Appl. Phys. Lett.* **22** 562

Journal of Physics: Conference Series **2207** (2022) 012039 doi:10.1088/1742-6596/2207/1/012039

Optical Properties of OLED Materials by TDDFT

N Akino and Y Zempo

Faculty of Computer and Information Sciences, Hosei University
3-7-2 Kajino Koganei, Tokyo, 184-8584, JAPAN

E-mail: akino@hosei.ac.jp

Abstract For the material design of organic light emitting diodes, it is highly desired to simulate the spectral profile, that is, not only the peak wavelength but also its spectrum shape. Time dependent density functional theory in real-space and real-time formalism is applied to simulate the spectrum profile of typical organic materials, poly(9,9'-dioctyl-fluorene) and $Ir(ppy)_3$. Although the some red-shift in the peak wavelength has been observed due to the use of the local density approximation, the overall simulated profiles agree well with the experiments. Although the study is in preliminary stage, the results show that this approach may be able to simulate the spectrum profile reasonably well within a reasonable computation time.

1. Introduction

Recently, organic light emitting diodes (OLEDs) have been utilized in smartphones, large-size displays and solid-state lightings due to their desired characteristics such as self-emission, vivid color, light weight, and so on.

Excitons, which are generated by the recombination of injected electrons and holes, consist of singlet excitons (25%) and triplet excitons (75%) according to the rules of spin statistics. Since the light emission in usual organic compounds is normally only from the singlet exciton (fluorescence), the triplet excitons, which are 75% of generated excitons, do not contribute to the light emission and are heat dissipated. In other words, only 25% of the excitons obtained by recombination are utilized in light emission. Hence, many efforts have been devoted to the utilization of the triplet excitons for years in order to make an internal quantum efficiency (IQE) 100%. One route is a use of metal complex such as an iridium complex and a platinum complex where the triplet excitons could be utilized as emission (phosphorescence) [1,2]. Due to the strong spin-orbit interaction of a heavy atom such as an iridium, the intersystem crossing (ISC) becomes possible, thus singlet excited state is converted into triplet excited state, and theoretically 100% of the generated excitons can be extracted as phosphorescence. Thus, theoretically IQE=100% can be achieved [3,4]. However, in the use of metal complex, the broad spectrum shape has been one of the issues for the display applications. Furthermore, it is very difficult to develop stable (deep-)blue phosphorescent emitters, and it has been a challenging field in material development [4].

In this study, we have focused on the spectrum profile of emitting materials, which is one of the most important material properties as it determines not only the emission color, but also the device efficiency. Most of theoretical studies have been performed so far were focused on the prediction of the peak wavelength, not the spectrum profile, due to the computational difficulty. However, the prediction of spectral profile is really desired in material development as it relates with the device performance. Especially the material with sharp spectral profile is necessary for display application to avoid any loss by the energy filter.

In order to study the spectrum property of materials, we have employed the time dependent density functional theory (TDDFT) [5], which is a well-recognized method to calculate the properties related to the excited states such as the excitation energy of molecules. We apply this technique to study the optical properties of OLED materials, especially focusing on their spectral profiles.

In the next section we describe the theory with our numerical details. Then, the results of its application to poly-(9,9'-dialkyl-fluorene) as an example of polymer (oligomer) fluorescent material, and tri(2-phenylpyridinato)iridium(III), Ir(ppy)$_3$, as a phosphorescence material, are given with brief concluding remarks.

2. Theory and numerical details

In this section, the TDDFT technique used in this study is briefly explained [6,7]. The basis is the density functional theory (DFT) [8], which is expressed by the Kohn-Sham (KS) equation [9],

$$\left\{-\frac{1}{2}\nabla^2 + V_{ion}(\boldsymbol{r}) + V_H(\boldsymbol{r}) + V_{XC}[\rho(\boldsymbol{r})]\right\}\psi_j(\boldsymbol{r}) = \varepsilon_j\psi_j(\boldsymbol{r}) \tag{1}$$

where $V_{ion}(\boldsymbol{r})$, $V_H(\boldsymbol{r})$, and $V_{XC}[\rho(\boldsymbol{r})]$ are the ionic potential, the Hartree potential, and the exchange correlation potential, respectively. The electronic density is given by $\rho(\boldsymbol{r}) = \sum_j |\psi_j(\boldsymbol{r})|^2$ and the summation is taken over all occupied orbitals $\psi_j(\boldsymbol{r})$.

Since DFT is a theory for the ground state property, it is not suited to study the excited property. The time-dependent extension of the time-independent DFT becomes necessary and it is called the TDDFT [5]. The equation of motion for TDDFT is a time dependent extension of Eq. (1) and is given by

$$\left\{-\frac{1}{2}\nabla^2 + V_{ion}(\boldsymbol{r}) + V_H(\boldsymbol{r},t) + V_{XC}[\rho(\boldsymbol{r},t)] + V_{ext}(\boldsymbol{r},t)\right\}\psi_j(\boldsymbol{r},t) = i\frac{\partial}{\partial t}\psi_j(\boldsymbol{r},t) \tag{2}$$

where $V_{ion}(\boldsymbol{r})$, $V_H(\boldsymbol{r},t)$, $V_{XC}[\rho(\boldsymbol{r},t)]$, and $V_{ext}(\boldsymbol{r},t)$ are the ionic potential and the Hartree potential, the exchange correlation potential, and the external field, respectively. The potential of the external field, $V_{ext}(\boldsymbol{r},t)$, is included as it is used as a perturbation at $t = 0$ in our real time calculations. For the ionic potential, the pseudopotential $V_{ion}(\boldsymbol{r})$ in the separable form was employed so that only the valence electrons were considered [10]. The Hartree and exchange-correlation potentials are given as functionals of the electronic charge density, $\rho(\boldsymbol{r},t) = \sum_j |\psi_j(\boldsymbol{r},t)|^2$ where the summation is taken over all the occupied states j as in DFT. The Hartree potential is determined by solving the Poisson's equation $\nabla^2 V_H = -4\pi\rho$. For the exchange-correlation potential V_{XC}, the local density approximation (LDA) is used in this study [8].

In our calculations, the real-time and real-space technique was adopted in solving Eq.(1) and (2) by means of the finite difference approach [11] without using explicit bases such as plane waves [12] and gaussians [13]. Within this approach, we divide the physical system into the space grids (cells) and solve for the wave functions on each grid. Thus, the linear dimension of grid is related to the accuracy in energy. In this study, the uniform grid with the linear dimension $\Delta x = 0.3$ (Å) which the KS equation, Eq.(1), converges well [6,7] was used for simplicity.

Prior to the calculation of optical responses, we need to solve the time-independent DFT to obtain the optimized electronic structure [14]. After obtaining the equilibrium electronic structure, an external field, $V_{ext}(\boldsymbol{r},t)$, is applied at $t = 0$ to the system as a perturbation and follow the linear responses of the system in real time. The wave functions are evolved by the time evolution operator,

$$\psi(t) = e^{-iHt}\tilde{\psi}(0) \tag{3}$$

where H is the Hamiltonian of the system, and $\tilde{\psi}(0)$ is the initial wave function at $t = 0$. The latter is given by $\tilde{\psi}(0) = e^{ik_z z}\psi(0)$ where k_z is a small wave number corresponding to the external perturbation in the z direction. In the linear response, the time-dependent polarizability is proportional to the dipole matrix element

$$\mu_\xi(t) = \langle\psi(t)|\xi|\psi(t)\rangle \tag{4}$$

where ξ represents x, y, and z. The polarizability $\alpha_\xi(\omega)$ is numerically calculated by the Fourier transform of $\mu_\xi(t)$ and the average over three directions gives the polarizability of the molecule, $\alpha(\omega) = (\alpha_x + \alpha_y + \alpha_z)/3$. The strength function $S(\omega)$ is related to the imaginary part of the polarizability and given by

$$S(\omega) = \frac{2\omega}{\pi}\mathrm{Im}[\alpha(\omega)] \tag{5}$$

In the time evolution, the time step width Δt and the total number of time steps N_{time} are important parameters. The former has to be small enough so that the Hamiltonian can be regarded as static, and the latter is related to the energy resolution in physical quantity in interest [6,7].

3. Results and discussions

3.1. Poly(9-9'-dioctyl-fluorene)

Poly(9,9'-dioctyl-fluorene) has been a well-known and well-studied material as a candidate for the blue light emitting polymer in polymer OLEDs [15,4]. Its molecular structure is shown in Figure 1(a). Since a real polymer is too large to handle in numerical simulations, the simulations have been limited to oligomers with small number of repeating units [16,6]. The lowest unoccupied molecular orbital (LUMO) and the highest occupied molecular orbital (HOMO) of dimer of 9-9'-dimethyl-fluorene are shown in Figure 1(b) and 1(c) calculated at B3LYP/6-31g(d) level by Gaussioan09 [13]. From Figure 1(b) and 1(c), one can see that the orbitals are well delocalized on the backbone, but little on the alkyl sidechain. This supports the shortening of alkyl chain length in our simulations. The main contribution to absorption and emission spectra of this molecule comes from the $\pi - \pi^*$ transition between HOMO and LUMO.

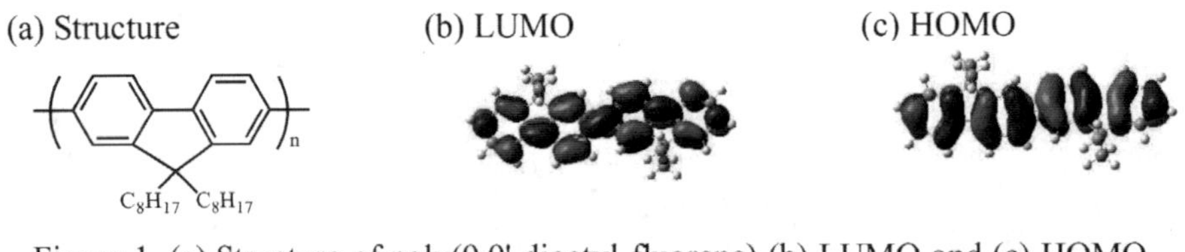

Figure 1. (a) Structure of poly(9,9'-dioctyl-fluorene) (b) LUMO and (c) HOMO orbitals of dimer of 9,9'-dimethyl-fluorene calculated at B3LYP/6-31g(d) level.

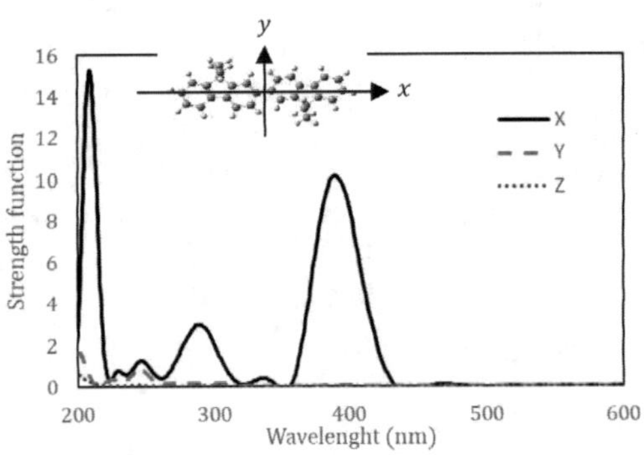

Figure 2. Strength function of dimer of 9,9'-dimethyl-fluorene by RSRT-TDDFT. Black solid line, red dashed line, and blue dotted line correspond to the perturbation in x, y, and z direction, respectively. The x axis is taken in the direction of conjugation extension in the dimer.

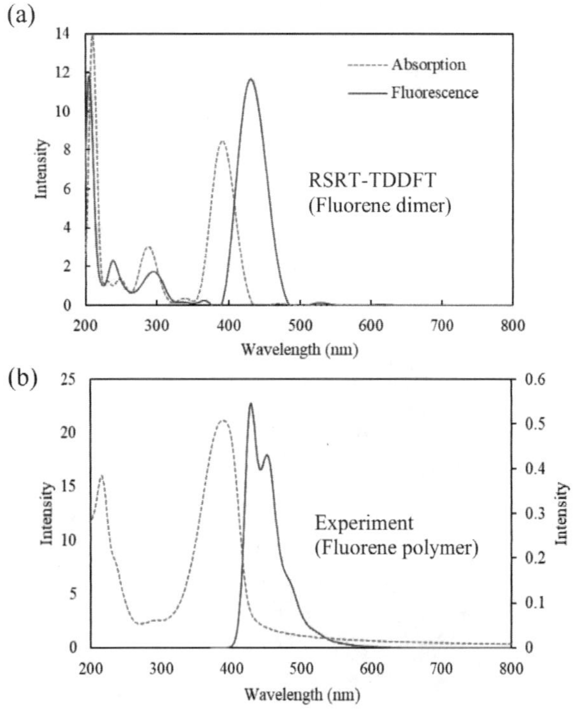

Figure 3. (a) Absorption and fluorescence spectra of fluorene dimer by RSRT-TDDFT and (b) experimental absorption and fluorescence of fluorene polymer.

In order to study the spectral profile, RSRT-TDDFT was applied to dimer of 9,9-dimethyl-fluorene whose geometry is optimized at B3LYP/6-31g(d) level for the ground state S_0 and TD-B3LYP/6-31G(d) for the first excited state S_1. As an example, the simulated strength functions $S_\xi(\omega) = \frac{2\omega}{\pi}\mathrm{Im}[\alpha_\xi(\omega)]$ for the absorption, where for ξ represents x (black solid line), y (red dashed line), and z (blue dotted line) are shown in Figure 2. The coordinate is taken so that the direction of x axis is parallel to the direction of backbone (the direction of conjugation extension) of the dimer as also shown in Figure 2. As one can see, the lowest energy absorption at ~390nm is given by the case with the perturbation in x direction, $S_x(\omega)$, and there seems almost no absorption in the case of perturbation in y and z directions in this energy range. These are reasonable as the x axis is parallel to the direction of conjugation, thus the long time response can occur, on the other hand, the y and z axis are perpendicular to the molecular backbone, thus the electron transfer is limited compared to the x direction.

The simulated spectrum profile of dimer of 9,9-dimethyl-fluorene and the experimentally observed spectrum of the poly(9-9'-dioctyl-fluorene), absorption (red dashed line) and emission (blue solid line), are shown in Figure 3(a) and 3(b), respectively. Although the red shift of the peak wavelength in simulation has been observed, two spectral profiles are quite similar in shape. One reason of this red shift is considered due to the use of LDA. Aside from this red-shift, the agreement of spectral profile between the simulation and the experiment seems to be very encouraging and shows the possibility of the spectrum profile prediction by this RSRT-TDDFT technique.

3.2. *Fac-tri(2-phenylpyridine)iridium(III), fac-Ir(ppy)$_3$*

Fac-tri(2-phenylpyridine)iridium(III), fac-Ir(ppy)$_3$, is the well-known efficient green phosphorescent material. Its structure is shown in Figure 4(a). HOMO and LUMO orbitals are shown in Figure 4(c) and 4(b). The HOMO orbital is mostly localized on the iridium atom and LUMO is evenly distributed onto three ligands. This iridium complex is known to form the metal to ligand charge transfer (MLCT) excited states. There exists the absorbance of MLCT singlet (^{1}MLCT) at ~385nm and MCLT triplet (^{3}MLCT) at ~450nm in its absorption spectrum as shown in Figure 5(b) [3]. Due to the charge transfer nature of orbitals, its spectrum profile becomes broad.

The RSRT-TDDFT results is shown in Figure 5(a). The agreement of the simulated and experimentally observed spectral profiles are reasonable aside from the red shift in simulation due to the use of LDA. Not only the agreement of ^{1}MLCT, but also the agreement of high energy $\pi - \pi^*$ type transition below ~300nm can be observed in the comparison. This agreement is suggesting that the technique in this study may be able to simulate charge transfer type as well as $\pi - \pi^*$ type transitions.

The geometry used in the RSRT-TDDFT are optimized by B3LYP level for the ground states geometry. Basis used for the geometry optimization is 6-31G(d) for all atoms except for Ir atom where LANL2DZ is used.

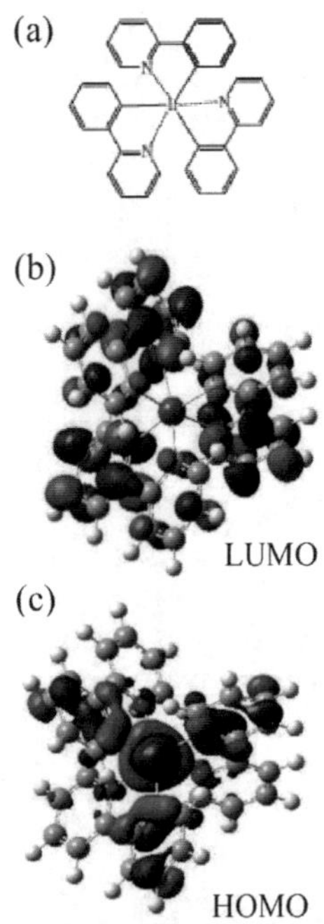

Figure 4. (a) Structure of fac-Ir(ppy)$_3$ (b) LUMO and (c) HOMO calculated at B3LYP.

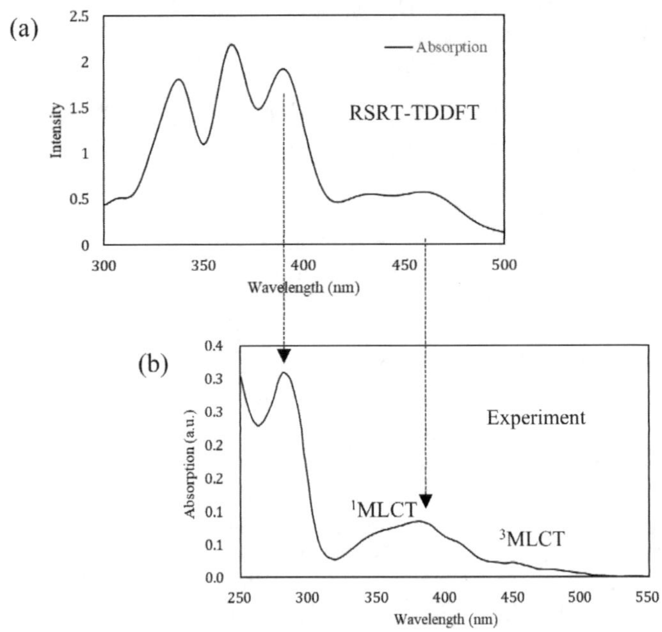

Figure 5. Absorption spectrum of Ir(ppy)3 (a) by RSRT-TDDFT and (b) by the experiment at 9K.

3.3. Concluding remarks

TDDFT in real-space and real-time formalism has been applied to OLED materials, poly(9,9'-dioctyl-fluorene) and Ir(ppy)$_3$. The former is a fluorescent material of $\pi - \pi^*$ type transition with narrow width and the latter is a phosphorescent material of MLCT type transition with broad width. Although the study is limited and in preliminary stage at this moment, the good agreement of spectrum profiles between RSRT-TDDFT calculation and the experiment has been observed for two types of transitions, $\pi - \pi^*$ and MLCT. Thus, these results suggest that this approach may be able to simulate the spectrum profile reasonably well within a reasonable computation time. Further confirmation including the application to TADF materials is in progress.

References

[1] Baldo M A, O'Brien D F, You Y, Shoustikov A, Sibley S, Thompson M E, and Forrest S R, 1998 *Nature* **395**, 151
[2] Baldo M A, Lamansky S, Burrows P E, Thomson M E, and Forrest S R, 1999 *Appl. Phys. Lett.* **75**, 4
[3] Goushi K, Kawamura Y, Sasabe H, Adachi C, 2004 *Jpn. J. Appl. Phys.* **43** L937
[4] Akino N, Tsubata Y, and Yamada T, 2018 *SUMITOMO KAGAKU* **2018**, 1
[5] Runge E and Gross E K U, 1984 *Phys. Rev. Lett.* **52**, 997

[6] Zempo Y, Akino N, Ishida M, Ishitobi M and Kurita Y, 2008 *J. Phys. Condens. Matter* **20**, 064231
[7] Toogoshi M, Kato M, Kano S S, and Zempo Y, 2014 *J. of Phys. Conf. Seri.*, **510**, 012027
[8] Hohenberg P and Kohn W, 1964 *Phys. Rev.* **136**, B864
[9] Kohn W and Sham L. J, 1965 *Phys. Rev.* **140**, A1133
[10] Troullier N and Martins J L, 1991 *Phys. Rev.* B**43**, 1993
[11] Chelikowsky J, Troullier N, Wu K and Saad Y, 1994 *Phys. Rev.* B**50**, 11355
[12] Kresse G and Purthmueller J, 1996 *Phys. Rev.* B**54**, 11169
[13] Gaussian 09, Revision E.01, Frisch M J, Trucks G W, Schlegel H B, Scuseria G E, Robb M A, Cheeseman J R, Scalmani G, Barone V, Petersson G A, Nakatsuji H, Li X, Caricato M, Marenich A, Bloino J, Janesko B G, Gomperts R, Mennucci B, Hratchian H P, Ortiz J V, Izmaylov A F, Sonnenberg J L, Williams-Young D, Ding F, Lipparini F, Egidi F, Goings J, Peng B, Petrone A, Henderson T, Ranasinghe D, Zakrzewski V G, Gao J, Rega N, Zheng G, Liang W, Hada M, Ehara M, Toyota K, Fukuda R, Hasegawa J, Ishida M, Nakajima T, Honda Y, Kitao O, Nakai H, Vreven T, Throssell K, Montgomery J A, Peralta Jr., J E, Gliaro F, Bearpark M, Heyd J J, Brothers E, Kudin K N, Staroverov V N, Keith T, Kobayashi R, Normand J, Raghavachari K, Rendell A, Burant J C, Iyengar S S, Tomasi J, Cossi M, Millam J M, Klene M, Adamo C, Cammi R, Ochterski J W, MartinR L, Morokuma K, Farkas O, Foresman J B, and Fox D J, Gaussian, Inc., Wallingford CT, 2016.
[14] Yabana K and Bertisch G F, 1996 *Phys. Rev.* B**54**, 4484
[15] Bernius M, Inbasekaran M, O'Brien J, and Wu W, 2000 *Adv. Mater.* **12**, 1737 (2000)
[16] Akino N and Zempo Y, 2004 *MRS Proceedings* **846**

Journal of Physics: Conference Series **2207** (2022) 012040 doi:10.1088/1742-6596/2207/1/012040

Phenomenological Simulations of Intermediate States in a Type I Superconducting Wire

Yohitaka mizuhara and Masaru Kato

Department of Physics and Electronics, Osaka Prefecture University, 1-1 Gakuencho, Naka-ku, Sakai 599-8531, Japan

sab01160@edu.osakafu-u.ac.jp, kato@ms.osakafu-u.ac.jp

Abstract. We have investigated superconducting state of a type I superconducting wire solving the Ginzburg Landau equation with finite element method. When a critical current is applied to a type I superconducting wire, superconductivity becomes inhomogeneous. We find superconductivity shows oscillatory structures which are similar to London proposed intermediate state. In the simulations, we impose two types of boundary conditions, phase-fixed and vector potential-fixed. Under both boundary conditions, we have obtained similar periodic structures.

1. Introduction

Since the discovery of the superconductivity by Kamerlingh Onnes in 1911 [1], superconductors have been studied for applications. Especially, magnetic properties have been studied extensively, because superconductors are intended to be used under the magnetic field. Depending on response to the magnetic field, superconductors are classified into two types. First one shows a perfect diamagnetism, which is called the Messner effect. We call this type of superconductor type-I superconductor. For higher field that is larger than the critical field H_c, superconductivity in the type-I superconductor disappears due to large magnetic energy. Second one shows a vortex state, where magnetic field enters into the superconductor as a quantized magnetic flux. We call this type of superconductor type- II superconductor. In the type- II superconductor, the vortex state appears between the lower critical field H_{c1} and the upper critical field H_{c2} [3]. For applications of superconductors, the type- II superconductors are usually used, because of large value of H_{c2}. However, magnetic properties of type-I superconductor are not still clearly understood. In a thin film of the type-I superconductor, a normal and a superconducting regions can coexist and this state is called an intermediate state [4].

In 1950s, London [5] proposed a periodic structure of intermediate state in a type I superconducting wire carrying current that is larger than the critical current [3,6]. The critical current is determined as follows. When a current induced magnetic field becomes H_c at the surface of the superconducting region, the current is the critical current $I_c = \frac{H_c r_1 c}{2}$ (Silsbee's rules) [3]. Here, r_1 is a radius of a superconducting region. If the current is larger than I_c, magnetic field exceeds the critical field H_c and a surface layer become normal. Then, the current flows only inner superconducting region, and the current density increases. This causes magnetic field at a surface of inner region to become larger than H_c. Therefore, this structure with a superconducting core surrounded by normal region is not stable. Then London proposed a periodic structure of intermediate as shown in figure 1 [2]. We require that for

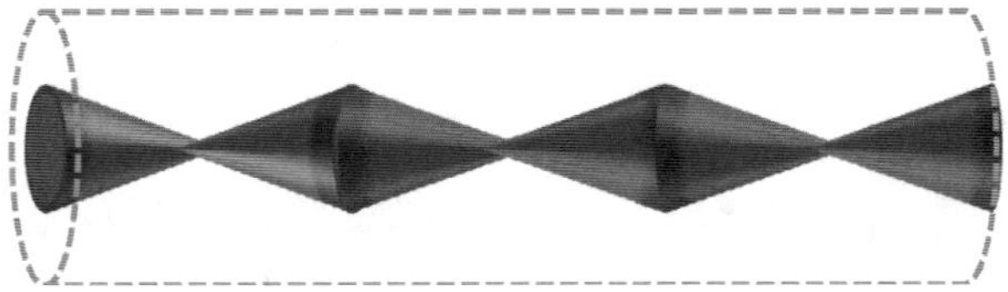

Figure 1. London's model of the intermediate state of a type I superconducting wire carrying an external current that is larger than I_c.

$r < r_1$, superconducting and normal regions coexist. Then, the magnetic field $H(r) = H_c$ and then the total current $I(r)$ inside radius r,

$$I(r) = \frac{crH_c}{2}. \quad (1)$$

We obtain a current density $J(r)$,

$$J(r) = \frac{1}{2\pi r}\frac{dI(r)}{dr} = \frac{cH_c}{4\pi r}. \quad (2)$$

Since the electric field $\boldsymbol{E}$ is independent of r, the fractional path of resistive material must be proportional to r. So, the structure in figure 1 may become stable.

Previously, Kramer and Baratoff studied phase slip states in a quasi-one-dimensional superconducting wire model. They found that phase-slip presumably occur at weak spots[7]. Also, Bronsard and Stoth studied the time dependent Ginzburg-Landau(TDGL) equations for a type I cylindrical superconducting wire under a parallel magnetic field. They showed analytically that the problem is well-posed[8]. Unlike previous studies, we study superconductivity in a square shaped wire, using the three-dimensional finite element method, in order to clarify whether the London's intermediate state actually exists and what kind of structure the intermediate state can be stable and whether the phase slips occur at weak points.

2. Model

In order to treat spatial variation of the superconducting order parameter, we solve the Ginzburg-Landou (GL) equations,

$$\alpha\psi + \beta|\psi|^2 + \frac{1}{m^*}\left(\frac{\hbar}{i}\nabla - \frac{e^*}{c}\boldsymbol{A}\right)^2 \psi = 0 \ , \quad (3)$$

$$\boldsymbol{J} = \frac{c}{4\pi} curl\boldsymbol{h} = \frac{e^*\hbar}{2m^*i}(\psi^*\nabla\psi - \psi\nabla\psi^*) - \frac{e^{*2}}{m^*c}\psi^*\psi. \quad (4)$$

Here, α and β is expansion coefficients, ψ is the order parameter, m^* and e^* are an effective mass and an elective charge of a superconducting electron pair, which is called superconducting Cooper pair, respectively. In this study, we solve this equation self-consistently using the three-dimensional finite element method (3D-FEM). In 3D-FEM, we divide a three-dimensional system into tetrahedrons. Then, we expand ψ and $\boldsymbol{A}$ using volume coordinates N_i as,

$$\psi^e(x,y,z) = \textstyle\sum_i \psi_i^e N_i^e(x,y,z) \ , \quad (5)$$

$$\boldsymbol{A}^e(x,y,z) = \textstyle\sum_i \boldsymbol{A}_i^e N_i^e(x,y,z) \ , \quad (6)$$

where ψ_i^e and $\boldsymbol{A}_i^e$ are values of the order parameter and vector potential at i-th node in e-th element, respectively.

In this study, we consider two types of boundary conditions for constant external current at both ends of the wire. First, we impose following boundary conditions at both ends,

$$\Delta\theta = -\frac{4\pi}{c}\frac{2\pi\lambda(T)^2}{\Phi_0}\frac{\mathbf{j}_{input}}{|\psi|^2}. \quad (7)$$

Second, we impose following boundary condition with the fixed vector potential,

$$\boldsymbol{A} = -\frac{4\pi}{c}\frac{2\pi\lambda(T)^2}{\Phi_0} \times \frac{\boldsymbol{j}_{input}}{|\psi^{new}|^2} - \nabla\theta \quad (8)$$

Here $\Delta\theta$ is difference of the phase $(\boldsymbol{\nabla}\theta \cdot \boldsymbol{n})\Delta x$ and $\boldsymbol{n}$ is a normal vector at the end, $\Phi_0 = \frac{hc}{2e}$ is a flux quantum, and $\lambda(T) = \left[\frac{mc^2}{4\pi e^2 \cdot 2\psi_0^2}\right]^{\frac{1}{2}}$ is a penetration depth.
Inserting equations (5) and (6) into equations (3) and (4), we obtain

$$\sum_j \left[P_{ij}(\{\boldsymbol{A}\}) + P_{ij}^{2R}(\{\psi\})\right] Re\,\psi_j^e + \sum_j \left[P_{ij}(\{\boldsymbol{A}\}) + P_{ij}^{2R}(\{\psi\})\right] Im\,\psi_j^e = V_i^R(\{\psi\}), \tag{9}$$

$$\sum_j \left[-Q(\{\boldsymbol{A}\}) + Q_{ij}^2(\{\psi\})\right] Re\,\psi_j^e + \sum_j \left[P_{ij}(\{\boldsymbol{A}\}) + P_{ij}^{2I}(\{\psi\})\right] Im\,\psi_j^e = V_i^I(\{\psi\}), \tag{10}$$

$$\sum_j R_{ij}(\{\psi\})\ A_{jx} + \sum_j S_{ij}^{xy} A_{jy} + \sum_j S_{ij}^{xz} A_{jy} = T_i^x - U_i^x, \tag{11}$$

$$\sum_j R_{ij}(\{\psi\})\ A_{jy} + \sum_j S_{ij}^{yx} A_{jx} + \sum_j S_{ij}^{yz} A_{jz} = T_i^y - U_i^y, \tag{12}$$

$$\sum_j R_{ij}(\{\psi\})\ A_{jz} + \sum_j S_{ij}^{zx} A_{jx} + \sum_j S_{ij}^{zy} A_{jy} = T_i^z - U_i^z. \tag{13}$$

Coefficients are defined as,

$$P_{ij}(\{\boldsymbol{A}\}) = \sum_\alpha K_{ij}^{\alpha\alpha} + \sum_{i_1,i_2} I_{i_1 i_2 ij} \sum_\alpha A_{i_1\alpha}^e A_{i_2\alpha}^e - \frac{1}{\xi(T)^2} I_{ij}, \tag{14}$$

$$P_{ij}^{2R}(\{\psi\}) = \frac{1}{\xi(T)^2} \sum_{i_1,i_2} I_{i_1 i_2 ij} \left(3Re\psi_{i_1}^e Re\psi_{i_2}^e + Im\psi_{i_1}^e Im\psi_{i2}^e\right), \tag{15}$$

$$P_{ij}^{2R}(\{\psi\}) = \frac{1}{\xi(T)^2} \sum_{i_1,i_2} I_{i_1 i_2 ij} \left(Re\psi_{i_1}^e Re\psi_{i_2}^e + 3Im\psi_{i_1}^e Im\psi_{i2}^e\right), \tag{16}$$

$$Q_{ij}(\{\boldsymbol{A}\}) = \sum_{i_1} \sum_\alpha (J_{ji_1 i}^\alpha - J_{ii_1 j}^\alpha) A_{i_1}^\alpha, \tag{17}$$

$$Q_{ij}^2(\{\psi\}) = \frac{2}{\xi(T)^2} \sum_{i_1 i_2} I_{i_1 i_2 ij}\, Re\psi_{i_1}^e Re\psi_{i_2}^e, \tag{18}$$

$$R_{ij}(\{\psi\}) = \kappa^2 \xi(T)^2 \sum_\alpha K_{ij}^{\alpha\alpha} + \sum_{i_1,i_2} I_{i_1 i_2 ij}\, \psi_{i_1}^{e*} \psi_{i_2}^e, \tag{19}$$

$$S_{ij}^{\alpha\beta} = \kappa^2 \xi(T)^2 \left(K_{ij}^{\alpha\beta} - K_{ij}^{\beta\alpha}\right), \tag{20}$$

$$T_i^\alpha = \sum_{i_1,i_2} J_{i_1 i_2 i}^\alpha\, Im\left(\psi_{i_1}^{e*} \psi_{i_2}^e\right), \tag{21}$$

$$U_i^\alpha = \kappa^2 \xi(T)^2 \frac{2\pi}{\Phi_0} \left(J_i^\beta H_\gamma - J_i^\gamma H_\beta\right), \tag{22}$$

$$V_i^R(\{\psi\}) = \frac{2}{\xi(T)^2} \sum_{i_1 i_2 i_3} I_{i_1 i_2 i_3}\, Re\left(\psi_{i_1}^e \psi_{i_2}^{e*}\right) Re\psi_{i_3}^e, \tag{23}$$

$$V_i^{I}(\{\psi\}) = \frac{2}{\xi(T)^2} \sum_{i_1 i_2 i_3} I_{i_1 i_2 i_3}\, Re\left(\psi_{i_1}^e \psi_{i_2}^{e*}\right) Re\psi_{i_3}^e. \tag{24}$$

The integrals are defined as,

$$I_{ij} = \int_{V_e} N_i^e N_j^e\, dV, \tag{25}$$

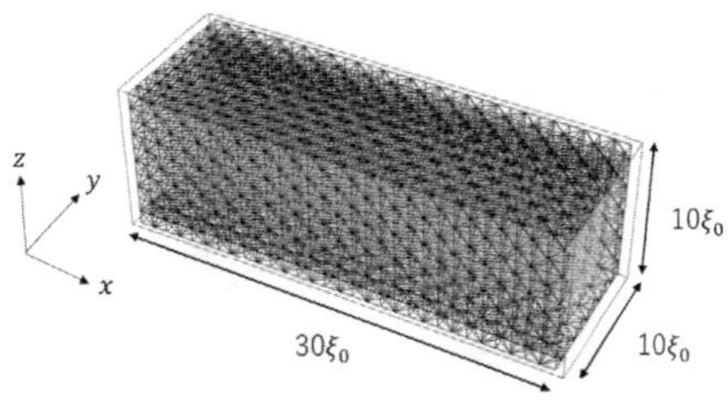

Figure 2. Finite elements of simulation system. Green and black region are the superconductor and the vacuum, respectively

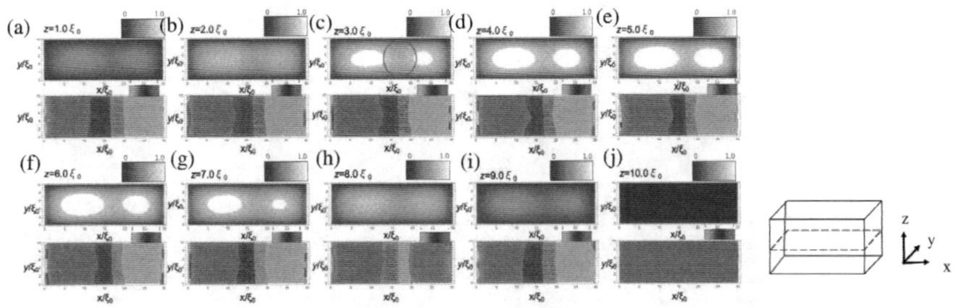

Figure 3. Two dimensional distributions of amplitude (upper) and phases (lower) of order parameter at z=1.0 ξ_0(a), 2.0 ξ_0(b) , 3.0 ξ_0(c) , 4.0 ξ_0(d) , 5.0 ξ_0(e) , 6.0 ξ_0(f) , 7.0 ξ_0(g) , 8.0 ξ_0(h) , 9.0 ξ_0(i) and 10.0 ξ_0(j).

$$I_{i_1 i_2 i_3} = \int_{V_e} N_{i_1}^e N_{i_2}^e N_{i_3}^e dV, \tag{26}$$

$$I_{i_1 i_2 i_3 i_4} = \int_{V_e} N_{i_1}^e N_{i_2}^e N_{i_3}^e N_{i_4}^e dV, \tag{27}$$

$$J_i^{x_i} = \int_{V_e} \frac{\partial N_i}{\partial x_i} dV, \tag{28}$$

$$J_{i_1 i_2 i_3}^{x_i} = \int_{V_e} \frac{\partial N_{i_1}^e}{\partial x_i} N_{i_2}^e N_{i_3}^e dV, \tag{29}$$

$$K_{i_1 i_2}^{x_i x_j} = \int_{V_e} \frac{\partial N_{i_1}^e}{\partial x_i} \frac{\partial N_{i_2}^e}{\partial x_j} dV, \tag{30}$$

$$J_{i_1 i_2}^{x_i} = \int_{V_e} \frac{\partial N_{i_1}}{\partial x_i} N_{i_2} dV. \tag{31}$$

We start initial random states, and solve above equations self-consistently. We show simulation results for both boundary conditions.

In this study $30\xi_0 \times 8\xi_0 \times 8\xi_0$ superconductor surrounding by a vacuum layer with thickness1 ξ_0 (figure 2). Here ξ_0 is a coherence length of absolute zero. We set GL parameter κ=0.7.

3. Results and discussion

3.1. Numerical results with the fixed-phase boundary condition

First, we show an order parameter structure for T/T_c =0.4 and $J/J_0 = 1.0$ with the fixed-phase boundary condition in figure 3. This structure is similar to that was proposed by London. In this structure, the phase varies along the x direction, so current flows along x-direction. Also, there is a weak superconducting region.

3.2. Numerical results with the fixed-magnetic vector potential boundary condition

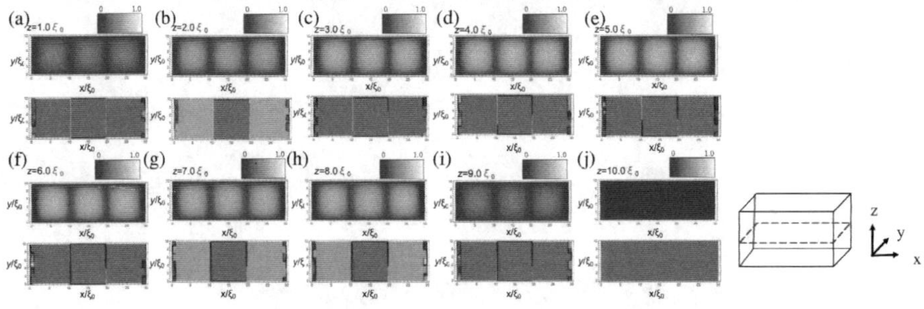

Figure 4. Two dimensional distributions of amplitude (upper) and phases (lower) of order parameter at z=1.0 ξ_0(a) , 2.0 ξ_0(b) , 3.0 ξ_0(c) , 4.0 ξ_0(d) , 5.0 ξ_0(e) , 6.0 ξ_0(f) , 7.0 ξ_0(g) , 8.0 ξ_0(h) , 9.0 ξ_0(i) and 10.0 ξ_0(j).

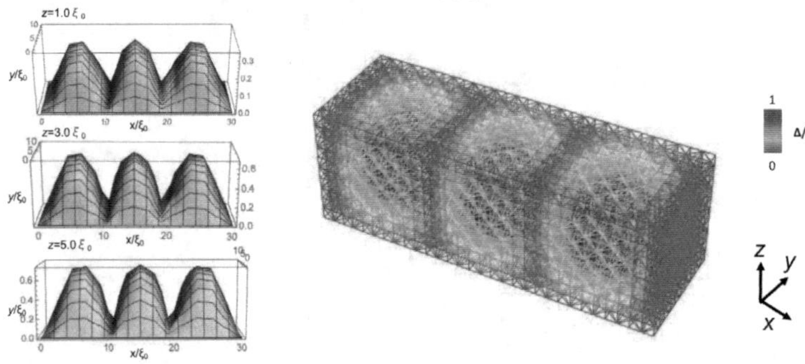

Figure 5. Distributions of amplitude of order parameter in each z cross section(left). Three dimensional order parameter structure(right)

Next, we show a numerical result for T/T_c =0.3 and $J/J_0 = 0.1$ with the fixed-magnetic vector potential boundary condition in figure 4. This structure is also similar to that was proposed by the London. Convergence of this state during iteration, is rapid compared with previous structure with the fixed phase boundary condition, and oscillation of superconductivity becomes clear. In figure 5, we show three dimensional distribution of the order parameter. We can see clean oscillation of order parameter amplitude. The superconductor is separated into three regions. The phase of order parameter is almost constant within these regions. The abrupt change of the phase is called phase slip. In this result, we found that phase slips may occur at weak superconducting regions.

4. Summary

We have studied superconducting states for a type I superconducting wire under an applied current, based on the GL theory. Under the fixed phase boundary condition and the fixed magnetic vector potential boundary condition, we have found oscillating superconducting structure. For the fixed magnetic vector potential boundary condition, there appear phase slip at weak superconducting region. Convergence of this state during iteration, is rapid compared with the structure with the fixed phase boundary condition, and oscillation of superconductivity becomes clear.

Acknowledgements
We would like to thank T. Aoki, J. Yamanaka, T. Wachi, T. Kato, M. Kawabata for useful discussions.

References

[1] Onnes H K 1911 *Leiden Comm* 120-124
[2] Abrikosov A A 1957 *Zh. Eksperim Teor. Fiz* **32** 1442
[3] Tinkham M 1996 *Introduction To Superconductivity* 2nd ed. (New York : McGraw-Hill)
[4] London F 1950 *Superfluids* vol. I (New York : Wiley)
[5] Huebener R P and Gallus D E 1973 Phys. Rev. B **7** 4089
[6] Baird D C and Mukherjee B K 1970 Phys. Rev. B **3** 1043
[7] Kramer L and Baratoff A 1977 Phys. Rev. L **38** 518-521
[8] Bronsard L and Stoth B 1998 Anal. Non. L **15** 371-397

Journal of Physics: Conference Series **2207** (2022) 012041 doi:10.1088/1742-6596/2207/1/012041

Dynamics of photoinduced entanglement generation between remote electron-phonon systems

Kunio Ishida[1], **Hiroaki Matsueda**[2]

[1] School of Engineering and Center for Optical Research and Education, Utsunomiya University, Utsunomiya, Tochigi 321-8585, Japan

[2]Department of Applied Physics, Graduate School of Engineering, Tohoku University, Sendai 980-8579, Japan

E-mail: `ishd_kn@cc.utsunomiya-u.ac.jp`

Abstract. We numerically study the dynamics of quantum entanglement generation between remote electron-phonon systems by photoirradiation employing a model of multiple spins and boson modes. The Schmidt decomposition on the time-dependent wavefunction enables us to discuss the mechanism of the entanglement generation as dynamical behaviors of the singular vectors. The calculated results show that an effective entanglement source will be provided by an appropriate measurement on the phonon state.

1. Introduction

The generation of quantum entanglement is a key to realize various quantum information devices. As shown in a previous experimental study on phonons in diamond[1], the irradiation of light induces quantum entanglement between systems that do not directly interact with each other. If we treat the electromagnetic field classically, initially decomposable states remain fully decomposable by irradiation of light, and no entanglement is generated. In other words, irradiation of "classical light" to remote systems corresponds to "local operations and classical communication" in the terminology of quantum information science, and no entanglement generation is obtained between them[2, 3]. Furthermore, when light is regarded as classical electromagnetic field, simultaneous treatment of the Maxwell equations and the Schrödinger equation is possible to describe the dynamics of the system. However, light-induced quantum correlation(entanglement) cannot be taken into account there, and we need a theory with quantized light to study the dynamics of entanglement in spite of the huge computational resources required.

The paper aims to reveal the dynamics of entanglement generation between phonons in remote material systems. As shown in our previous study[3], the quantum mutual information[4, 5] is an appropriate measure for the entanglement, since we treat multipartite systems composed of electrons, phonons, and photons in the present case. Since the quantum mutual information is derived from the entanglement entropy between subsystems, the singular vectors of the Schmidt decomposition of the entire system reflect the entanglement dynamics. In this paper, we divide the entire system into the phonon states and the electron-photon states, and performed the Schmidt decomposition to obtain the entangled phonon states as singular vectors. Since those phonon singular vectors compose a set of projection-valued measures (PVM)[6], measurement of the phonon states will provide an entangled pure phonon state.

2. Model and method

We employ a model of two remote(non-interacting) electron-phonon systems irradiated by photons, of which the Hamiltonian is described by[2, 3]

$$\mathcal{H} = \sum_{i=1}^{3} \Omega_i c_i^\dagger c_i + \sum_{j=1}^{M} \left[\omega a_j^\dagger a_j + \sigma_x^j \left\{ \sum_{i=1}^{3} \mu_i (c_i^\dagger + c_i) + \lambda \right\} + \hat{n}_j \{ \nu (a_j^\dagger + a_j) + \varepsilon \} \right], \quad (1)$$

where a_j and c_i denote the annihilation operators of optical phonons in the j-th material system (system j) and the photons of the i-th mode, respectively. σ_ρ^j ($\rho = x, y, z$) describes the Pauli matrix that operates on the electronic states of the j-th material system, $|g\rangle_j$(ground state) and $|e\rangle_j$(excited state). $\hat{n}_j = (\sigma_z^j + 1)/2$ is the excited state electronic population of the j-th system, and μ_i denotes the dipole interaction between electrons and photons. We take into account the nonadiabatic interaction between the electronic states $\lambda \sigma_x^j$, which contributes to the entanglement generation as well as the other parameters.

We take $\hbar = 1$ and the values of the parameters are chosen in the following way.

First, we set $\omega = 1$ and $\varepsilon = 13.5$. When $\hbar\omega$ is $\sim$ 150meV, the light absorption of the system is observed in near infrared or visible region.

The number of irradiated photons is much smaller than that in an optical pulse due to the limitation of the numerical calculation. We choose $\mu_i = 0.5 (i = 1, 2, 3)$ in order to make the effect of light irradiation clear under such a weak excitation condition.

As for the electron-phonon coupling ν, it lies between 0 and 6 in various materials[7, 8]. In the present study it is set to 3.5, which is a strong coupling case. By setting $\lambda = 1.5$, we discuss the dynamics of structural change accompanied by nonadiabatic transition, e.g, photoisomerization, though we do not restrict ourselves to discuss molecular systems.

Since Raman scattering is one of the primary processes in the excitation of electron-phonon systems, three-photon modes corresponding to the pump mode ($\Omega_1 = 13.5$), Stokes mode ($\Omega_2 = 12.5$), and anti-Stokes mode ($\Omega_3 = 14.5$) are taken into account. In this case, the pump mode frequency is tuned to that for the electronic transition.

The time-dependent Schrödinger equation for Hamiltonian (1) was numerically solved to obtain the wavefunction $|\Phi(t)\rangle$, which is described by

$$|\Phi(t)\rangle = \sum_{\sigma_1\sigma_2 n_1 n_2 m_1 m_2 m_3} p_{\sigma_1\sigma_2 n_1 n_2}^{m_1 m_2 m_3}(t) |\sigma_1 \sigma_2 n_1 n_2 m_1 m_2 m_3\rangle, \quad (2)$$

where σ_j and n_j denote the indices for electronic states and the number of phonons in system j, respectively, while m_i is the number of photons in the i-th mode. Then, the coefficients $p_{\sigma_1\sigma_2 n_1 n_2}^{m_1 m_2 m_3}(t)$ obey a set of differential equations

$$\begin{aligned}
\frac{dp_{\sigma_1\sigma_2 n_1 n_2}^{m_1 m_2 m_3}}{dt} &= (\sum_i^3 m_i \Omega_i + \omega \sum_j^2 n_j) p_{\sigma_1\sigma_2 n_1 n_2}^{m_1 m_2 m_3} + \lambda (p_{-\sigma_1\sigma_2 n_1 n_2}^{m_1 m_2 m_3} + p_{\sigma_1 -\sigma_2 n_1 n_2}^{m_1 m_2 m_3}) \\
&+ \mu_1 \sqrt{m_1} (p_{-\sigma_1\sigma_2 n_1 n_2}^{m_1-1 m_2 m_3} + p_{\sigma_1 -\sigma_2 n_1 n_2}^{m_1-1 m_2 m_3}) + \mu_2 \sqrt{m_2} (p_{-\sigma_1\sigma_2 n_1 n_2}^{m_1 m_2-1 m_3} + p_{\sigma_1 -\sigma_2 n_1 n_2}^{m_1 m_2-1 m_3}) \\
&+ \mu_3 \sqrt{m_3} (p_{-\sigma_1\sigma_2 n_1 n_2}^{m_1 m_2 m_3-1} + p_{\sigma_1 -\sigma_2 n_1 n_2}^{m_1 m_2 m_3-1}) + \mu_1 \sqrt{m_1+1} (p_{-\sigma_1\sigma_2 n_1 n_2}^{m_1+1 m_2 m_3} + p_{\sigma_1 -\sigma_2 n_1 n_2}^{m_1+1 m_2 m_3}) \\
&+ \mu_2 \sqrt{m_2+1} (p_{-\sigma_1\sigma_2 n_1 n_2}^{m_1 m_2+1 m_3} + p_{\sigma_1 -\sigma_2 n_1 n_2}^{m_1 m_2+1 m_3}) + \mu_3 \sqrt{m_3+1} (p_{-\sigma_1\sigma_2 n_1 n_2}^{m_1 m_2 m_3+1} + p_{\sigma_1 -\sigma_2 n_1 n_2}^{m_1 m_2 m_3+1}) \\
&+ \delta_{\sigma_1 1} \{ \nu (\sqrt{n_1} p_{\sigma_1\sigma_2 n_1-1 n_2}^{m_1 m_2 m_3} + \sqrt{n_1+1} p_{\sigma_1\sigma_2 n_1+1 n_2}^{m_1 m_2 m_3}) + \varepsilon p_{\sigma_1\sigma_2 n_1 n_2}^{m_1 m_2 m_3} \} \\
&+ \delta_{\sigma_2 1} \{ \nu (\sqrt{n_2} p_{\sigma_1\sigma_2 n_1 n_2-1}^{m_1 m_2 m_3} + \sqrt{n_2+1} p_{\sigma_1\sigma_2 n_1 n_2+1}^{m_1 m_2 m_3}) + \varepsilon p_{\sigma_1\sigma_2 n_1 n_2}^{m_1 m_2 m_3} \}. \quad (3)
\end{aligned}$$

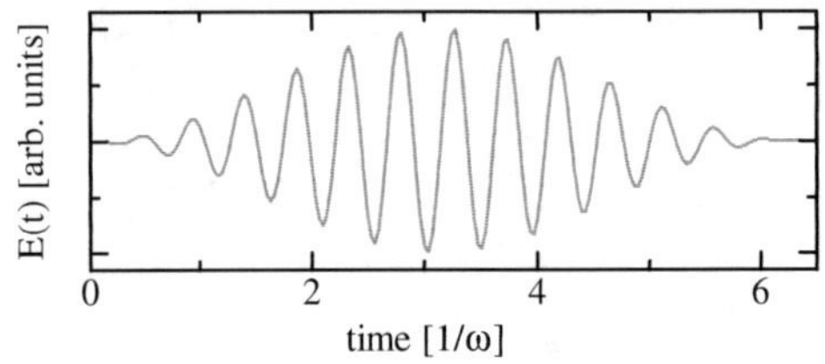

Figure 1. Electric field amplitude of the incident optical pulse.

Equation (3) is solved by the fourth-order Runge-Kutta method, where the time step τ is set to 5×10^{-5}. We tested and confirmed numerical convergence by using other values of τ ($5 \times 10^{-6} \leq \tau \leq 5 \times 10^{-4}$). The dimension of the photon and phonon Hilbert spaces are truncated to 64 and 32 per mode, respectively, and thus the dimension of the entire Hilbert space is 2^{30}. In this case hybrid MPI-OpenMP programming is effective for parallel computing and 64 MPI processes with 9 threads are used in all the calculations.

The initial condition is given by $|\Phi(0)\rangle = |\alpha_1\rangle \otimes |\alpha_2\rangle \otimes |\alpha_3\rangle \otimes |G\rangle_1 \otimes |G\rangle_2$, where $|\alpha_i\rangle$ denotes a coherent state parameterized by α_i for the i-th photon mode and $|G\rangle_j$ is the ground state of the j-th electron-phonon system. In the present study, we set $\alpha_1 = 5$ and $\alpha_2 = \alpha_3 = -2.5$ to describe a train of optical pulses. The expectation value of the electric field of the incident optical pulse is described by $E(t) \propto \sum_{n=1}^{3} i\langle \alpha_n e^{-i\omega_n t}|(c_n - c_n^\dagger)|\alpha_n e^{-i\omega_n t}\rangle \propto \sum_{n=1}^{3} \alpha_n \sin \omega_n t$, and shown in Fig. 1.

The entanglement properties are discussed by the entanglement entropy and the quantum mutual information, since the entanglement between phonons should be referred to as multipartite or mixed-state entanglement[9] in the present case. When the system is divided into two subsystems A and B, its wavefunction is described by

$$|\phi\rangle = \sum_{mn} q_{mn}|m\rangle|n\rangle, \tag{4}$$

where $|m\rangle$ and $|n\rangle$ denote the basis of A and B, respectively. The reduced density matrix is defined by

$$\rho_r = \mathrm{Tr}_A|\phi\rangle\langle\phi| = \sum_n q'_n|n\rangle\langle n|, \tag{5}$$

where $q'_n = \sum_m q^*_{mn}q_{mn}$. The entanglement entropy is calculated as the von Neumann entropy of ρ_r defined by

$$S = -\mathrm{Tr}\rho_r \log \rho_r = -\sum_k r_k \log r_k, \tag{6}$$

where r_k is the eigenvalue of ρ_r.

When the entire system is divided into three subsystems A, B, and C, it is necessary to consider multipartite entanglement to study the entanglement between two of them. The quantum mutual information between A and B is an appropriate measure for it, and defined by

$$I_M = S_A + S_B - S_{A\otimes B}, \tag{7}$$

where S_α is the von Neumann entropy of the reduced density matrix obtained by tracing out the degrees of freedom other than α. It is also interpreted as the relative entropy of the reduced density matrix in the phonon subspace to its closest separable state[4, 5]. In the present case, A, B, and C correspond to the phonon and the electrons in systems 1 and 2, and the irradiated photons, respectively.

Journal of Physics: Conference Series **2207** (2022) 012041 doi:10.1088/1742-6596/2207/1/012041

3. Calculated results

I_M as a function of t shows the entanglement generation dynamics as the light irradiation proceeds[3]. In order to study the mechanism of the entanglement generation, we performed the Schmidt decomposition to $|\Psi(t)\rangle$ and discuss the dynamical behavior of the phonon states. Precisely, the entire system is divided into the phonon states and the electron-photon states, and $|\Psi(t)\rangle$ is rewritten by

$$|\Psi(t)\rangle = \sum_n \lambda_n(t)|\eta_n(t)\rangle|\theta_n(t)\rangle, \tag{8}$$

where $|\eta_n(t)\rangle$ belongs to the phonon space and $|\theta_n(t)\rangle$ is in the Hilbert space of the electrons and the photons, respectively. $\lambda_n(t)$ is sorted in descending order, and the variable t will be omitted for brevity in the rest of the paper.

We note that the entanglement entropy in Eq. (7) is described by the singular values of the Schmidt decomposition, e.g., $S_{A\otimes B}$ is described by λ_n in Eq. (8) as

$$S_{A\otimes B} = -\sum_n \lambda_n^2 \log \lambda_n^2, \tag{9}$$

which shows that the entanglement entropy and/or the quantum mutual information is determined by the distribution of the singular values.

Tracing out the photon and electron degrees freedom in the density matrix $\rho = |\Psi\rangle\langle\Psi|$, we obtain a reduced density matrix Ξ as

$$\Xi = \sum_n \lambda_n^2 |\eta_n\rangle\langle\eta_n|. \tag{10}$$

Equation (10) shows that $|\eta_n\rangle$ corresponds to experimentally observed phonon states, if experiments are designed to detect the phonons in both systems simultaneously.

Since $\mathcal{M}_n \equiv |\eta_n\rangle\langle\eta_n|$ satisfies the closure relation $\sum_n \mathcal{M}_n = 1$, they construct a set of PVM. When we consider a measurement of the phonon states corresponding to $\mathcal{M}_n$, we expect to obtain a single $|\eta_n\rangle$ with probability λ_n^2. Thu, though the accumulated data in pump-probe experiments, for example, bring the averaged properties of the phonon states, it is worth while discussing the entanglement properties of individual singular vectors $|\eta_n\rangle$ to reveal the capacity of the phonon states as an entanglement source in the present configuration. In other words, the averaged value regarding the phonon states contains the classical correlation that originates from the multiple terms in Eq. (10), and the quantum correlation will be distilled by the measurement process represented by $\mathcal{M}_n$. Thus, we analyze the entanglement properties of the phonon states. For this purpose we calculated the entanglement entropy of $|\eta_n\rangle$ by the reduced density matrix ρ_{r2}^n for $|\eta_n\rangle$

$$\rho_{r2}^n = \sum_{ij}\sum_k u_{ik}^n \bar{u}_{kj}^n |i\rangle\langle j| \equiv \sum_{ij} v_{ij}^n |i\rangle\langle j|, \tag{11}$$

where $|ij\rangle$ denotes the phonon Fock states. $|\eta_n\rangle$ is expanded as $|\eta_n\rangle = \sum_{ij} u_{ij}|ij\rangle$ and $v_{ij}^n = \sum_k u_{ik}^n \bar{u}_{kj}$. The entanglement entropy between phonons in systems 1 and 2 is calculated as $S_n = -\mathrm{Tr}\rho_{r2}^n \log \rho_{r2}^n$.

Figures 2 and 3 show the time-evolution of the singular values λ_n^2 and the entanglement entropy S_n for $n = 0, 1$ and 2, respectively. Initially, only S_0 appears in the Schmidt decomposition of $|\Phi\rangle$, and thus the phonons are decomposable. When the excitation of phonon starts, S_0 gradually grows and becomes a weakly entangled state as we show in Fig. 3. As the photoexcitation process proceeds, λ_1 and λ_2 become to take finite values, which is synchronized to the generation of phonons[3]. We calculated the corresponding entanglement entropy S_1 and S_2 when λ_1, $\lambda_2 > 0.01$. As shown in Fig. 3, they are larger than S_0, which shows that these states primarily contribute to the phonon entanglement. Hence, if the projection onto $|\eta_1\rangle$ or

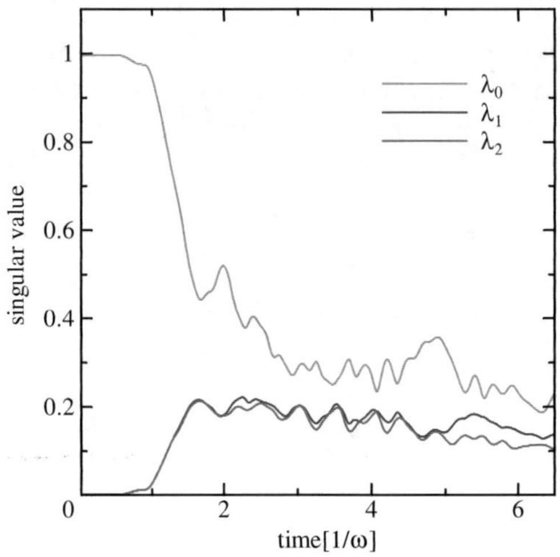

Figure 2. λ_0, λ_1, and λ_2 as functions of t.

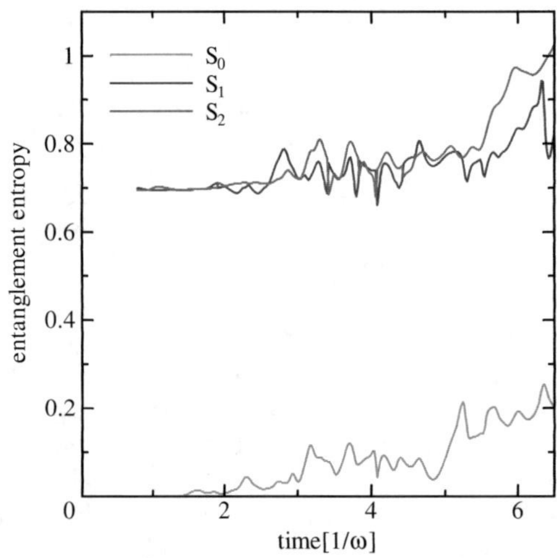

Figure 3. Entanglement entropy between phonons in systems 1 and 2, S_0, S_1, and S_2 as functions of t. S_1 and S_2 are not shown when corresponding singular values $\lambda_1,\ \lambda_2 < 0.01$.

$|\eta_2\rangle$ is realized by some measurement procedure, more strongly entangled phonon states are obtained, which is advantageous to apply to quantum memories and/or entanglement sources. Since Fig. 2 shows that λ_1, and λ_2 are as large as λ_0 after the photoirradiation ($t = 6.5/\omega$), $|\eta_1\rangle$ and $|\eta_2\rangle$ can be obtained by the measurement of phonon states with high probability comparable to that for $|\eta_0\rangle$.

4. Summary

We calculated the phonon entanglement generation dynamics between remote electron-phonon systems induced by irradiated photons. We focused on the phonon properties by dividing the entire electron-phonon-photon state $|\Phi\rangle$ into phonon states $|\eta_n\rangle$ and electron-photon states $|\theta_n\rangle$

by the Schmidt decomposition. Numerical calculations on the time-evolution of the singular values λ_n and $|\eta_n\rangle$ show that light irradiation induces weak entanglement between different electron-phonon systems at first, and that the creation of entangled phonons follows to construct strongly entangled phonon states. By realizing a measurement procedure of the phonon states corresponding to $\mathcal{M}_n = |\eta_n\rangle\langle\eta_n|$, we will obtain an efficient entanglement source.

K.I. is grateful to K. G. Nakamura for fruitful discussion. This work was partly supported by JSPS KAKENHI Grant Numbers JP21H01752 and the Collaborative Research Project of Laboratory for Materials and Structures, Institute of Innovative Research, Tokyo Institute of Technology, Japan. Numerical calculations were performed on the facilities of the Supercomputer Center, Institute for Solid State Physics, the University of Tokyo, Japan, and the computer resources offered under the category of General Projects by Research Institute for Information Technology, Kyushu University, Japan.

References

[1] K. C. Lee, M. R. Sprague, B. J. Sussman, J. Nunn, N. K. Langford, X.-M. Jin, T. Champion, P. Michelberger, K. F. Reim, D. England, D. Jaksch, and I. A. Walmsley, Science **334**, 1253 (2011).
[2] K. Ishida, Eur. Phys. J. **D73**, 117 (2019).
[3] K. Ishida and H. Matsueda, J. Phys. Soc. Jpn. **90**, 104714 (2021).
[4] B. Groisman, S. Popescu, and A. Winter, Phys. Rev. **A72**, 032317 (2005).
[5] A. Kumar, Phys. Rev. **A96**, 012332 (2017).
[6] M. A. Nielsen and I. L. Chuang, *"Quantum Computation and Quantum Information"* (Cambridge University Press, Cambridge, 2010).
[7] A. Hoffmann, B. K. Meyer, and E. Malguth, in *"Zinc Oxide: From Fundamental Properties Towards Novel Applications"* , volume 120, Springer Series in Materials Science, 233 (Springer, Berlin, Heidelberg, 2010).
[8] R. Valiev, A. Freidzon, and A. Berezhnoy, RSC Adv. **4**, 79 (2014).
[9] C. H. Bennett, D. P. DiVincenzo, J. A. Smolin, and W. K. Wootters, Phys. Rev. **A54**, 3824 (1996).

The Phonon Quasiparticle Approach for Anharmonic Properties of Solids

Zhen Zhang[1], Dong-Bo Zhang[2,3], Tao Sun[4], and Renata M. Wentzcovitch[1,5,6,*]

[1]Department of Applied Physics and Applied Mathematics, Columbia University, New York, NY 10027, USA
[2]College of Nuclear Science and Technology, Beijing Normal University, Beijing 100875, People's Republic of China
[3]Beijing Computational Science Research Center, Beijing 100193, People's Republic of China
[4]Key Laboratory of Computational Geodynamics, Chinese Academy of Sciences, Beijing 100049, People's Republic of China
[5]Department of Earth and Environmental Sciences, Columbia University, New York, NY 10027, USA
[6]Lamont–Doherty Earth Observatory, Columbia University, Palisades, NY 10964, USA

[*]rmw2150@columbia.edu

Abstract. Knowledge of lattice anharmonicity is essential to elucidate distinctive thermal properties in crystalline solids. Yet, accurate *ab initio* investigations of lattice anharmonicity encounter difficulties owing to the cumbersome computations. Here we introduce the phonon quasiparticle approach and review its application to various materials. This method efficiently and reliably addresses lattice anharmonicity by combining *ab initio* molecular dynamics and lattice dynamics calculations. Thus, in principle, it accounts for full anharmonic effects and overcomes finite-size effects typical of *ab initio* molecular dynamics. The validity and effectiveness of the current approach are demonstrated in the computation of thermodynamic and heat transport properties of weakly and strongly anharmonic systems.

1. Introduction
In crystalline materials, anharmonicity describes the deviation of vibrations from harmonic oscillations and is caused by phonon-phonon interactions. Lattice anharmonicity is essential in understanding phonon-driven phenomena, e.g., structural phase transitions [1-5], thermal expansion [6-8], and lattice thermal conductivity [9]. Lattice anharmonicity can also be characterized by the intrinsic temperature dependence of phonon frequencies. In terms of *ab initio* calculations, there are two prevailing ways to deal with lattice anharmonicity. The first is the interatomic force-constant-based methods, which can evaluate anharmonic phonon properties and material's thermal properties with dense **q**-meshes including essential long-wavelength phonons. However, the higher-order (3rd, 4th, etc.) interatomic force constants are complicated and computationally heavy to obtain. Inevitably, it suffers from computational errors or truncation issues. The second is the molecular-dynamics-based methods. Free energies and thermodynamic properties can be obtained by thermodynamic integration (TI). However, to approach

the thermodynamic limit ($N \to \infty$), conducting TI using *ab initio* molecular dynamics (AIMD) simulations with a sufficiently large supercell is beyond the current computational capability.

The present phonon quasiparticle approach deals with lattice anharmonicity by extracting phonon quasiparticle properties, i.e., renormalized frequencies and phonon lifetimes, from AIMD simulations without explicitly expressing the higher-order interatomic force constants. This is done by computing the normal-mode-projected velocity autocorrelation function (VAF) obtained by AIMD. Renormalized frequencies are next Fourier interpolated over the whole Brillouin zone (BZ) to overcome finite-size effects. Then thermodynamic properties in the thermodynamic limit ($N \to \infty$) can be obtained within the framework of the phonon gas model (PGM). Since the phonon quasiparticle approach is based on AIMD, in principle, anharmonicity is accounted for to all orders in perturbation theory. This approach can be used to study various thermal properties, e.g., structural stability, thermodynamic properties, phase boundaries, and lattice thermal conductivities.

2. Method

We define a phonon quasiparticle numerically by the normal-mode-projected velocity autocorrelation function (VAF) [4, 10],

$$\langle V_{\mathbf{q}s}(0) \cdot V_{\mathbf{q}s}(t) \rangle = \lim_{\tau \to \infty} \frac{1}{\tau} \int_0^\tau V_{\mathbf{q}s}^*(t') V_{\mathbf{q}s}(t'+t) dt', \quad (1)$$

where $V_{\mathbf{q}s}(t) = \sum_{i=1}^{N} \sqrt{M_i} \mathbf{v}_i(t) e^{i\mathbf{q}\cdot\mathbf{R}_i} \cdot \hat{\mathbf{e}}_{\mathbf{q}s}$ is the mass-weighted mode-projected velocity for normal mode ($\mathbf{q}$, s). $\mathbf{q}$ is the phonon wave vector, and s indexes the $3n$ phonon branches of an n-atom primitive cell. M_i, $\mathbf{R}_i$, and $\mathbf{v}_i$ ($i = 1, \ldots, N$) are the atomic mass, the atomic equilibrium coordinate, and the atomic velocity computed by AIMD simulations of an N-atom supercell, respectively. $\hat{\mathbf{e}}_{\mathbf{q}s}$ is the harmonic phonon polarization vector determined by the harmonic phonon calculations. $\mathbf{q}$ is commensurate with the supercell size. For a well-defined phonon quasiparticle, the VAF can be phenomenologically described as an exponentially decaying cosine function,

$$\langle V_{\mathbf{q}s}(0) \cdot V_{\mathbf{q}s}(t) \rangle = A_{\mathbf{q}s} \cos(\tilde{\omega}_{\mathbf{q}s} t) e^{-\Gamma_{\mathbf{q}s} t}, \quad (2)$$

where $A_{\mathbf{q}s}$ is the oscillation amplitude, $\tilde{\omega}_{\mathbf{q}s}$ is the renormalized phonon frequency, and $\Gamma_{\mathbf{q}s}$ is the phonon linewidth inversely proportional to the lifetime, $\tau_{\mathbf{q}s} = 1/(2\Gamma_{\mathbf{q}s})$ [10, 11]. The corresponding power spectrum,

$$G_{\mathbf{q}s}(\omega) = \left| \int_0^\infty \langle V_{\mathbf{q}s}(0) \cdot V_{\mathbf{q}s}(t) \rangle e^{i\omega t} dt \right|^2 \quad (3)$$

should have a Lorentzian line shape with a single peak at $\tilde{\omega}_{\mathbf{q}s}$ and a linewidth of $\Gamma_{\mathbf{q}s}$ [10, 11]. $\tilde{\omega}_{\mathbf{q}s}$ can be used to calculate the anharmonic thermodynamic properties. To overcome the finite-size effects on the vibrational contribution to the free energy and to obtain the free energy in the thermodynamic limit, the anharmonic phonon spectrum is further obtained via Fourier interpolation. At each $\mathbf{q}$, we construct the effective harmonic dynamical matrix [4, 10],

$$\tilde{D}(\mathbf{q}) = [\hat{\mathbf{e}}_{\mathbf{q}}] \Omega_{\mathbf{q}} [\hat{\mathbf{e}}_{\mathbf{q}}]^\dagger, \quad (4)$$

where the diagonal matrix $\Omega_{\mathbf{q}} = \mathrm{diag}[\tilde{\omega}_{\mathbf{q}1}^2, \tilde{\omega}_{\mathbf{q}2}^2, \ldots, \tilde{\omega}_{\mathbf{q}3n}^2]$ contains $\tilde{\omega}_{\mathbf{q}s}^2$ in the diagonal, and the matrix $[\hat{\mathbf{e}}_{\mathbf{q}}] = [\hat{\mathbf{e}}_{\mathbf{q}1}, \hat{\mathbf{e}}_{\mathbf{q}2}, \ldots, \hat{\mathbf{e}}_{\mathbf{q}3n}]$ contains harmonic polarization vectors in columns. Next, the effective harmonic force constant matrix is obtained by Fourier transform,

$$\tilde{\Phi}(\mathbf{r}) = \sum_{\mathbf{q}} \tilde{D}(\mathbf{q}) \cdot e^{i\mathbf{q}\cdot\mathbf{r}}, \quad (5)$$

where anharmonic interactions are effectively captured. Then, $\tilde{\omega}_{\mathbf{q}'s}$ at arbitrary wave vector $\mathbf{q}'$ in the Brillouin zone (BZ) can be obtained by diagonalizing

$$\tilde{D}(\mathbf{q}') = \sum_{\mathbf{r}} \tilde{\Phi}(\mathbf{r}) \cdot e^{-i\mathbf{q}'\cdot\mathbf{r}}, \quad (6)$$

from which the T-dependent phonon quasiparticle spectrum can be computed. Within the PGM, the vibrational entropy formula with T-dependent phonon quasiparticle spectra is still applicable [4, 10, 11],

$$S_{\mathrm{vib}}(T) = k_{\mathrm{B}} \sum_{\mathbf{q}s} [(n_{\mathbf{q}s} + 1) \ln(n_{\mathbf{q}s} + 1) - n_{\mathbf{q}s} \ln n_{\mathbf{q}s}], \quad (7)$$

where $n_{\mathbf{q}s} = [\exp(\hbar\tilde{\omega}_{\mathbf{q}s}(T)/k_{\mathrm{B}}T) - 1]^{-1}$. Then the Helmholtz free energy $F(T)$ at any T can be obtained by definition [12, 13],

$$F(T) = U(T) - TS_{\mathrm{el}}(T) - TS_{\mathrm{vib}}(T), \quad (8)$$

where $U(T)$ is the ensemble average of internal energy, i.e., potential (total) energy plus kinetic energy, of the system at finite T obtained from AIMD. $S_{\rm el}$ is the electronic entropy for a metallic system obtained using the Mermin functional [14]. Alternatively, with $F(T_0)$ at a reference temperature T_0 is known, F at any other T can also be obtained by integrating the entropy [5, 15],

$$F(T) = F(T_0) - \int_{T_0}^{T} S_{\rm el}(T')dT' - \int_{T_0}^{T} S_{\rm vib}(T')dT'. \quad (9)$$

The reference $F(T_0)$ can be obtained by equation (8) at T_0. Both approaches (equations (8) and (9)) have been tested to be effective [5, 12, 13, 15]. In contrast to $S_{\rm vib}$, $S_{\rm el}$ and U depend only weakly on simulation cell size [13]. Therefore, $S_{\rm el}$ and U can be readily obtained from the direct output of AIMD calculations in finite-size cells. Apart from $\tilde{\omega}_{\mathbf{q}s}$, $\tau_{\mathbf{q}s}$ can be used to calculate the lattice thermal conductivity within the relaxation-time approximation (RTA) of the linearized Boltzmann transport equation (LBTE) [16-18],

$$\kappa_{\rm lat} = \frac{1}{3}\sum_{\mathbf{q}s} c_{\mathbf{q}s} v_{\mathbf{q}s}^2 \tau_{\mathbf{q}s}, \quad (10)$$

where $c_{\mathbf{q}s}$ is the mode heat capacity, and $v_{\mathbf{q}s}$ is the mode group velocity. The latter is obtained by computing **q**-gradients of T-dependent phonon quasiparticle dispersions. The details of the phonon quasiparticle approach and code to perform phonon quasiparticle calculations can be found in [19].

3. Applications

3.1. Thermodynamics

The method has been successfully applied to weakly [10] and strongly [4, 5, 13, 15] anharmonic systems, nonmetallic [4, 10, 15, 19] and metallic [5, 12, 13] anharmonic systems, and used to study structural stability [4, 5, 20-22], anharmonic phonon dispersions [4, 5, 10, 12, 13, 19], anharmonic free energies [4, 10], phase transitions [5, 12, 13], and anharmonic thermodynamic quantities [15].

3.1.1. Weakly anharmonic insulating $MgSiO_3$ perovskite. *Pbnm* $MgSiO_3$ perovskite (MgPv) is the most abundant phase of the Earth's mantle. Despite the successful application of quasiharmonic approximation (QHA) to such weakly anharmonic systems, intrinsic anharmonic effects have been identified in the Raman spectrum of MgPv at ambient pressure (P) [23-25]. By conducting phonon quasiparticle calculations, we obtained frequency shifts $\Delta\omega_{\mathbf{q}s} = \tilde{\omega}_{\mathbf{q}s} - \omega_{\mathbf{q}s}$ at constant volume (V) for MgPv, where $\omega_{\mathbf{q}s}$ is the harmonic phonon frequency, as shown in figures 1(a)–(c). Then $\Delta\omega_{\mathbf{q}s}$ at constant volume were converted into $\Delta\omega_{\mathbf{q}s}$ at constant P by using the quasiharmonic thermal expansivity and mode Grüneisen parameter, which are shown in figures 1(d)–(f). We successfully reproduced the irregular temperature induced frequency shifts [23-25] observed in the Raman spectrum of this complex system [10].

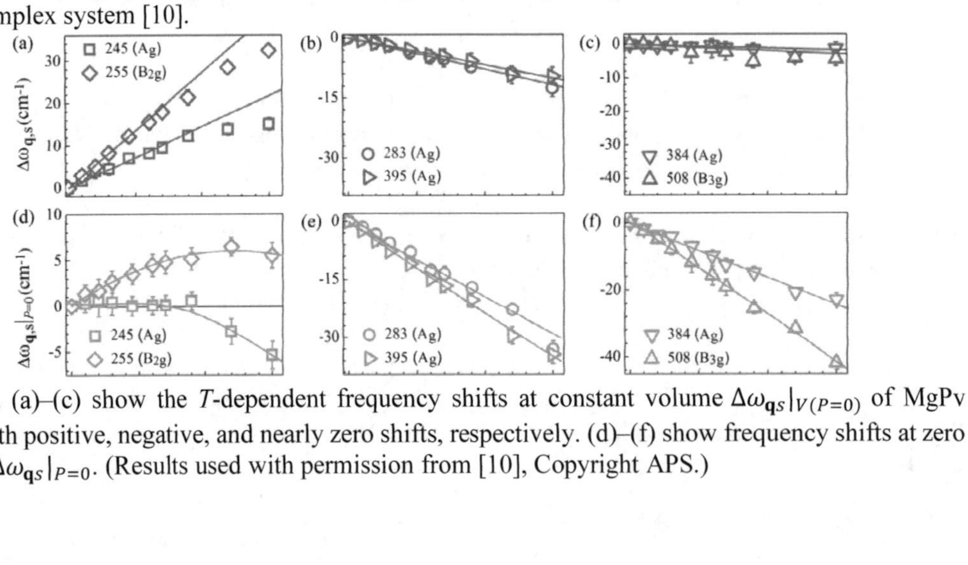

Figure 1. (a)–(c) show the T-dependent frequency shifts at constant volume $\Delta\omega_{\mathbf{q}s}|_{V(P=0)}$ of MgPv modes with positive, negative, and nearly zero shifts, respectively. (d)–(f) show frequency shifts at zero pressure $\Delta\omega_{\mathbf{q}s}|_{P=0}$. (Results used with permission from [10], Copyright APS.)

As long as phonon quasiparticles exist, i.e., as long as anharmonicity does not mix harmonic modes, and the power spectra of all modes have well-defined single peaks, one can also obtain anharmonic dispersions over the entire BZ, as shown in figure 2(a). The anharmonic phonon dispersion at 300 K was also compared with the harmonic one obtained by density-functional perturbation theory (DFPT) [26]. The frequency shifts are not significant, in line with the fact that MgPv is weakly anharmonic. Their corresponding vibrational densities of states (VDoS) are shown in figure 2(b). Combining these thoroughly sampled quasiparticle dispersions with the phonon gas model can offer first-principles free energy and thermodynamics properties in the thermodynamic limit at temperatures beyond the limit of validity of the QHA [10]. Figure 2(c) shows that free energy obtained from the phonon quasiparticle approach plus PGM agrees well with that obtained from thermodynamic integration (TI) with the same coarse **q**-mesh, which validates the present methodology. Fourier interpolation also allowed free energy computation with a sufficiently dense **q**-mesh, which is discernible from the one obtained with the coarse **q**-mesh. Therefore, the present method overcomes finite-size effects on computations of thermodynamic properties using AIMD simulations. This is especially important for calculations of thermal expansivity and phase boundaries, which are fundamental properties in geodynamics simulations [10].

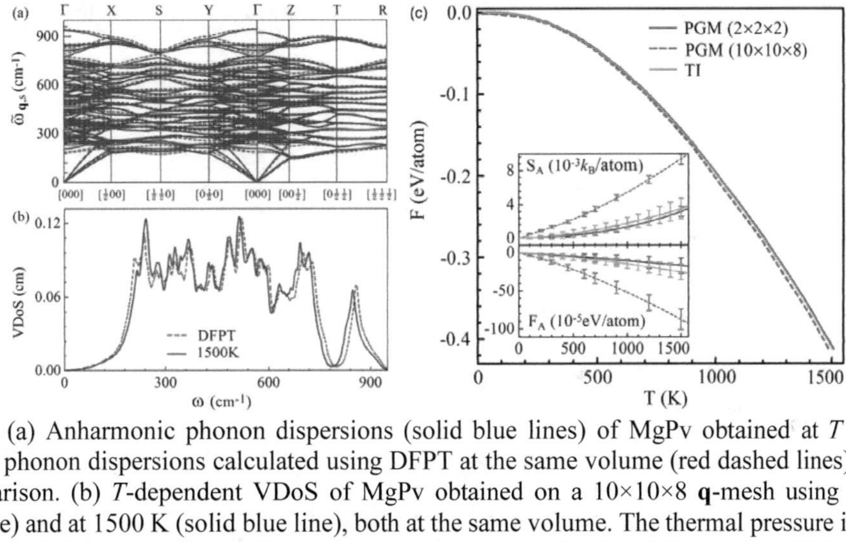

Figure 2. (a) Anharmonic phonon dispersions (solid blue lines) of MgPv obtained at T = 1500 K. Harmonic phonon dispersions calculated using DFPT at the same volume (red dashed lines) are shown for comparison. (b) T-dependent VDoS of MgPv obtained on a 10×10×8 **q**-mesh using DFPT (red dashed line) and at 1500 K (solid blue line), both at the same volume. The thermal pressure is ~10 GPa. (c) Vibrational free energy F and (insert) excess anharmonic vibrational free energy (F_A) and entropy (S_A) obtained with the PGM and the TI approach using an MD run sampling a coarse **q**-mesh (2×2×2). Results with a finely interpolated **q**-mesh (10×10×8) use the PGM. Error bars are evaluated from 5 parallel replicas (60 ps each) of Born-Oppenheimer molecular dynamics (BOMD) [27]. (Results used with permission from [10], Copyright APS.)

3.1.2. Strongly anharmonic insulating $CaSiO_3$ perovskite. $CaSiO_3$ perovskite (CaPv) is the third most abundant mineral of the Earth's lower mantle (LM) (~7 vol %). It is also the last major lower mantle phase whose structural and elastic properties have not been well characterized. It is widely believed that under the LM conditions, $23 < P < 135$ GPa and $2000 < T < 4000$ K, the cubic phase with $Pm\bar{3}m$ space group is adopted by CaPv. However, static calculations for cubic CaPv yield double-well potential and harmonic phonon calculations yield imaginary phonon frequencies around **R** (1/2, 1/2, 1/2) and **M** (0, 1/2, 1/2). At low temperatures, e.g., $T < 500$ K, CaPv adopts a variety of tetragonal or orthorhombic phases. We used the phonon quasiparticle approach to study cubic CaPv's structural stability and the corresponding phonon spectra. AIMD simulations were carried out for over 50 ps, and the local-density approximation (LDA) was chosen for the exchange-correlation functional. Figure 3(a) shows oxygen atom displacements distributions from AIMD at various T. Figure 3(b) shows the phonon spectra for the normal mode with imaginary harmonic frequency at **R** at the same T as in figure 3(a). At $T < 600$ K, the

O exhibits double-peak distributions with respect to the cubic equilibrium position, meaning the cubic phase is distorted. The corresponding power spectra display double peaks, indicating the breakdown of the phonon quasiparticle. At T = 600 K and above, the O shows centered distribution at the cubic equilibrium position, meaning the cubic phase is dynamically stabilized by anharmonic interactions. The corresponding phonon spectrum is well-defined with a single peak [4].

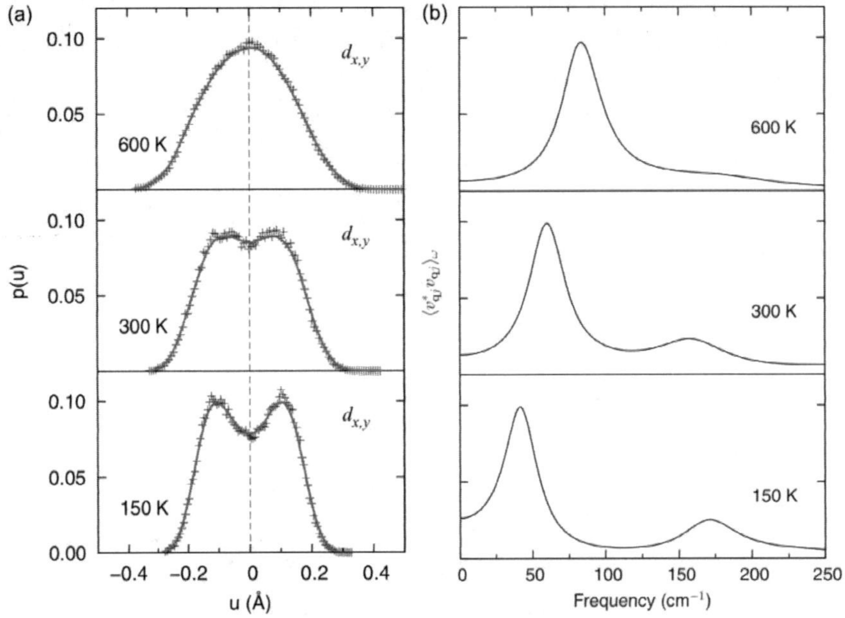

Figure 3. (a) Distributions of the O displacements in the x or y direction (octahedral tilting) with respect to the ideal cubic structure ($a^0a^0a^0$) at various T. (b) Power spectra of the VAF of R_{25} with **q** vector **R** (1/2, 1/2, 1/2) at low T. (Results used with permission from [4], Copyright APS.)

Using a 2×2×2 supercell, the anharmonic entropy calculated using the present phonon quasiparticle spectrum plus the PGM approach and the one by TI are in good agreement [4]. By Fourier interpolation, the high-T anharmonic phonon dispersion for cubic CaPv is free of imaginary frequencies, as shown in figure 4(a), which enables us to compute the vibrational entropy and free energy in the thermodynamic limit. The anharmonic phonon dispersion obtained by the present method also compares well with that obtained by the effective harmonic potential method [3], which further validates the present approach. The advantage of the present method is that complete information on the dynamics of the phonon quasiparticle is given [4]. At high T, where the cubic phase is adopted, we carried out phonon quasiparticle calculations for a series of V and T. Vibrational entropy and Helmholtz free energy were obtained via equations (7) and (9) and are shown in figures 4(b) and 4(c), respectively [15].

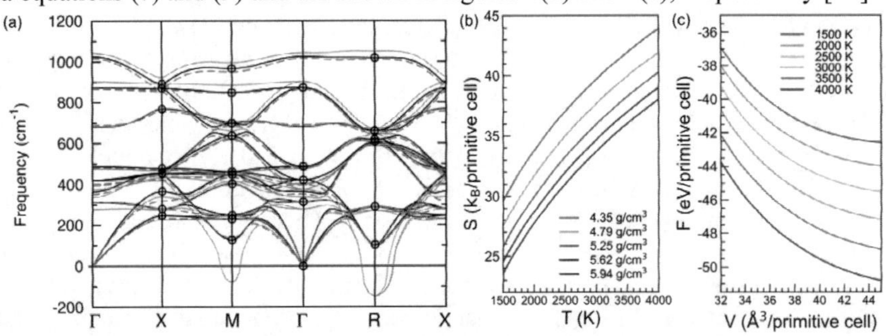

Figure 4. (a) Phonon dispersions at 1000K. The circles represent the frequencies extracted from the normal-mode-projected VAF. Fourier interpolating these frequencies yields the renormalized phonon

dispersions, shown in solid black curves. Dashed curves are the phonon dispersions determined by fitting an effective harmonic force constant matrix [3]. Harmonic phonon dispersions calculated with DFPT are shown in orange (light) curves. (Results used with permission from [4], Copyright APS.) (b) Vibrational entropy and (c) Helmholtz free energy of cubic CaPv. (Results used with permission from [15], Copyright APS.)

With Helmholtz free energy obtained, various thermodynamic properties such as thermal expansivity, thermodynamic Grüneisen parameter, isobaric heat capacity, and adiabatic bulk modulus were computed and are shown in figure 5. Results were also compared with a previous work using the Mie-Grüneisen-Debye (MGD) formulation [28], which may not account for full anharmonicity. The differences between the two methods are discernible for all thermodynamic quantities, suggesting the importance of full anharmonic effects. The difference is significant for the Grüneisen parameter, which can be used as a straightforward indicator of anharmonic effects [15]. Note the MGD formulation is based on the fitting for the Mie-Grüneisen equation of state. Its obtained Grüneisen parameter (γ_{mg}) and the thermodynamic Grüneisen parameter (γ) of this study are not strictly equal. The two quantities coincide with each other [29, 30] when satisfying three criteria [15]. First, the system is within the framework of QHA. Second, all mode Grüneisen parameters (γ_i) are equal. Third, at sufficiently high temperature, all mode isochoric heat capacities (C_{Vi}) are equal, and all mode thermal energies ($E_{\mathrm{th}i}$) are equal. Here for strongly anharmonic cubic CaPv at finite temperatures, none of the criteria is satisfied, resulting in a difference between γ and γ_{mg}. A detailed discussion on the difference and the relation among the three criteria can be found in the last paragraph of section III of [15].

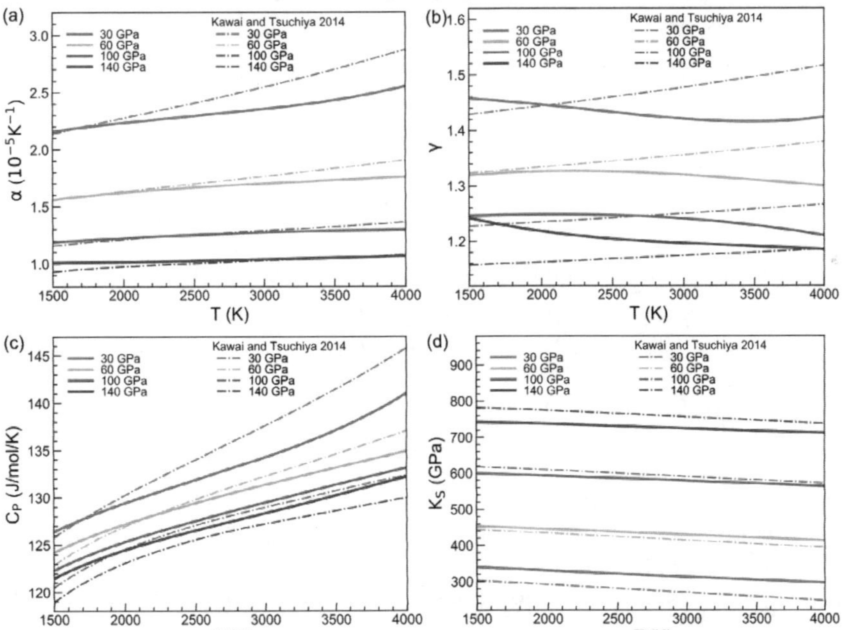

Figure 5. (a) Thermal expansivity (α), (b) thermodynamic Grüneisen parameter (γ), (c) isobaric heat capacity (C_P), and (d) adiabatic bulk modulus (K_S) versus T at a series of P of cubic CaPv, compared with results obtained by the MGD formulation [28]. (Results used with permission from [15], Copyright APS.)

3.1.3. Metallic Fe. The method has also been applied to determine the melting properties of hexagonal close-packed (*hcp*) Fe at conditions of the Earth's inner core boundary (330 GPa, 5500 to 6500 K). These properties have great significance in Earth sciences. Yet, precise determination of melting properties is challenging as it requires highly accurate free energies: an error of 10 meV/atom shifts the predicted melting temperature by ~100 K. Accordingly, melting properties are often evaluated via the formally exact TI approach. However, performing TI involves a series of extra AIMD simulations along the integration path, making it computationally very demanding. By contrast, the phonon quasiparticle method requires only one post-processing of the AIMD trajectory and is much more efficient. Figure 6 summarizes the results of the phonon quasiparticle calculations. The mode-projected VDoS exhibit nice Lorentzian line shapes (figure 6(a)), indicating well-defined phonon quasiparticles. These phonon quasiparticles can exhibit either red or blue frequency shifts with respect to their harmonic counterparts. However, the red-shifts dominate (figures 6(b) and 6(c)), leading to an increase of the vibrational entropy, from the harmonic value of 10.15 k_B/atom to 10.48 k_B/atom. We also find that Fourier interpolation is critical to get the vibrational entropy in the thermodynamic limit: the ionic entropy associated with the 64-atom supercell (without interpolation) is 10.25 k_B/atom, 0.23 k_B/atom lower than the converged value. As the entropy change at melting is ~1 k_B/atom, such differences would cause profound changes (~20%, 1000 K) in the predicted melting temperature. Remarkably, while the computational procedures of the phonon quasiparticle method and TI are quite different, the Gibbs free energies predicted from the two approaches are nearly identical (figure 6(d), within ±10 meV/atom) [12].

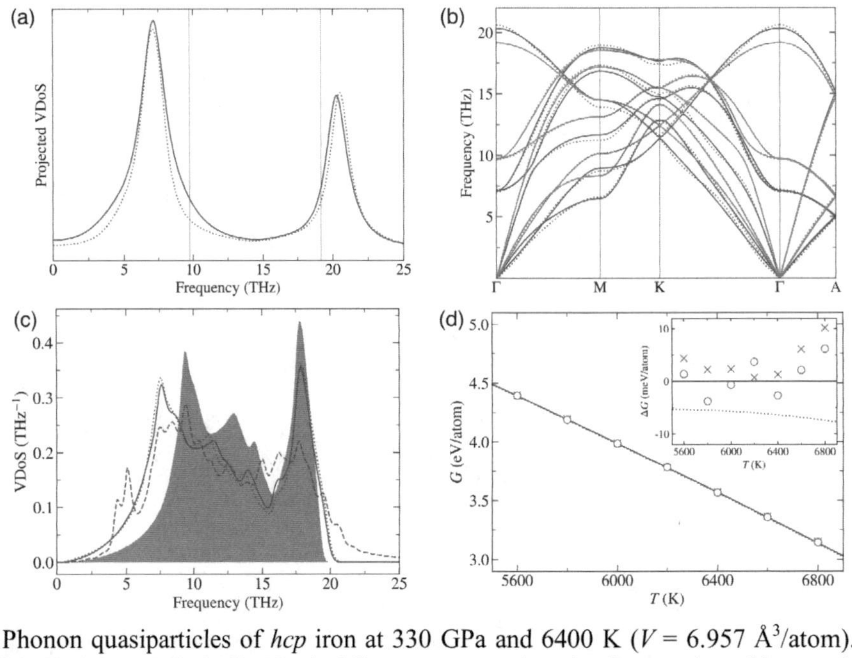

Figure 6. Phonon quasiparticles of *hcp* iron at 330 GPa and 6400 K (V = 6.957 Å^3/atom). (a) Mode projected VDoS at $\mathbf{q} = \mathbf{\Gamma}$ (0, 0, 0). Vertical lines correspond to the frequencies of harmonic TO (left) and LO (right) phonons. (b) Dispersions of phonon quasiparticles (red) and harmonic phonons (grey). (c) Complete VDoS of phonon quasiparticles, where the effective harmonic force constant matrices are from MD involving 64 atoms (4×2×2 supercell) (solid red curve) and 180 atoms (5×3×3 supercell) (red dotted curve). The grey area denotes the VDoS of harmonic phonons, where the harmonic force constant matrix is from the 64-atom supercell. Fourier interpolations are performed on a dense **q**-mesh equivalent to a supercell containing 64000 atoms for both phonon quasiparticles and harmonic phonons. The dashed blue curve denotes the VDoS of MD involving 64 atoms without interpolation. (d) The Gibbs free energy per atom (G) in the thermodynamic limit as a function of T. Red circles (crosses) denote results from the phonon quasiparticle approach using simulation cells with 64 (180) atoms. Blue solid (dotted) lines denote results from TI using 64 (180) atoms. (Results used with permission from [12], Copyright APS.)

3.1.4. Strongly anharmonic metallic Be. Beryllium is an important material with wide applications ranging from aerospace components to x-ray equipment. Yet, a precise understanding of its phase diagram remains elusive. Be assumes an *hcp* structure at relatively low *T*. However, unlike other metals, the stability of the body-centered cubic (*bcc*) phase at high *T* and the associated *hcp/bcc* phase transition are not well understood yet. We have investigated the phase stability of Be using the phonon quasiparticle method that accounts for anharmonic effects. We find that an *hcp* to *bcc* transition occurs near the melting curve at low pressures of $0 < P < 11$ GPa. This also demonstrates the validity of this theoretical framework based on the phonon quasiparticle to study the structural stability and phase transitions in strongly anharmonic metallic materials. Phonon quasiparticles of *bcc* are shown in figures 7(a)–(d). At low *T*, the VAF of the chosen mode does not decay exponentially, and the corresponding power spectrum exhibits a non-Lorentzian line shape, indicating the breakdown of the phonon quasiparticle. At high *T*, e.g., 1000 K and above, the phonon quasiparticle becomes well-defined with exponentially decaying cosine function-shaped VAF and Lorentzian-shaped power spectrum, meaning the *bcc* phase is dynamically stabilized. Accordingly, figure 7(e) displays the anharmonic phonon dispersion at 1000 K free of imaginary frequencies, which is not the case in the harmonic dispersion. Anharmonic phonon dispersion was also obtained for the *hcp* phase and is shown in figure 7(f), where anharmonic effects are nonnegligible. Vibrational entropy and Helmholtz free energies were obtained via equations (7) and (9), and Gibbs free energies were then calculated by $G = F + PV$. By comparing the *G* of both phases, a pre-melting *hcp* to *bcc* phase transition was discovered [5]. Under high *P-T* conditions, the *hcp/bcc* phase boundary computed with the phonon quasiparticle approach [13] was confirmed by another study using a different approach [31].

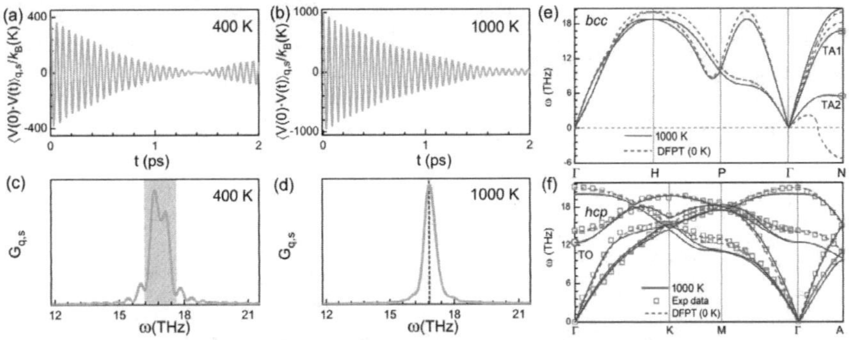

Figure 7. Mode-projected VAF of the TA1 acoustic mode at $\mathbf{q} = \mathbf{N}$ with a harmonic frequency of 18.2 THz of *bcc* Be at (a) 400 and (b) 1000 K, respectively. (c) and (d) show the corresponding power spectra. In (c), the shaded area between 16.2 and 17.6 THz covers two major peaks, indicating the breakdown of the phonon quasiparticle picture. In (d), the vertical dashed line at 16.8 THz indicates the frequency of the well-defined phonon quasiparticle. (e) Anharmonic phonon dispersion at 1000 K (solid blue curves) and harmonic phonon dispersion calculated using the DFPT (gray dashed curves) both at $V = 7.81$ Å^3/atom, the static *bcc* Be equilibrium volume. The two transverse branches are labeled TA1 and TA2, respectively. (f) Anharmonic phonon dispersion calculated at 1000 K (solid blue curves) and harmonic phonon dispersion calculated using the DFPT (gray dashed curves) both at $V = 7.89$ Å^3/atom, the static *hcp* Be equilibrium volume. The experimental data are shown for comparison [32]. (Results used with permission from [5], Copyright APS.)

3.2. Lattice Thermal Conductivity

The phonon quasiparticle method has also been successfully applied to study the lattice thermal conductivity of weakly [18, 33] and strongly [20, 22, 34] anharmonic systems.

3.2.1. Weakly anharmonic $MgSiO_3$ perovskite and postperovskite. At low *T*, the prediction of κ_{lat} by equation (10) is usually reliable. However, it fails to recur the high *T* saturation of thermal conductivity, a phenomenon where κ_{lat} violates the *1/T* law as discovered in many materials [35-37]. In order to address the issue, the minimal phonon mean free path (MFP) hypothesis was proposed [6, 36, 38]. It states that the MFP of a phonon, $l_{\mathbf{q}s}$, cannot be shorter than the lattice constants. Otherwise, the phonon picture breaks. With this hypothesis as an amendment to the Peierls-Boltzmann theory, equation (10) yields a minimal lattice thermal conductivity when $l_{\mathbf{q}s}$ approaches the lattice constants at high *T*. As a phenomenology, the minimal phonon MFP has been regarded as valid and useful in various areas. Here we provide a validity check of the minimal MFP hypothesis using the phonon quasiparticle approach and taking MgPv as an example. In our simulation, we have carried out Born-Oppenheimer molecular dynamics (BOMD) [27] on a 2×2×2 supercell of MgPv for a series of *T* from 300 to 1500 K. The initial atomic structure is first fully relaxed to $P = 0$ GPa with variable-cell-shape molecular dynamics [39]. The resulting volume is 24.1 cm^3/mol.

Figures 8(a) and 8(b) show the obtained mode projected VAF for one optical mode at $\mathbf{q} = (0, 0, 1/2)$ with a harmonic frequency 884 cm^{-1} at $T = 300$ and 1500 K, respectively. Both display an oscillating behavior. Figures 8(c) and 8(d) show that the corresponding power spectra both display a single peak. From the power spectra, we can also identify the frequency shift $\Delta\omega_{\mathbf{q}s}$ and line width $\Gamma_{\mathbf{q}s}$. At 300 K (1500 K), $\Delta\omega_{\mathbf{q}s} \approx 2$ cm^{-1} (10 cm^{-1}) and $\Gamma_{\mathbf{q}s} \approx 7.3$ cm^{-1} (34.5 cm^{-1}). Therefore, the criterion for well-defined phonon quasiparticles, $|\Delta\omega_{\mathbf{q}s} - i\Gamma_{\mathbf{q}s}| \ll \omega_{\mathbf{q}s}$, is satisfied both at low *T* and at high *T*. The good characterization of phonon quasiparticles enables us to investigate the vibrational properties with the concept of phonon. Figure 8(e) shows that the obtained $l_{\mathbf{q}s}$ for the same mode, which broadly follows a *1/T* law as guided by the solid curve. This is not surprising given a weakly anharmonic system like MgPv. More importantly, beyond certain temperatures, i.e., $T = 500$, 672, and 680 K, $l_{\mathbf{q}s}$ can be shorter than lattice constants a_0, b_0, and c_0, respectively. Since phonon quasiparticle is valid, i.e., lattice vibrations can be properly depicted in terms of phonon throughout the temperature range considered here, this result suggests that it is unphysical to presume a lower bound in MFP, i.e., the minimal MFP phenomenology does not exist [18].

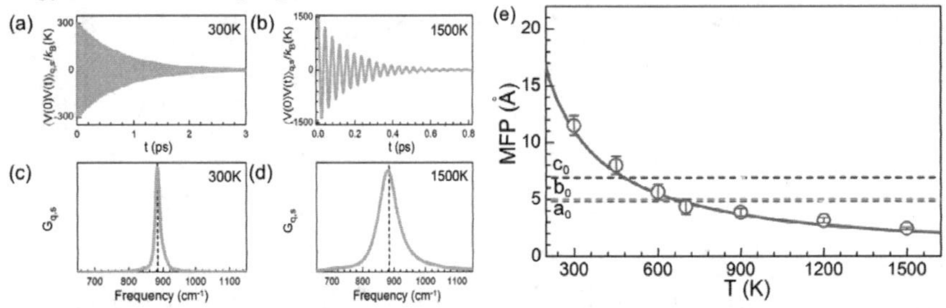

Figure 8. The mode-projected VAF (a) at 300 K and (b) 1500 K, respectively, of one optical mode at $\mathbf{q} = (0, 0, 1/2)$. (c) and (d) are the corresponding power spectra. The vertical dashed lines at 884 cm^{-1} indicate the harmonic frequency of this phonon mode. (e) MFP, $l_{\mathbf{q}s}$, of the same mode. Scatters denote the calculated data, and continuous curves having the *1/T* dependence are the fitted results. The horizontal dashed lines represent the lattice parameters of MgPv, a_0, b_0, and c_0, which define the orthogonal primitive cell with a volume of 24.1 cm^3/mol. (Results used with permission from [18], Copyright APS.)

The phonon quasiparticles calculation was also conducted for *Cmcm* $MgSiO_3$ post-perovskite (MgPPv), which is also weakly anharmonic. To overcome the finite-size effects on the lattice thermal conductivity, we relied on the frequency dependence of phonon linewidth [18, 33, 34]. The obtained phonon lifetimes, group velocities (figure 9(a)), and lattice thermal conductivities were compared between the MgPv and MgPPv. κ_{lat} of MgPv as functions of *T* and *P* are displayed in figures 9(b) and 9(c), respectively. MgPv's κ_{lat} at 300 K agrees well with a previous theoretical report using perturbative method accounting for up to three-phonon scattering processes [40], which is in line with the weakly anharmonic nature of MgPv. κ_{lat} of MgPPv as functions of *T* and *P* are displayed in figures 9(d) and 9(e), respectively. MgPPv's κ_{lat} agrees well with previous experimental measurements at 300 K [41]. Experimental and theoretical discrepancies on MgPv's and MgPPv's κ_{lat} were also discussed [33]. At each temperature, both MgPv's and MgPPv's κ_{lat} vary linearly with pressure. It was also found that MgPPv's κ_{lat} is ~25% higher than that of MgPv. This is attributed to MgPPv's smaller primitive cell

(10 atoms) compared to that of MgPv (20 atoms). Thus, MgPPv has generally larger $v_{\mathbf{q}s}$, as shown in figure 9(a), and higher κ_{lat} [33].

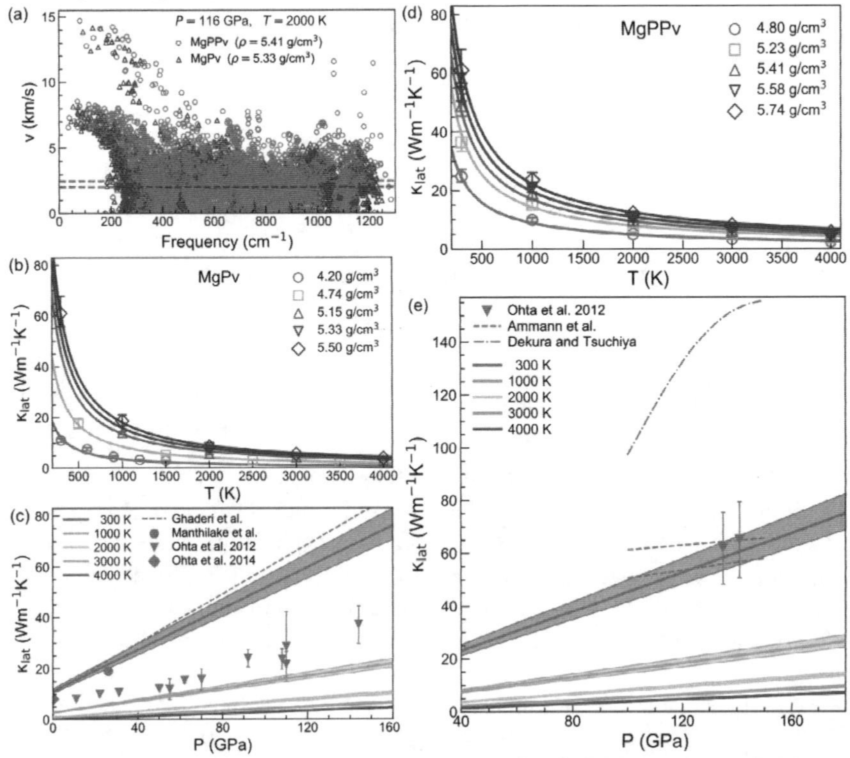

Figure 9. (a) $v_{\mathbf{q}s}$ versus $\tilde{\omega}_{\mathbf{q}s}$ sampled by a 10×10×10 **q**-mesh of MgPPv at P = 116 GPa and T = 2000 K are shown as red circles. $v_{\mathbf{q}s}$ versus $\tilde{\omega}_{\mathbf{q}s}$ sampled by an 8×8×8 **q**-mesh of MgPv at the same *P-T* conditions are shown as blue triangles for comparison. Dashed lines indicate the average phonon velocities. (b) κ_{lat} versus T of MgPv (symbols) for a series of densities. Error bars show the computational uncertainties. (c) κ_{lat} versus P of MgPv at a series of T. Shaded areas indicate the same computational uncertainties as in (b). Symbols with error bars show previous experimental measurements at 300 K [41-43]. The dashed line shows the previous *ab initio* perturbative calculations at 300 K [40]. (d) κ_{lat} versus T of MgPPv (symbols) for a series of densities. Error bars show the computational uncertainties. (e) κ_{lat} versus P of MgPPv at a series of T. Shaded areas indicate the same computational uncertainties as in (d). Symbols with error bars show the previous experimental measurements at 300 K [41]. Dashed lines show the previous semiclassical nonequilibrium MD calculations by adopting two different sets of interatomic potentials at 300 K [44]. The dashed-dotted line shows the previous *ab initio* perturbative calculations at 300 K [45]. (Results used with permission from [33], Copyright APS.)

3.2.2. Strongly anharmonic $CaSiO_3$ perovskite. CaPv is the last primary lower mantle phase whose thermal conductivity has not been determined, either experimentally or theoretically, due to its strong anharmonicity. The measurements must be carried out at high *P-T* because the cubic phase is unquenchable [4, 46, 47]. Calculations by perturbative methods without renormalization of interatomic force constants are also invalid for cubic CaPv. The static double-well potential deviates from the harmonic one substantially, and harmonic phonons exhibit imaginary frequencies. The phonon quasiparticle approach can overcome such difficulty by extracting well-defined quasiparticle lifetimes at high T. Only when phonon quasiparticles exist can equation (10) be used to calculate κ_{lat} within the RTA of the LBTE [34]. Figure 10(a) shows the obtained MFP. The sub-minimal MFP was also observed in such a strongly anharmonic system. Figures 10(b) and 10(c) report the obtained κ_{lat} of cubic CaPv as functions of T and P, respectively. The linear dependence of κ_{lat} on pressure was once again observed. Our theoretical predictions for κ_{lat} were further confirmed by the high P-T thermoreflectance measurements [34, 48, 49].

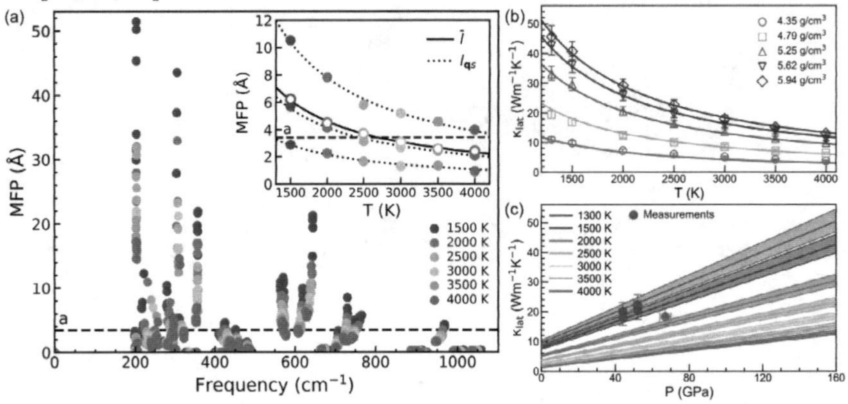

Figure 10. (a) MFP, $l_{\mathbf{q}s}$, from phonon quasiparticles sampled by MD simulations using a 3×3×3 supercell (135 atoms) with $\rho = 4.79$ g/cm^3 of CaPv at a series of T. The horizontal dashed black lines in both the main and the inserted figures represent the primitive cell lattice parameter, a. Insert: Three $l_{\mathbf{q}s}$ (solid circles) versus T are showcased. One remains longer than a (top), one remains shorter than a (bottom), and the other crosses the lattice parameter (middle) in the T range investigated. Average MFP, $\bar{l}$ (hollow circles), is also found to cross a as T increases. We find no lower-bound limit on phonon MFP in CaPv. The dotted and solid black curves show the $1/T$ fits [18]. (b) κ_{lat} versus T of CaPv (symbols) at several densities. Error bars show the computational uncertainties. (c) κ_{lat} versus P of CaPv at a series of T. Shaded areas indicate the same computational uncertainties as in (b). Solid symbols with error bars are experimental data by the high P-T thermoreflectance measurements [34, 48, 49]. (Results used with permission from [34], Copyright APS.)

4. Conclusion

The phonon quasiparticle approach addresses lattice anharmonicity by extracting renormalized frequencies and phonon lifetimes from AIMD simulations without explicitly expressing the higher-order interatomic force constants. In principle, it captures full anharmonicity and can overcome finite-size effects typical of AIMD simulations by offering complete phonon quasiparticle dispersions through Fourier interpolation. It has been demonstrated [4, 5, 10, 12, 13, 15, 18-22, 33, 34] to be valid and powerful in studying thermodynamics and lattice thermal conductivities of both weakly and strongly anharmonic materials when phonon quasiparticles are well defined.

References

[1] Souvatzis P, Eriksson O, Katsnelson M I and Rudin S P 2008 *Phys. Rev. Lett.* **100** 095901
[2] Errea I, Rousseau B and Bergara A 2011 *Phys. Rev. Lett.* **106** 165501
[3] Hellman O, Abrikosov I A and Simak S I 2011 *Phys. Rev. B* **84** 180301(R)
[4] Sun T, Zhang D-B and Wentzcovitch R M 2014 *Phys. Rev. B* **89** 094109
[5] Lu Y, Sun T, Zhang P, Zhang P, Zhang D-B and Wentzcovitch R M 2017 *Phys. Rev. Lett.* **118** 145702
[6] Kittel C 1949 *Phys. Rev.* **75** 972
[7] Wallace D C 1965 *Phys. Rev.* **139** A877
[8] Carrier P, Wentzcovitch R M and Tsuchiya J 2007 *Phys. Rev. B* **76** 064116
[9] Klemens P G 1958 *Solid State Phys.* **7** 1
[10] Zhang D-B, Sun T and Wentzcovitch R M 2014 *Phys. Rev. Lett.* **112** 058501
[11] Sun T, Shen X and Allen P B 2010 *Phys. Rev. B* **82** 224304

[12] Sun T, Brodholt J P, Li Y and Vočadlo L 2018 *Phys. Rev. B* **98** 224301
[13] Xian J-W, Yan J, Liu H-F, Sun T, Zhang G-M, Gao X-Y and Song H-F 2019 *Phys. Rev. B* **99** 064102
[14] Mermin N D 1965 *Phys. Rev.* **137** A1441
[15] Zhang Z and Wentzcovitch R M 2021 *Phys. Rev. B* **103** 104108
[16] Ziman J M 2001 *Electrons and Phonons: The Theory of Transport Phenomena in Solids.* (Oxford: Oxford Univ. Press)
[17] Sun T and Allen P B 2010 *Phys. Rev. B* **82** 224305
[18] Zhang D-B, Allen P B, Sun T and Wentzcovitch R M 2017 *Phys. Rev. B* **96** 100302(R)
[19] Zhang Z, Zhang D-B, Sun T and Wentzcovitch R M 2019 *Comput. Phys. Commun.* **243** 110
[20] Lu Y, Sun T and Zhang D-B 2018 *Phys. Rev. B* **97** 174304
[21] Lu Y, Zheng F, Yang Y, Zhang P and Zhang D-B 2019 *Phys. Rev. B* **100** 054304
[22] Lu Y, Zheng F, Yang Y, Zhang P and Zhang D-B 2021 *Phys. Rev. B* **103** 014304
[23] Gillet P, Guyot F and Wang Y 1996 *Geophys. Res. Lett.* **23** 3043
[24] Gillet P, Daniel I, Guyot F, Matas J and Chervin J C 2000 *Phys. Earth Planet. Inter.* **117** 361
[25] Durben D J and Wolf G H 1992 *Am. Mineral.* **77** 890
[26] Baroni S, Gironcoli S d, Corso A D and Giannozzi P 2001 *Rev. Mod. Phys.* **73** 515
[27] Wentzcovitch R M and Martins J L 1991 *Solid State Commun.* **78** 831
[28] Kawai K and Tsuchiya T 2014 *J. Geophys. Res. Solid Earth* **119** 2801
[29] Jackson I and Rigden S M 1996 *Phys. Earth Planet. Inter.* **96** 85
[30] Anderson O L 1995 *Equations of State of Solids for Geophysics and Ceramic Science.* (New York: Oxford University Press)
[31] Wu J, González-Cataldo F and Militzer B 2021 *Phys. Rev. B* **104** 014103
[32] Stedman R, Amilius Z, Pauli R and Sundin O 1976 *J. Phys. F* **6** 157
[33] Zhang Z and Wentzcovitch R M 2021 *Phys. Rev. B* **103** 144103
[34] Zhang Z, Zhang D-B, Onga K, Hasegawa A, Ohta K, Hirose K and Wentzcovitch R M 2021 *Phys. Rev. B* **104** 184101
[35] Auerbach A and Allen P B 1984 *Phys. Rev. B* **29** 2884
[36] Cahill D G, Braun P V, Chen G, Clarke D R, Fan S, Goodson K E, Keblinski P, King W P, Mahan G D, Majumdar A, Maris H J, Phillpot S R, Pop E and Shi L 2014 *Appl. Phys. Rev.* **1** 011305
[37] Roufosse M C and Klemens P G 1972 *J. Geophys. Res.* **79** 703
[38] Grimvall G 1999 *Thermophysical Properties of Materials.* (Amsterdam: North-Holland)
[39] Wentzcovitch R M 1991 *Phys. Rev. B* **44** 2358
[40] Ghaderi N, Zhang D-B, Zhang H, Xian J, Wentzcovitch R M and Sun T 2017 *Sci. Rep.* **7** 5417
[41] Ohta K, Yagi T, Taketoshi N, Hirose K, Komabayashi T, Baba T, Ohishi Y and Hernlund J 2012 *Earth Planet. Sci. Lett.* **349–350** 109
[42] Manthilake G M, Koker N d, Frost D J and McCammon C A 2011 *Proc. Natl. Acad. Sci. U.S.A.* **108** 17901
[43] Ohta K, Yagi T and Hirose K 2014 *Am. Mineral.* **99** 94
[44] Ammann M W, Walker A M, Stackhouse S, Wookey J, Forte A M, Brodholt J P and Dobson D P 2014 *Earth Planet. Sci. Lett.* **390** 175
[45] Dekura H and Tsuchiya T 2019 *Geophys. Res. Lett.* **46** 12919
[46] Komabayashi T, Hirose K, Sata N, Ohishi Y and Dubrovinsky L S 2007 *Earth Planet. Sci. Lett.* **260** 564
[47] Thomson A R, Crichton W A, Brodholt J P, Wood I G, Siersch N C, Muir J M R, Dobson D P and Hunt S A 2019 *Nature* **572** 643
[48] Hasegawa A, Yagi T and Ohta K 2019 *Rev. Sci. Instrum.* **90** 074901
[49] Okuda Y, Ohta K, Hasegawa A, Yagi T, Hirose K, Kawaguchi S I and Ohishi Y 2020 *Earth Planet. Sci. Lett.* **547** 116466

Journal of Physics: Conference Series **2207** (2022) 012043 doi:10.1088/1742-6596/2207/1/012043

The Unstable Analysis of Soliton Compression in Silicon Photonic Crystal

Sarun Phibanchon

Faculty of Education, Burapha Univerity
169 Longhardbangsaen road, Saensuk, Muang, Chonburi, 20130, Thailand

E-mail: sarunp@buu.ac.th

Abstract. Soliton solutions to the perturbed nonlinear Schrödinger equation, for describing the wave in the photonic crystal, can be found by the sine-Gordon expansion method. This method provides both bright and dark soliton solutions. The instability of the perturbed bright soliton in the perpendicular direction to the wave motion cab be expressed in terms of the growth rate. The time evolution of the unstable soliton can be visualized from the transformation to the higher dimensional solitons by the spectral method.

1. Introduction

The cubic nonliear Schrödinger (cNLS) equation,

$$i\phi_t + \phi_{xx} + |\phi|^2\phi = 0,$$

where subscripts x and t denote the partital derivatives with respect to the space and time, respectively, used to describe the pulse in fibre optics [1, 2], nonlinear radio pulsar emission [3] and laser physics [4]. However, for the photonic crystal [5], the perturbed cNLS will be used to describe some behaviours of the wave pulse but there are too many perturbed terms. To simplify the problem, we consider the 1D photonic crystals [6] where it has the only one perturbed term.

$$i\phi_t + \phi_{xx} + |\phi|^2\phi + \phi = 0. \tag{1}$$

In this work, we will apply the sine-Gordon (sG) expansion method [7] (the series of hyperbolic functions) to obtain both bright and dark solitons. There are many methods that have the same idea to the sG method, for example, tanh method [8], Exp-function method [9]. These two methods work very well for the Korteweg-de Vries (KdV) equation [10]. For all possible solitons of the perturbed cNLS equation, the sG expansion is more suitable method because of the hyperbolic functions.

After We obtain the bright soliton, we will apply the long-wavelength function to the perpendicular direction to the soliton propagation in order to investigate the all possible wavenumbers to make bright soliton unstable. We also study the time evolution of the unstable bright soliton to observe some results. We begin with the calculation of the soliton solution.

2. sine-Gordon Expansion

The sine-Gordon (sG) equation [7],

$$u_{xx} - u_{tt} = m^2 \sin(u),$$

where m is a constant. We consider the traveling wave solution, $\xi = \eta(x - ct)$, in which c is wave speed. The sG can now be written as

$$\frac{d^2u}{d\xi^2} = \frac{m^2}{\eta^2(1-c^2)} \sin u$$

where $a^2 = m^2/(\eta^2(1-c^2))$. By integration and change a new variable, $w = u/2$, the solutions can be expressed in terms of the hyperbolic functions such that

$$\begin{aligned} \sin w &= \operatorname{sech} a(\xi - \xi_0) \\ \cos w &= \tanh a(\xi - \xi_0). \end{aligned}$$

We can compose the series from these solutions for the ansatz solution to the nonlinear evolution equation,

$$\phi(w) = \sum_{j=1}^{n} \cos^{j-1} w \, [B_j \sin w + A_j \cos w] + A_0, \tag{2}$$

where A_j, B_j and A_0 are constant. To determine n, the number of terms in the series, the balance of the degrees between the highest power of the nonlinear term and the higher derivative must be required. The operator D is used to get the highest degree,

$$D\left(\frac{d^2}{dw^2}(\sin^2 w + 3\cos w)\right) = 4.$$

We now can solve (1) using (2).

3. The solutions of the perturbed cNLS Equation

We will start with the bright soliton. The traveling wave solution, $\phi(x,t) = u(\xi)e^{i(kx-\omega t)}$ with $\xi = \eta(x - ct)$,

$$\eta^2 u_{\xi\xi} + (1 + |u|^2)u - (\omega + k^2)u = 0. \tag{3}$$

To get the value of n, we have

$$D(u_{\xi\xi}) = n + 2 \quad \text{and} \quad D(u^3) = 3n,$$

and make it equal

$$n + 2 = 3n,$$

therefore $n = 1$, or

$$u(\xi) = B_1 \sin(w(\xi)) + A_1 \cos(w(\xi)) + A_0. \tag{4}$$

Substituting (4) into (3) and consider all possible degrees of a polynomial, the first one is a constant term,

$$A_0 + A_0^3 + 3A_0B_1^2 - A_0k^2 + A_0\omega = 0.$$

The $\cos(w(\xi))$ term,

$$-A_1 + 3A_0^2A_1 + 3A_1B_1^2 - A_1k^2 - 2A_1\eta^2 + A_1\omega = 0.$$

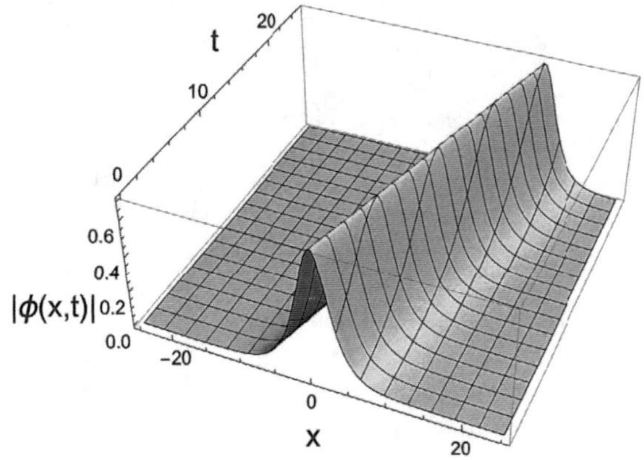

Figure 1. A bright solitary wave (5) with $c = \eta = 0.5$ and $t \in [0, 25]$

The $\sin(w(\xi))$ term,

$$B_1 + 3A_0^2 B_1 + B_1^3 - B_1 k^2 - B_1 \eta^2 + B_1 \omega = 0.$$

The $\sin(w(\xi))\cos(w(\xi))$ term,

$$6A_0 A_1 B_1 = 0.$$

The $\cos^2(w(\xi))$ term,

$$3A_0 A_1^2 - 3A_0 B_1^2 = 0.$$

The $\sin(w(\xi))\cos^2(w(\xi))$ term,

$$3A_1^2 B_1 - B_1^3 + 2B_1 \eta^2 = 0,$$

and then $\cos^3(w(\xi))$ term,

$$A_1^3 - 3A_1 B_1^2 + 2A_1 \eta^2 = 0.$$

Consider the case of $A_0 = 0$, $B_1 = \eta\sqrt{2}$, $\omega = c^2/4 - \eta^2 - 1$, the bright soliton solution can be obtained as

$$\phi(x,t) = \eta\sqrt{2}\,\mathrm{sech}\,\eta(x - ct)e^{i(cx/2 - \{\frac{c^2}{4} - \eta^2 - 1\}t)}, \tag{5}$$

the solution profile can be shown in Figure 1.

We next consider the dark solution, where $A_0 = 0$, $A_1 = i\eta\sqrt{2}$, $\omega = (c^2/4 + 2\eta^2 - 1)$, the solution can be written as

$$\phi(x,t) = i\eta\sqrt{2}\tanh(\eta(x - ct))e^{i(cx/2 - \{\frac{c^2}{4} + 2\eta^2 - 1\}t)}. \tag{6}$$

This type of solution is known as the dark solution solution as shown in Figure 2. In the next section, the growth rate for the long-wavelength perturbation to the bright soliton will be determined.

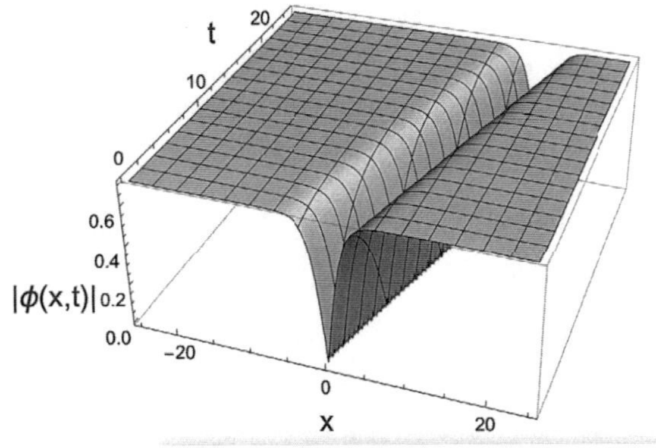

Figure 2. A dark solitary wave (6) with $c = \eta = 0.5$ and $t \in [0, 25]$

4. The long-wavelength perturbation

To study the instability of (5) to the long-wavelength perturbation in the perpendicular direction of the wave propagation [10], the perturbed cNLS will be extended into 2-D form,

$$i\phi_t + \phi_{xx} + \phi_{yy} + (1 + |\phi|^2)\phi = 0. \tag{7}$$

The perutrbation of (5) is written as

$$\phi = \left(u(\xi) + \epsilon n(\xi)e^{i\bar{k}y+\gamma t}\right) e^{i(kx-\omega t)}, \tag{8}$$

where n is any real or complex functions, ϵ is the small parameter,γ denotes the growth rate and $\bar{k}$ is the wave number with the long-wavelength. Eq.(8) is substituted into into (7) and consider $n = r(\xi, y, t) + iv(\xi, y, t)$, we then have

$$\begin{aligned} \eta^2 r_{\xi\xi} + (3u^2 + \omega + 1 - k^2 - \bar{k}^2)r(\xi) &= \gamma v(\xi) \\ \eta^2 v_{\xi\xi} + (u^2 + \omega + 1 - k^2 - \bar{k}^2)v(\xi) &= -\gamma r(\xi) \end{aligned}$$

The action [11] of these equations can be written as

$$\mathcal{S} = \int_{-\infty}^{\infty} \mathcal{L} dt$$

with

$$\begin{aligned} \mathcal{L} &= -\frac{\eta^2 r_\xi^2}{2} + \frac{\eta^2 v_\xi^2}{2} - \gamma v r \\ &+ \left[(6\eta^2 \operatorname{sech}^2 \xi + 1 + \omega - k^2 - \bar{k}^2)\right] \frac{r^2}{2} \\ &- \left[(2\eta^2 \operatorname{sech}^2 \xi + 1 + \omega - k^2 - \bar{k}^2)\right] \frac{v^2}{2}. \end{aligned}$$

We consider the trial functions as,

$$r(\xi) = \beta \operatorname{sech}^2(\xi) \quad \text{and} \quad v(\xi) = \alpha \operatorname{sech}(\xi).$$

To maximize the action with these trial functions, the growth rate can be obtained by eliminating α and β,

$$\gamma^2 = \frac{32\bar{k}^2(3-\bar{k}^2)}{3\pi^2}. \tag{9}$$

The growth rate curve is shown in Figure 3. The result of the time evolution will be shown in the next section.

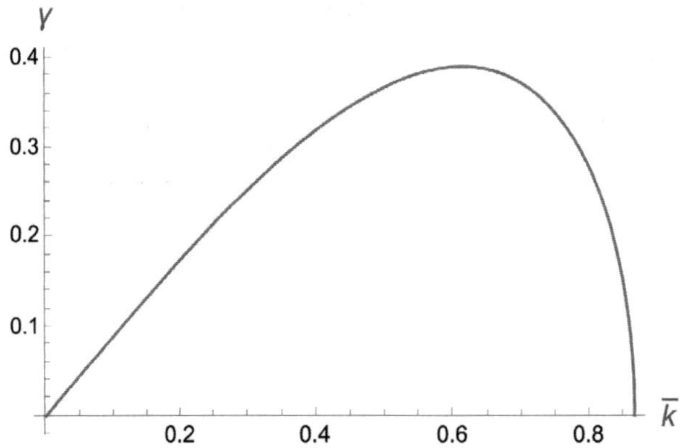

Figure 3. The growth rate (9) for the wavenumber, $\bar{k} \in [0, \sqrt{3}]$

5. Time Evolution of the Perturbed Bright Soliton

The spectral method [12] will be applied for the time evolution of the instability bright soliton, We take the Fourier transformation on the spatial coordinates of (7),

$$i\hat{\phi}_t - \xi^2\hat{\phi} - \chi^2\hat{\phi} + FT\left(|\phi|^2\phi\right) = 0,$$

where

$$\hat{\phi}_{p,q} = FT(\phi_{l,m}) = \sum_{l=0}^{N_x-1}\sum_{m=0}^{N_y-1} \phi_{l,m} e^{i(\xi_p x_l + \chi_q y_m)},$$

N_x and N_y are the number of mesh points of x and y, $(x_l, y_y) = (lL_x/N_x, mL_y/N_y)$, and L_x and L_y are lengths of the domain in the x and y, respectively, $(\xi_p, \chi_q) = 2\pi(p/L_x, qL_y)$. The initial condition is chosen as

$$\phi = \sqrt{2}\eta\,\mathrm{sech}(\eta x)e^{icx/2}\left(1 + \epsilon\cos\left(\frac{2\pi y}{L_y}\right)\right).$$

The numerical parameters for the calcuation are $L_x = 200$, $L_y = 70$, $N_x = 256$, $N_y = 64$, $\eta = 0.15$, $\epsilon = 0.01$ and $dt = 0.001$ The perturbed bright soliton will change into 2D soliton as shown in Figure 4(c)-(f).

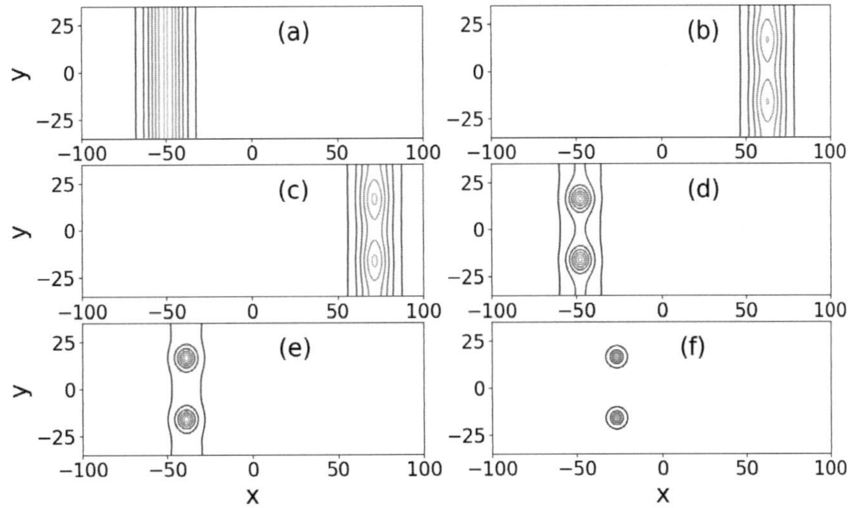

Figure 4. The unstable soliton with $\bar{k} = 0.1257$ (a) t=0, (b) t=100, (c) t=125, (d) t=150, (e) t=175, (f) t=200

6. Conclusion

The sine-Gordon expansion method is successfully applied to determine the solitary waves of perturbed cNLS. The result of the long-wavelength perturbation to the bright soliton produce the higher dimensional soliton which travels along the fibre optics faster than the plane soliton.

7. Acknowledgments

SP wishes to acknowledge support from NRCT 222/2561.

8. References

[1] Agrawal G P 2007 *Nonlinear Fiber Optics* 4th ed (New York: Academic Press)
[2] Cohen G 1995 *Phys. Rev. E* **52** 5565–73
[3] Asseo E, Pelletier G and Sol H 1990 *Mon. Not. Roy. Astron. Soc.* **247** 529–48
[4] Bullough R K, Jack P M, Kitchenside P W and Saunders R 1979 *Phys. Scr.* **20** 364–381
[5] Blanco-Redondo A, Husko C, Eades D, Zhang Y, Li J, Krauss T and Eggleton B 2014 *Nature Communications* **5**
[6] Mayteevarunyoo T and Malomed B A 2008 *Journal of the Optical Society of America B* **25** 1854
[7] Yan C 1996 *Physics Letters A* **224** 77–84
[8] Malfliet W and Hereman W 1996 *Phys. Scr.* **54** 563–8
[9] He J H and Wu X H 2006 *Chaos Soliton. Fract.* **30** 700–708
[10] Infeld E and Rowlands G 2000 *Nonlinear Waves, Solitons and Chaos* 2nd ed (Cambridge: Cambridge University Press)
[11] Anderson D, Bondeson A and Lisak M 1979 *J. Plasma Phys.* **21** 259–66
[12] Press W H, Teukolsky S A, Vetterling W T and Flannery B P 1992 *Numerical Recipes in C* 2nd ed (Cambridge: Cambridge University Press)

Journal of Physics: Conference Series **2207** (2022) 012044 doi:10.1088/1742-6596/2207/1/012044

DFT-NEGF Study of Electron Transfer Property in Graphene Nanoribbon Double Barrier System

S Sakamoto, M Tomiya and R Iida

Faculty of Science and Technology, Seikei University, Musashino, Tokyo 180-8633, Japan

sakamoto@st.seikei.ac.jp

Abstract. Current-voltage ($I - V$) characteristics in armchair graphene nanoribbon (AGNR) double barrier systems (DBSs) are investigated by the density functional theory (DFT) and the non-equilibrium Green's function (NEGF) methods. The hexagonal boron nitride (h-BN) which has an analogous structure with a graphene shows wide bandgap and becomes an electrical barrier in DBSs. The heterojunction structure of AGNR and two h-BNs with one six-membered ring width has properties as a resonant tunnelling device. Due to the resonance, sharp peaks are formed in the transmission function around the Fermi energy E_{F}. Our previous studies have explained the origins of some peaks by solving the one-dimensional Dirac equation for massless particles. Here, by using the two-dimensional Dirac equation, we show that the remaining sharp peaks depend on the transverse nodal pattern of the wavefunction as well as the anisotropy of the Fermi velocity. Moreover, in the $I - V$ characteristics, a flat part is formed without the negative differential resistance appearing. By partial density of state (PDOS) analysis, we reveal that the flat part in the current is caused by the fact that the peak positions in the resonance region are stable against the bias voltage.

1. Introduction

While various two-dimensional electronic devices have been proposed [1-13], the graphene nanoribbon (GNR) [14-17] is one of the devices expected for the next generation [18-22]. The hexagonal boron nitride (h-BN) which has an analogous structure with GNR shows wide bandgaps (~5eV) and becomes electrical insulators or wide bandgap semiconductors [23]. For bond lengths of C-C in GNR and B-N in h-BN are close to each other, GNR and h-BN combine and form heterojunctions [24-28]. The h-BN nanoribbon stripe becomes an electrical barrier when it is embedded in the conductive armchair graphene nanoribbon (AGNR). By adjusting the width and spacing of the two h-BN stripes, it is expected to be a double barrier system (DBS) of AGNR. We have investigated AGNRs with an interest in the electrical properties, especially the resonance tunnel phenomenon of the DBS. So far, the $I - V$ characteristics and the transmission function of the DBS have been obtained by calculation using the density functional theory (DFT) and the non-equilibrium Green's function (NEGF) method [25,26]. Unlike the macro system of the resonant tunneling diode, the nanocarbon system of AGNR has no clear negative differential resistance region in the $I - V$ characteristics, and the flat current region is observed. Considering the DBS of AGNR as a one-dimensional system, we have also explained some peaks of the transmission function around the Fermi energy (E_{F}) by solving the one-dimensional Dirac equation of massless particles for AGNR. The

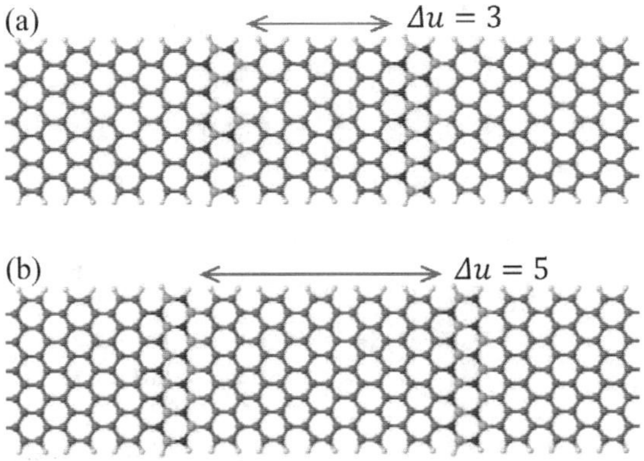

Figure 1. AGNR DBS models. Two rows of h-BN six-membered rings are embedded separately in AGNR. (a) The graphene spacing $\Delta u = 3$ and (b) $\Delta u = 5$. The unit of Δu is the width of the six-membered ring. Both ends of the system are connected to semi-infinite AGNR electrodes.

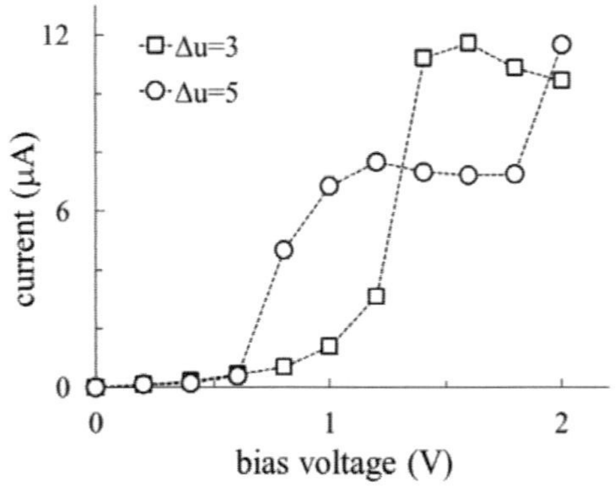

Figure 2. $I - V$ Characteristics of the DBS in figure 1. Corresponding to the interval of resonance peaks in figure 3, the current rises faster in the $\Delta u = 5$ than in the $\Delta u = 3$. In both cases, the current flattens in the middle. There is no clear negative differential resistance.

purpose of this paper is to evaluate remained other transmission peaks by the two-dimensional Dirac equation and to explain the flat current region by the partial density of states (PDOS). We introduce our methods for DFT and NEGF numerical calculations and the analysis of the transmission peaks in the following section. We show the results of the transverse nodal pattern analysis and the PDOS analysis with transport properties obtained previously [25,26] to clarify the purpose. We discuss the anisotropy of electron transport and the flat region of the $I - V$ characteristics as summary.

2. Method

2.1. Computational details

Computations of geometrical optimizations of our DBS of 11-AGNR (figure 1) are performed by using the SIESTA package based on the DFT method [29,30]. A total of 390 carbon, boron, nitrogen and hydrogen atoms are arranged in a $3.2 \times 0.8 \times 6.5$ nm supercell. The generalized gradient approximation (GGA) of the Perdew, Burke and Ernzerhof (PBE) form is used for the exchange-correlation term, and the Kleinman–Bylander (KB) type pseudo-potential is attached to each atom through all our calculations. We employ an energy cutoff for a real space mesh size of 350Ry and the Monkhorst-Pack k-sampling of $4 \times 4 \times 8$ points in reciprocal space. A double-zeta basis set orbital is adopted. Before calculation of electronic properties, using the conjugate gradient (CG) method for the structural optimization, all atomic positions in the super cell have been fully relaxed.

We have also computed transport properties, the DOS and the electronic quantum transmission as a function of energies, by using the TRANSIESTA code [31] which is based on the NEGF method combined with the DFT technique. In these calculations, both left and right edges of the central DBS are connected to the semi-infinite AGNR leads. Under a non-equilibrium situation, the bias voltage V_b between both leads is defined as $eV_b = \mu_R - \mu_L$ where μ_L and μ_R are chemical potentials in the left and right electrodes connecting to the central part, respectively. From the transmission function $T(E)$ under the bias voltage, the current $I(V_b)$ should be evaluated by the Landauer formula [32].

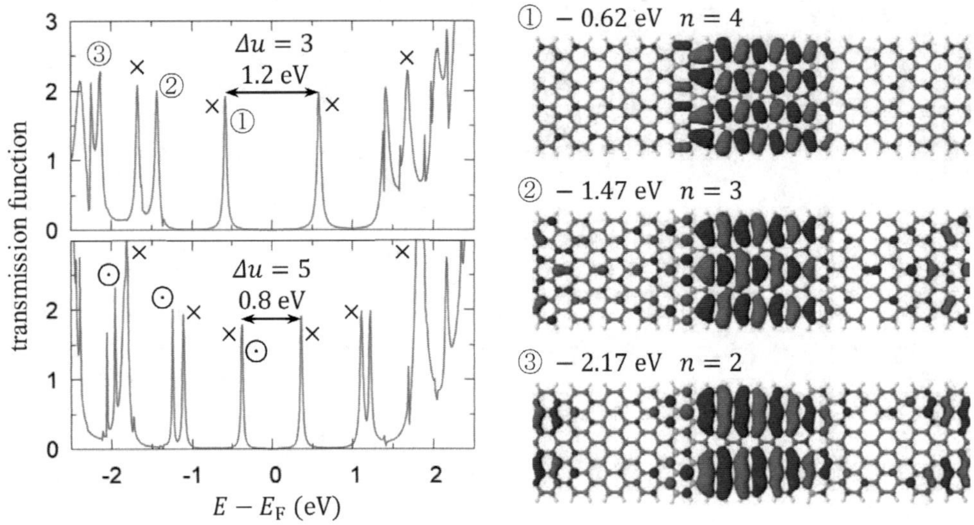

Figure 3. The left side is the transmission function with bias voltage $V_b = 0$ V. There are some sharp peaks due to the double barrier resonance. The narrower the width of the AGNR central region between the h-BN barriers, the larger the peak spacing around the Fermi energy E_F. The peaks indicated by × correspond to the results obtained by the analysis with the one-dimensional Dirac equation. Other peaks indicated by 1, 2, 3 and ⊙ can be explained by the two-dimensional Dirac equation. The right side is the wave function of peaks 1, 2 and 3 of the $\Delta u = 3$. The signs $\pm$ of the wavefunction value are indicated in red and blue. The eigenenergy depends on the number of the antinodes n in the transverse direction.

2.2. Two-dimensional Dirac equation of massless particle

Some of the peak positions of the transmission function are evaluated by considering the DBS as one-dimensional [25,26]. The particle mass m is zero for AGNR and is the bandgap for h-BN. Using the Fermi velocity v_F and momentum p_x, the Dirac equation is as follows.

$$(v_F \sigma_x p_x + m\beta + U(x) - E)\Phi(x) = 0 ,$$

where

$$\sigma_x = \begin{pmatrix} 0 & 1 \\ 1 & 0 \end{pmatrix}, \quad \beta = \begin{pmatrix} 1 & 0 \\ 0 & -1 \end{pmatrix}, \quad \Phi(x) = (\phi(x), \psi(x)) .$$

Assuming the potential $U =$ const. we have

$$\psi(x) = -\frac{i\hbar v_F}{m v_x^2 + E - U} \frac{d\phi(x)}{dx} .$$

Representing the wave function by the superposition of the traveling and backward wave, then the transmittance can be obtained from the solution that satisfies the boundary conditions.

We here propose a simple two-dimensional Dirac equation of the massless particle to evaluate the influence of the transverse mode in the resonance region between the barriers.

$$(v_x \sigma_x p_x + v_y \sigma_y p_y + U)\Phi(x, y) = E\Phi(x, y) ,$$

where

$$\sigma_y = \begin{pmatrix} 0 & -i \\ i & 0 \end{pmatrix}, \quad \Phi(x, y) = (\phi(x, y), \psi(x, y)) .$$

Then, we have

$$\left((v_x p_x)^2 + (v_y p_y)^2\right)\phi(x, y) = (E - U)^2 \phi(x, y) .$$

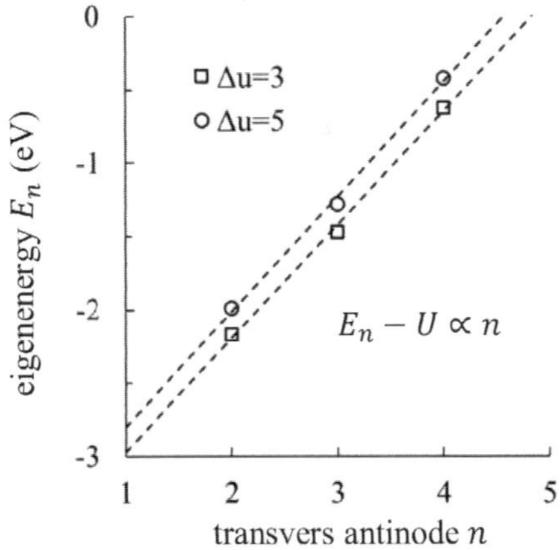

Figure 4. Plots of the eigenenergy against the transvers antinode number n. The eigenenergy of the massless particle described by the Dirac equation is proportional to the quantum number n, not n^2.

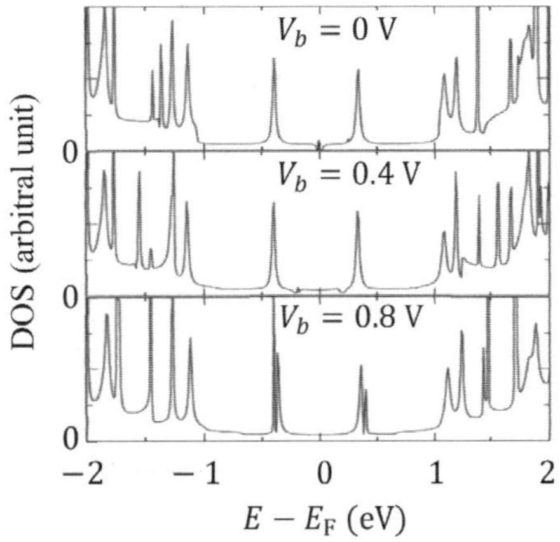

Figure 5. DOS plot at bias voltages V_bs, 0, 0.4 and 0.8 V in the $\Delta u = 5$. The resonance peak positions have no change with the bias voltage. DOS around the Fermi energy E_F has a finite value, but its contribution to current is small (figure 2).

If the longitudinal direction and the transverse direction are considered separately, the transverse eigenenergy E_y is obtained as

$$E_y - U_y = v_y p_y = v_y \hbar k_y$$

where U_y is a certain constant. The certainty of this approximation is guaranteed when the eigenenergy obtained from the DFT method is proportional to the wave number.

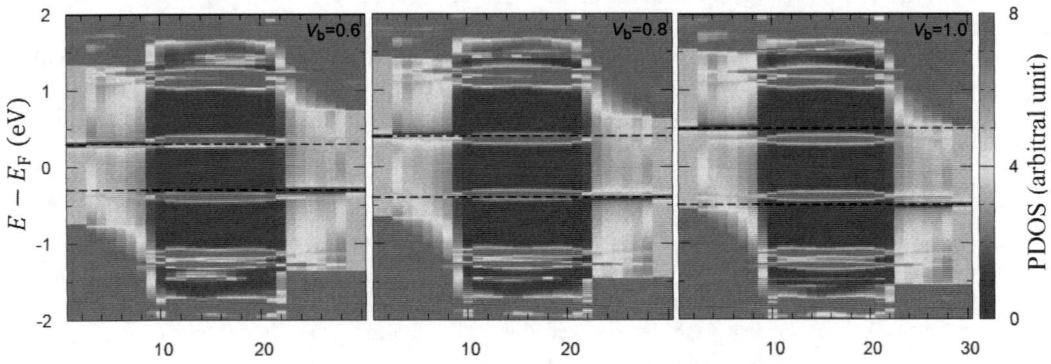

longitudinal position number of sliced zigzag chain

Figure 6. PDOS map for slices in the longitudinal direction of the $\Delta u = 5$ DBS at the $V_b = 0.6, 0.8$ and 1.0 V. Each slice is a zigzag chain of C or h-BN in the transverse direction. The horizontal axis of the map is the position number of slices. Numbers 9, 10 and numbers 21, 22 are the left and right h-BN barriers, and numbers 11 to 20 are the central resonance region. The black dashed lines indicate V_b centered on the E_F. In the central region, the DOS other than the resonance peak is almost zero. The effect of increasing the bias voltage is blocked by the h-BN barriers and does not affect the resonance region.

Journal of Physics: Conference Series **2207** (2022) 012044 doi:10.1088/1742-6596/2207/1/012044

3. Results

In the DBS with $\Delta u = 3$ and 5 (figure 1), the $I - V$ characteristics obtained from the Landauer formulation are shown in figure 2. In both cases, the current flattens in the middle, and there is no clear negative differential resistance region like the macro resonant tunneling diode [33]. The voltage at which the current rises corresponds to the peak interval around E_{F} in the transmission function (figure 3). The transmission function at $V_{\mathrm{b}} = 0$ V has sharp peaks of the double barrier resonance. Some peaks marked with × can be evaluated by the one-dimensional Dirac equation, and the then the Fermi velocity is about $v_{\mathrm{F}} \simeq c/350$. Other peaks indicated by 1, 2, 3 and ⨀ are related to the eigenstate of the central resonance region. The peak position of the transmission function and the DFT eigenenergy are the same. The eigen wavefunctions (at the Γ point) of peaks 1, 2 and 3 are shown on the right side of figure 3, and the number of antinodes in the transverse direction is 4, 3 and 2, respectively. Even in the two-dimensional Dirac equation, the wave function in the longitudinal direction needs to consider the boundary condition between AGNR and h-BN. While the wavefunction in the transverse direction of AGNR is connect to the vacuum, so the wave number can be easily evaluated by the transverse width and the antinode number. In order to confirm that the eigenstate can be described by the Dirac equation, the eigenenergy E_n is plotted against the antinode number n (figure 4). The result is that E_n is proportional to n, and the slope is about 0.78 for both DBSs of $\Delta u =$ 3 and 5. Note that the eigenenergy described by Schrödinger equation is proportional to n^2. From the width of AGNR (figure 1), 1.24 nm, and the slope 0.78, the transverse Fermi velocity $v_y \simeq c/630$.

Figure 5 shows the DOS of the $\Delta u = 5$ DBS at $V_{\mathrm{b}} = 0, 0.4$ and 0.8 V. The positions of the resonance peak, especially near the Fermi energy, hardly change with voltage. Although DOS has a non-zero finite value around E_{F}, it does not contribute to the magnitude of the current. We compute PDOS to see how DOS is reflected in the transmission function. The PDOS map here is the DOS of the sliced part of the DBS divided in the longitudinal direction. Each slice is a C or h-BN zigzag chain in the transverse direction. In figure 6, we show PDOSs with $V_{\mathrm{b}} = 0.6, 0.8$ and 1.0 V near the current rise in the $\Delta u = 5$ DBS. The horizontal axis is the slice number, 9 and 10 are the left h-BN barrier, 11 to 20 are the central resonance region, 21 and 22 are the right barrier, and the rest are the AGNR connected to the electrodes. The PDOS shifts in the regions outside of the barriers due to the difference in chemical potential, but the central resonance region is blocked by the barrier and is not affected by the chemical potential. There is no state other than the resonance peaks in the central resonance region, and the current conducts only in these stable peaks. The black dashed lines in figure 6 show the integration range of the current, i.e. the value of V_{b}. No current flows unless the integration region reaches the resonance peaks. When V_{b} increases and reaches the resonance peaks, the current begins to flow and increases until it exceeds the peaks. As spreading V_{b} beyond the first peaks, then the current becomes flat.

4. Discussion and Conclusion

From our numerical investigations of the graphene nanoribbon DBS, it was found that there are cases where the $I - V$ characteristics show no negative resistance and become flat. This feature, which is different from the macro resonant tunneling diode, is derived from the nano-sized property peculiar to AGNR. The energy gap of the h-BN barrier stabilizes the isolated peaks in the resonance region, and the chemical potential of the electrodes does not affect the resonance state. Analysis using PDOS showed that these resonance peaks play an important role in electron transfer in the DBS. The position of the resonance peak of the transmission function can also be evaluated from the boundary condition of the wave function obtained by solving the Dirac equation. The linear relationship between the eigenenergy and the number of nodes supports that the description of the Dirac equation is correct. In the narrow AGNR DBS, the anisotropy of fermi velocity in the longitudinal and transverse directions is confirmed.

It can be concluded that these properties are peculiar to nanocarbon resonance devices and their behavior as Dirac particles is important.

References

[1] Cui T, Lv R, Huang Z-H, Chen S, Zhang Z, Gan X, Jia X, Wang K, Wu D, and Kang F 2013 *J. Mater. Chem.* A **1** 5736
[2] Nepal N, Wheeler V D, Anderson T J, Kub F J, Mastro M A, Myeys-Ward R L, Qadri S B, Freitas J A, Hernandez S C, Nykiti, Walton S G, Gaskill K, and Eddy C R Jr. 2013 *J. Appl. Phys. Express* **6** 061003
[3] Lin F, Chen S-H, Meng J, Tse G, Fu X-W, Xu F-J, Shen B, Liao Z-M, and Yu D-P 2014 *Appl. Phys. Lett.* **106** 073103
[4] Cuo G-C, Wang D, Wei X-L, Zhang Q, Liu H, Lau W-M, and Liu L-M 2015 *J. Phys. Chem. Lett.* **6** 5002
[5] Cai Y, Zhang G, and Zhang Y-W 2015 *J. Phys. Chem.* C **119** 13929
[6] Padilha J E, Fazzio A, and da Silva A J R 2015 *Phys. Rev. Lett.* **114** 066803
[7] Hu W, Lin L, Yang C, Dai J, and Yang J 2016 *Nano Lett.* **16** 1675
[8] Zhang Y-Y, Pei Q-X, Mai Y-W, and Lai S-K 2016 *J. Phys.* D **49** 465301
[9] Kumar A, Kashid R, Ghosh A, Kumar V, and Singh R 2016 *ACS Matter. Interfaces* **8** 8213
[10] Yang Q, Tan C-J, Meng R-S, Jiang J-K, Liang Q-H, Sun X, Yang D-G, and Chen X-P 2016 *IEEE Electron Device Lett.* **38** 145
[11] Zhu J, Zhang J, and Hao Y 2016 *Jpn. J. Appl. Phys.* **55** 080306
[12] Kalita G, Shaarin M D, Paudei B, Mahyavanshi R, and Tanemura M 2017 *Appl. Phys. Lett.* **111** 013504
[13] Hu X-R, Zheng J-M, Ren Z-Y 2018 *Front. Phys.* **13** 137302
[14] Han M Y, Özyilmaz B, Zhang Y, and Kim P, 2007 *Phys. Rev. Lett.* **98** 206805
[15] Berger C, Song Z, Li X, Wu X, Brown N, Naud C, Mayou D, Li T, Hass J, Marchenkov A N, Conrad E H, First P N, and de Heer W A 2006 *Science* **312** 1191
[16] Yang L, Park C -H, Son Y -W, Cohen M L, and Louie S G 2007 *Phys. Rev. Lett.* **99** 186801
[17] Neto A H C, Guinea F, Peres N M R, Novoselov K S, and Geim A K 2009 *Rev. Mod. Phys.* **81** 109
[18] Djavid N, Khaliji K, Tabatabaei S M and Pourfath M 2014 *IEEE TRANSACTIONS ON ELECTRON DEVICES* **61** 23
[19] Zoghi M and Goharrizi A Y 2017 *IEEE TRANSACTIONS ON ELECTRON DEVICES* **64** 4322
[20] Hammam A M M, Schmidt M E, Muruganathan M, Suzuki S and Mizuta H 2018 *Carbon* **126** 588
[21] Martini L, Chen Z, Mishra N, Barin G B, Fantuzzi P, Ruffieux P, Fasel R, Feng X, Marita A, Coletti C, Müllen K and Candini A 2019 *Carbon* **146** 36
[22] Kheirabadi S J, Ghayour R and Sanaee M 2020 *Diamond & Related Materials* **108** 107970
[23] Liu L, Park J, Siegel D A, McCarty K F, Clark K W, Deng W, Basile L, Idrobo J C, Li A -P, and Gu G 2014 *Science* **343** 163
[24] Goharrizi A Y, Zoghi M and Saremi M 2016 *IEEE TRANSACTIONS ON ELECTRON DEVICES* **63** 3761
[25] Wakai T, Sakamoto S and Tomiya M 2018 *J. Phys.: Condens. Matter* **30** 265302A
[26] Tomiya M, Sakamoto S and Wakai T 2019 *J. Phys.: Conf. Ser.* **1290** 012010
[27] Yami F M and Hosseini S E 2019 *27th Iranian Conference on Electrical Engineering* 315
[28] Sanaeepur M, Jafari M R and Esmaeili M 2020 *IEEE TRANSACTIONS ON ELECTRON DEVICES* **67** 725
[29] Sánchez-Portal D, Ordejón P, Artacho E and Soler J M 1997 *Int. J. Quantum Chem.* **65** 453
[30] Soler J M, Artacho E , Gale J D, García A, Junquera J, Ordejón P and Sánchez-Portal D 2002 *J. Phys.: Condens. Matter* **14** 2745
[31] Brandbyge M, Mozos J-L, Ordejón P, Taylor J and Stokbro K 2002 *Phys. Rev.* B **65** 165401
[32] Datta S 1997 *Electronic Transport In Mesoscopic Systems* (Cambridge: Cambridge University Press) p 57
[33] Tsu R and Esaki L 1973 *Appl. Phys. Lett.* **22** 562

Excitation and transport of bound magnon clusters in frustrated ferromagnetic chain

Hiroaki Onishi

Advanced Science Research Center, Japan Atomic Energy Agency, Tokai, Ibaraki 319-1195, Japan

E-mail: onishi.hiroaki@jaea.go.jp

Abstract. We investigate magnetic excitations and spin transport properties of a frustrated spin chain with ferromagnetic nearest-neighbor and antiferromagnetic next-nearest-neighbor exchange couplings in a magnetic field by density-matrix renormalization group methods. In the field-induced quadrupole and octupole regimes, the low-energy excitation is governed by two- and three-magnon bound states, respectively, so that bound magnon clusters would contribute to the spin transport. We show that spin current correlations decrease when the system goes from the quadrupole regime into the octupole regime. This indicates that the spin transport is suppressed in the octupole regime, although bound three-magnon clusters carry a larger amount of angular momentum than bound two-magnon clusters.

1. Introduction

Spin-related transport phenomena in magnetic materials have been attracting a growing interest in the fields of condensed-matter physics and spintronics. In magnetic insulators, magnons, which are elementary excitations in magnetic ordered states, carry spin current without electric current, expected to be a key feature to realize efficient devices without Joule heating. Recently, novel types of carriers in quantum disordered states have been discussed, such as spinons [1, 2], bound magnon pairs [3, 4], and Majorana fermions [5, 6], not only to understand fundamental properties of exotic spin states from the viewpoint of transport but also to explore a new mechanism for the efficient spin current generation.

Here we focus on spin multipole states emerging in a frustrated spin chain with ferromagnetic nearest-neighbor and antiferromagnetic next-nearest-neighbor exchange couplings in a magnetic field [7, 8, 9]. It is well established that the ground state is a vector chiral state in low magnetic fields, while in high magnetic fields there appear a series of spin multipole liquid ground states such as quadrupole, octupole, and hexadecapole, which are two-, three-, and four-magnon bound states, respectively. Since the low-energy excitation is governed by multi-magnon bound states, we expect that bound multi-magnon clusters would carry spin current.

We have previously studied the dynamical spin and quadrupole structure factors [10, 11, 12] and the spin Drude weight [3] with a particular focus on the quadrupole regime. In this paper, to gain further insight into the relation between the low-energy excitation and the spin transport property, we numerically investigate the multi-magnon excitation energy, the dynamical spin, quadrupole, and octupole structure factors, and the spin current correlation function, in the both quadrupole and octupole regimes. We show that spin current correlations are suppressed in the octupole regime, although bound three-magnon clusters carry a larger amount of angular momentum than bound two-magnon clusters.

2. Model and method

We consider a spin-1/2 J_1-J_2 Heisenberg model with competing ferromagnetic J_1 (<0) and antiferromagnetic J_2 (>0) in a magnetic field h on an N-site chain, described by

$$H = J_1 \sum_i \boldsymbol{S}_i \cdot \boldsymbol{S}_{i+1} + J_2 \sum_i \boldsymbol{S}_i \cdot \boldsymbol{S}_{i+2} - h \sum_i S_i^z, \tag{1}$$

where $\boldsymbol{S}_i$ are spin-1/2 operators at site i. We set $J_2 = 1$ and take it as the energy unit. Note that the total magnetization $m = M/N = \sum_i S_i^z/N$ is conserved, so that it can be used to block-diagonalize the Hamiltonian. In the magnetization curve, M changes by one with h in the vector chiral phase, two in the quadrupole liquid phase, three in the octupole liquid phase, etc. For the analysis of a given m, we set h to the midpoint of the magnetization plateau of M in the N-site system, and use the saturation field h_s for $m = 1/2$.

We investigate magnetic excitations and spin transport properties at zero temperature by density-matrix renormalization group (DMRG) methods in the open boundary condition [13, 14]. To clarify the property of the low-energy excitation, we examine the dependence of the p-magnon excitation energy, defined by

$$\Delta_p(N, M) = [E_0(N, M+p) + E_0(N, M-p) - 2E_0(N, M)]/2, \tag{2}$$

where $E_0(N, M)$ is the lowest energy of the N-site system in the subspace of M at $h = 0$. Note that excitation energies for adding p magnons and removing p magnons are averaged. At the saturation $M = N/2$, we evaluate the p-magnon excitation energy as

$$\Delta_p(N, N/2) = E_0(N, N/2 - p) - E_0(N, N/2) + p h_\mathrm{s} \tag{3}$$

Moreover, we calculate the dynamical spin, quadrupole, and octupole structure factors, which are, respectively, given by

$$S^-(q,\omega) = -\mathrm{Im}\frac{1}{\pi}\langle\psi_\mathrm{G}|(S_q^-)^\dagger \frac{1}{\omega + E_\mathrm{G} - H + \mathrm{i}\eta} S_q^- |\psi_\mathrm{G}\rangle, \tag{4}$$

$$Q^{--}(q,\omega) = -\mathrm{Im}\frac{1}{\pi}\langle\psi_\mathrm{G}|(Q_q^{--})^\dagger \frac{1}{\omega + E_\mathrm{G} - H + \mathrm{i}\eta} Q_q^{--} |\psi_\mathrm{G}\rangle, \tag{5}$$

$$O^{---}(q,\omega) = -\mathrm{Im}\frac{1}{\pi}\langle\psi_\mathrm{G}|(O_q^{---})^\dagger \frac{1}{\omega + E_\mathrm{G} - H + \mathrm{i}\eta} O_q^{---} |\psi_\mathrm{G}\rangle, \tag{6}$$

where S_q^-, Q_q^{--}, and O_q^{---} are the Fourier transforms of S_i^-, $Q_i^{--} = S_i^- S_{i+1}^-$, and $O_i^{---} = S_i^- S_{i+1}^- S_{i+2}^-$, respectively, $|\psi_\mathrm{G}\rangle$ is the ground state, E_G is the ground-state energy, and η is a small broadening factor, set to 0.1 in the present calculations. To calculate the dynamical structure factors, we use a dynamical DMRG method [14], multi-targeting $|\psi_\mathrm{G}\rangle$, $A|\psi_\mathrm{G}\rangle$, and $[\omega + E_\mathrm{G} - H + \mathrm{i}\eta]^{-1} A|\psi_\mathrm{G}\rangle$, where A is S_q^-, Q_q^{--}, or O_q^{---}.

Regarding the transport property, we investigate the spin current correlation function as a measure of the degree of the flow of the spin current, defined by

$$I_\mathrm{s} = \langle\psi_\mathrm{G}|j_\mathrm{s}^\dagger j_\mathrm{s}|\psi_\mathrm{G}\rangle, \tag{7}$$

where $j_\mathrm{s} = \sum_i J_1(\boldsymbol{S}_i \times \boldsymbol{S}_{i+1})^z + \sum_i 2J_2(\boldsymbol{S}_i \times \boldsymbol{S}_{i+2})^z$ is the spin current operator.

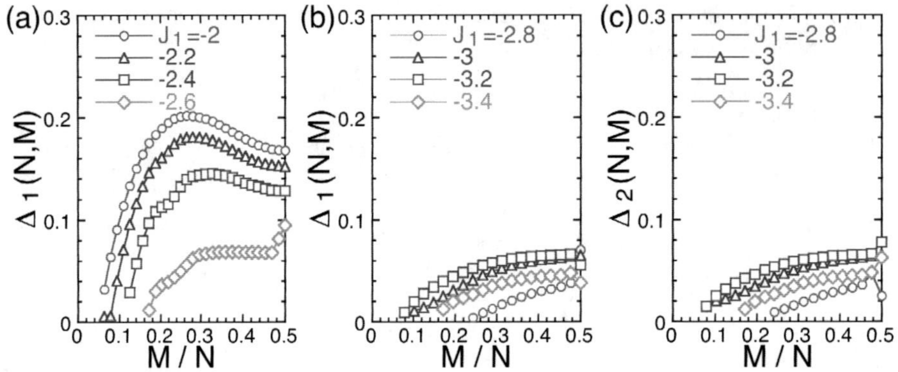

Figure 1. The excitation energy (a) $\Delta_1(N, M)$ for several values of J_1 in the quadrupole regime, (b) $\Delta_1(N, M)$ and (c) $\Delta_2(N, M)$ for several values of J_1 in the octupole regime. The system size is $N = 128$.

3. Results

Let us start with the discussion on magnetic excitations. In Fig. 1(a), we show the excitation energy $\Delta_1(N, M)$ for several values of J_1 in the quadrupole regime. Note that the two-magnon bound state appears for $J_1 \gtrsim -2.7$ and the ground state turns to the three-magnon bound state for smaller J_1 [7, 8, 9]. We find that $\Delta_1(N, M)$ decreases with decreasing J_1, indicating that the two-magnon bound state becomes unstable as the system approaches the phase boundary. We note that $\Delta_1(N, M)$ increases with decreasing J_1 for larger J_1 (> -2), indicating a reentrant behavior [10].

Figures 1(b) and 1(c) present the excitation energies $\Delta_1(N, M)$ and $\Delta_2(N, M)$, respectively, in the octupole regime. We find that $\Delta_1(N, M)$ and $\Delta_2(N, M)$ exhibit similar dependences. This is because we obtain both single-spin-flip and two-spin-flip states when we break a bound three-magnon cluster. With decreasing J_1, $\Delta_1(N, M)$ and $\Delta_2(N, M)$ first increase for $J_1 = -2.8$, -3, and -3.2, indicating that the ferromagnetic exchange interaction stabilizes the three-magnon bound state, while they turn to decrease with further decreasing J_1. This reentrant behavior resembles that of $\Delta_1(N, M)$ in the quadrupole regime. Here we point out that the excitation energies in the octupole regime are much reduced from those in the quadrupole regime. This is indicative that the three-magnon bound state is easily broken comparing with the two-magnon bound state.

To obtain more deep insight into the low-energy excitation, we investigate the dynamical spin, quadrupole, and octupole structure factors. In Fig. 2, we show numerical results for $J_1 = -2$ and -2.4 with fixed $m = 0.35$, where the ground state is in the quadrupole regime. For $S^-(q, \omega)$, we find gapped spin excitations, composed of a dispersive mode extending to high energy. The momentum position of the gap, corresponding to the bottom of the dispersion at a little below $q = \pi/2$, is incommensurate. With decreasing J_1, the momentum position of the gap gradually moves toward small momentum, while the gap decreases, as observed in Fig. 1(a). In addition, the dispersion for small momentum below the gap position is pushed down to the low-energy side. For $Q^{--}(q, \omega)$, we find gapless quadrupole excitations at $q = \pi$ due to quasi-long-range antiferro-quadrupole correlations. The overall structure of the intensity distribution does not change with J_1 significantly, while it moves to the low-energy side, so that the slope of the dispersion becomes gentle. For $O^{---}(q, \omega)$, we find gapped octupole excitations. The intensity distribution moves to the low-energy side with decreasing J_1, which is similar to the findings for $S^-(q, \omega)$ and $Q^{--}(q, \omega)$.

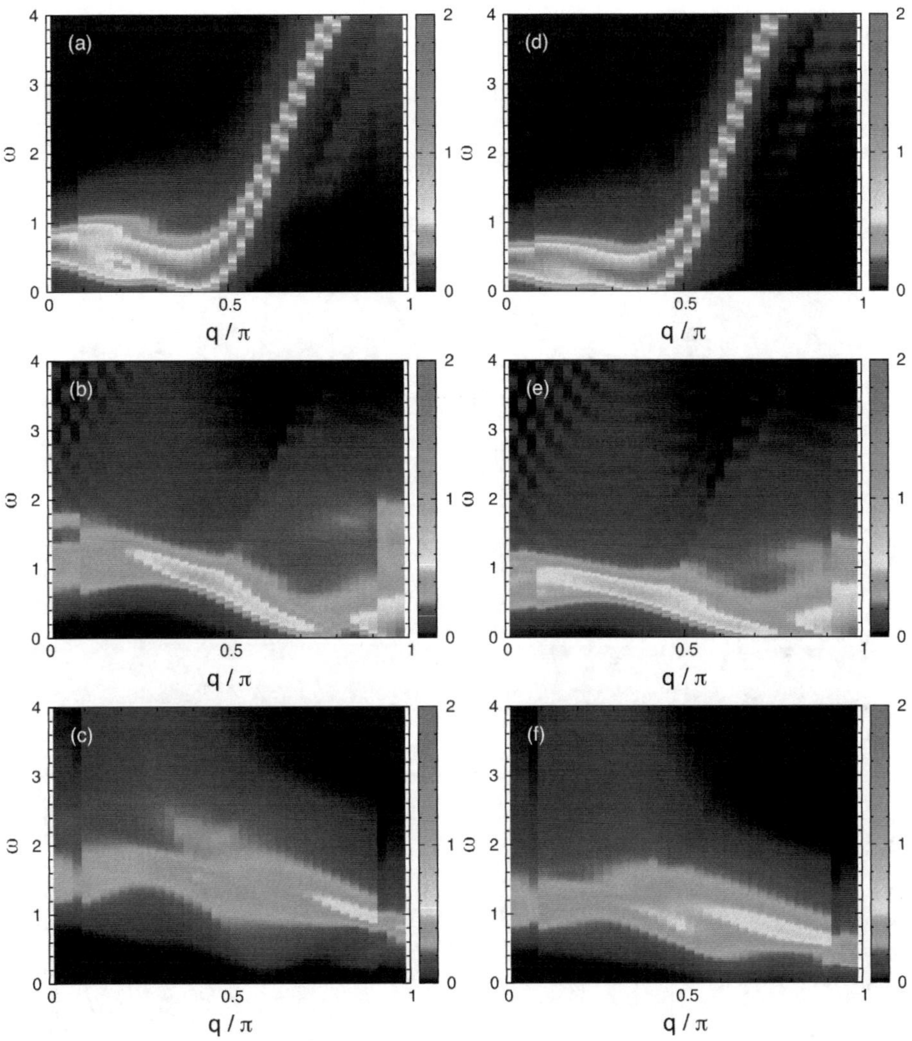

Figure 2. Intensity plots of the dynamical structure factors for two values of J_1 with fixed $m = 0.35$ in the quadrupole regime. Left panels: (a) $S^-(q,\omega)$, (b) $Q^{--}(q,\omega)$, and (c) $O^{---}(q,\omega)$ for $J_1 = -2$. Right panels: (d) $S^-(q,\omega)$, (e) $Q^{--}(q,\omega)$, and (f) $O^{---}(q,\omega)$ for $J_1 = -2.4$. The system size is $N = 40$.

Figure 3 presents the dynamical structure factors for $J_1 = -3$ and -3.4 with fixed $m = 0.35$, where the ground state is in the octupole regime. For $S^-(q,\omega)$, we find gapped spin excitations. Note that the lowest-energy peak is found at a finite energy, but the gap is not clearly identified in the color plot due to large intensity near zero energy. With decreasing J_1, the dispersion in small momentum goes further down to the low-energy side, leading to a flat structure of the dispersion. For $Q^{--}(q,\omega)$, we find gapped quadrupole excitations. The momentum position of the gap is away from $q = \pi$. That is, dominant quadrupole correlations change from quasi-long-range antiferro-quadrupole correlations in the quadrupole liquid to short-range incommensurate quadrupole correlations in the octupole liquid. The intensity is concentrated in a low-energy dispersion, in contrast to the highly dispersive mode of $S^-(q,\omega)$. For $O^{---}(q,\omega)$, we find

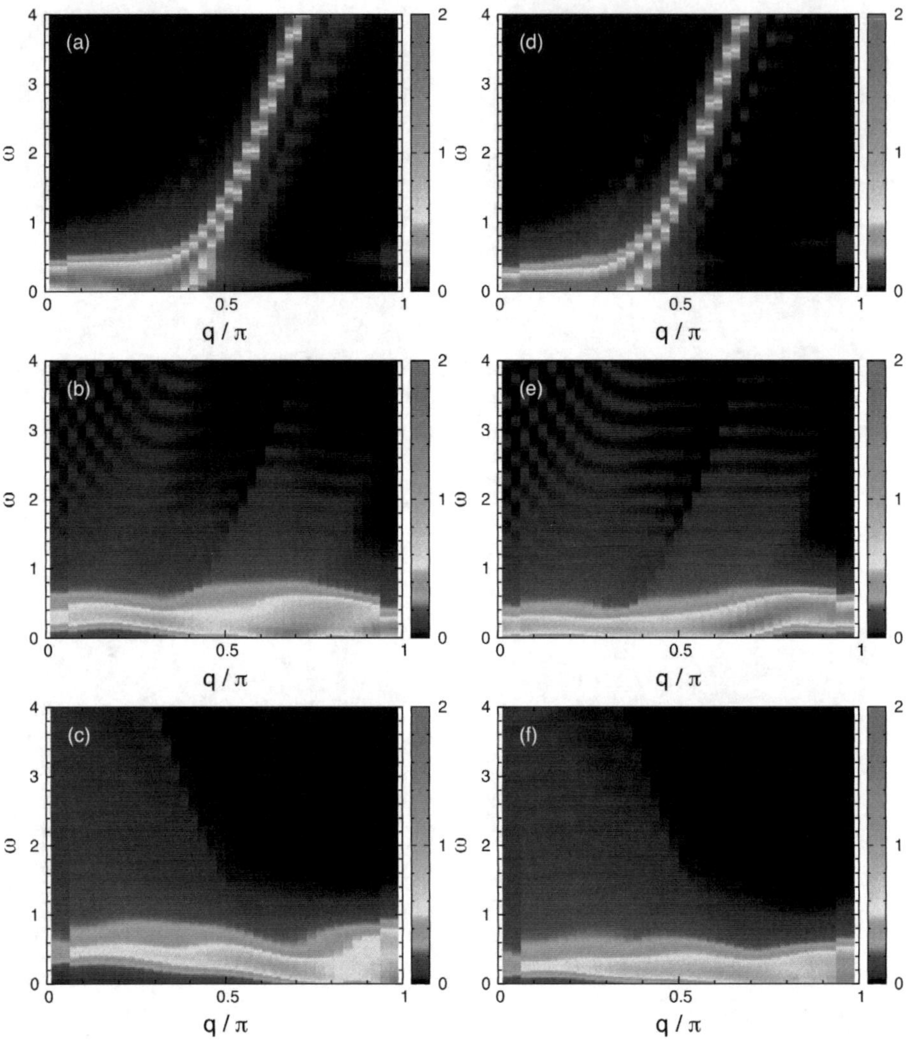

Figure 3. Intensity plots of the dynamical structure factors for two values of J_1 with fixed $m = 0.35$ in the octupole regime. Left panels: (a) $S^-(q,\omega)$, (b) $Q^{--}(q,\omega)$, and (c) $O^{---}(q,\omega)$ for $J_1 = -3$. Right panels: (d) $S^-(q,\omega)$, (e) $Q^{--}(q,\omega)$, and (f) $O^{---}(q,\omega)$ for $J_1 = -3.4$. The system size is $N = 40$.

gapless octupole excitations at $q = \pi$ due to quasi-long-range antiferro-octupole correlations. The intensity is concentrated in a low-energy dispersion in similar to $Q^{--}(q,\omega)$. Thus the slope of the low-energy dispersion becomes gentler than that in the quadrupole regime.

Now we move on to the transport property. In Fig. 4(a), we show typical results of the spin current correlation function I_s/N. Plots for various system sizes collapse, indicating that the finite-size effect is small. In Fig. 4(b), we present I_s/N at $J_1 = -2$ and -2.4 for the quadrupole regime, and at $J_1 = -3$ and -3.4 for the octupole regime. For the quadrupole regime, we observe that the spin current correlation is enhanced with decreasing J_1, implying that the system changes to be favorable for the flow of bound two-magnon clusters. With further decreasing J_1, we find that the spin current correlation in the octupole regime is much reduced

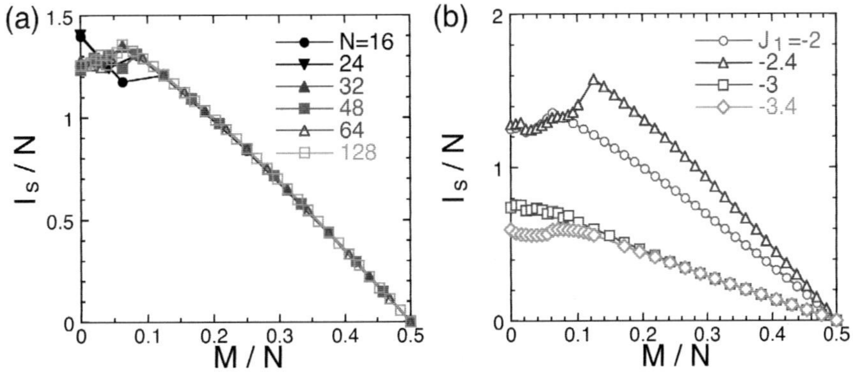

Figure 4. The spin current correlation function I_s/N (a) at $J_1 = -2$ with various system sizes, and (b) the dependence on J_1 with $N = 128$.

from that in the quadrupole regime, indicating that the spin transport is suppressed. This is contrary to a naive expectation, because bound three-magnon clusters carry a larger amount of angular momentum than bound two-magnon clusters. This behavior would be attributed to the dependences of the low-energy dispersion. That is, the slope of the dispersion, corresponding to the velocity of propagating magnon clusters, in the octupole regime is gentler than that in the quadrupole regime, leading to the reduction of the flow of the spin current.

4. Summary

We have studied magnetic and transport properties of the frustrated ferromagnetic chain by numerical methods. The spin current correlation in the octupole regime is reduced from that in the quadrupole regime, although bound three-magnon clusters carry a larger amount of angular momentum than bound two-magnon clusters.

Acknowledgments

This work was supported by JSPS KAKENHI Grant Number JP19K03678. Computations were done using supercomputers at the Japan Atomic Energy Agency and the Institute for Solid State Physics, the University of Tokyo.

References

[1] D. Hirobe, M. Sato, T. Kawamata, Y. Shiomi, K. Uchida, R. Iguchi, Y. Koike, S. Maekawa, and E. Saitoh, Nat. Phys. **13**, 30 (2017).
[2] B. Bertini, F. Heidrich-Meisner, C. Karrasch, T. Prosen, R. Steinigeweg, and M. Žnidarič, Rev. Mod. Phys. **93**, 025003 (2021).
[3] H. Onishi, J. Magn. Magn. Mater. **479**, 88-90 (2019).
[4] D. Hirobe, M. Sato, M. Hagihala, Y. Shiomi, T. Masuda, and E. Saitoh, Phys. Rev. Lett. **123**, 117202 (2019).
[5] A. Koga, T. Minakawa, Y. Murakami, and J. Nasu, J. Phys. Soc. Jpn. **89**, 033701 (2020).
[6] T. Minakawa, Y. Murakami, A. Koga, and J. Nasu, Phys. Rev. Lett. **125**, 047204 (2020).
[7] L. Kecke, T. Momoi, and A. Furusaki, Phys. Rev. B **76**, 060407(R) (2007).
[8] T. Hikihara, L. Kecke, T. Momoi, and A. Furusaki, Phys. Rev. B **78**, 144404 (2008).
[9] J. Sudan, A. Lüscher, and A. M. Läuchli, Phys. Rev. B **80**, 140402(R) (2009).
[10] H. Onishi, J. Phys. Soc. Jpn. **84**, 083702 (2015).
[11] H. Onishi, J. Phys.: Conf. Ser. **592**, 012109 (2015).
[12] H. Onishi, Physica B **536**, 346-349 (2018).
[13] S. R. White, Phys. Rev. Lett. **69**, 2863 (1992).
[14] E. Jeckelmann, Phys. Rev. B **66**, 045114 (2002).

Plasma propagation via radiation transfer in millimeter-wave discharge under subcritical condition

S Suzuki[1], C Kato[2], M Takahashi[1], and N Ohnishi[1]

[1]Department of Aerospace Engineering, Tohoku University, Sendai, 980-8579, Japan
[2]Department of Science and Technology, Tokyo University of Science, Noda, 278-8510, Japan

E-mail: soichiro.suzuki.r3@dc.tohoku.ac.jp

Abstract. An improved model was proposed to reduce a computational cost for subcritical millimeter-wave discharge. The proposed model was able to reproduce the plasma-front propagation via radiation transport as similar to the conventional model, and the plasma-front propagation speed was in agreement with the previous simulation. An electron transport effect by neutral fluid advection, which has been introduced into the conventional model, does not affect the propagation speed. By using the presented model, a computational time was reduced by 35%, which was suitable for a multi-dimensional simulation in the future.

1. Introduction

Experiments and numerical calculations have been carried out to study discharge phenomena caused by a high-power millimeter-wave irradiation [1–14]. A millimeter-wave discharge can be applied to a microwave rocket (Fig. 1) which can dramatically reduce a transportation cost to an outer space [3, 4]. On the microwave rocket, a high-power millimeter wave beam with the pulse width τ is irradiated from a ground facility to the rocket. A gas discharge is induced in the rocket nozzle when the incident beam is focused by a parabolic mirror because an electron absorbs the electromagnetic wave energy and ionization is kept. Finally, the discharge region elongates toward the beam source direction at the propagation speed u_{ion}, while forming a shock wave. A high-pressure region behind the shock wave gives the propulsive force to the microwave rocket. The plasma elongation distance $u_{\text{ion}}\tau$ can be an important parameter which determines a thrust efficiency because the past experiment using a model rocket indicated that a thrust performance is changed depending on $u_{\text{ion}}\tau/L$ [3]. Therefore, a prediction of propagation speed u_{ion} is necessary to design the microwave rocket.

The millimeter-wave discharge is classified into subcritical and overcritical conditions depending on a power density of the incident millimeter-wave beam. The microwave rocket can be driven under the subcritical condition because its slower propagation speed can give a higher thrust in comparison with the overcritical condition. A past simulation for the subcritical discharge by Takahashi *et al.* revealed that a plasma-front propagation was affected by an interaction between the neutral fluid and plasma [11]. Under the subcritical condition, They performed two-dimensional (2D) simulation by coupling the Euler equation which describes inviscid compressible fluid dynamics with the millimeter-wave discharge model to introduce the expansion effects of neutral fluid. Their simulation qualitatively reproduced a branching

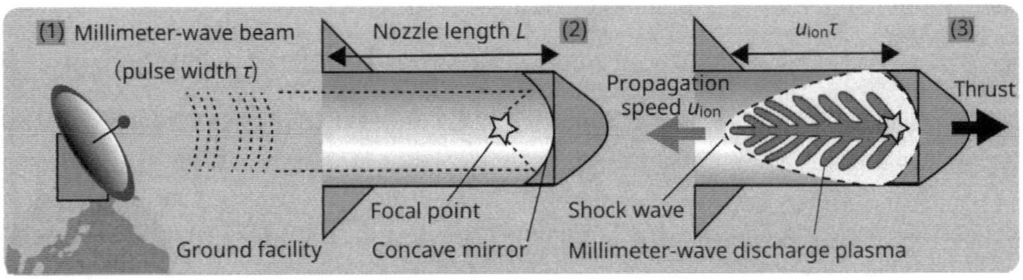

Figure 1. The microwave-rocket concept

structure which has been observed in the experiments [3], but a quantitative reproduction for the propagation speed and breakdown threshold was not obtained.

On the other hand, in our previous 1D simulation, detailed chemical reaction and radiation transfer modules were added to Takahashi's conventional model [11] to consider ionization processes ignored in the conventional model [14]. In our previous simulation, the breakdown process above 0.2 MV/m was successfully reproduced because the thermal ionization, associative ionization, and ionization via radiation transfer sustained a sufficient electron number density. Moreover, the propagation speed obtained by their model was in good agreement with the experimental results below 0.3 MV/m, but it was underestimated above 0.4 MV/m [1, 3, 6, 14]. It can be considered that ignoring a multi-dimensional effect causes a disagreement of the plasma front propagation speed above 0.4 MV/m; however, a computational cost of the multi-dimensional simulation must be huge, and cost reduction is required.

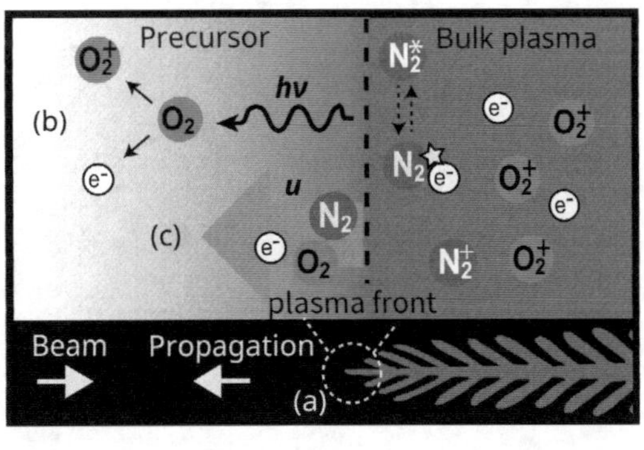

Figure 2. The schematics view of millimeter-wave plasma. Here, u is the flow velocity of neutral particles. (a) The head of branched plasma structure, which is called plasma front, is numerically reproduced in our the 1D simulation. (b) The ionization via radiation transfer, which was discussed in detail in our previous paper [14]. In this paper, we focus on (c) the electron transport via neutral fluid advection.

The electron transport via neutral fluid advection was considered in our previous simulation model, but there is a possibility that the effect of neutral fluid advection on the plasma front propagation speed is negligibly small, and computational cost can be reduced by simplifying the electron transport model. Here, in Fig. 2, the process (c) is the electron transport through the advection. Moreover, the explicit Euler method was used to integrate the electron' s diffusion equation in our conventional model, but implicit method can be effective to decrease the computational cost. Therefore, in this study, we propose the simplified transport model and change the temporal integration manner, aiming to the cost reduction for the multi-dimensional simulation in the future. The plasma structure and propagation speed, which are captured by the improved model, are compared with our conventional model.

2. Numerical model

Numerical models A and B are used in this paper to check the effect of electron transport via neutral fluid advection. Model A is the same as our previous simulation [14], which combines electromagnetic wave propagation, reaction-diffusion of the plasma, neutral gas dynamics, radiation transfer, and detailed chemical reaction processes. In model B, the modules of reaction-diffusion of plasma are improved.

The Maxwell's equations are solved by finite-difference time-domain (FDTD) method [15] to reproduce the electromagnetic wave propagations. The calculation procedure is the same as past simulations of millimeter-wave discharge [2, 11, 14].

The number density of electron n_e, positive ion n_p, and negative ion n_n are obtained by the following reaction diffusion equation:

$$\frac{\partial n}{\partial t}+\frac{\partial \Gamma}{\partial x}=S, \tag{1}$$

where n is the number density ($n=n_e,n_p,n_n$) and S is the source term. Here, S is calculated using the detailed chemical reaction solver [14]. Γ is the numerical flux, which is changed depending on model A or B as follows:

$$\Gamma=\begin{cases}-D\frac{\partial n}{\partial x}+un & \text{in model A,}\\ -D\frac{\partial n}{\partial x} & \text{in model B,}\end{cases} \tag{2}$$

where D is the diffusion coefficient, and the effective diffusion coefficient D_{eff} is used for electrons [2]. The diffusion coefficients of ions are assumed to be 0 [2]. un is the electron-transport term through neutral fluid advection. The effect of the advection on millimeter-wave plasma is investigated by comparing models A with B because the advection term is ignored in model B.

The transportation process of translational energy of electron $\varepsilon_e=3/2k_BT_e$ is calculated by the following reaction-diffusion equation:

$$\frac{\partial\left(n_e\varepsilon_e\right)}{\partial t}+\frac{\partial \Gamma_\varepsilon}{\partial x}=\frac{e^2n_e}{m_e}\frac{\nu_m}{\omega^2+\nu_m^2}E_{\text{rms}}^2+W_\varepsilon, \tag{3}$$

where k_B is the Boltzmann constant and T_e is the electron temperature, $\omega=2\pi f$ is the angular frequency of millimeter-wave with frequency f, E_{rms} is the root-mean-square (RMS) value of local electric field intensity that is obtained using the FDTD solver, ν_m is the elastic collision frequency, and W_ε is the translational energy of electron which is lost by all electron impact processes. ν_m, W_ε, and other transport coefficients are calculated using the Boltzmann equation solver Bolsig+ [16]. Γ_ε is the energy flux given by:

$$\Gamma_\varepsilon=\begin{cases}-\frac{5}{3}D_{\text{eff}}\frac{\partial(n_e\varepsilon_e)}{\partial x}+u\left(n_e\varepsilon_e\right) & \text{in model A,}\\ -\frac{5}{3}D_{\text{eff}}\frac{\partial(n_e\varepsilon_e)}{\partial x} & \text{in model B.}\end{cases} \tag{4}$$

The time integration of Eqs. (1) and (3) are performed using the explicit Euler method using a time step $\frac{1}{10f}$ to maintain an accuracy in model A. On the other hand, in model B, they are discretized using the Crank-Nicolson method using the time step $\frac{1}{f}$ to reduce the total computational cost. A decrease in calculation loops effectively reduces calculation cost because the transport coefficients are updated in the time step of Eqs. (1) and (3). The numerical flux in model A is calculated using the Scharfetter-Gummel scheme [17].

Journal of Physics: Conference Series **2207** (2022) 012046 doi:10.1088/1742-6596/2207/1/012046

The following Euler equation is used to consider the effect of the expansion of neutral fluid and shock waves driven by a gas heating effect:

$$\frac{\partial}{\partial t}\begin{bmatrix}\rho \\ \rho u \\ \varepsilon\end{bmatrix}+\frac{\partial}{\partial x}\begin{bmatrix}\rho u \\ \rho u^2+p \\ (\varepsilon+p)u\end{bmatrix}=\begin{bmatrix}0 \\ 0 \\ W_g\end{bmatrix}, \tag{5}$$

where ρ is the mass density of the neutral gas, ε is the total energy per unit mass, and W_g is the heat source evaluated by the reaction rate and reaction heat for all chemical reactions, which is calculated using the detailed chemical reaction solver. The time integration of Eq. (5) is performed using the explicit Euler method, and the equation is spatially discretized using the cell-centered finite volume method. The AUSM-DV scheme and the MUSCL interpolation are used to evaluate the numerical flux [18, 19].

The detailed chemical reaction process is evaluated using the zero-dimensional plasma kinetics solver ZDPlasKin in this study [20]. The ZDplasKin can generate a subroutine that solves rate equations for each species by inputting the data file of the chemical species, the chemical equation, and the corresponding rate coefficients. The reaction system of 53 species and 337 reactions are used in this study, which is similar to those of our previous study [14].

The ionization via radiation transfer is also considered. The calculation method is the same as that discussed in our conventional model [14].

3. Calculation Condition

The 1D calculation is performed in 300 cells. The cell size is $\lambda/100$, where λ is the wavelength of millimeter-wave. The incident millimeter-wave frequency is set to 170 GHz ($\lambda = 1.76$ mm). The electric field intensity of incident millimeter-wave is 0.2 MV/m (10.6 kW/cm^2). These values are determined by the past experiments [3]. The millimeter-wave is irradiated from the left boundary (x=0 mm). Other calculation conditions are detailed in [14]. The simulation code is parallelized using the message passing interface (MPI). The calculations were conducted using Intel(R) Xeon(R) X5670 4 cores in Mac Pro (2010 model).

4. Results and Discussion

Numerical calculations using models A and B reproduced the discharge structure under the subcritical condition in which Takahashi's model could not reproduce breakdown. The propagation mechanism is explained using the result of model B because the propagation mechanism is the same for models A and B.

Figure 3 shows the electron number density n_e, the gas temperature T_g (Fig. 3(a)), and the reaction rate of principal ionization processes (Fig. 3(b)). The plasma front is formed at x/λ=1.6, and Fig. 3(a) shows that the precursor electron with $n_e = 10^{11}$–10^{13} cm^{-3} distributes in $0.6<x/\lambda<1.6$. The most dominant ionization process is the photoionization as shown in Fig. 3(b); therefore, the precursor electron is generated by the photoionization process. The precursor electron heats neutral gas and increases T_g to 8,000 K by the vibrational-translational (V-T) relaxation process. Subsequently, the thermal ionization process becomes dominant (at x/λ=1.6 in Fig. 3(b)) because the high gas temperature promotes the thermal ionization process. Such high gas temperature also promotes the dissociation process. Therefore, the associative ionization process is dominant at x/λ=1.7 in Fig. 3(b). Because of the weak electric field intensity, the electron impact ionization is not enhanced as shown in Fig. 3(b)), unlike previous studies which focus on the condition of stronger electric field intensity [2, 11]. Furthermore, the plasma-front propagation via radiation transfer did not occur in the previous studies [2, 11]. These results indicate that the radiation transfer, thermal ionization, and associative ionization process play an important role on the subcritical discharge with low electric field intensity.

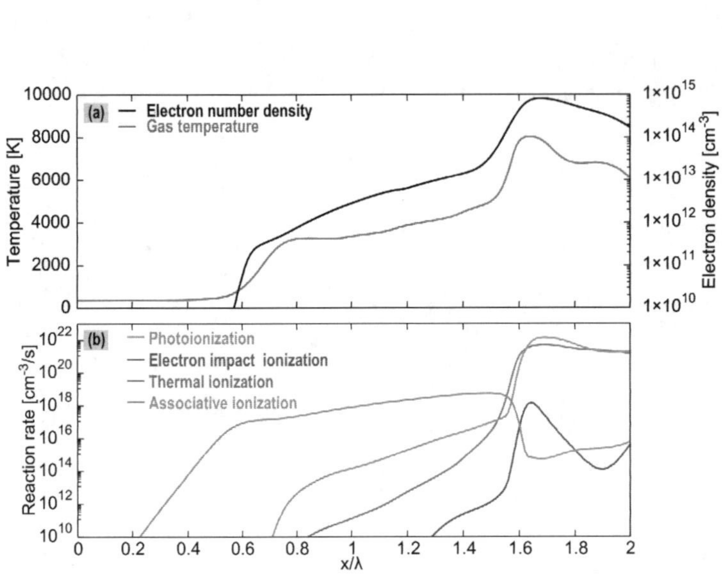

Figure 3. The plasma structures on millimeter-wave discharge: (a) the electron number density n_e and the gas temperature T_g; (b) the reaction rate of principal ionization processes such as photoionization, thermal ionization, associative ionization, and electron impact ionization.

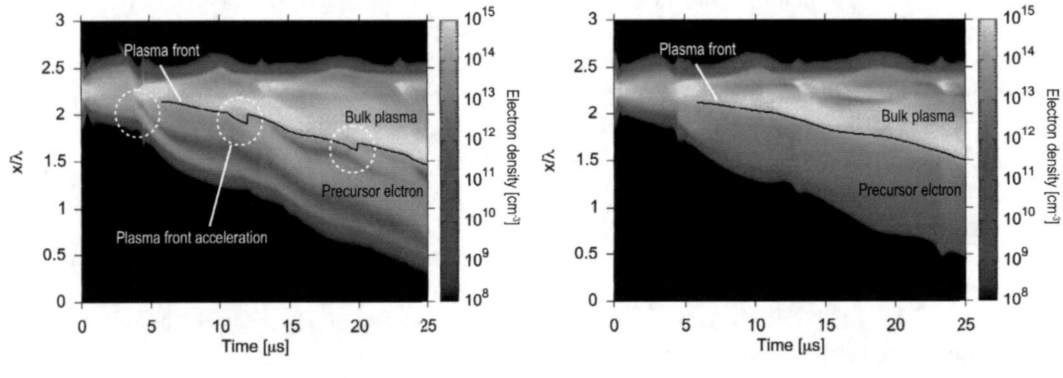

(a) model A

(b) model B

Figure 4. The discharge structures obtained using two models. The color maps show the electron number density in the x-t space and black lines show the position of the plasma front.

Here, we conduct a comparison between models A and B. Figure 4 is the time evolution of the electron number density in models A and B. The millimeter-wave beam is irradiated from $x/\lambda = 0$, which is bottom of Fig. 4. In both models, the precursor electron region is formed ahead of the plasma front (underside of black lines). However, an acceleration of plasma front occurs at $t = 11$ and 19 μs in Fig. 4 (a) when the shock waves pass through the plasma front because an advection of neural fluid carries the electron from the bulk to precursor regions. This phenomenon does not occur in model B as shown in Fig. 4 (b) because the advection term is ignored in model B as in Eq. (2). Table 1 shows plasma-front propagation velocity u_{ion} and computational time required to calculate up to 25 μs in models A and B. The u_{ion}, which is evaluated as the slope of the black line obtained using the least-squares method are 63 and 58 m/s

Journal of Physics: Conference Series **2207** (2022) 012046 doi:10.1088/1742-6596/2207/1/012046

Table 1. Comparison between models A and B

Model	u_{ion} [m/s]	Computational time [days]
A	63	4.9
B	58	3.2

in models A and B, respectively. From Table 1, it is found that the electron transport through neutral fluid advection does not affect the plasma-propagation speed significantly. While the change in u_{ion} is negligibly small between models, model B can reduce the computational time by 35%; therefore, model B is suitable for multi-dimensional calculations.

5. Conclusion

The simulation model was improved to reduce the computational cost of subcritical millimeter-wave discharge simulation. The explicit Euler method was used to solve the diffusion equation with the advection term. The integration was repeated ten times during the one-wave period, which caused the slower computational time. However, in the improved model, the Crank-Nicolson method was used to integrate the diffusion equations and the advection term was ignored. In both models, the plasma-front propagation was reproduced because of the combination between radiation transfer, thermal ionization, and associative ionization. Furthermore, a comparison of both models indicated that the electron transport through neutral fluid advection was not important because the propagation speed did not change significantly. The improved model was able to reduce computational time by 35% than the conventional model. By using the proposed model, multi-dimensional calculations which require a large amount of calculation are possible in the future.

References

[1] Bogatov N A, Bykov Y, Venediktov N P, Golubev S, Zorin V G, Eremeev A and Semenov V E 1986 *Fizika Plazmy* **12** 725–732
[2] Boeuf J P, Chaudhury B and Zhu G Q 2010 *Physical Review Letters* **104** 015002
[3] Oda Y, Komurasaki K, Takahashi K, Kasugai A and Sakamoto K 2006 *Journal of Applied Physics* **100** 113307
[4] Fukunari M, Arnault A, Yamaguchi T and Komurasaki K 2014 *Applied Optics* **53** 17972
[5] Fukunari M, Tanaka S, Shinbayashi R, Yamaguchi Y, Tatematsu Y and Saito T 2019 *Scientific Reports* **9** 17972
[6] Tabata K, Harada Y, Nakamura Y, Komurasaki K, Koizumi H, Kariya T and Minami R 2020 *Journal of Applied Physics* **127** 063301
[7] Artem'ev K V, Batanov G M, Berezhetskaya N K, Borzosekov V D, Davydov A M, Konchekov E M, Kossyi I A, Sarksyan K A, Stepakhin V D and Kharchev N K 2021 *Plasma Physics Reports* **47** 86–91
[8] Shimamura K, Yamasaki J, Miyawaki K, Minami R, Kariya T, Yang J and Yokota S 2021 *Physics of Plasmas* **28** 033505
[9] Nakamura Y, Komurasaki K, Fukunari M and Koizumi H 2018 *Journal of Applied Physics* **124** 033303
[10] Nakamura Y and Komurasaki K 2020 *Plasma Sources Science and Technology* **29** 105017
[11] Takahashi M, Kageyama Y and Ohnishi N 2017 *AIP Advances* **7** 055206
[12] Takahashi M and Komurasaki K 2018 *Advances in Physics: X* **3** 1417744
[13] Takahashi M and Ohnishi N 2019 *Journal of Applied Physics* **125** 163303
[14] Suzuki S, Hamasaki K, Takahashi M, Kato C and Ohnishi N *Physics of Plasmas (To be submitted)*
[15] Kunz K S and Luebbers R J 1983 *The Finite Difference Time Domain Method for Electromagnetics* (CRC Press)
[16] Hagelaar G J M and Pitchford L C 2005 *Plasma Sources Science and Technology* **14** 722–733
[17] Scharfetter D L and Gummel H K 1969 *IEEE Transactions on Electron Devices* **64** 64–77
[18] van Leer B 1979 *Journal of Computational Physics* **32** 101–136
[19] Wada Y and Liou M S 1994 *AIAA Paper 1994-83*
[20] Pancheshnyi S, Eismann B, Hagelaar G and Pitchford L *Computer code ZDPlasKin* http://www.zdplaskin.laplace.univ-tlse.fr (University of Toulouse, LAPLACE, CNRS-UPS-INP, Toulouse, France, 2008)

Journal of Physics: Conference Series **2207** (2022) 012047 doi:10.1088/1742-6596/2207/1/012047

Coupling Simulation on Two-dimensional Axisymmetric Beaming Propulsion System

Masayuki Takahashi

Department of Aerospace Engineering, Tohoku University, Sendai 980-8579, Japan

E-mail: masayuki.takahashi.c8@tohoku.ac.jp

Abstract.
A two-dimensional (2D) axisymmetric fluid modeling for plasma dynamics was coupled with an electromagnetic wave propagation to capture the electromagnetic wave and plasma structures on the microwave-driven in-tube accelerator (MITA) concept. The electromagnetic wave injected into a waveguide was focused on at the vehicle's rear side through reflection processes by the front mirror and the waveguide. An overcritical plasma was created at the high-intensity electric field region because of enhancement of an electron-impact ionization. The incident microwave was reflected by the plasma with overcritical density and a standing wave was induced in front of the dense plasma because the reflected wave overlapped with the incident microwave. In addition to wave enhancement by standing process, the strong electric field region was obtained at edges of the overcritical plasma because of wave diffraction, which can affect a shock wave strength and thrust performance of the MITA.

1. Introduction

Concept of beaming propulsion has been proposed to reduce a transportation cost to the outer space and promote a space usage in the fields of science, business, and national defence. On the air-breakdown type beaming rocket, an intense pulse beam is irradiated from a ground-based oscillator to a flying vehicle. An electric field concentration occurs inside the rocket nozzle when the incident beam is focused on by using a parabolic mirror putted on the vehicle, which induces gas breakdown and dense plasma. A strong shock wave is then generated from the focal point of the mirror because an energy of the high density plasma is transferred to neutral particles. The beaming rocket can obtain an impulsive thrust when a positive pressure of the shock wave interacts with a vehicle surface. A launch cost can be saved because a fuel for obtaining the thrust is not equipped on the vehicle. Moreover, a reusability of the vehicle can be increased because a complex gas combustion engine is not required to acquire a continuous thrust, which can contribute to cost reduction.

On the beaming rocket, microwave rockets are especially attracting attention for achieving a drastic decrease of the launch cost because a construction cost of high-power microwave oscillator can be lower than that of laser beam [1–15]. The intense microwave beam is used to generate the propulsion force; however, a beam power per unit area tends to be lower than that in laser beam when a long-distance beam transmission of the microwave is conducted because of its larger divergence angle. Therefore, it is necessary to increase a size of the incident beam receiver (mirror) on the vehicle to capture the microwave and induce a stronger shock wave, which means that a thrust-mass ratio of the rocket is decreased. Furthermore, the past experiments for the microwave rocket described that the thrust performance became lower with time when the repetitive pulses beam was irradiated to the rocket because the plasma, which is generated by the previous pulse, did not disappear and plasma screening for the incident electromagnetic wave occurred inside the rocket nozzle [2, 3]. Additionally, in free flight, a flight stability of the microwave rocket cannot be maintained as simulated in the previous paper [12]. These characteristics of the microwave rocket prevent system establishment and growth of this field.

The microwave in-tube accelerator (MITA) concept was proposed by us to avoid the divergence of microwaves, decrease in thrust performance during operation with the repetitive pulses, and in-air flight instability [13]. Figure 1 describes the schematic view of the MITA. In the MITA concept, the vehicle is installed in a waveguide, and the intense microwave beam is transmitted inside the waveguide to avoid the beam divergence. The microwave propagates from the front to rear sides of the vehicle, and the gas breakdown is induced at the rear side owing to beam-focusing process. A strong shock wave is then obtained at the focal point of the rear side, which gives a propulsive force to the vehicle and accelerates the vehicle toward the beam source. The plasma screening for the next microwave pulse does not occur because the dense plasma is exhausted to the rear side of the vehicle. Moreover, the flight control is not required on this system because rotational and non-axial direction motions of the vehicle are suppressed by the waveguide.

Simulations for the electromagnetic wave propagation around the vehicle were conducted in the previous study for the MITA [13], but the plasma structure and interaction with the electromagnetic wave were not discussed. Therefore, in this study, by developing a coupling code between the plasma and electromagnetic wave behaviors, a two-dimensional (2D) axisymmetric simulation is done to capture the electromagnetic wave and plasma structures in the MITA.

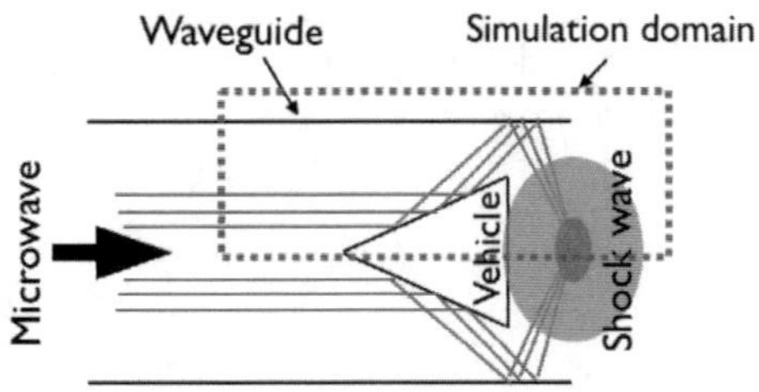

Figure 1. Schematic view of the MITA. The simulation domain in this study is shown by the blue dashed line.

2. Numerical methods

2.1. Axisymmetric electromagnetic wave propagation

In simulation of the MITA, the 2D axisymmetric model for the electromagnetic wave propagation is coupled with transport of the plasma because the waveguide and vehicle have an axisymmetric geometry. There is a possibility that the electromagnetic wave with 3D mode is used in an actual launch system, but only the 2D axisymmetric wave mode was treated in this paper for simplification. The following 2D axisymmetric Maxwell's equations are numerically solved using the finite-difference time-domain (FDTD) method to reproduce propagation of the electromagnetic wave around the vehicle:

$$\nabla \times \mathbf{E} = -\frac{\partial \mathbf{B}}{\partial t}, \tag{1}$$

$$\nabla \times \mathbf{H} = \epsilon_0 \frac{\partial \mathbf{E}}{\partial t} + \mathbf{J}, \tag{2}$$

where $\mathbf{B}$ is the magnetic flux density, $\mathbf{H}$ is the magnetic field, $\mathbf{E}$ is the electric field, t is the time, ϵ_0 is the electric permittivity in vacuum, and $\mathbf{J}$ is the current density. Here, $\mathbf{B} = \mu\mathbf{H}$ with the magnetic permeability μ. The vector of $\mathbf{E}$ has the r- and z-components and the vector of $\mathbf{B}$ has the θ-component, where r, z, and θ are the radial, axial, and azimuthal directions. The discretization form for the axisymmetric Maxwell's equations is shown as follows:

$$H^{n+1}_{\theta,i,j} = H^{n}_{\theta,i,j} + \frac{\Delta t}{\mu_0}\left\{-\frac{E^{n+1/2}_{r,i+1/2,j} - E^{n+1/2}_{r,i-1/2,j}}{\Delta z} + \frac{E^{n+1/2}_{r,i,j+1/2} - E^{n+1/2}_{r,i,j-1/2}}{\Delta r}\right\}, \quad (3)$$

$$E^{n+1/2}_{r,i+1/2,j} = E^{n-1/2}_{r,i+1/2,j} - \frac{\Delta t}{\epsilon_0 \Delta z}\{H^{n}_{\theta,i+1,j} - H^{n}_{\theta,i,j}\} - \frac{\Delta t}{\epsilon_0} J^{n-1/2}_{r,i+1/2,j}, \quad (4)$$

$$E^{n+1/2}_{z,i,j+1/2} = E^{n-1/2}_{z,i,j+1/2} + \frac{2\Delta t}{\epsilon_0 (r_{i,j+1} + r_{i,j})\Delta r}\{H^{n}_{\theta,i,j+1} - H^{n}_{\theta,i,j}\} - \frac{\Delta t}{\epsilon_0} J^{n-1/2}_{z,i,j+1/2}, \quad (5)$$

where $H^{n+1}_{\theta,i,j}$ is the magnetic field component in the θ-direction at the grid index $(i,\ j)$ and time step $n+1$, $E^{n+1/2}_{r,i+1/2,j}$ is the electric field component in the r-direction at the grid index $(i+1/2,\ j)$ and time step $n+1/2$, $E^{n+1/2}_{z,i,j+1/2}$ is the electric field component in the z-direction at the grid index $(i,\ j+1/2)$ and time step $n+1/2$, $J^{n-1/2}_{r,i+1/2,j}$ is the current density component in the r-direction at the grid index $(i+1/2,\ j)$ and time step $n-1/2$, $J^{n-1/2}_{z,i,j+1/2}$ is the current density component in the z-direction at the grid index $(i,\ j+1/2)$ and time step $n-1/2$, $r_{i,j}$ is the r-directing position, and Δz and Δr are the grid interval in the z and r-directions (Fig. 2). An interaction between the electromagnetic wave and dense plasma can be considered by feedback of the plasma motion via $\mathbf{J}$ in Eq. (2). Here, the current density $\mathbf{J}$ is evaluated based on the electron's bulk velocity $\mathbf{u}$, which is obtained from the fluid model of plasma transport. This coupling manner is based on the past numerical simulations for microwave breakdown [11, 14–20]. The time step of the electromagnetic wave propagation is determined in terms of the Courant-Friedrichs-Lewy (CFL) condition for light speed.

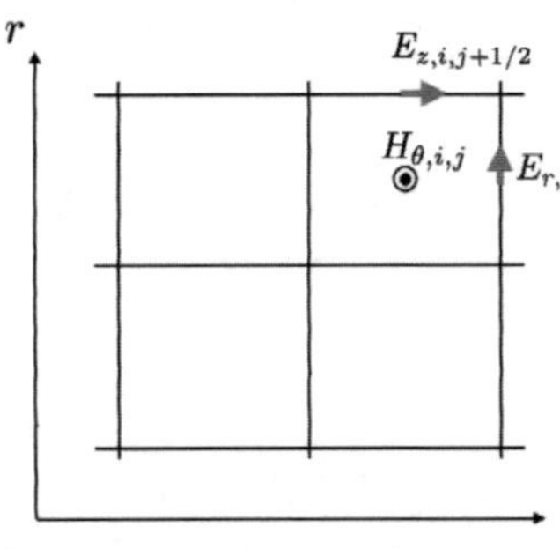

Figure 2. The grid index and definitions of H_θ, E_r, and E_z.

2.2. Fluid model of plasma transport

The fluid model is used to describe diffusion-reaction process of the plasma because the computational cost is lower than that of the particle description. A simple diffusion equation based on a quasi-neutral assumption is integrated as discussed in the past studies [11, 14, 15, 18–20], but the diffusion-reaction equation is changed to the 2D axisymmetric configuration in this paper. The 2D axisymmetric simulation is done in the r-z domain.

$$\frac{\partial n_e}{\partial t} - D_{\text{eff}} \left(\frac{\partial^2 n_e}{\partial z^2} + \frac{\partial^2 n_e}{\partial r^2} \right) - \frac{D_{\text{eff}}}{r} \frac{\partial n_e}{\partial r} = \nu_i n_e, \tag{6}$$

where n_e is the electron density, D_{eff} is the effective diffusion coefficient, and ν_i is the ionization frequency. The third term of left hand side $-\frac{D_{\text{eff}}}{r}\frac{\partial n_e}{\partial r}$ is the axisymmetric term. The effective diffusion coefficient is evaluated as $D_{\text{eff}} = (\alpha D_e + D_a)/(\alpha + 1)$, having the ambipolar diffusion coefficient $D_a = (\mu_i D_e + \mu_e D_i)/(\mu_i + \mu_e)$ and $\alpha = \tau_M \nu_i$. Here, $\tau_M = \epsilon_0/[en(\mu_i + \mu_e)]$ is Maxwell's relaxation time, where μ_e is the electron's mobility, μ_i is the ion's mobility, D_e is the electron's diffusion coefficient, and D_i is the ion's diffusion coefficient. The ambipolar diffusion coefficient is used in the bulk plasma region because a quasi-neutral condition is satisfied, but the diffusion coefficient is changed to free at the plasma front. Switching of the diffusion coefficient is realized using the parameter α. The transport coefficients of D_e and ν_i are evaluated using the data table made by us [14, 15]. The table of the transport coefficients was made by repeating the particle-in-cell simulation with collisional reactions The ion's diffusion coefficient D_i is set as 0 because the time scale of ion motion is much longer than that of electron motion. The time step for the electron's diffusion equation is set as one period of the incident wave, which is longer than that for the FDTD simulation. The Euler explicit method is used to integrate the 2D axisymmetric diffusion equation. The finite difference manner is used for the spatial discretization, and the second-order central difference method is used to evaluate the diffusion term of Eq. (6).

To feed back to Maxwell's equations, the electron's current density is evaluated as $\mathbf{J} = -en_e\mathbf{v_e}$ in the fluid modeling, where $\mathbf{v_e}$ is the bulk velocity of electron. The electron velocity is updated by integrating an equation of motion, which has the electric force and loss owing to elastic collision;

$$\frac{\partial \mathbf{v_e}}{\partial t} = -\frac{e\mathbf{E}}{m_e} - \nu_m \mathbf{v_e}, \tag{7}$$

where m_e is the mass of electron and ν_m is the elastic collision frequency. The elastic collision frequency ν_m is evaluated by the previous simulation with the fully kinetic model [14, 15]. The time step for the equation of motion is the same as for the electromagnetic wave simulation. The temporal integration for the equation of motion for electrons is conducted by utilizing the Euler explicit method. The current density $\mathbf{J} = -en_e\mathbf{v_e}$ is fed back to Maxwell's equation in order to capture interactions between the microwave beam and the dense plasma. The root-mean-square (RMS) electric field is locally evaluated by solving Maxwell's equation, which is used to obtain the local transport coefficients such as D_e, ν_i, and ν_m. A linear interpolation on the data map is used to calculate the transport coefficients.

3. Simulation condition

The 2D coupling simulation is conducted in background gas of pure nitrogen in order to examine the electromagnetic wave and plasma structures when the microwave is injected from the front side of the vehicle. The simulation domain is shown by the blue dashed line in Fig. 1. The microwave with axisymmetric TM_{01} mode is injected from the left to right boundaries. The microwave frequency is set as 110 GHz. This value follows Hidaka's breakdown experiment [21]. On an actual rocket, the subcritical wave (the beam intensity is lower than the breakdown threshold) can be selected on an actual rocket, but here, the overcritical microwave beam with the maximum wave amplitude of 10 MV/m is injected for speeding-up of coupling simulation. The pulse duration is set as 100 ns. The axial length of the simulation domain is 6.2 cm, which is sufficiently large to capture the microwave wavelength. The left boundary is set as Mur absorbing boundary [22]. The perfect matched layer (PML) boundary is used at the right boundary because the electromagnetic wave has a strong two-dimensionality at the rear side of the vehicle. A shape of the vehicle is same as the model used in the past experiment for laser propulsion [23]. The radius of the waveguide is 1.2 cm. The reflection mirror for microwave focusing is equipped on the vehicle's front side. The propagation direction of the incident microwave is changed by the front mirror, and the microwave is then focused at the rear side of the vehicle. An initial electron with a Gaussian distribution is set at the rear side of the vehicle to reproduce a gas breakdown process at the focal point. A background gas is pure nitrogen with a pressure of 1 atm.

4. Results

4.1. Electromagnetic wave propagation and plasma structure on the MITA

The breakdown process on the MITA is numerically captured by conducting the coupling simulation between the electromagnetic wave and plasma propagations (Fig. 3(a)). The direction of the beam's wave vector is changed because of an interaction between the TM_{01} microwave and the front mirror, while obtaining a beam focusing at the rear side. The electric field intensity becomes higher at the rear side of the vehicle because of focusing process. Here, there is a possibility that the axisymmetric mode of the incident microwave is changed into the 3D mode when reflection occurs on the mirror and waveguide surfaces, but mode transition is ignored in this simulation for simplification. It is considered that an outline of beam focusing process and wave-plasma interaction is captured by our simple modeling.

An electron number density is drastically increased at the vehicle's rear side because an electron-impact ionization process is enhanced owing to the high electric field (Fig. 3(b)). The incident microwave beam is reflected by the dense plasma because the electron number density reaches the cut-off density for wave reflection by repeating the electron-impact ionization. The low electric field region is then created inside the plasma because the microwave cannot penetrate into the overcritical plasma. A standing wave is induced in front of the overdense plasma because the incident microwave is reflected by the dense plasma and reflected wave overlaps with the incident wave. A mechanism for standing-wave generation is same as that captured in the past experiment and numerical simulation for the overcritical microwave breakdown [18, 21]. In addition to field enhancement by the standing wave, the electric field becomes higher at edges of the high density plasma ((r, z)=(0, 4) and (0, 5) cm) because the electromagnetic wave is diffracted by the electron with high number density. This wave diffraction does not occur in the case of laser propulsion because the wavelength of the incident laser beam is shorter than the scale length of bulk plasma. An enhancement by wave diffraction can increase a thrust performance of the vehicle because a strong shock wave can be induced, but it is necessary to consider a shock wave propagation to evaluate the thrust performance of the MITA, in addition to the electromagnetic wave and plasma propagations. A coupling simulation model will be updated to capture the shock wave dynamics and discuss mechanical performance of the MITA in the forthcoming paper.

5. Conclusion

A coupling simulation model was developed to reproduce electromagnetic wave and plasma structures on the microwave-driven in-tube accelerator (MITA) concept. An axisymmetric electromagnetic wave was focused on at the rear side of the vehicle through reflections on the front mirror and the waveguide wall. An overcritical plasma was created at the rear side of the vehicle because an intense electric field promoted an electron-impact ionization process, which reflected the incident wave. A standing wave was created in front of the dense plasma because the reflected wave overlapped with the incident wave. In addition to field enhancement by the standing wave, the high electric field region was induced at edges of the overcritical plasma owing to wave diffraction effect, which can affect a thrust performance of the MITA.

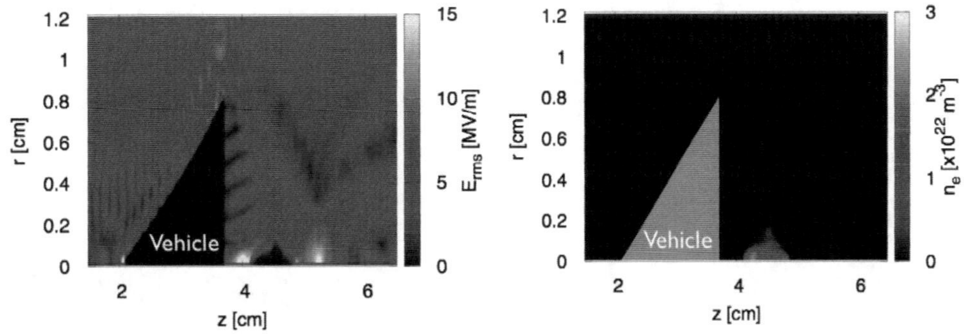

(a) RMS electric field (b) Electron number density

Figure 3. The root-mean-square (RMS) electric field and the electron number density on the MITA. The images were captured at 2 ns.

Acknowledgements

This work was supported by JSPS KAKENHI Grant Number JP20K14947. The simulations in this work were done on a FUJITSU PRIMEHPC FX100 at the Japan Aerospace Exploration Agency and a FUJITSU Server PRIMERGY CX2550M4 at the Advanced Fluid Information Research Center, Institute of Fluid Science, Tohoku University.

References

[1] T. Nakagawa, Y. Mihara, and K. Komurasaki, J. Spacecraft and Rockets **41** (1), 151–153 (2004).
[2] Y. Oda and K. Komurasaki, J. Appl. Phys. **100**, 113307 (2006).
[3] Y. Oda, Doctor Thesis (The University of Tokyo, Tokyo, 2008).
[4] M. Fukunari, R. Komatsu, A. Arnault, T. Yamaguchi, K. Komurasaki, and Y. Arakawa, Vacuum **88**, 155–159 (2013).
[5] M. Fukunari, A. Arnault, T. Yamaguchi, and K. Komurasaki, Appl. Optics **53** (31), 16–22 (2014).
[6] M. Takahashi and N. Ohnishi, Appl. Phys. Lett. **105**, 223504 (2014).
[7] M. Takahashi and N. Ohnishi, J. Appl. Phys. **124**, 173301 (2018).
[8] M. Takahashi and N. Ohnishi, J. Appl. Phys **125**, 163303 (2019).
[9] M. Takahashi and N. Ohnishi, J. Prop. Power **34**, 762–771 (2018).
[10] M. Takahashi and N. Ohnishi, J. Appl. Phys. **120**, 063303 (2016).
[11] M. Takahashi and N. Ohnishi, AIP Advances **7**, 055206 (2017).
[12] M. Takahashi and N. Ohnishi, AIAA Journal **50**, 2600-2608 (2011).
[13] M. Takahashi and N. Ohnishi, AIAA Paper 2019-1242 (2019).
[14] M. Takahashi and N. Ohnishi, J. Phys. Conf. Series **905** 012024 (2017).
[15] M. Takahashi and N. Ohnishi, Jpn. J. Appl. Phys. **55**, 07LD02 (2016).
[16] K. S. Nam and P. J. Verboncoeur, Phys. Rev. Lett. **103**, 055004 (2009).
[17] Q. Zhou and Z. Dong, Appl. Phys. Lett. **98**, 161504 (2011).
[18] J. P. Boeuf, B. Chaudhury, and G. Q. Zhu, Phys. Rev. Lett. **104**, 015002 (2010).
[19] B. Chaudhury, J. P. Boeuf, and G. Q. Zhu, Phys. Plasmas **17**, 123505 (2010).
[20] K. Kourtzanidis, J. P. Boeuf, and F. Rogier, Phys. Plasmas **21**, 123513 (2014).
[21] Y. Hidaka, E. M. Choi, I. Mastovsky, M. A. Shapiro, J. R. Sirigiri, and R. J. Temkin, Phys. Plasmas **16**, 055702 (2009).
[22] G. Mur, IEEE Trans. EMC **EMC-23** (4), 377–382 (1981).
[23] A. Sasoh, Rev. Sci. instruments **72**, 1893–1898 (2001).

Journal of Physics: Conference Series **2207** (2022) 012048 doi:10.1088/1742-6596/2207/1/012048

3D hydrodynamics simulations of massive stars with the PROMPI code

F Rizzuti[1], R Hirschi[1,2], C Georgy[3], W D Arnett[4], C Meakin[4,5], M Mocák[1]

[1] Astrophysics Group, Lennard-Jones Laboratories, Keele University, Keele ST5 5BG, UK
[2] Kavli IPMU (WPI), University of Tokyo, 5-1-5 Kashiwanoha, Kashiwa 277-8583, Japan
[3] Geneva Observatory, Geneva University, CH-1290 Sauverny, Switzerland
[4] Steward Observatory, University of Arizona, 933 N. Cherry Avenue, Tucson AZ 85721, USA
[5] Pasadena Consulting Group, 1075 N Mar Vista Ave, Pasadena, CA 91104 USA

E-mail: f.rizzuti@keele.ac.uk

Abstract. Evolutionary models of massive stars are affected by uncertainties related to three-dimensional processes like convection. However, it is not computationally possible to simulate at the same time both the entire lifetime of a star and its fluid motions. 3D hydrodynamics simulations can be employed to study the fluid behaviour in very high detail for a short time-scale, which can be used to improve 1D models. The code `PROMPI` is a multi-dimensional hydrodynamics software developed specifically for simulations of the late phases in massive star interiors. In these proceedings, we present results from a new `PROMPI` simulation of the neon-burning shell in a 15 $M_{\odot}$ star, focusing on the interplay between convective turbulence and nuclear reactions. In particular, we show how it is possible to use the chemical composition to study the time evolution of the convective zone, which is subjected to turbulent entrainment.

1. Introduction

Stellar evolution models in one dimension are an excellent tool to simulate the life of a full star for an extended period of time, possibly from its birth (zero-age main sequence, ZAMS) to the final instants before its death. Nevertheless, due to their simplicity and effectiveness, these models largely rely on assumptions and prescriptions to reproduce the most important physical processes that take place in the star. These prescriptions are usually calibrated by comparing results from specific 1D or multi-D simulations to observational data. One distinguished example is the mixing length theory (MLT) by [1], extensively used in 1D models to reproduce the fluid mixing in convective regions of a star.

In this context, it is essential to have a good understanding of such complex phenomena, like convection, rotation and magnetic activity. This can be obtained employing multi-dimensional models, where these processes can be naturally reproduced without assuming any prescription. Unfortunately, multi-D models require a computing power much larger than simpler 1D models, so they are restricted to short time-scales and limited spatial extensions. Therefore, it is the interplay between fine-tuning prescriptions from multi-D and their use in 1D that brings about progress in understanding stellar evolution.

We present here results from a new set of hydrodynamics models simulating convection in the neon-burning shell of a 15 $M_{\odot}$ star, obtained with the `PROMPI` code. Similar studies of oxygen-

Journal of Physics: Conference Series **2207** (2022) 012048 doi:10.1088/1742-6596/2207/1/012048

and carbon-burning shells have already been produced with the same code (see e.g. [2, 3, 4]). We focus here on the developing and evolution of the convective zone in the neon shell, with particular attention to the kinetics and chemical composition of the fluid.

2. Initial conditions and input physics

For these simulations, initial conditions were taken from a 1D stellar evolution simulation of a 15 $M_\odot$ star obtained with the GENEC code ([5]). To prepare our neon-shell initial model, radial profiles in 1D of fundamental variables (e.g. density, temperature, abundances) enclosing the second neon-burning shell were mapped to a cubic box of side 0.64×10^8 cm. The simulation then evolves from this initial setup.
Concerning the computational domain, we decided to run a set of four simulations with different numerical resolution. Four grids of $128^3, 256^3, 512^3$ and 1024^3 cells have been used, resulting in a cell length from $\sim$ 6 to 50×10^4 cm. We made this choice to investigate how strongly the results depend on the chosen resolution, and if physical quantities converge to a stable value with increasing resolution. A similar study has been presented in [6] for carbon shell.
Another parameter we included in our simulations is the so-called "boosting factor", i.e. a constant factor which multiplies the nuclear energy generation rate. There are many reasons for introducing this factor: to compare different burning layers equalizing their luminosity, to boost convection and make the computing time more accessible, and to study the dependence of key quantities on the energy driving the convection. Focusing on the latter, we produced a set of simulations with different boosting factors, namely $\times 1, 10, 100, 1000$ the nominal 1D case, in a similar way to what has been done in the work of [4] for carbon shell.
Naturally, changing the resolution and boosting factor has a strong impact on the computing power required to run the simulations. In particular, with a fixed computing budget we are able to simulate a longer time-range and a faster convection for models with lower resolution and larger boosting factor, respectively.
In these proceedings, we present results from the 512^3 run with boosting factor $\times 10$. This is one of most complete simulations in our set, since the resolution is high enough to study the variables in detail, and the boosting factor allows to follow the evolution of convection without requiring an excessively long physical time-scale.

3. Results

The 1D model used for initial conditions cannot reproduce the turbulent motions of the flow, but rely on prescriptions to define the extension of the convective zone (MLT, see Introduction). Therefore, in our 3D simulations there is an intermediate phase, which we call "initial transient", between the beginning of the simulation and when convection is fully developed, showing eddies and plumes. The length of the transient depends on the boosting factor, since a stronger nuclear energy generation results in faster convection. After this transient, the simulation is in a "quasi-steady" state and we can start analysing the results.
In Figure 1 we display a vertical cross-section across the domain with velocity magnitude in colour scale, taken toward the end of the simulation. The convective zone is clearly visible where the flow has higher speed and forms plumes and eddies. Above and below the convective zone, the stable region is dominated by low-speed internal gravity waves.
In a similar way, we show in Figure 2 the mass fraction in colour scale of the four key nuclides used for the nuclear burning network: ^{16}O, ^{20}Ne, ^{24}Mg, ^{28}Si. The turbulent fluid elements, enriched by the nuclear burning at the bottom of the cell, can be seen filling the convective zone. It is interesting to notice that for ^{16}O, ^{24}Mg and ^{28}Si the plumes are overabundant (toward the red) compared to the initial composition of the convective zone (white), while ^{20}Ne is underabundant (toward the blue). This is an effect of the nuclear network of our model, which is burning neon to produce oxygen and magnesium, and burning magnesium to produce silicon.

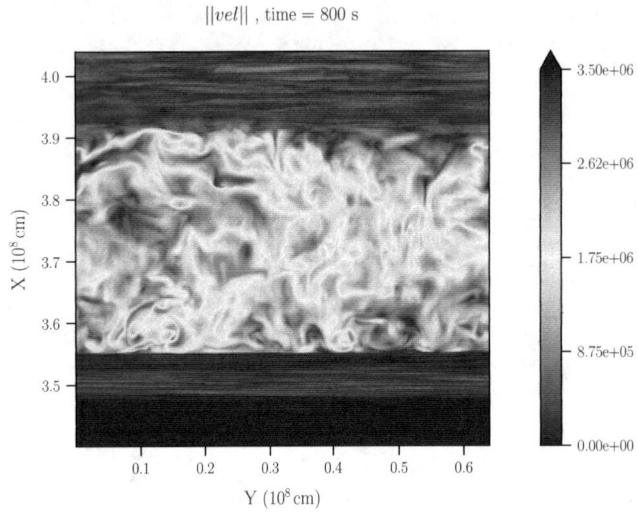

Figure 1. Vertical cross section of the velocity magnitude (values in colour scale, in cm s^{-1}) taken at 800 seconds from the beginning of the simulation. It is possible to see the plumes and eddies which characterize the convective zone.

The nuclides explicitly included in our simulations also provide a way of studying the time evolution of the convective zone. This is evident from Figure 3, where we show the horizontally-averaged mass fraction profiles of the four elements, at the beginning (dashed) and at the end (solid) of the simulation. The central plateaus, common to all the profiles, correspond to the convective zone, which is chemically well mixed. The locations where these plateaus are interrupted represent the edges of the convective zone, and they can be used to define the convective boundaries.

It is interesting to notice in Figure 3 that the upper boundary is shifting rightward with time, as indicated by all the four elements. This phenomenon is well-known in the literature (see [7, 8]) and is usually called "turbulent entrainment". It is a result of the shear mixing near the convective boundaries, so that the turbulent flow entrains some material from the stable region into the convective zone, resulting in more fluid becoming convective. In this way, the convective zone itself grows, both upward and downward, with time. Entrainment in stellar hydrodynamics simulations has already been studied with `PROMPI` for oxygen-burning ([2]) and carbon-burning ([4]) shells.

In our case, we study entrainment for the neon-burning shell in the following way. First, we define the boundaries of the convective zone as the location where the horizontally-averaged mean atomic mass is equal to the mid-point between the value in the convective and in the stable zone. Using this definition, we are able to follow how the convective boundary locations evolve with time, as shown in Figure 4. The first part of the simulation, up to 200 seconds, is the initial transient, when convection is not fully developed. After this time, the plot shows the convective zone growing both upward and, less rapidly, also downward, due to turbulent entrainment as previously explained. Entrainment in stars is expected from theory and previous hydrodynamics models. Our new simulations confirm these expectations.

4. Conclusions and outlook

In these proceedings, we presented a new 3D hydrodynamics simulation of the neon-burning shell in a 15 $M_\odot$ star, employing the `PROMPI` code. The simulation follows the development and evolution of a convective zone, which has been studied in terms of kinetics and chemical composition. The results presented show the turbulent motions of the fluid in the convective region and the time evolution of the chemical species, which fully reflects the nuclear network assumed in the model. Finally, it is shown that the mixing which takes place near the convective

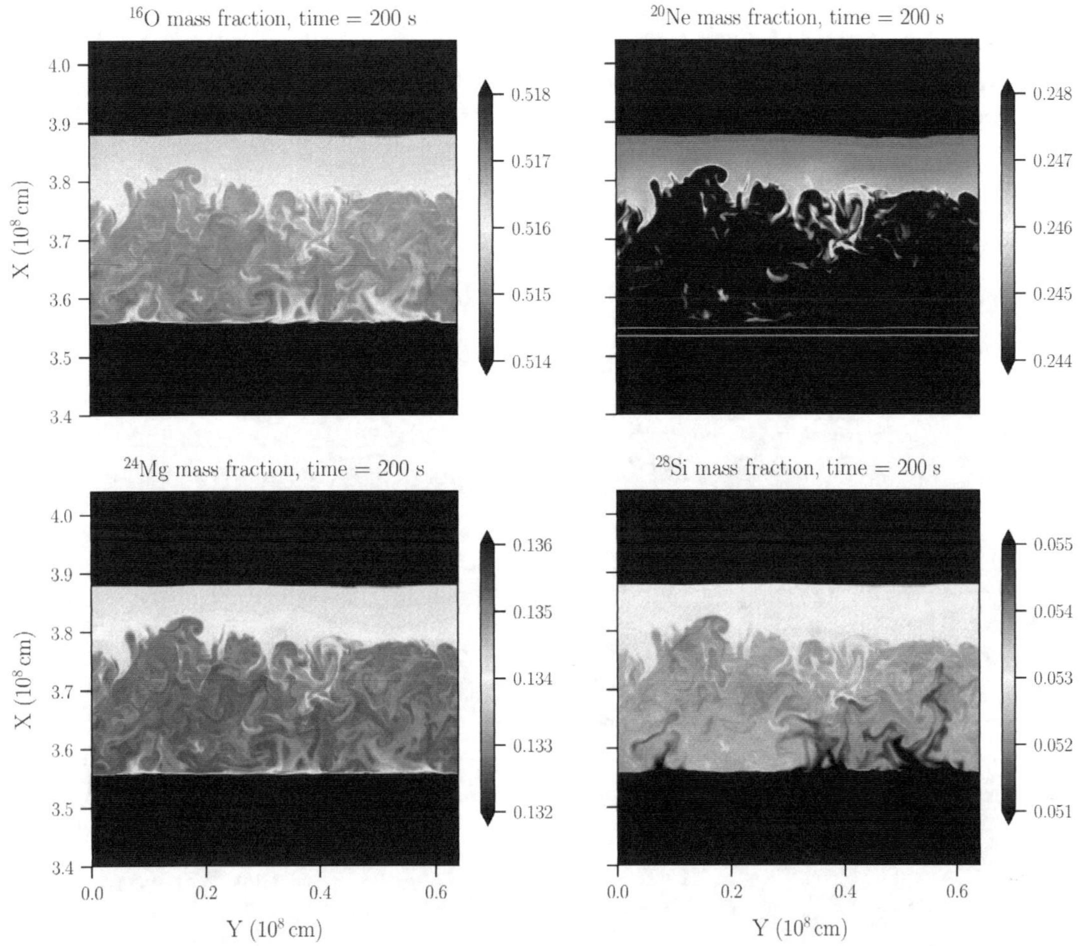

Figure 2. Vertical cross sections of the mass fraction (values in colour scale) of the different isotopic nuclei ^{16}O, ^{20}Ne, ^{24}Mg, ^{28}Si, taken at 200 seconds from the beginning of the simulation, during the initial transient.

boundaries entrains material for the surrounding stable regions, leading to a growth of the convective zone with time.

These preliminary results prepare the ground for a more extensive study that shall enrich our knowledge about convection in the neon-burning shell. The comparison between simulations with different resolution and boosting factor, as explained in the Introduction, shall allow us to evaluate the impact of these parameters on the physics of the simulations. The convergence of results with increasing resolution is expected, according to the similar study on a carbon-burning shell in [6]. Studying the dependence of certain quantities, like the entrainment rate, on the boosting factor shall enable us to produce a parametrized law with general validity, most useful for 1D stellar evolution models, which cannot reproduce multi-dimensional processes but rely on simplified assumptions. Different entrainment laws have already been obtained from hydrodynamics models in several studies (see e.g. [2, 4, 9, 10]): our complete set of neon-burning shell simulations shall expand and refine this repository.

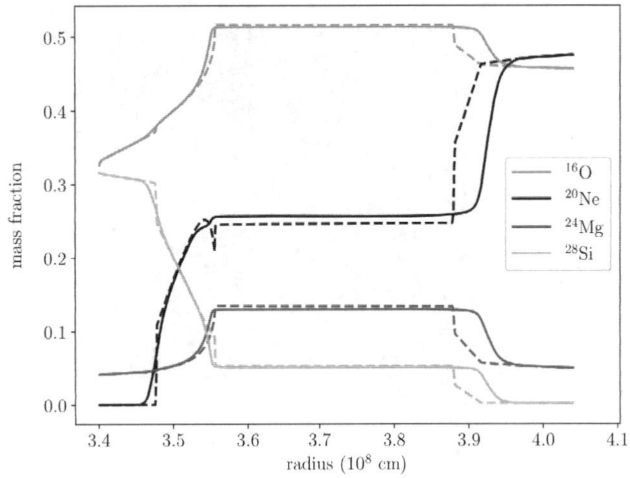

Figure 3. Radial profiles of the horizontally-averaged mass fractions for the four key isotopes, at the beginning of the simulation (dashed) and after 800 seconds (solid).

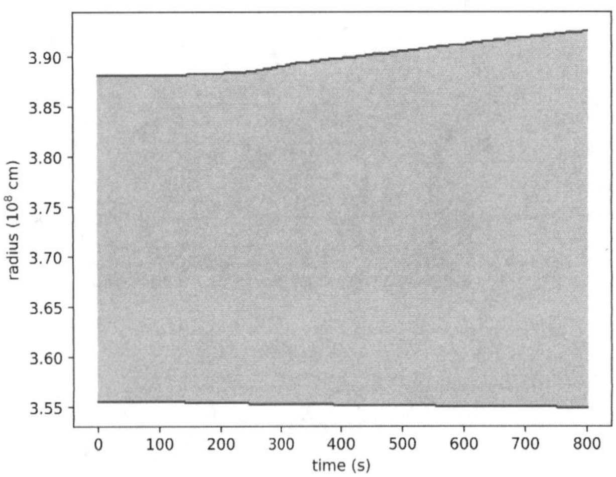

Figure 4. Time evolution of the upper and lower boundaries enclosing the convective zone (in magenta). After the initial transient, lasting about 200 seconds, the growth of the convective zone due to entrainment can be clearly seen.

Acknowledgments
RH acknowledges support from the World Premier International Research Centre Initiative (WPI Initiative), MEXT, Japan and the IReNA AccelNet Network of Networks, supported by the National Science Foundation under Grant No. OISE-1927130. CG has received funding from the European Research Council (ERC) under the European Union's Horizon 2020 research and innovation program (Grant Agreement No. 833925). WDA acknowledges support from the Theoretical Astrophysics Program (TAP) at the University of Arizona and Steward Observatory. This article is based upon work from the ChETEC COST Action (CA16117), supported by COST (European Cooperation in Science and Technology). This project has received funding from the European Union's Horizon 2020 research and innovation programme under grant agreement No. 101008324 (ChETEC-INFRA). The authors acknowledge PRACE for awarding access to the resource MareNostrum 4 based in Spain at Barcelona Supercomputing Center (project ID 2017174145). The support of David Vicente and Janko Strassburg from Barcelona Supercomputing Center, Spain, to the technical work is gratefully acknowledged. This work used the DiRAC@Durham facility managed by the Institute for Computational Cosmology on

behalf of the STFC DiRAC HPC Facility (www.dirac.ac.uk). The equipment was funded by BEIS capital funding via STFC capital grants ST/P002293/1 and ST/R002371/1, Durham University, and STFC operations grant ST/R000832/1. This work also used the DiRAC Data Centric system at Durham University, operated by the Institute for Computational Cosmology on behalf of the STFC DiRAC HPC Facility. This equipment was funded by BIS National E Infrastructure capital grant ST/ K00042X/1, STFC capital grants ST/H008519/1 and ST/ K00087X/1, STFC DiRAC Operations grant ST/K003267/1, and Durham University. DiRAC is part of the National E Infrastructure.

References

[1] Böhm-Vitense E 1958 *ZAp* **46** 108
[2] Meakin C A and Arnett D 2007 *ApJ* **667** 448
[3] Viallet M, Meakin C, Arnett D and Mocák M 2013 *ApJ* **769** 1
[4] Cristini A, Hirschi R, Meakin C, Arnett D, Georgy C and Walkington I 2019 *MNRAS* **484** 4645
[5] Eggenberger P, Meynet G, Maeder A, Hirschi R, Charbonnel C, Talon S and Ekström S 2008 *Ap&SS* **316** 43
[6] Cristini A, Meakin C, Hirschi R, Arnett D, Georgy C, Viallet M and Walkington I 2017 *MNRAS* **471** 279
[7] Maeder A 1976 *A&A* **47** 389
[8] Zahn J P 1991 *A&A* **252** 179
[9] Jones S, Andrassy R, Sandalski S, Davis A, Woodward P and Herwig F 2017 *MNRAS* **465** 2991
[10] Horst L, Hirschi R, Edelmann P V F, Andrassy R and Roepke F K 2021 *A&A* **653** A55

The formation of dark-matter-deficient galaxies through galaxy collisions

Koki Otaki and Masao Mori

Graduate School of Pure and Applied Sciences, University of Tsukuba, 1-1-1 Tennodai, Tsukuba, Ibaraki 305-8577 Japan

Center for Computational Sciences, University of Tsukuba, 1-1-1 Tennodai, Tsukuba, Ibaraki 305-8577 Japan

E-mail: otaki@ccs.tsukuba.ac.jp

Abstract. In the standard galaxy formation model, cold dark matter (CDM) drives the hierarchical structure formation. However, recently it is suggested that the satellite galaxy NGC1052-DF2, a member of the elliptical galaxy NGC1052 group, has a very little dark matter component compared to the theoretical predictions. 27 dark-matter-deficient galaxies have been discovered, but it is difficult to explain their formation process in the CDM model. We investigate the formation process through dark matter subhalo (DMSH) collisions using three-dimensional numerical simulations. As a result, we show the bifurcation sequence of the formation of dark-matter-dominated galaxies and dark-matter-deficient galaxies. In the case of low-speed collisions, a merger of the two DMSHs produces dark-matter-dominated galaxies. However, the two DMSHs penetrate each other, but the gaseous mediums collide for the middle-speed encounter. The enhancement of the gas density induces a burst of star formation, and then a dwarf galaxy without a dark matter halo forms at the interface of the DMSH collision.

1. Introduction

In the current standard model of galaxy formation, cold dark matter (CDM) has played an essential role in the universe. Based on the CDM model, cosmic structures are believed to form in a bottom-up fashion in which small dark matter halos repeatedly collide and merge, growing into larger structures. The gas falls into the deep gravitational potential created by the dark matter halo, where stars and galaxies are formed. Many dark matter subhalos (DMSHs) inside a massive dark matter halo are gravitationally condensed smaller structures composed of dark matter. Satellite galaxies may form in some subhalos. So far, observations of galaxy surveys and cosmological simulations in the CDM model have shown that the dark matter fraction in galaxies is expected to be more than 90% [1].

However, recently it has been suggested that the satellite galaxy NGC1052-DF2, a member of the elliptical galaxy NGC1052 group, has a very little dark matter component compared to the theoretical predictions [2]. Thus far, 27 dark-matter-deficient galaxies have been discovered [3–5]. It is still challenging to explain the formation of dark-matter-deficient galaxies in the CDM model.

Ogiya (2018) investigated the formation of dark-matter-deficient galaxies by tidal interaction between a host galaxy and a satellite galaxy using N-body simulations [6]. They demonstrated that an NGC1052-DF2-like galaxy successfully reproduces, assuming that the dark halo of the satellite galaxy has a cored density profile with a tightly bound and extremely radial orbit.

Recently, it has been pointed out that high-velocity collisions between gas-rich dwarf galaxies are also capable of forming dark-matter-deficient galaxies[7,8]. Shin et al. (2020) showed that dark-matter-deficient galaxies formed when two galaxies collide with each other at a relative velocity of $300\,\mathrm{km\,s^{-1}}$ using self-gravitating hydrodynamics simulations.

In this study, we focus on the head-on collision process between DMSHs and investigate the possibility of the formation of dark-matter-deficient galaxies using our original simulation code.

2. Simulation method

In this section, we describe our original N-body/Smoothed Particle Hydrodynamics (SPH) code. Two types of particles are adopted in the simulation: collisionless particles such as stars and dark matter, and gas particles. Only the gravitational interaction is calculated for collisionless particles, while the gravity and hydrodynamic interactions are calculated for gas particles. In the gravitational N-body algorithm for a collisionless system, the acceleration of particle i at position $\boldsymbol{r}_i$ is given by

$$\frac{\mathrm{d}\boldsymbol{v}_i}{\mathrm{d}t} = -\sum_j \frac{Gm_j\left(\boldsymbol{r}_i - \boldsymbol{r}_j\right)}{\left(\left|\boldsymbol{r}_i - \boldsymbol{r}_j\right|^2 + \epsilon^2\right)^{3/2}}, \tag{1}$$

where G is the gravitational constant, $\boldsymbol{v}_i$ is the velocity of particle i, m_j is the mass of a particle j and ϵ is gravitational softening. Tree algorithm[9] is adopted to calculate the gravitational interaction.

In SPH formulation, the gas density of particle i at position $\boldsymbol{r}_i$ is given by

$$\rho_i = \sum_{j=1}^{N_{\mathrm{neigh}}} m_j W(r_{ij}, h_i), \tag{2}$$

where $r_{ij} = |\boldsymbol{r}_i - \boldsymbol{r}_j|$, $W(r,h)$ is the kernel function adopting the Wendland C^4 function[10] to avoid the clumping instability[11,12], h is the smoothing length and $N_{\mathrm{neigh}} = 200$ is the number of neighbour particles. We implemented the algorithm to calculate the smoothing length to conserve the mass enclosed in a smoothing volume[13]. The smoothing length h_i of particle are determined by

$$\frac{4\pi}{3} h_i^3 \rho_i = \overline{m} N_{\mathrm{neigh}}, \tag{3}$$

where $\overline{m}$ is an average mass of gas particles. These equations (2) and (3) need to be solved implicitly for ρ_i and h_i. We follow the entropy formulation for the calculation of the momentum equations and energy equations[13] and adopt Monaghan's artificial viscosity[14] combined with Balsara's switch[15]. In order to handle strong shock waves at the DMSH collisions, we set the parameter, $\alpha = 5$, which is a measure of the strength of viscosity.

In the dissipationless case, only the entropy equation is solved for the thermal evolution, but since radiative cooling is important in galaxy formation, it is needed to add a dissipation term to the entropy equation:

$$\left(\frac{\mathrm{d}A}{\mathrm{d}t}\right)_{\mathrm{cool}} = \frac{\gamma - 1}{\rho^\gamma}\left(\frac{\mathrm{d}u}{\mathrm{d}t}\right)_{\mathrm{cool}}, \quad \left(\frac{\mathrm{d}u}{\mathrm{d}t}\right)_{\mathrm{cool}} = -\frac{n^2\Lambda(u,Z)}{\rho} = -\frac{\rho\Lambda(u,Z)}{\mu^2 m_{\mathrm{p}}^2}, \tag{4}$$

where n, μ and m_{p} is the number density of gas, the mean molecular weight and the proton mass, respectively. Λ is the cooling function of the specific internal energy u and the metallicity

Z. In order to solve the cooling equation (4), the Exact Integration (EI) scheme[16] is introduced. Discretizing the time evolution at the time step Δt, we get

$$\int_{u_i^n}^{u_i^{n+1}} \frac{\mu(u)^2}{\Lambda(u, Z)} \mathrm{d}u = -\frac{\rho_i}{m_\mathrm{p}^2} \Delta t. \tag{5}$$

Then, using the temporal evolution function which is defined by

$$Y(u) = \frac{\Lambda_\mathrm{ref}}{\mu_\mathrm{ref}^2 u_\mathrm{ref}} \int_u^{u_\mathrm{ref}} \frac{\mu(u)^2}{\Lambda(u, Z)} \mathrm{d}u, \tag{6}$$

the cooling equation (4) becomes

$$u_i^{n+1} = Y^{-1}\left[Y(u_i^n) + \frac{\Lambda_\mathrm{ref}}{\mu_\mathrm{ref}^2 u_\mathrm{ref}} \frac{\rho_i}{m_\mathrm{p}} \Delta t\right], \tag{7}$$

where $Y(u)$ can be obtained as a table by fitting the cooling function Λ with a piecewise power-law.

To parallelize our simulation code, Framework for Developing Particle Simulator (`FDPS`)[17,18] is installed. This is a framework to support the development of particle simulation codes. Users can implement their code using this library, and it will be executed efficiently on a supercomputer without spending much time for parallelization.

3. Numerical simulation

We consider head-on collisions between two DMSHs as the formation process of dark-matter-deficient galaxies. The two colliding DMSHs each have a mass of $M_\mathrm{tot} = M_\mathrm{DM} + M_\mathrm{gas} = 10^9\,\mathrm{M}_\odot$ and the mass ratio between dark matter and gas is 5.36. The density profile of the dark matter is assumed to be the Navarro-Frenk-White (NFW) profile[19], and the density profile of the gas is assumed to be in hydrostatic equilibrium with the potentials of the dark matter. The c-M relation[20] is used to determine the concentration c of NFW profile. We use the Many-component Galaxy Initializer (`MAGI`)[21] to create the initial conditions. The total number of particles is about 1.36 million.

The DMSH centres are initially at $(x, y, z) = (\pm 5, 0, 0)$ kpc, and the initial bulk-velocities of the DMSHs are $(v_x, v_y, v_z) = (\mp v_\mathrm{col}, 0, 0)$. In this paper, we show the results of two collision simulations, one with $v_\mathrm{col} = 20\,\mathrm{km\,s^{-1}}$ and the other with $100\,\mathrm{km\,s^{-1}}$.

Fig. 1 and Fig. 2 show the time evolution of the mass density sliced at z = 0 plane with $v_\mathrm{col} = 20\,\mathrm{km\,s^{-1}}$ and $100\,\mathrm{km\,s^{-1}}$, respectively. In the case of low-speed collisions, a merger of the two DMSHs produces dark-matter-dominated galaxies. However, the two DMSHs penetrate each other, but the gaseous mediums collide for the high-speed encounter.

As the gas density increases and the gas collapses gravitationally, star formation will be triggered. The enhancement of the gas density induces a burst of star formation, and then a dwarf galaxy without a dark matter halo may form at the interface of the DMSH collision. To test this idea, we install the algorithms for the star formation and the resulting energetic "feedback" from young, massive stars. The simulation of the galaxy formation has usually insufficient resolution to resolve these processes directly and instead adopt sub-grid physics tuned to match large-scale observational constraints. The star formation is realized by the conversion of the SPH particle into a stellar particle with an initial mass function (IMF). In this study, we assume that a stellar particle has Salpeter IMF[22]. For a single SPH particle, it is converted to a stellar particle when the conditions are satisfied[23]:

$$n_\mathrm{H} > 10\,\mathrm{cm^{-3}}, \tag{8}$$

$$r < 1 - \exp\left(-C_\star \Delta t / t_\mathrm{ff}\right), \tag{9}$$

where $n_{\rm H}$ is the number density of hydrogen, r is the random value between 0 and 1, $C_\star = 1$ is the constant value corresponding to the star formation efficiency and $t_{\rm ff}$ is the local free-fall time.

When a star with a mass of more than $8\,{\rm M}_\odot$ reaches the end of its life, it undergoes a supernova explosion, releasing about 10^{51} erg per star into the surrounding gas. This supernova feedback heats up the gas, leading to a decrease in the star formation rate of the galaxy. Assuming a Salpeter IMF, the number of massive stars ($> 8\,{\rm M}_\odot$) in a stellar particle of mass $m_\star$ is $N_{\rm SN} \simeq 7.31\left(m_\star/10^3\,{\rm M}_\odot\right)$. We set the upper mass limit at $60\,{\rm M}_\odot$. Of all the stars that undergo supernova explosions, the shortest lifetime is 5.4 Myr ($60\,{\rm M}_\odot$) and the longest is 43 Myr ($8\,{\rm M}_\odot$). During this period, the feedback energy is released from the star to the gas. If we assume that the energy is equally distributed to neighbour SPH particles of the star particle, the energy received by the SPH particle per time step Δt is

$$\Delta E = \frac{L_{\rm SN} N_{\rm SN} \Delta t}{N_{\rm neigh}}, \tag{10}$$

where $L_{\rm SN}$ is the average energy rate per star during the explosion period. SPH particles that receive energy turn off radiative cooling calculations and evolve adiabatically[24]. The time step Δt of the particles is determined by the CFL conditions.

In Fig. 1, a dark-matter-dominated galaxy with dark matter mass of $7.7 \times 10^8\,{\rm M}_\odot$, gas mass of $1.1 \times 10^8\,{\rm M}_\odot$ and stellar mass of $1.2 \times 10^8\,{\rm M}_\odot$ is formed by the merger of two DMSHs. Here we calculate the mass enclosed within the half-mass radius of the dark matter halo of the initial condition. In the case of $v_{\rm col} = 100\,{\rm km\,s^{-1}}$, a starburst is triggered at the time of the collision, and the dark matter component penetrate each other. Finally, a dark-matter-deficient galaxy with dark matter mass of $0\,{\rm M}_\odot$, gas mass of $8.6 \times 10^8\,{\rm M}_\odot$ and stellar mass of $1.2 \times 10^8\,{\rm M}_\odot$ is formed.

4. Conclusion

We developed a three-dimensional simulation code with radiative cooling, star formation and supernova feedback, and performed numerical simulations of head-on collisions between dark matter subhalos with each mass of $10^9\,{\rm M}_\odot$. The simulation results show that a dark-matter-dominated galaxy is formed when the collision velocity is $20\,{\rm km\,s^{-1}}$, and a dark-matter-deficient galaxy is formed when the collision velocity is $100\,{\rm km\,s^{-1}}$. In the forthcoming paper, we will show critical collision-velocities for the bifurcation sequence of the formation of dark-matter-dominated galaxies and dark-matter-deficient galaxies through the DMSH collision by analytical model coupling energy balancing and one-dimensional hydrodynamics problems[25].

5. Acknowledgments

Numerical computations were performed with computational resources provided by Multidisciplinary Cooperative Research Program in Center for Computational Sciences, University of Tsukuba, Oakforest-PACS operated by the Joint Center for Advanced High-Performance Computing (JCAHPC). This work was supported by JSPS KAKENHI Grant Numbers JP21J21888, JP20K04022.

References

[1] Behroozi P S, Wechsler R H and Conroy C 2013 *ApJ* **770** 57 ISSN 0004-637X

[2] van Dokkum P, Danieli S, Cohen Y, Merritt A, Romanowsky A J, Abraham R, Brodie J, Conroy C, Lokhorst D, Mowla L, O'Sullivan E and Zhang J 2018 *Nature* **555** 629–632 ISSN 1476-4687

[3] Mancera Piña P E, Fraternali F, Adams E A K, Marasco A, Oosterloo T, Oman K A, Leisman L, di Teodoro E M, Posti L, Battipaglia M, Cannon J M, Gault L, Haynes M P, Janowiecki S, McAllan E, Pagel H J, Reiter K, Rhode K L, Salzer J J and Smith N J 2019 *ApJL* **883** L33 ISSN 2041-8205

[4] van Dokkum P, Danieli S, Abraham R, Conroy C and Romanowsky A J 2019 *ApJL* **874** L5 ISSN 2041-8205

[5] Guo Q, Hu H, Zheng Z, Liao S, Du W, Mao S, Jiang L, Wang J, Peng Y, Gao L, Wang J and Wu H 2020 *Nat Astron* **4** 246–251 ISSN 2397-3366

[6] Ogiya G 2018 *Mon Not R Astron Soc Lett* **480** L106–L110 ISSN 1745-3925

[7] Silk J 2019 *Mon Not R Astron Soc Lett* **488** L24–L28 ISSN 1745-3925

[8] Shin E j, Jung M, Kwon G, Kim J h, Lee J, Jo Y and Oh B K 2020 *ApJ* **899** 25 ISSN 0004-637X

[9] Barnes J and Hut P 1986 *Nature* **324** 446–449 ISSN 1476-4687

[10] Wendland H 1995 *Adv Comput Math* **4** 389–396 ISSN 1572-9044

[11] Dehnen W and Aly H 2012 *Mon Not R Astron Soc* **425** 1068–1082 ISSN 0035-8711

[12] Zhu Q, Hernquist L and Li Y 2015 *ApJ* **800** 6 ISSN 0004-637X

[13] Springel V and Hernquist L 2002 *Monthly Notices of the Royal Astronomical Society* **333** 649–664 ISSN 0035-8711

[14] Monaghan J J 1997 *Journal of Computational Physics* **136** 298–307 ISSN 0021-9991

[15] Balsara D S 1995 *Journal of Computational Physics* **121** 357–372 ISSN 0021-9991

[16] Townsend R H D 2009 *ApJS* **181** 391–397 ISSN 0067-0049

[17] Iwasawa M, Tanikawa A, Hosono N, Nitadori K, Muranushi T and Makino J 2016 *Publ Astron Soc Jpn Nihon Tenmon Gakkai* **68** ISSN 0004-6264

[18] Namekata D, Iwasawa M, Nitadori K, Tanikawa A, Muranushi T, Wang L, Hosono N, Nomura K and Makino J 2018 *Publications of the Astronomical Society of Japan* **70** ISSN 0004-6264

[19] Navarro J F, Frenk C S and White S D M 1996 *The Astrophysical Journal* **462** 563 ISSN 0004-637X

[20] Prada F, Klypin A A, Cuesta A J, Betancort-Rijo J E and Primack J 2012 *Mon Not R Astron Soc* **423** 3018–3030 ISSN 0035-8711

[21] Miki Y and Umemura M 2018 *Mon Not R Astron Soc* **475** 2269–2281 ISSN 0035-8711

[22] Salpeter E E 1955 *The Astrophysical Journal* **121** 161 ISSN 0004-637X

[23] Katz N 1992 *ApJ* **391** 502 ISSN 0004-637X, 1538-4357

[24] Mori M, Yoshii Y, Tsujimoto T and Nomoto K 1997 *ApJ* **478** L21 ISSN 1538-4357

[25] Otaki K and Mori M (in preparation)

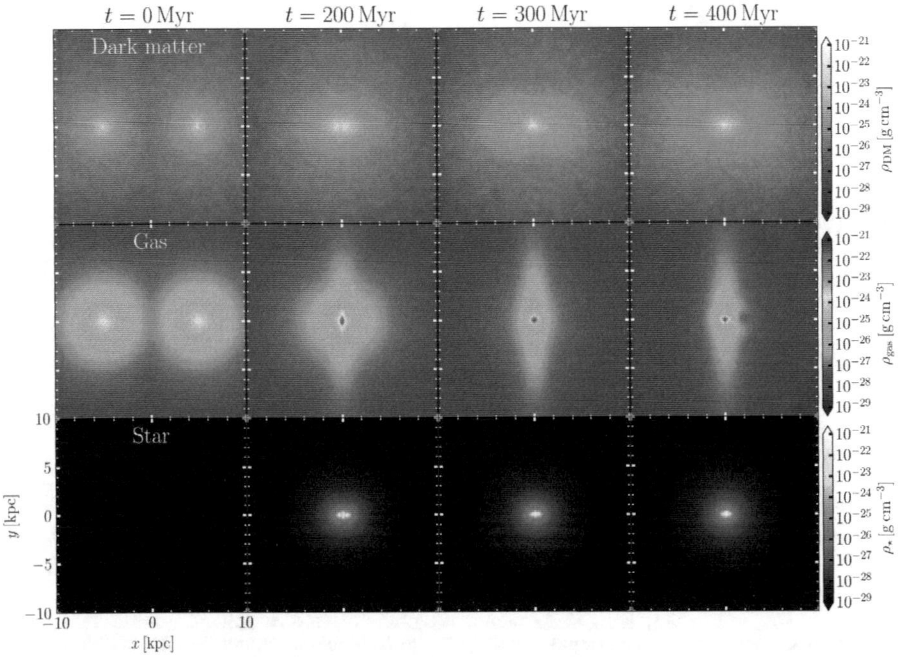

Figure 1. Snapshots of dark matter density (top), gas density (middle) and stellar density (bottom) of DMSH collision simulation at collision velocity of $20\,\mathrm{km\,s^{-1}}$. From left to right, $t = 0$, 200, 300, 400 Myr, respectively. A dark-matter-dominated galaxy forms for this velocity.

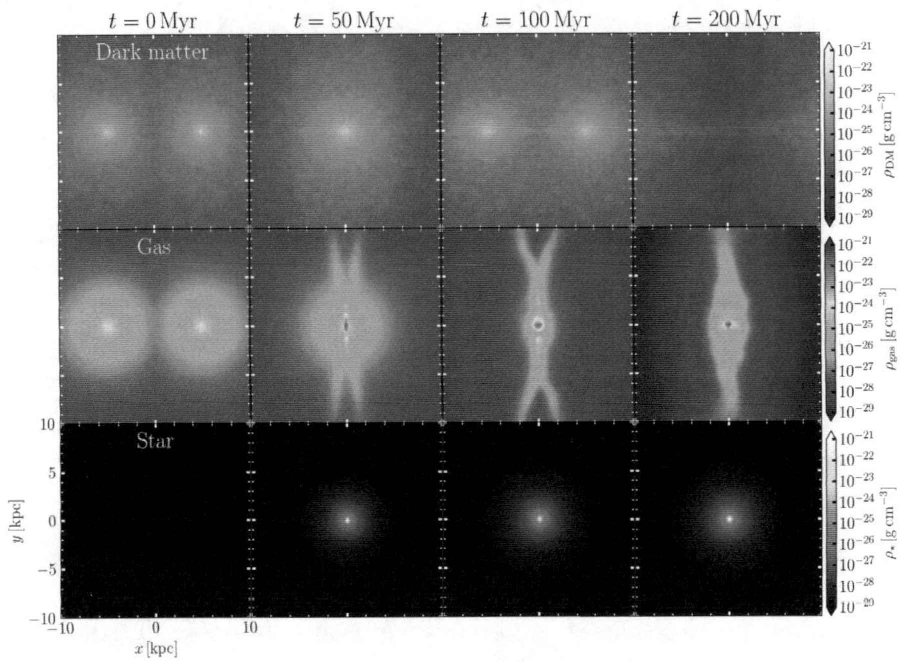

Figure 2. Same as Figure 1, but collision velocity is $100\,\mathrm{km\,s^{-1}}$. From left to right, $t = 0$, 50, 100, 200 Myr, respectively. A dark-matter-deficient galaxy forms for this velocity.

Journal of Physics: Conference Series **2207** (2022) 012050 doi:10.1088/1742-6596/2207/1/012050

Predicting the Expansion of Supernova Shells for High-Resolution Galaxy Simulations Using Deep Learning

K. Hirashima[1], K. Moriwaki[2], M. S. Fujii[1], Y. Hirai[3,4,5], T. Saitoh[6], and J. Makino[6,5]

[1]Department of Astronomy, Graduate School of Science, The University of Tokyo, 7-3-1 Hongo, Bunkyo-ku, Tokyo 113-0033, Japan
[2]Department of Physics, Graduate School of Science, The University of Tokyo, 7-3-1 Hongo, Bunkyo-ku, Tokyo 113-0033, Japan
[3]Department of Physics, University of Notre Dame, 225 Nieuwland Science Hall, Notre Dame, IN 46556, USA
[4]Astronomical Institute, Tohoku University, 6-3, Aramaki, Aoba-ku, Sendai, Miyagi 980-8578, Japan
[5]RIKEN Center for Computational Science, 7-1-26 Minatojima-minami-machi, Chuo-ku, Kobe, Hyogo 650-0047, Japan
[6]Department of Planetology, Graduate School of Science, Kobe University, 1-1 Rokkodai-cho, Nada-ku, Kobe, Hyogo 657-8501, Japan

E-mail: hirashima.keiya@astron.s.u-tokyo.ac.jp

Abstract. Small integration timesteps for a small fraction of the particles become a bottleneck for future galaxy simulations with a higher resolution, especially for massively parallel computing. As we increase the resolution, we must resolve physics on a smaller timescale while the total integration time is fixed as the universe age. The small timesteps for a small fraction of the particles worsen the scalability. More specifically, the regions affected by supernovae (SN) have the smallest timestep in the whole galaxy. Using a Hamiltonian splitting method, we calculate the SN regions with small timesteps using a few thousand CPU cores but integrate the entire galaxy using a shared timestep. For this approach, we need to pick up particles in regions, which will be affected by SN (the target particles) by the next global step (the integration timestep for the entire galaxy) in advance. In this work, we developed the deep learning model to predict the region where the shell due to a supernova explosion expands during one global step. In addition, we identify the target particles using image processing of the density distribution predicted by our deep learning model. Our algorithm could identify the target particles better than the method based on the analytical solution. This particle selection method using deep learning and the Hamiltonian splitting method will improve the performance of galaxy simulations with extremely high resolution.

1. Introduction

Since the timescale of the universe is much longer than that of human beings, numerical simulations are one of the ways to understand the formation history of the universe. Galaxies are a system consisting of billions of stars embedded in a dark matter halo, and the sun is a star in the Milky-Way galaxy, which has been evolved for $\sim 10\,\mathrm{Gyr}$. To simulate galaxy formation, we need to treat multiple complex phenomena such as gravitational and hydrodynamic forces, radiative cooling and heating, star formation, supernova explosions, and chemical evolution.

One commonly used method for the hydrodynamics of galaxy formation is smoothed particle hydrodynamics (SPH) [1, 2], a particle-based method using gas particles smoothed with a kernel size depending on the local density. Stars in galaxies are also modeled using particles (N-body simulations).

The resolution of N-body/SPH simulations depends on the number of star and gas particles. The number of stars in the Milky-Way galaxy exceeds 10^{10}. We need more than 10^{10} particles to simulate galaxies resolving individual stars (star-by-star simulation). In addition, we need the mass resolution for gas particles similar to that of stars.

Thanks to the development of supercomputers, a higher mass resolution has been achieved. IllustrisTNG [3, 4] is one of the simulations with the highest resolution. The mass resolution is $8 \times 10^4 M_\odot$ (about 10^{10} particles), and the simulation was performed using 25,000 CPU cores. However, the number of particles is three order-of-magnitude smaller than the number of particles necessary to resolve individual stars, even for Milky-Way-sized galaxies.

The communication overhead is a crucial problem of galaxy simulations using massively parallel computers. In parallel computing with more than thousands of CPU cores, the communication takes longer than the calculations (see Figure 63. in [5]). As we increase the resolution, we have to resolve physics on a smaller timescale, although the total integration time (the universe's age) does not change. The smaller timesteps for a small fraction of the particles worsen the scalability. Thus, we cannot reach star-by-star galaxy simulations without solving this problem.

The regions affected by supernovae (SN) have the smallest timestep in the whole galaxy. Using a Hamiltonian splitting method (e.g., Fujii *et al.* [6] and Saitoh and Makino [7]), we calculate the SN regions with small timesteps using a few thousand CPU cores but integrate the entire galaxy using a shared timestep.

For this approach, we need to pick up particles in regions, which will be affected by the subsequent global step (the integration timestep for the entire galaxy), in advance. We use a deep learning method to predict the region in which the shell due to a supernova explosion expands during one global step.

In this study, we propose a method of deep learning to identify the particles whose future timesteps will be small by using the density distribution just before a supernova explosion as the input data.

2. Methods

2.1. Extended Memory In Memory Network

We employ the Memory In Memory network (MIM) [8], which utilizes differential signals effectively to learn the non-stationary spatiotemporal changes of the video. MIM is composed of two main sequential sectors: the Convolutional Neural Network (CNN) and MIM block. CNNs have been widely used to learn correlations in images [9]. The CNN sector extracts spatial features of the image. MIM block is an extension of Recurrent Neural Network (RNN), in which hidden states are propagated recurrently to lean changes in sequential data. The MIM block makes use of a module that efficiently captures non-stationary changes as well as stationary changes.

We make two improvements to the vanilla MIM to apply for our simulation data. First, we expand the internal dimension of MIM by one dimension so that our model can deal with the three-dimensional physical quantity distribution represented by voxel. We extend the convolutional layer from two dimensions to three. Second, we improve the length of the prediction sequence. While the original MIM [8] performs a many-to-many prediction, in particular, the last ten frames from the initial ten frames, we build a one-to-many prediction

Journal of Physics: Conference Series **2207** (2022) 012050 doi:10.1088/1742-6596/2207/1/012050

model that is commonly used for, for example, the generation of music and video. Our model predicts all the subsequent gas distribution only from the initial one. The schematic diagram of our model is shown in Figure We call our model **Extended-MIM**.1.

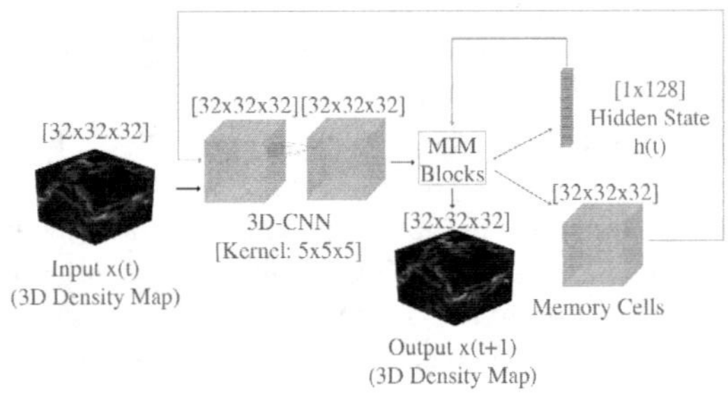

Figure 1. The architecture of our deep learning model.

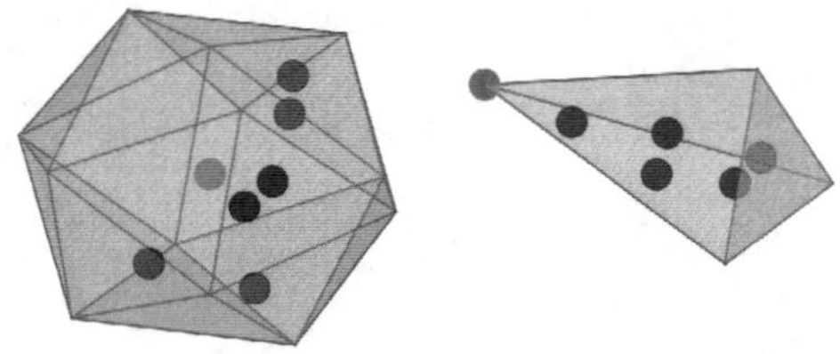

(a) Icosahedron (b) A piece of domain

Figure 2. Domain decomposition using icosahedron. Red particle represents the center of the explosion. Darkgreen particles represent SPH particles.

2.2. Data Preparation
As training data, we employ the outputs of simulations of SN explosions in non-uniform gas distributions. We set a density typical for the galactic disc, where the SN explosions occur. Each supernova explosion is simulated using SPH code, ASURA-FDPS (Saitoh et al. in press). By converting the particle data using the SPH kernel (like the Gaussian kernel), we obtain a time series of voxel data composed of 32x32x32 grids. A single training data is composed of time series of 3D density distribution. We generate 14,400 training data with 20 frames by taking 48 different viewing angles for 300 simulations.

2.3. Image Processing
We predict the area affected by a supernova explosion by processing the predicted future density distribution. First, we compute the ratio of the predicted density of each voxel divided by the initial density of the same voxel and collect voxels where the ratio exceeds a certain threshold. Among the regions where four or more of those voxels are connected, we pick up the one whose center of gravity is closest to the voxel where the SN explodes. We assigned a "1" to the voxels in this blob and a "0" to the others.

Another three types of image processing are performed: "Dilate", "Erode", and "Gradient". Dilate assigns 1 to a pixel if at least one of its neighbors has a value of 1. Conversely, Erode assigns 0 to the pixel if it is surrounded by at least one pixel with a value of 0. "Gradient" compares the results of "Dilate" and "Erode" on an image and assigns 1 if the pixel has different values.

2.4. Defining Particles with Domain Decomposition Using Icosahedron
To evaluate our new method to predict the regions affected by SN explosion, we also test a method to determine the region using an analytic solution for the evolution of SN shells. The expansion of the supernova shell is described as a self-similar solution. It approximates the SN as a spherical point explosion in a uniform medium. Here, we derive the radius R of a SN's shell with the released energy E in the uniform density ρ at some time t. Introducing the

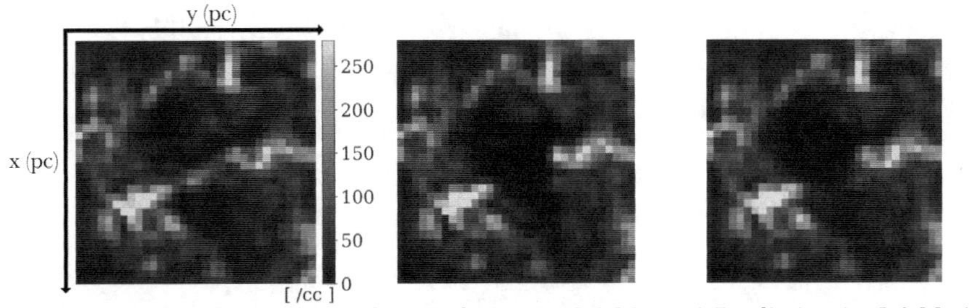

Figure 3. This figure shows an example of predicted density distributions after a SN using our deep learning model. One side of each panel corresponds to 60 pc. Color maps show the density distribution. The color bar and scale are the same in all panels. Panel (a) shows the initial condition just before the supernova explosion in the center. Panel (b) shows the result of the SPH simulation up to a hundred thousand years after the supernova explosion. Panel (c) shows the results predicted using our deep learning model until the same time as panel (b).

dimensionless similarity variable ξ into which the parameters of E and ρ, and the valuables of R and t are combined, the radius R is written as the following;

$$R(t) = \xi \left(\frac{E}{\rho}\right)^{4/5} t^{2/5}. \tag{1}$$

This is known as the Sedov-Taylor solution [10].

Since SN explosions rarely occur in situations under the assumption of uniform density and isotropy, we considered the anisotropy of the density distribution by dividing the region using an icosahedron. First, we align the center of the icosahedron with the center of the explosion. Next, we calculate the average density of the 20 tetrahedral regions formed by connecting each vertex of the icosahedron to the center and calculate the radius of the shell at a specific time in each direction using the 20 average densities and Equation 1. The particle inside each shell radius is identified as a particle with a small timestep in the future and is acquired.

3. Result

3.1. Predicting the shell of the Supernova Explosion

Figure 3 shows the density distributions of the input (panel (a)), simulated using the SPH code until 0.1 Myr (panel (b)), and the predicted using deep learning (panel (c)). We evaluate the performance of the trained model using Mean Structural SIMilarity (MSSIM) [11] and Mean Absolute Percentage Error (MAPE). MSSIM is the index for quality assessment based on the degradation of structural information. MAPE is the index for forecasting accuracy in statistics. When we increased training epochs, these indices converged on a good value. Our model can predict the density distribution sufficiently well based on these indices.

3.2. Identification of Target Particles and Comparison with Analytic Method

We use image processing to determine the voxels in which we expect particles with small timesteps to be included. The table 1 shows the list of parameters we used.

In Fig. 4, we present the relation between the identification rate and the ratio of the number of 'target particles' to the number of 'non-target particles'. The identification rate is defined as detected particles divided by the particles with small timesteps, which is given from the results

Table 1. The list of parameters used for image processing. After binarization, images are processed using three types of morphological operators: "Dilate", "Erode", and "Gradient". Dilate assigns 1 to the pixel if it is surrounded by at least one pixel with a value of 1. Conversely, Erode assigns 0 to the pixel if it is surrounded by at least one pixel with a value of 0. "Gradient" compares the results of "Dilate" and "Erode" on an image, and assigns 1 if the pixel has different values. In the column of iterations, D_x and E mean the number of iterations.

Morphological Operators	kernel or threshold	iterations
Dilate	(3,3)	1
Gradient	(3,3)	-
Dilate	(3,3)	D_3 (e.g. 1)
Dilate	(5,5)	D_5 (e.g. 3)
Majority	≥ 2	-
Erode	(3,3)	E (e.g.1)
Dilate	(3,3)	$E+1$ (e.g.2)
Majority	≥ 2	-

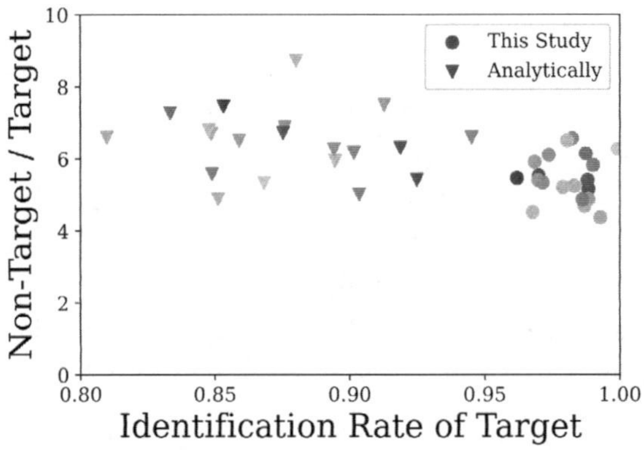

Figure 4. Identification rate of target particles versus non-target-to-target particle rate in the selected region. Circles and triangles indicate the methods using deep learning and analytical solution, respectively. The colors indicate the individual simulation data. The vertical axis shows the ratio of the number of non-target particles to the number of target particles in the enclosed region. The horizontal axis shows the ratio of non-target particles to the target particles in the enclosed region. Here, target particles have a smaller timestep than global timestep and are hotter than 100 K.

of the SPH simulations. Hereafter, 'target' particles mean the particles that will have the small timestep due to a SN by the subsequent global step, whereas 'non-target' particles mean the particles incorrectly identified as target particles. This figure shows that we detected 95% of the target particles.

For comparison, we also show the results using the icosahedron method. If we set the non-target-to-target similar to that of the deep learning method, the identification rate is a maximum of 95%. In addition, the identification rate of the icosahedron method has a large scatter. The identification rate distributes between 80% and 95%. This is probably because the analytic solution fails when the gas density distribution is highly inhomogeneous. Thus, we conclude that our deep learning method is better and more stable compared to an analytic solution.

4. Conclusion

To improve the resolution of galaxy formation simulations using SPH code, we are developing a Hamiltonian splitting scheme, in which only SN regions are integrated with timesteps smaller than that for the entire region (global step). For this method, we developed a new algorithm with a deep learning model to select particles that will have timesteps smaller than the global timestep. Our new deep learning model successfully predicted the region which is affected by a SN explosion in 0.1 Myr. By performing image processing of the predicted density distributions, our new algorithm can identify particles that will have small timesteps. We confirmed that our method could select particles with small timesteps better than that based on a self-similar analytic solution.

We in the future will include this method in our N-body/SPH code, ASURA-FDPS [12, 13], and perform a star-by-star galaxy formation simulation using a massively parallel computer such as Fugaku.

Acknowledgements

Numerical computations were carried out on Cray XC50 CPU-cluster at the Center for Computational Astrophysics (CfCA) of the National Astronomical Observatory of Japan. This work was supported by JSPS KAKENHI Grant Number 19H01933, 19J21379, 20K14532, 21J00153, 21K03614, 21K03633, 21H04499 and Initiative on Promotion of Supercomputing for Young or Women Researchers, Information Technology Center, The University of Tokyo, and MEXT as "Program for Promoting Researches on the Supercomputer Fugaku" (Toward a unified view of the universe: from large scale structures to planets, Revealing the formation history of the universe with large-scale simulations and astronomical big data). MF was supported by The University of Tokyo Excellent Young Researcher Program.

References

[1] Gingold R A and Monaghan J J 1977 *MNRAS* **181** 375–389
[2] Lucy L B 1977 *AJ* **82** 1013–1024
[3] Weinberger R, Springel V, Hernquist L, Pillepich A, Marinacci F, Pakmor R, Nelson D, Genel S, Vogelsberger M, Naiman J and Torrey P 2017 *MNRAS* **465** 3291–3308 (*Preprint* 1607.03486)
[4] Pillepich A, Springel V, Nelson D, Genel S, Naiman J, Pakmor R, Hernquist L, Torrey P, Vogelsberger M, Weinberger R and Marinacci F 2018 *MNRAS* **473** 4077–4106 (*Preprint* 1703.02970)
[5] Springel V, Pakmor R, Zier O and Reinecke M 2021 *MNRAS* **506** 2871–2949 (*Preprint* 2010.03567)
[6] Fujii M, Iwasawa M, Funato Y and Makino J 2007 *PASJ* **59** 1095 (*Preprint* 0706.2059)
[7] Saitoh T R and Makino J 2010 *PASJ* **62** 301 (*Preprint* 0908.1460)
[8] Wang Y, Zhang J, Zhu H, Long M, Wang J and Yu P S 2018 *arXiv e-prints* arXiv:1811.07490 (*Preprint* 1811.07490)
[9] Krizhevsky A, Sutskever I and Hinton G E 2012 Imagenet classification with deep convolutional neural networks *Advances in Neural Information Processing Systems 25* ed Pereira F, Burges C J C, Bottou L and Weinberger K Q (Curran Associates, Inc.) pp 1097–1105 URL http://papers.nips.cc/paper/4824-imagenet-classification-with-deep-convolutional-neural-networks.pdf
[10] Sedov L I 1959 *Similarity and Dimensional Methods in Mechanics*
[11] Wang Z, Bovik A C, Sheikh H R and Simoncelli E P 2004 *IEEE Transactions on Image Processing* **13** 600–612
[12] Saitoh T R, Daisaka H, Kokubo E, Makino J, Okamoto T, Tomisaka K, Wada K and Yoshida N 2008 *PASJ* **60** 667–681 (*Preprint* 0802.0961)
[13] Iwasawa M, Tanikawa A, Hosono N, Nitadori K, Muranushi T and Makino J 2016 *PASJ* **68** 54 (*Preprint* 1601.03138)

ISTEDDAS: a new direct N-Body code to study merging compact-object binaries

Mattia Mencagli[1], **Natalia Nazarova**[1], **Mario Spera**[1,2,3]

[1] SISSA, Via Bonomea 265, I-34136, Trieste, Italy
[2] INFN, Sezione di Trieste, I-34127 Trieste, Italy
[3] INFN, Sezione di Padova, Via Marzolo 8, I–35131, Padova, Italy

E-mail: mattia.mencagli@sissa.it

Abstract. On September 14, 2015, the two detectors of the Laser Interferometer Gravitational wave Observatory (LIGO) reported the first detection of gravitational waves, a signal generated from the coalescence of two stellar-mass black holes. The discovery represented the beginning of an entirely new way to investigate the Universe. From the theoretical point of view, the formation and evolution of compact-object binaries are still very uncertain. One of the main issues is that most stars form in dense stellar environments, and numerical simulations of merging compact-object binaries in such crowded stellar systems are very challenging. In this work, we review the main numerical bottlenecks that hamper our knowledge on merging binaries in dense environments and we present a new GPU-accelerated N-body code, which is currently under development, called ISTEDDAS, that can overcome most of the obstacles.

1. Introduction

Gravitational waves (GWs) are ripples in space-time caused by accelerated masses such as two compact objects orbiting around each other. These perturbations are strong close to the source and become weak when they reach earth, making them difficult to detect. GWs were predicted by Einstein in 1916 in the context of his theory of general relativity but they were never observed until September 14 2015, when the LIGO-Virgo collaboration (LVC) announced, for the first time, the detection of a GW signal coming from the coalescence of two stellar-mass black holes, a few seconds before the two objects merged [1, 2]. After the first detection, 89 mergers of compact-object binaries have been detected during the first, second, and third LIGO-Virgo-KAGRA observing run [3, 4, 5, 6, 7, 8, 9, 10, 11, 12, 13].

The astrophysical interpretation of GW detections is crucial to shed light on the evolutionary history of compact-object binaries. We still do not know how heavy stellar black holes form, how two black holes can merge within a Hubble time, and what are the distinguishing features that would allow us to piece-together the formation pathways of compact-object binaries [14, 15].

From the theoretical point of view, there are two main channels to explain the formation of merging compact-object binaries: the isolated evolution of binary stars and the dynamical scenario. In the first case, compact-object binaries form from the isolated evolution of massive-star binaries. Investigating this scenario is complex because the astrophysical processes that drive the evolution of tight binaries (e.g. stable mass transfer, common-envelope evolution, chemical homogeneous evolution, tidal effects) are very uncertain [16, 17, 18, 19, 20, 21, 22, 23, 24, 25, 26, 73]. In contrast, in the dynamical formation scenario, compact-object binaries form via multiple gravitational encounters. Thus, in this scenario, collisional dynamics is crucial for the formation, evolution and merger of binaries, especially if compact objects form in a very dense environment, such as globular clusters (GCs), young massive star clusters (YMCs), or open star clusters (OCs) [27, 28]. Different dynamical processes can take place in such dense environments. For instance, an encounter between a binary and an object, where one of the binary members is replaced by the intruder (dynamical exchange), are expected to lead to the formation of tighter compact-object binaries [44]. The semi-major axis of hard binaries can shrink further via non-exchanging, close three-body encounters (i.e. hardening [46, 45]). Moreover, black holes (BHs), which are among the most massive objects we can find in dense stellar environments, tend to segregate towards the clusters' innermost region, forming subsystems kinematically dominated by their interactions [29, 30, 31]. Therefore, three-body interactions might easily pair BHs in tight binaries that can merge within a Hubble time [32, 33, 31, 34, 35, 36, 37, 38, 39, 40, 41, 42, 43].

Thus, studying the formation and evolution of merging compact-objects in dense star clusters is challenging because it involves the complex interplay between different spatial and time scales [27]. For a typical GC, with half-mass radius of ~1pc, the two-body relaxation time (t_{relax}) is of the order of Gyrs and the crossing time is of the order of Myrs [48, 49, 50, 51]. At the same time, the orbital period of tight binaries can be as short as ~days [46, 47]. Furthermore, it is also crucial to consider time-scales related to binary stellar evolution, such as the typical timescale for stable mass transfer (SMT) $O\left(10^3\right)$ yr, or the common envelope (CE) time scale, that is $O\left(10^3 - 10^5\right)$ yr [15]. N-body simulations that aim at investigating merging compact objects in GCs need to cover, at least, t_{relax} (~Gyrs) of evolution and to resolve all the mentioned time-scales simultaneously. Therefore, such simulations are challenging because they require the use of small time steps, significantly smaller compared to the total integration time. This means that such simulations are slow, and they accumulate significant numerical error, in the form of both truncation and round-off errors.

The dynamical evolution of stars is governed by the gravitational N-body problem, that can be solved numerically using a direct or non-direct approach. In the first one, all the forces are computed between star pairs, with a computational complexity of $O(N^2)$, where N is the total number of stars. In the second approach, either an approximated mathematical expression is used instead of the Newton's gravitational force, or only the force from close stars is computed directly, softening the $O(N^2)$ computational complexity. Non-direct approaches are faster but less accurate than the direct approach. Besides stellar dynamics, both direct and non-direct N-body codes generally include fitting formulas or look-up tables to treat stellar evolution of single and binary stars.

Given the high computational complexity, special-purpose hardware components, such as GRAPE, were developed to speed up direct N-body simulation of massive star clusters [52, 53, 54]. Today, graphics processing units (GPUs) have replaced the GRAPE system as main computing accelerator for N-body simulations. Various codes with only Central Processing Unit (CPU) support exist (e.g. MYRIAD, STARLAB/KIRA [55, 56]) but they can take advantage of external libraries to speed-up the force calculation process through GPUs (e.g. SAPPORO [57, 58]). The hybrid CPU-GPU approach represents the state-of-the-art for N-body codes and simulations [59, 60, 61, 62, 63, 64, 65, 66, 67, 68, 69, 70]. Finally, stellar evolution of single and binary stars is implemented, in most cases, through population-synthesis codes that use, in general, polynomial fitting formulas [71, 72, 73, 74, 75, 22, 76, 77].

In this work, we present ISTEDDAS, a new direct N-body code designed to run entirely on GPUs, to avoid any possible CPU-GPU communication bottleneck, and on multi-node GPU clusters through Message Passing Interface (MPI). The main integrator of the code is the 6th-order Hermite scheme, coupled with the block-time-step method and the Ahmad-Cohen neighbour scheme [78, 79, 80, 81]. Binaries and hierarchical few-body systems are integrated in ISTEDDAS through an interface to the TSUNAMI code, which implements a high-accuracy, few-body integrator based on the Algorithmic Regularization Chain , and that includes general relativity corrections as post-Newtonian terms up to 3.5pN [84, 86, 88]. Currently, ISTEDDAS does not include single and binary stellar evolution prescriptions, but it will be coupled with the new, up-to-date population-synthesis code SEVN [90, 91, 92]. To evolve single and binary stars, SEVN implements a large set of look-up tables at different ages, masses and metallicities, which are interpolated on-the-fly. The default look-up tables included in the SEVN code have been generated through the PARSEC stellar evolution code [93, 94, 95]. The main advantage of the ISTEDDAS-SEVN coupling is that the look-up tables in SEVN can be easily switched with another set of look-up tables. Therefore, we will be able to investigate the impact of different stellar evolution prescriptions on the dynamics of star clusters, without changing anything in the codes.

In Section 2 we describe into details the integrators of ISTEDDAS: the Hermite integrator with its features, and the Algorithmic Regularization chain. In Section 3 we briefly describe SEVN and how it is coupled with ISTEDDAS. Finally, in Section 4 we summarize our results.

2. The direct N-body code ISTEDDAS

In this section, we will give an overview of the strategies adopted in our code to numerically solve the N-body problem using GPUs. ISTEDDAS (the word for “stars” is Sardinian language) is a work-in-progress direct N-body code that implements the Hermite 6th-order time integrator combined with the block time steps method, the Ahmad-Cohen neighbours scheme, and the Algorithmic Regularization Chain integrator coupled with the Bulirsch-Stoer algorithm. ISTEDDAS is written combining utilities of C and C++ and it uses CUDA, MPI and OpenMP to exploit GPU workstations as well as GPU clusters. The main features of ISTEDDAS are described in the following sub-sections.

2.1. The Hermite Integrator

The Hermite time integrators (4th, 6th and 8th order) represent the state-of-the-art for direct N-body simulations [78, 79]. The 4th-order scheme is, by far, the most used algorithm in this context [59, 60, 61, 62, 63, 65, 56], but the 6th-order can be more accurate and almost as fast as the 4th-order when implemented on GPUs [78]. The Hermite integrators are based on Taylor series of positions, velocities and accelerations and their most important feature is that they need to evaluate the distances between particles just once per time-step. For instance, this is a huge advantage compared to a classic 4th-order Runge-Kutta method, which needs to evaluate accelerations 4 times per integration step.

To evolve a stellar system from time t_0 to time t_1 ($\Delta t = t_1 - t_0$, 1 integration step), the Hermite 6th-order integrator takes advantage of a predictor-evaluation-corrector scheme. The prediction step performs a Taylor expansion on stars' positions ($\mathbf{r}$), velocities ($\mathbf{v}$) and accelerations ($\mathbf{a}$), using time derivatives up to $\dddot{\mathbf{a}}$[1]:

$$\mathbf{r}_p = \mathbf{r}_0 + \mathbf{v}_0\Delta t + \mathbf{a}_0\frac{\Delta t^2}{2} + \dot{\mathbf{a}}_0\frac{\Delta t^3}{6} + \ddot{\mathbf{a}}_0\frac{\Delta t^4}{24} + \dddot{\mathbf{a}}_0\frac{\Delta t^5}{120}, \tag{1}$$

$$\mathbf{v}_p = \mathbf{v}_0 + \mathbf{a}_0\Delta t + \dot{\mathbf{a}}_0\frac{\Delta t^2}{2} + \ddot{\mathbf{a}}_0\frac{\Delta t^3}{6} + \dddot{\mathbf{a}}_0\frac{\Delta t^4}{24}, \tag{2}$$

$$\mathbf{a}_p = \mathbf{a}_0 + \dot{\mathbf{a}}_0\Delta t + \ddot{\mathbf{a}}_0\frac{\Delta t^2}{2} + \dddot{\mathbf{a}}_0\frac{\Delta t^3}{6}, \tag{3}$$

where the quantities labelled with p are the predicted ones and the quantities labelled with 0 come from the corrector performed in the previous step, or from initial conditions.

[1] Throughout the manuscript, we use bold letters to refer to three-dimensional vectors

During the evaluation step, we compute the total force on the star i from all the other stars in the system, using the classical Newtonian formula and the predicted values $\mathbf{p}$, $\mathbf{v}$, and $\mathbf{a}$:

$$\mathbf{a}_{i,1} = \sum_{j\neq i}^{N} \frac{m_j}{|\mathbf{r}_{ij,p}|^3}\mathbf{r}_{ij,p}\,, \tag{4}$$

$$\dot{\mathbf{a}}_{i,1} = \sum_{j\neq i}^{N} \frac{m_j}{|\mathbf{r}_{ij,p}|^3}\left(\mathbf{v}_{ij,p} - 3\frac{\mathbf{r}_{ij,p}\cdot\mathbf{v}_{ij,p}}{|\mathbf{r}_{ij,p}|^2}\mathbf{r}_{ij,p}\right), \tag{5}$$

$$\ddot{\mathbf{a}}_{i,1} = \sum_{j\neq i}^{N} \frac{m_j}{|\mathbf{r}_{ij,p}|^3}\left(\mathbf{a}_{ij,p} - 6\frac{\mathbf{r}_{ij,p}\cdot\mathbf{v}_{ij,p}}{|\mathbf{r}_{ij,p}|^2}\mathbf{v}_{ij,p} + \frac{3}{|\mathbf{r}_{ij,p}|^2}\left(5\frac{(\mathbf{r}_{ij,p}\cdot\mathbf{v}_{ij,p})^2}{|\mathbf{r}_{ij,p}|^2} - \mathbf{r}_{ij,p}\cdot\mathbf{a}_{ij,p} - |\mathbf{v}_{ij,p}|^2\right)\mathbf{r}_{ij,p}\right), \tag{6}$$

where the quantities with subscript "ij, p" are the relative predicted values between the stars i and j, and we have assumed the gravitational constant $G = 1$.

Finally, in the corrector step the predicted quantities are corrected to obtain a more accurate solution using the newly evaluated accelerations. Positions and velocities are corrected as:

$$\mathbf{v}_c = \mathbf{v}_0 + (\mathbf{a}_1 + \mathbf{a}_0)\frac{\Delta t}{2} - (\dot{\mathbf{a}}_1 - \dot{\mathbf{a}}_0)\frac{\Delta t^2}{10} + (\ddot{\mathbf{a}}_1 + \ddot{\mathbf{a}}_0)\frac{\Delta t^3}{120}\,, \tag{7}$$

$$\mathbf{r}_c = \mathbf{r}_0 + (\mathbf{v}_c + \mathbf{v}_0)\frac{\Delta t}{2} - (\mathbf{a}_1 - \mathbf{a}_0)\frac{\Delta t^2}{10} + (\dot{\mathbf{a}}_1 + \dot{\mathbf{a}}_0)\frac{\Delta t^3}{120}\,. \tag{8}$$

During the correction phase, a new time-step for each star is computed using the new accelerations, as in [78]:

$$\Delta t = \eta_h \sqrt[6]{\frac{|\mathbf{a}_1||\ddot{\mathbf{a}}_1| + |\dot{\mathbf{a}}_1|^2}{|\mathbf{a}_1^{(5)}||\dddot{\mathbf{a}}_1| + |\mathbf{a}_1^{(4)}|^2}}\,, \tag{9}$$

where η_h is an accuracy parameter that we set to 0.4 [64, 65]. $\dddot{\mathbf{a}}_1$, $\mathbf{a}_1^{(4)}$ and $\mathbf{a}_1^{(5)}$ are the third, fourth and fifth derivatives of the acceleration, and they are estimated analytically during the corrector step using $\mathbf{a}_0$, $\mathbf{a}_1$, $\dot{\mathbf{a}}_0$, $\dot{\mathbf{a}}_1$, $\ddot{\mathbf{a}}_0$ and $\ddot{\mathbf{a}}_1$ [78]. Furthermore, $\dddot{\mathbf{a}}_1$ is saved for the next iteration since it is needed in the predictor for the Taylor expansion.

2.2. Block time-steps method

Stars in an N-body system can have very different accelerations. This corresponds to having a large variety of evolutionary dynamical time-scales. In this context, it is convenient to assign to all stars their individual time step, which becomes a function of the physical parameters that describe the kinematic state of stars. To avoid time-synchronization issues between the stars, and to simplify the parallelization process, the time-step of the i-th star is quantized to powers of two: $\Delta t_i^{quantized} = 2^{-n}$, where n is an integer. For each star, we use Eq.9 to calculate the ideal time step Δt_i^{real}, but we always approximate n to the largest integer value, so the effective time step is $\Delta t_i^{quantized} \leq \Delta t_i^{real}$. Thus, particles are sub-divided into several groups (blocks) that share the same time-step [79] and the groups are indexed according to n. The latter strategy is very effective to select, for each step, the stars that need to be integrated, since we need to update positions and velocities only for m stars per time-step, where $m \leq N$. Specifically, stars with smaller time steps will be updated more often than stars with larger time steps, and for the latter, the kinematic state will be estimated using the predictor-step only. This implies that the computational complexity is reduced from $O(N^2)$ to $O(mN)$. This is a significant gain in terms of performance, with a negligible cost in terms of precision of the integrator.

2.3. The Ahmad-Cohen neighbours scheme

Star clusters may contain a large number of stars (up to $\sim 10^6$ for massive GCs), and this makes direct N-body simulations very challenging because of the high computational complexity of $O(N^2)$. In order to deal with such numerical challenges, a neighbour procedure that requires fewer total force calculations can be introduced. A very common strategy for direct N-body simulations is to adopt the Ahmad-Cohen (AC) neighbours scheme [80, 81]. In this method, the force on a star is splitted into two contributions that evolve over different time scales. The first contribution depends only on near stars and the other one is owed to distant stars. It is apparent that the latter force evolves over much slower time scales than the one from close stars, therefore it does not have to be recalculated with the same frequency. This implies less operations per step and a significant performance gain. The combination with the block time-steps method described above is straightforward: each star will be in two separate blocks, corresponding to two (quantized) time steps Δt_{close} and Δt_{far}, calculated for close and distant stars, respectively. For each star, we only compute the forces from its neighbours, proceeding with a time step Δt_{close} (neighbour steps), and the forces from distant particles are approximated using a Taylor expansion; in contrast, every k steps (with $k\Delta t_{close} = \Delta t_{far}$), we compute the force from both close and distant particles self-consistently. ISTEDDAS implements the AC neighbours scheme entirely on the GPU. This is a significant advantage in terms of performance since, for each step, there is no need to communicate the neighbours between the CPU and the GPU, reducing communication latencies.

To identify neighbours, we use a three-dimensional tree structure. To calculate the ideal number of neighbours, we evaluate a cost function by dividing the total number of operations performed with $(n_n\Delta t_f/\Delta t_c + (N - n_n))$ and without neighbours $(N\Delta t_f/\Delta t_c)$:

$$Cost(n_n) = \frac{n_n}{N} + \frac{\Delta t_c}{\Delta t_f}\left(1 - \frac{n_n}{N}\right), \tag{10}$$

where n_n is the number of neighbours. To compute both Δt_c and Δt_f we use the Hermite 4th-order time-step criterion [80]:

$$\Delta t = \eta_\ell \sqrt{\frac{|\mathbf{a}_1||\ddot{\mathbf{a}}_1| + |\dot{\mathbf{a}}_1|^2}{|\dot{\mathbf{a}}_1||\dddot{\mathbf{a}}_1| + |\ddot{\mathbf{a}}_1|^2}}, \tag{11}$$

where η_ℓ is a free parameter that we take as 0.05 for Δt_c and 0.1 for Δt_f. In Eq. 11, the time step depends on n_n through the accelerations. The cost function is computed for different values of n_n until a maximum of 25 neighbors is reached. The ideal value for n_n is the one that minimizes the cost function. For each star, we save the radius of its neighbour sphere, r_{neigh}, which corresponds to the distance of the star's n_n-th neighbour. To stabilize the number of neighbors per star, that is to avoid too frequent neighbors' changes, we have introduced a threshold value, so if the minimum cost is bigger then 0.6 no neighbours are selected ($r_{neigh} = 0$).

The tree structure and the number of neighbours are updated every time all stars are synchronized, therefore every time the code computes N^2 forces. However, r_{neigh} can change before one global synchronization and the next one. After the update of the tree structure, we calculate the local star density considering, for each star, the distance to its 6th nearest neighbor, and we use it to estimate the system's radial density profile, $\rho(r)$. Every Δt_f, we update r_{neigh} so the number of neighbours per star does not change significantly, that is:

$$r_{neigh,new} = r_{neigh,old}\left(\frac{\rho(r_{*,old})}{\rho(r_{*,new})}\right)^{\frac{1}{3}}, \tag{12}$$

where $r_{*,old}$ ($r_{*,new}$) is the position of the star in the system at the previous (current) time step.

Some complications arise using this scheme combined with the Hermite integrator. With the AC method, the Hermite 6th-order corrector writes as in Eq.7. During the neighbour integration steps (Δt_n) no issues are encountered, since the number of neighbours is constant. Though, during a distant step (i.e. every Δt_f), the neighbours might have changed, thus the Hermite corrector is not self-consistent anymore because in Eq.7 we are summing and subtracting accelerations and derivatives computed with potentially different neighbours. Therefore, we apply the following steps, as in [80] for the Hermite 4th-order integrator, to solve the problem:

- Compute the acceleration and its derivatives due to neighbors based on the list of neighbours calculated in the previous step ($\mathbf{a}_{close,old}$).
- Apply the corrector calculated for the acceleration due to neighbors
- Compute the acceleration and its derivatives due to neighbors and distant star based on the new list of neighbours calculated in this step ($\mathbf{a}_{close,new}$ and $\mathbf{a}_{far,new}$)
- Compute the acceleration and its derivatives due to distant particles based on the old neighbor list as: $\mathbf{a}_{far,old} = \mathbf{a}_{far,new} + \mathbf{a}_{close,new} - \mathbf{a}_{close,old}$.
- Apply the corrector for the acceleration due to distant particles using $\mathbf{a}_{far,old}$ and its derivatives.

It is worth noting that $\mathbf{a}$, $\dot{\mathbf{a}}$ and $\ddot{\mathbf{a}}$ are computed with the new neighbours list, during the evaluation step, while $\dddot{\mathbf{a}}$ (far and close) is estimated during the corrector step using the old neighbours list. Since in the Hermite 6th-order scheme $\dddot{\mathbf{a}}$ is needed in the predictor step and to calculate the correct individual time steps for stars, it is important to compute it self-consistently. To achieve this, we keep track of the incoming and outgoing neighbors with respect to r_{neigh}, and we retrieve $\dddot{\mathbf{a}}_{new}$ by subtracting and summing on $\dddot{\mathbf{a}}_{old}$ the contribution from outgoing and incoming neighbours:

$$\dddot{\mathbf{a}}_{new} = \dddot{\mathbf{a}}_{old} - \dddot{\mathbf{a}}_{outgoing} + \dddot{\mathbf{a}}_{incoming} \, . \tag{13}$$

We compute the outgoing and the incoming contributions using the explicit Newtonian formula for $\dddot{\mathbf{a}}$.

Finally, to estimate Δt_{close} and Δt_{far} we use Eq.9 only if the neighbours do not change; otherwise we use the low-order, Hermite 4th-order, criterion as in Eq.11. This is necessary because $\mathbf{a}^{(4)}$ and $\mathbf{a}^{(5)}$ would be inconsistent if the neighbor list changes. Furthermore, $\mathbf{a}^{(4)}$ and $\mathbf{a}^{(5)}$ are not needed in the predictor step and correcting them with the corresponding Newtonian formulas, as we do for $\dddot{\mathbf{a}}$, would have a significant computational cost without increasing the order of the integrator.

2.4. The TSUNAMI *code: the Algorithmic Regularization Chain*

While the 6th-order Hermite integrator is a powerful main integrator in ISTEDDAS, it might not be suitable for the long-term integration of stable, very tight systems (e.g. binaries, triples). A high-precision integration of tight systems is crucial in ISTEDDAS, especially if our aim is to investigate the evolution of merging compact-object binaries. Therefore, we have coupled ISTEDDAS with TSUNAMI, a code for the high-accuracy numerical integration of binaries and hierarchical systems [88]. TSUNAMI implements the Algorithmic Regularization Chain (AR Chain) integrator for high accuracy computation of particle motion in small N-body system [84, 85, 86, 87].

The method addresses very high accuracy and comparatively little computational effort combining: the use of chained coordinates, the Bulirsch-Stoer algorithm, a time-coordinate transformation (regularized time), and the 2nd-order symplectic Leapfrog integrator.

To evaluate the acceleration between close particles, the relative position between these particles is needed. Sometimes, especially when the two particles are far away from the origin of the coordinate system, this calculation implies a close-value subtraction between two floating-point numbers, which can lead to significant round-off error accumulation. To solve this problem the AR Chain uses a chained coordinate system, where the relative positions will be evolved instead of the absolute positions (relative-motion coordinate system). This transformation significantly reduces round-off errors.

To integrate the equations of motion, a time transformation is performed to obtain regular solutions even for collisional orbits (algorithmic regularization). For the details we refer the reader to [84, 85, 86, 87] .

TSUNAMI implements a symplectic Leapfrog. Though, the leapfrog method is only a 2nd-order method and it might not reach very high accuracy for very tight systems. Therefore, the Bulirsch-Stoer extrapolation is implemented to construct higher order results from a series of lower order results through an extrapolation table [82, 83]. The idea is to consider the final answer of a numerical calculation as itself being an analytic function of an adjustable parameter, which is the integration step dt. That analytic function can be probed by performing the calculation with various values of dt, none of them being necessarily small enough to yield the accuracy that we desire. When we know enough about the function, we fit it, and then evaluate it at the point $dt = 0$. The analytic fitting function can be either a rational function or a polynomial of variable degree. Furthermore, TSUNAMI also includes additional features such as perturbative forces, general relativity corrections through post-Newtonian terms up to 3.5 order, tidal interactions, and a collision-detection algorithm.

We are currently working on the interface between ISTEDDAS and TSUNAMI. In particular, we are developing the "decision making" part, that is we are working on effective criteria to decide whether a group of stars must be regularized with TSUNAMI, or it is better to leave it within ISTEDDAS. This is crucial, because TSUNAMI is a CPU-only code while ISTEDDAS is designed to run entirely on GPU. Thus, switching-on regularization would imply significant CPU-GPU data transfer, which might become a bottleneck if the decision-making is not carefully implemented. Moreover, we are working on a double-parallel scheme to evolve all the regularized systems simultaneously. This will lead to a significant speed-up with respect to others state-of-the-art direct N-body codes, especially when a large number of binaries are present in the system (e.g. high fractions of original binaries). The code will choose to use either the CPU or the GPU to parallelize the integration of the regularized systems, depending on the latter's count. We are currently working on a GPU version of the TSUNAMI code for this purpose.

3. Preliminary numerical tests

We have performed some preliminary tests to check the accuracy of the current ISTEDDAS+TSUNAMI machinery. All the tests are done on a laptop with a CPU Intel(R) Core(TM) i7-10750H 2.60GHz and a GPU NVIDIA GeForce GTX 1650 Ti, without using a softening factor[2] in the force calculations.

In Fig.1, ISTEDDAS is tested against the Pythagorean three-body problem. This system consists of three different masses initially located at rest at the corners of a right triangle whose sides have lengths 3, 4, and 5. Each of the three masses is located in such a way that its mass is equal to the length of the opposite side. In Fig.1 we show the trajectories of the bodies (left-hand panel) and the relative error on the total system energy as a function of simulation time (right-hand panel). In Fig.2 ISTEDDAS is tested against a cluster with 2048 stars, we show the initial and the final positions of the stars on the XY-plane together with the evolution of the total energy relative error of the cluster. The starting positions, velocities and masses were generated using the MCCLUSTER code [89]. For the first test shown in Fig.1, the AR chain of the TSUNAMI module is at work for the whole integrations. As expected, it reaches a quite high precision, with a maximum total energy relative error below $\sim 10^{-12}$. The N-body test in Fig.2 uses only the

[2] The softening factor is a small quantity added to the denominator in the newtonian gravitational potential. It is generally introduced to soften very close stellar encounters and to avoid very small integration time steps, at the expense of integration accuracy on the softening-comparable spatial scales.

AC scheme of ISTEDDAS and TSUNAMI is never called. As it is expected for a 6th-order Hermite integrator, the total energy relative error reaches a maximum of $\sim 10^{-6}$, after about 20 cluster's two-body relaxation times.

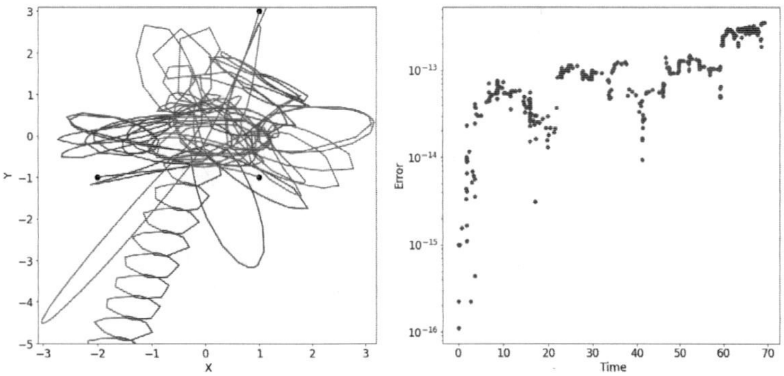

Figure 1. The Pythagorean three-body problem. On the left-hand panel the orbits of the three bodies are plotted in three different colours on the XY-plane, and the initial positions of the particles are marked with black circles. On the right-hand panel we show the relative energy error as a function of time. Time is N-body units.

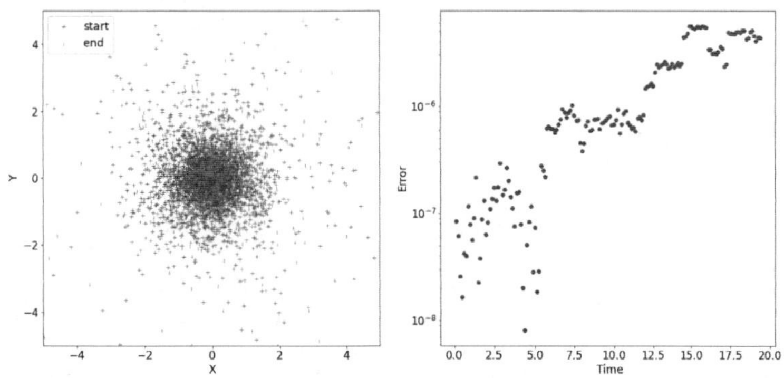

Figure 2. The simulation of a cluster with $N = 2048$ stars. On the left-hand panel, all stars of the cluster are plotted on the XY-plane at the beginning (blue) and at the end (red) of the simulation. On the right-hand side, we show the relative error on the total energy as a function of time. Time is in units of the cluster's two-body relaxation time.

4. The population-synthesis code SEVN

In this section, we will give an overview of the strategies adopted in our code to implement single and binary stellar evolutionary processes. ISTEDDAS is designed to be interfaced with the SEVN population-synthesis code. SEVN (acronym for "Stellar EVolution for N-body") is a population-synthesis tool that couples PARSEC evolutionary tracks for stellar evolution (as default option) with up-to-date models for core-collapse supernova explosions, and it also implements analytical prescriptions for pair-instability supernovae (PISNe) and pulsation pair-instability supernovae (PPISNe) [90, 91, 92, 93, 94, 95]. SEVN also includes analytic prescriptions to model binary-star evolutionary processes, such as wind mass transfer, Roche lobe overflow, common envelop evolution, stellar mergers, tidal evolution, gravitational wave decay, and magnetic braking.

In a N-body code, the evolution of single and binary stars can be implemented following different approaches. A possible strategy is to use a stellar evolution code (e.g. PARSEC) to evolve stars every time that is needed, but this is a very slow approach because stellar evolution codes are very sophisticated and need to integrate the equations for the internal structure of stars, at each step. Another strategy is to use fitting formulas for all the stellar parameters. This approach is very fast but the generation of such formulas is very complicated and time consuming since, each time a stellar prescription has to be updated, all the formulas must be re-derived from scratch. Therefore, it would be hard to keep up with the state-of-the-art stellar evolution prescriptions. In contrast, SEVN uses pre-evolved stellar evolution tracks that are provided in the form of look-up tables, for a grid of masses and metallicities, which are read and interpolated on the fly. This strategy has two crucial advantages: (i) it is computationally very fast, and (ii) users can easy switch to a different set of look-up tables, thus investigating the impact of different stellar evolution prescriptions is possible without changing any lines in the code. Finally, SEVN implements an adaptive time step to control the maximum per-step variations of physical stellar parameters.

4.1. Coupling ISTEDDAS *and* SEVN

SEVN can work as a stand-alone code (for fast population synthesis studies in the field), or can be linked to a N-body code, without having performance penalties. We plan to couple our dynamical evolution code ISTEDDAS with the new population-synthesis code SEVN. We are currently finalizing the interface between the two codes.

There are only three quantities that have to be communicated between the two codes, for each star: mass, radius, and the stellar evolution time-step. The radius is important to check whether two stars collide/merge or they are only passing close through each other. The new predicted masses are communicated to ISTEDDAS at the beginning of each step, since they are needed in the evaluation step. For each star, we have a maximum of three time-steps, two from ISTEDDAS, Δt_{close} and Δt_{far}, and one from SEVN, Δt_{se}. To simplify the synchronization/parallelization process, Δt_{se} is also quantized to powers of two. Thus, in ISTEDDAS, stars belong to three different block structures, corresponding to three different time steps (i.e. time step from neighbors, from distant particles and from stellar evolution processes).

For non-regularized systems, SEVN will have to communicate to ISTEDDAS its Δt_{se} for each star. Depending on the values of the time-steps, we will have three different cases:

- $\Delta t_{se} < \Delta t_{close} \leq \Delta t_{far}$: this could happen for single stars in a fast evolutionary phase, as the RGB (Red-Giant Branch), or in a tight binary with ongoing Roche-lobe overflow/stable mass transfer. In this case, SEVN performs the full stellar evolution step first, then, at the synchronization of Δt_{se} with Δt_{close}, ISTEDDAS will do a full dynamical step;
- $\Delta t_{close} < \Delta t_{se} \leq \Delta t_{far}$: this is similar to the prevoius case, could happen to a non-isolated star in a fast stellar evolutionary phase. In this case, SEVN always performs a full stellar evolution step. ISTEDDAS performs the dynamical evolution with neighbours and the relevant stellar evolution parameters (e.g. mass, radius) at $t < \Delta t_{se}$ are predicted through linear interpolations. At synchronization (i.e. $\Delta t_{close} = \Delta t_{se}$), SEVN performs the next stellar evolution step. We follow a similar procedure when $\Delta t_{far} = \Delta t_{close} = \Delta t_{se}$;
- $\Delta t_{close} \leq \Delta t_{far} < \Delta t_{se}$: the most common case, which happens for a generic star in a slow evolutionary phase (e.g. main sequence). This is similar to the previous case, but the next full stellar evolution step is performed only at the end of the full dynamical step, that is when $\Delta t_{far} = \Delta t_{close} = \Delta t_{se}$.

For regularized binaries, the procedures are similar to the ones described in the previous paragraph, but in this case the SEVN code communicates the physical stellar parameters directly to TSUNAMI, without going through the ISTEDDAS GPU interface. Furthermore, in this case, SEVN is always forced to go through the full binary evolution calculations.

Currently, for regularized multiples, SEVN evolves the hardest (inner) binary using the binary evolution prescriptions, while all the other stars in the subsystem are evolved as if they were single stars. We plan to improve this approach for hierarchical triples, using the SEVN binary routines to evolve both the inner binary and the outer binary.

5. Conclusions

An high-performance direct N-body code with up-to-date stellar evolution prescriptions and high-accuracy treatment of tight binaries is needed to interpret the present and future gravitational-wave detections. We are currently developing ISTEDDAS, a new direct N-body code which is designed to run natively on GPUs and that uses a combination of a high-order integrator (Hermite 6th order) with block time step the Ahmad-Cohen neighbours scheme to significantly speed up N-body simulations. Moreover, ISTEDDAS can accurately integrate very tight binaries, strong gravitational interactions and hierarchical few-body systems thanks to the state-of-the-art TSUNAMI code, which implements the Algorithmic Regularization Chain combined with the Bulirsch-Stoer method. This makes possible to resolve all the span of spacial and time scales inside a stellar cluster, and to treat with high accuracy those compact objects systems that are the final cause of the gravitational wave events observed by the LVC. ISTEDDAS will also be coupled with the population-synthesis code SEVN, a new, up-to-date, look-up table based population-synthesis code that will take care of the evolution of the physical stellar parameters of both single and binary stars.

A preliminary version of ISTEDDAS coupled with TSUNAMI is currently under testing, we have shown two tests on few-body systems (the Pythagorean and the ∞-shape three-body problems) and one test on a cluster with 2048 stars. All the tests are done on a laptop with a CPU Intel(R) Core(TM) i7-10750H 2.60GHz CPU and a GPU NVIDIA GeForce GTX 1650 Ti, without using a softening factor in the force calculations. The few-body tests have a quite high precision (relative error $\sim 10^{-13}$), while the N-body test, without regularization, has a relative error expected for a 6th-order Hermite integrator (maximum $\sim 10^{-6}$).

As next steps, we will continue to develop the "decision making" parts and corresponding interfaces: the one between ISTEDDAS and TSUNAMI, the one between ISTEDDAS and SEVN, and also the one between TSUNAMI and SEVN. We will then perform advanced tests with small and moderate-size star clusters, including all the physical ingredients at the same time. Then, we will implement the GPU version of TSUNAMI to speed-up the parallel integration of many regularized systems and we will perform some advanced profiling tests. Finally, we will investigate the impact of the new methods on the formation and evolution of merging compact-object binaries in star clusters.

6. Acknowledgments

We acknowledge Dr. Mauro Bisson, Dr. Josh Romero and Dr. Massimiliano Fatica for getting in touch during the conference and for joining the ISTEDDAS developing team, and to the anonymous referee for their helpful comments.

References

[1] B.P. Abbot, et al., Physical Review Letters 116 (2016)
[2] B.P. Abbot, et al., Physical Review D 93 (2016)
[3] B.P. Abbot, et al., Astrophysical Journal Letters 851 (2017)
[4] B.P. Abbot, et al., Physical Review Letters 119 (2017)
[5] B.P. Abbot, et al., Physical Review X 9 (2019)
[6] B.P. Abbot, et al., Astrophysical Journal Letters 892 (2020)
[7] R. Abbot, et al., Astrophysical Journal Letters 896 (2020)

[8] R. Abbot, et al., Astrophysical Journal Letters 900 (2020)
[9] R. Abbot, et al., Physical Review Letters 125 (2020)
[10] R. Abbot, et al., Physical Review D 102 (2020)
[11] R. Abbot, et al., Astrophysical Journal Letters 915 (2021)
[12] R. Abbot, et al., Physical Review X 11 (2021)
[13] R. Abbot, et al., arXiv:2111.03606 [gr-qc] (2021)
[14] M. Zevin, S.S. Bavera, C.P.L. Berry, V. Kalogera, T. Fragos, P. Marchant, C.L. Rodriguez, F. Antonini, D.E. Holz, C. Pankow, The Astrophysical Journal 910 (2021)
[15] M. Mapelli, Formation Channels of Single and Binary Stellar-Mass Black Holes, Handbook of Gravitational Wave Astronomy, Springer (2021)
[16] B. Paczynski, ed. P. Eggleton, S. Mitton, J. Whelan, 73 (1976)
[17] H.A. Bethe, G.E. Brown, The Astrophysical Journal 506 780 (1998)
[18] K. Belczynnski, V. Kalogera, T. Bulik, The Astrophysical Journal 527 407 (2002)
[19] M. Dominik, K. Belczynski, C. Fryer, et al., The Astrophysical Journal 759 52 (2012)
[20] J.J. Eldridge, E.R. Stanway, Monthly Notices of the Royal Astronomical Society 462 3302 (2016)
[21] S. Stevenson, A. Vigna-Gomez, I. Mandel, et al., Nature Communications 8 14906 (2017)
[22] N. Giacobbo, M. Mapelli, Monthly Notices of the Royal Astronomical Society 480 2011 (2018)
[23] C.J. Neijssel, A. Vigna-Gomez, S. Stevenson, et al., Monthly Notices of the Royal Astronomical Society 490 3740 (2019)
[24] S.E. De Mink, I. Mandel, Monthly Notices of the Royal Astronomical Society 460 3545 (2016)
[25] I. Mandel, S.E. De Mink, Monthly Notices of the Royal Astronomical Society 458 2634 (2016)
[26] P. Marchant, N. Langer, P. Podsiadlowski, T. Tauris, T. Moriya, Astronomy & Astrophysics 588 A50 (2016)
[27] S.F. Portegies Zwart, S.L.W. McMillan, M. Gieles, Annual Reviews 48 431-483 (2010)
[28] M. Mapelli M., Monthly Notices of the Royal Astronomical Society 459 3432 (2016)
[29] K. Kremer, C.S. Ye, S. Chatterjee, C.L. Rodriguez, F.A. Rasio, Cambridge University Press (2020)
[30] A.P. Lightman, S.L. Shapiro, Reviews of Modern Physics 50 437 (1978)
[31] S. Sigurdsson, L. Hernquist, Nature 364 423 (1993)
[32] S. McMillan, P. Hut, J. Makino, The Astrophysical Journal 372 111 (1991)
[33] P. Hut, S. McMillan, J. Goodman, et al., Publications of the Astronomical Society of the Pacific 104 981 (1992)
[34] M.C. Miller, D.P. Hamilton, The Astrophysical Journal 576 894 (2002)
[35] S.F. Portegies Zwart, S.L.W. McMillan, The Astrophysical Journal Letters 528 17 (2000)
[36] R.M. O'Leary, F.A. Rasio, J.M. Fregeau, N. Ivanova, R. O'Shaughnessy, The Astrophysical Journal 637 937 (2006)
[37] J.M.B. Downing, M.J. Benacquista, M. Giersz, R. Spurzem, Monthly Notices of the Royal Astronomical Society 407 1946 (2010)
[38] B.M. Ziosi, M. Mapelli, M. Branchesi, G. Tormen, Monthly Notices of the Royal Astronomical Society 441 3703 (2014)
[39] C.L. Rodriguez, M. Morscher, B. Pattabiraman, et al., Physical Review Letters 115 051101 (2015)
[40] F. Antonini, F.A. Rasio, The Astrophysical Journal 83, 1 (2016)
[41] A. Askar, M. Szkudlarek, M., D. Gondek-Rosinska, M. Giersz, T. Bulik, Monthly Notices of the Royal Astronomical Society 464 36 (2017)
[42] S. Banerjee, Monthly Notices of the Royal Astronomical Society 467 524 (2017)
[43] J. Samsing, E. Ramirez-Ruiz, The Astrophysical Journal Letters 840 L14 (2017)
[44] J.G. Hills, L.W. Fullerton, Astronomical Journal 85 1281–1291 (1980)
[45] P.C. Peters, Physics Review 136 1224–1232 (1964)
[46] D.C. Heggie, Monthly Notices of the Royal Astronomical Society 173 729 (1975)
[47] J.G. Hills Astronomical Journal 80 809 (1975)
[48] J. Binney, S. Tremaine S., Galactic Dynamics, Princeton University Press (1987)
[49] L. Wang, et al., Monthly Notices of the Royal Astronomical Society 458 1450 (2016)
[50] L. Spitzer, Dynamical Evolution of Globular Clusters, Princeton University Press (1987)
[51] S. Chandrasekhar, Stellar Structure and Stellar Atmospheres, The University of Chicago press (1942)
[52] J. Makino, E. Kokubo, M. Taiji, Publications of the Astronomical Society of Japan 45 349 (1993)
[53] J. Makino, M. Taiji, Scientific Simulations with Special-Purpose Computers: the GRAPE Systems, Wiley-VCH (1998)
[54] J. Makino, T. Fukushige, M. Koga, K. Namura, Publications of the Astronomical Society of Japan 55 1163 (2003)
[55] S. Konstantinidis, K.D. Kokkotas, Astronomy & Astrophysics 522 A70 (2010)
[56] S.F. Portegies Zwart, S.L.W. McMillan, P. Hut, J. Makino, Monthly Notices of the Royal Astronomical

Society 321 199-226 (2001)
[57] E. Gaburov, S.Harfst, S. Portegies Zwart, New Astronomy 14 630-637 (2009)
[58] E. Gaburov, S.Harfst, S. Portegies Zwart, arXiv:1510.04068 (2015)
[59] L. Wang, R. Spurzem, A. Aarseth, K. Nitadori, P. Berczik, M.B.N. Kouwenhoven, T. Naab, Monthly Notices of the Royal Astronomical Society 450 4070 (2015)
[60] S.J. Aarseth, Publications of the Astronomical Society of the Pacific 111 1333 (1999)
[61] S.J. Aarseth, Celestial Mechanics and Dynamical Astronomy 73 127–137 (1999)
[62] C. Maureira-Fredes, P. Amaro-Seoane, Monthly Notices of the Royal Astronomical Society 473 3113–3127 (2018)
[63] L. Wang, Ma. Iwasawa, K. Nitadori, J. Makino, Monthly Notices of the Royal Astronomical Society 497 536-555 (2020)
[64] R. Capuzzo-Dolcetta, M. Spera, D. Punzo, Journal of Computational Physics 236 580-593 (2013)
[65] P. Berczik, R. Spurzem, L. Wang, S. Zhong, O. Veles, I. Zinchenko, S. Huang, M. Tsai, G. Kennedy, S. Li, L. Naso, C. Li, arXiv:1312.1789 [astro-ph.IM] (2013)
[66] J. Bédorf, E. Gaburov, S. Portegies Zwart, arXiv:1204.2280 [astro-ph.IM] (2012)
[67] Y. Miki, M. Umemura, New Astronomy 52 65-81 (2017)
[68] C.L. Rodriguez, M. Morscher, L. Wang, S. Chatterjee, F.A. Rasio, R. Spurzem, Monthly Notices of the Royal Astronomical Society 463 2109–2118 (2016)
[69] A. Hypki, M. Giersz, Monthly Notices of the Royal Astronomical Society, 429 1221–1243 (2013)
[70] C.L. Rodriguez, B. Pattabiraman, S. Chatterjee, A. Choudhary, W. Liao, M. Morscher, F.A. Rasio, Computational Astrophysics and Cosmology 5 5 (2005)
[71] A.C. Tout, S.J. Aarseth, O.R. Pols, P.P. Eggleton, Monthly Notices of the Royal Astronomical Society 291 732-748 (1997)
[72] J.R. Hurley, O.R. Pols, C.A. Tout, Monthly Notices of the Royal Astronomical Society 315 543-569 (2000)
[73] J.R. Hurley, C.A. Tout, O.R. Pols, Monthly Notices of the Royal Astronomical Society 329 897-928 (2002)
[74] M. Mapelli, N. Giacobbo, E. Ripamonti, M. Spera, Monthly Notices of the Royal Astronomical Society 472 2422–2435 (2017)
[75] N. Giacobbo, M. Mapelli, M. Spera, Monthly Notices of the Royal Astronomical Society 474 2959–2974 (2018)
[76] Zhe Cui, Xiang-Dong Li, arXiv:2111.04496 [astro-ph.SR] (2021)
[77] J.E. Wall, S.L.W. McMillan, M.M.M. Low, R.S. Klessen, S. Portegies Zwart, The Astronomycal Journal 887 62 (2019)
[78] K. Nitadori, J. Makino, New Astronomy 13 498 (2008)
[79] S.J. Aarseth, Gravitational N-body simulations: tools and algorithms, Cambridge University Press (2003)
[80] J. Makino, S.J. Aarseth, Publications of the Astronomical Society of Japan 44 141-151 (1992)
[81] A. Ahmad, L. Cohen, Journal of computational physics 12 389-402 (1973)
[82] W.H. Press, S.A. Teukolsky, W.T. Vetterling, B.P. Flannery, Numerical Recipes in C : the Art of Scientific Computing, Cambridge University Press (1992)
[83] Y. Wang, N.W.C. Leigh, B. Liu, R. Perna, Monthly Notices of the Royal Astronomical Society 505 1053–1070 (2021)
[84] S. Mikkola, S.J. Aarseth, Celestial Mechanics and Dynamical Astronomy 57 439–459 (1993)
[85] S. Mikkola, K. Tanikawa, Monthly Notices of the Royal Astronomical Society 310 745–749 (1999)
[86] S. Mikkola, D. Merritt, Monthly Notices of the Royal Astronomical Society 372 219-223 (2006)
[87] S. Mikkola, D. Merritt, The Astronomical Journal 135 2398 (2008)
[88] A.A. Trani, M.S. Fujii, M. Spera, The Astrophysical Journal 875 42 (2019)
[89] A.H.W. Kupper, T. Maschberger, P. Kroupa, H. Baumgardt, Monthly Notices of the Royal Astronomical Society 417 2300–2317 (2011)
[90] M. Spera, M. Mapelli, A. Bressan, Monthly Notice of the Royal Astronomical Society 451 4086–4103 (2015)
[91] M. Spera, M. Mapelli, Monthly Notice of the Royal Astronomical Society 470 4739-4749 (2017)
[92] M. Spera, M. Mapelli, N. Giacobbo, A.A. Trani, A. Bressan, G. Costa, Monthly Notices of the Royal Astronomical Society 485 889–907 (2019)
[93] A. Bressan, P. Marigo, L. Girardi, B. Salasnich, C. Dal Cero, S. Rubele, A. Nanni, Monthly Notices of the Royal Astronomical Society 427 127-145 (2012)
[94] Y. Chen, L. Girardi, A. Bressan, P. Marigo, M. Barbieri, X. Kong, Monthly Notices of the Royal Astronomical Society 444 2525-2543 (2014)
[95] J. Tang, A. Bressan, P. Rosenfield, A. Slemer, P. Marigo, L. Girardi, L. Bianchi, Monthly Notices of the Royal Astronomical Society 445 4287-4305 (2014)

Efficient computations of continuous action densities of states for lattice models

Biagio Lucini[1], Olmo Francesconi[2], Markus Holzmann[3], David Lancaster[4] and Antonio Rago[4,5]

[1] Mathematics Department and Swansea Academy of Advanced Computing, Swansea University, Bay Campus, SA1 8EN, Swansea, Wales, United Kingdom
[2] IMADA and CP3-Origins, Univ. of Southern Denmark, Campusvej 55, DK-5230 Odense, Denmark
[3] Université Grenoble Alpes, CNRS, LPMMC, 3800 Grenoble, France
[4] Centre for Mathematical Sciences, University of Plymouth, Plymouth, PL4 8AA, United Kingdom
[5] CERN, Theoretical Physics Department, 1211 Geneva 23, Switzerland

E-mail: b.lucini@swansea.ac.uk, francesconi@imada.sdu.dk, markus.holzmann@grenoble.cnrs.fr, david.lancaster@plymouth.ac.uk, antonio.rago@plymouth.ac.uk

Abstract. The Logarithmic Linear Relaxation (LLR) algorithm is an efficient method for computing densities of states for systems with a continuous spectrum. A key feature of this method is exponential error reduction, which allows us to evaluate the density of states of a system over hundreds of thousands of orders of magnitude with a fixed level of relative accuracy. As a consequence of exponential error reduction, the LLR method provides a robust alternative to traditional Monte Carlo calculations in cases in which states suppressed by the Boltzmann weight play nevertheless a relevant role, e.g., as transition regions between dominant configuration sets. After reviewing the algorithm, we will show an application in U(1) Lattice Gauge Theory that has enabled us to obtain the most accurate estimate of the critical coupling with modest computational resources, defeating exponential tunneling times between metastable vacua. As a further showcase, we will then present an application of the LLR method to the decorrelation of the topological charge in SU(3) Lattice Gauge Theory near the continuum limit. Finally, we will review in general applications of the LLR algorithm to systems affected by a strong sign problem and discuss the case of the Bose gas at finite chemical potential.

1. Introduction

Markov Chain Monte Carlo (MCMC) are particularly well suited in calculations in which ensemble averages of extensive quantities that can be expressed explicitly as a function of the fields need to be computed. While cases in this category cover a good cross-section of relevant quantities in physics, they are by no means exhaustive. In fact, there are various scenarios, ranging from computations of partition functions to the sign problem, in which MCMC are inefficient. In this contribution, we will review the Logarithmic Linear Relaxation (LLR) algorithm, which enables us to overcome the limitations of MCMC methods in the presence of small overlaps between domains of relevant configurations. The superior performance of this algorithm is provided by exponential error reduction[1], which is a central feature of the method.

This proceeding is organised as follows. We will present the LLR algorithm for real action systems and provide some example applications in Sect. 2. In Sect. 3, we will formulate the algorithm for complex action systems and discuss its application in a numerical study of the Bose gas at finite chemical potential, which is a popular benchmark model for assessing the efficiency of an algorithm at circumventing the sign problem. Sect. 4 will summarise our contribution and outline some further directions of investigation.

2. The LLR algorithm for real action systems

Let us start by considering an Euclidean quantum field theory[2] written in a general form, whose path integral is given by the expression

$$Z(\beta) = \int [D\phi] e^{-\beta S[\phi]} \ . \tag{1}$$

Here, ϕ is a field configuration over a lattice Λ, β is the coupling and S the action. The integral is performed over all possible values of the fields. The density of states associated to a value E of the action S of the model is defined as

$$\rho(E) = \int [D\phi] \delta(S[\phi] - E) \ . \tag{2}$$

Using this expression for ρ, we can reformulate the path integral as an integral over all possible energy values weighted by the density of states and the Boltzmann measure:

$$Z(\beta) = \int dE \rho(E) e^{-\beta E} = e^{-\beta F} \ . \tag{3}$$

If the density of states is known, free energies and expectation values are accessible via simple numerical integrations. For instance, for an observable that depends only on E,

$$\langle O \rangle = \frac{\int dE \rho(E) O(E) e^{-\beta E}}{\int dE \rho(E) e^{-\beta E}} \ . \tag{4}$$

Therefore, an algorithm that enables us to compute $\rho(E)$ will straightforwardly give us access, through a numerical integration, to the free energy F and to values of thermodynamic ensemble averages of observables depending on E as a function of β. For systems with discrete energy levels, an algorithm of this type, the celebrated Wang-Landau algorithm, was provided in [1]. The LLR method, which was inspired by the latter work, is an algorithm for the calculation of densities of states in systems with a continuous spectrum [2, 3]. This algorithm is implemented as follows. We start by dividing the (continuum) energy interval of the system in N sub-intervals of amplitude δ_E, with the interval n centered at the value E_n. We then define a piecewise continuous local linear approximation of $\log \rho(E)$ as

$$\log \tilde{\rho}(E) = a_n (E - E_n) + c_n \qquad \text{for } E_n - \delta_E/2 \le E \le E_n + \delta_E/2 \ , \tag{5}$$

which is valid for sufficiently small width δ_E of the energy sub-intervals. We obtain a_n as the root of the stochastic equation

$$\langle\langle \Delta E \rangle\rangle_{a_n} = \int_{E_n - \frac{\delta_E}{2}}^{E_n + \frac{\delta_E}{2}} (E - E_n) \rho(E) e^{-a_n E} dE = 0 \tag{6}$$

[1] We use the expression *exponential error reduction* to indicate that the relative error of a quantity is independent from the value of the latter.

[2] Although for the sake of definiteness in this work we use the language of Euclidean quantum field theories, a translation of the method to a statistical mechanics context is immediate.

using the Robbins-Monro iterative method [4]

$$\lim_{m\to\infty} a_n^{(m)} = a_n \; , \qquad a_n^{(m+1)} = a_n^{(m)} - \frac{\alpha}{m} \frac{\langle\langle \Delta E \rangle\rangle_{a_n^{(m)}}}{\langle\langle \Delta E^2 \rangle\rangle_{a_n^{(m)}}} \; , \qquad \alpha \text{ constant} \; . \tag{7}$$

In Eqs. (6) and (7) we have used the double-angle notation for the expectation of an observable $O(E)$, which is defined as

$$\langle\langle O(E) \rangle\rangle_{a_n} = \frac{1}{\mathcal{N}} \int_{E_n - \frac{\delta_E}{2}}^{E_n + \frac{\delta_E}{2}} O(E) \rho(E) e^{-a_n E} dE \; , \qquad \mathcal{N} = \int_{E_n - \frac{\delta_E}{2}}^{E_n + \frac{\delta_E}{2}} \rho(E) e^{-a_n E} dE \; . \tag{8}$$

These energy-restricted integrals can be easily reformulated as integrals over the all spectrum with sharp cut-offs provided by appropriate Heaviside functions. These sharp cut-offs can be replaced with a smooth Gaussian cut-off [5]. The double-angle expectations are computed with a MCMC restricted to the relevant energy interval. Note that if we repeat the Robbins-Monro algorithm starting from different random numbers, asymptotically, at fixed number of iterations m, the $a_n^{(m)}$ are gaussianly distributed around the root $a_n = a_n^{(\infty)}$ with a variance that goes to zero as m increases. This property provides an immediate strategy for converting potential systematic errors into statistical errors in the root finding procedure.

After computing the a_n for all n, setting $c_1 = 0$, the piecewise continuity of $\log \tilde{\rho}(E)$ gives

$$c_n = \frac{\delta}{2} a_1 + \delta \sum_{k=2}^{n-1} a_k + \frac{\delta}{2} a_n \; , \qquad n \geq 2 \; . \tag{9}$$

The numerically determined $\tilde{\rho}(E)$ can be used in path integral calculations *in lieu* of $\rho(E)$. This procedure result in a systematic approximation error in δ_E that is however controlled, since it scales quadratically with the width of the sub-intervals. Remarkably, our procedure for determining $\tilde{\rho}$ provides exponential error suppression [3]. It is worth noting that the restricted sampling discussed above is non-ergodic. Ergodicity can be recovered using the replica exchange method, as discussed in [6].

Exponential error suppression has been proved to be spectacularly implemented in the method in [2], where, using the piecewise continuous density of states that has been reconstructed over 250000 orders of magnitude, it has been shown that the LLR-determined SU(3) lattice gauge theory plaquette agrees with MCMC calculations over a wide range of β values that interpolate between the strong coupling and the weak coupling regime of the theory.

Due to the presence of a first-order deconfinement phase transition, a notoriously hard to simulate system is compact U(1) lattice gauge theory. The best available MCMC calculations used a large amount of computer time on then state-of-the-art supercomputers, reaching a maximum lattice size of 18^4 when periodic boundary conditions were imposed [7]. Using the LLR algorithm, we were able to perform accurate and robust calculations with moderate computational resources on a 20^4 lattice [3]. A reconstruction of the probability distribution of the energy at the pseudocritical value of β obtained on our largest system with the LLR algorithm is shown in Fig. 1.

More recently, we have shown in [8] that the LLR algorithm mitigates significantly the problem of the topological freezing [9] by performing a calculation of the correlation time of the topological charge in SU(3) pure gauge theory near the continuum limit. Fig. 2 shows that, compared to MCMC methods, the LLR algorithm reduces this correlation time by about one order of magnitude.

Other successful applications of the LLR method include the calculation of the order-disorder interface tension in the Potts model [6] and the calculation of the renormalisation constant for the energy-momentum tensor in SU(2) lattice gauge theory using the method of shifted boundary conditions, which in turn requires computations of free energies from partition functions [5].

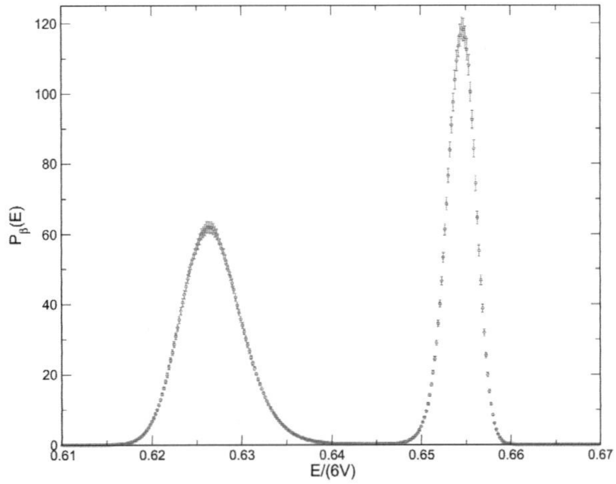

Figure 1. Probability distribution of the energy E for compact U(1) lattice gauge theory at $\beta = 1.011006$ (central estimate for the critical value using the latent heat peak) on a 20^4 lattice.

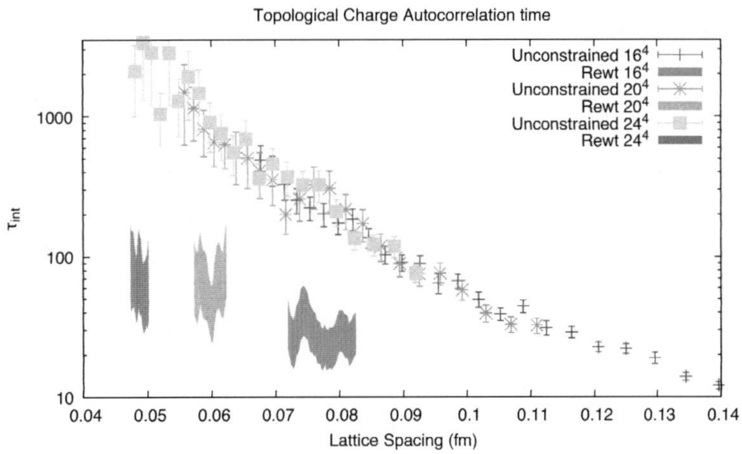

Figure 2. Correlation time of the topological charge for SU(3) gauge theory at the lattice volumes shown as a function of the lattice spacing. The "unconstrained" points are obtained with a standard heat bath method, while the "rewt" points come from LLR simulations.

3. The LLR algorithm for complex action systems

We now consider an Euclidean quantum field theory with a complex action, whose path integral we write as

$$Z(\mu) = \int [D\phi] e^{-\beta S[\phi] + i\mu Q[\phi]} \ . \tag{10}$$

The generalised density of states is defined as

$$\rho(q) = \int [D\phi] e^{-\beta S[\phi]} \delta(Q[\phi] - q) \ . \tag{11}$$

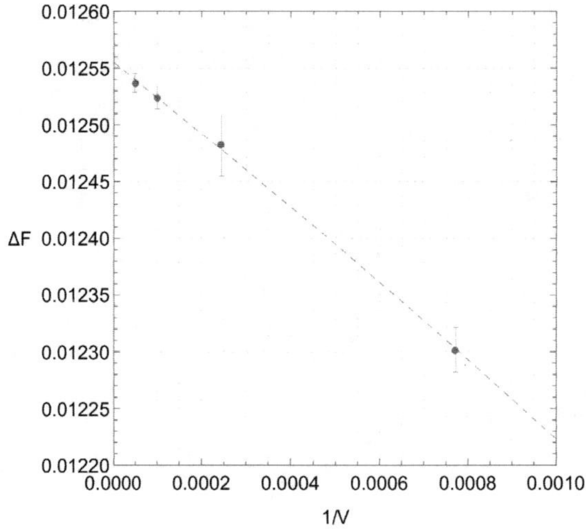

Figure 3. Computation of the overlap free energy for the Bose gas at finite chemical potential for various volumes. A linear extrapolation in $1/V$ to the thermodynamic limit is also shown.

In terms of $\rho(q)$, we rewrite Z as

$$Z(\mu) = \int dq \rho(q) e^{i\mu q} \ . \tag{12}$$

In general, the above integral is strongly oscillating, These oscillations generate the large cancellations typical of a sign problem scenario. Because of the latter, when performing a calculation $\rho(q)$ in general needs to be known with an extraordinary degree of accuracy.

The severity of the sign problem is indicated by the *vev* of the phase factor in the phase quenched ensemble:

$$\langle e^{i\mu q} \rangle = \frac{Z(\mu)}{Z(0)} = e^{-V\Delta F} \to 0 \qquad \text{exponentially in } V \ . \tag{13}$$

Eq. (13) shows that the sign problem can be reformulated as an *overlap problem*, as the ensemble average of the phase factor depends on the overlap of relevant configurations for two different partition functions, which is exponentially suppressed with the volume. Since the LLR method has proved to be very efficient at computing regions with suppressed densities, it is natural to explore how it performs in this case[3]. We note though that, while the computation of the density is in the scope of the LLR algorithm, integration is a separate problem. In particular, in the complex case, it has been shown in [11] that a piecewise approximation of ρ does not provide sufficient precision for computing relevant observables. In the same work, using a $Z(3)$ spin model as a prototype system, the authors showed that a polynomial interpolation of ρ can be performed that is stable against the order of the interpolation and at the same time provides the required accuracy[4].

More recently, the method has been applied to the four-dimensional Bose gas system at finite chemical potential [13]. Fig. 3 shows the computation of the overlap free energy ΔF as a function

[3] A method for calculations of densities of states for systems with complex actions that is similar in spirit to the LLR algorithm is the FFA method, first proposed in [10].

[4] See [12] for a related proposal for the interpolation of the density based on a cumulant expansion.

of the volume V for the latter system at chemical potential $\mu = 0.8$. Volumes up to the size $V = 16^4$ have been used. A linear extrapolation in $1/V$ (also reported in the figure) gives

$$\Delta F = (0.012557 \pm 0.000004) - \frac{(0.329 \pm 0.008)}{V} , \tag{14}$$

whose infinite volume limit is reassuringly close to the mean-field result $\Delta F_{MF} \simeq 0.012522$ computed in [14]. At the same time, the LLR result is sufficiently accurate to expose the expected deviations from the mean-field value.

4. Conclusions

The LLR algorithm enables accurate computations of density of states that can be used for precise calculations in scenarios in which MCMC are known to be inefficient. In this contribution, we have reviewed in particular some applications to first-order phase transitions, to the decorrelation of the topological charge and to the sign problem. A natural future direction is an extension to the algorithm to systems with fermions [15].

Acknowledgements

This work has been partially supported by the ANR project ANR-15-IDEX-02e. The work of BL is supported in part by the Royal Society Wolfson Research Merit Award WM170010 and by the Leverhulme Foundation Research Fellowship RF-2020-461\9. BL received funding also from the European Research Council (ERC) under the European Union's Horizon 2020 research and innovation programme under grant agreement No 813942. AR is supported by the STFC Consolidated Grant ST/P000479/1. Numerical simulations have been performed on the Swansea SUNBIRD system, provided by the Supercomputing Wales project, which is part-funded by the European Regional Development Fund (ERDF) via the Welsh Government, and on the HPC facilities at the HPCC centre of the University of Plymouth.

References

[1] Wang F and Landau D P 2001 *Phys. Rev. Lett.* **86** 2050 (*Preprint* cond-mat/0011174)
[2] Langfeld K, Lucini B and Rago A 2012 *Phys. Rev. Lett.* **109** 111601 (*Preprint* 1204.3243)
[3] Langfeld K, Lucini B, Pellegrini R and Rago A 2016 *Eur. Phys. J.* **C76** 306 (*Preprint* 1509.08391)
[4] Robbins H and Monro S 1951 *The Annals of Mathematical Statistics* **22** 400–407 ISSN 00034851 URL http://www.jstor.org/stable/2236626
[5] Pellegrini R, Lucini B, Rago A and Vadacchino D 2017 *PoS* **LATTICE2016** 276
[6] Lucini B, Fall W and Langfeld K 2016 *PoS* **LATTICE2016** 275 (*Preprint* 1611.00019)
[7] Arnold G, Bunk B, Lippert T and Schilling K 2003 *Nucl. Phys. B Proc. Suppl.* **119** 864–866 (*Preprint* hep-lat/0210010)
[8] Cossu G, Lancaster D, Lucini B, Pellegrini R and Rago A 2021 *Eur. Phys. J. C* **81** 375 (*Preprint* 2102.03630)
[9] Schaefer S, Sommer R and Virotta F (ALPHA) 2011 *Nucl. Phys. B* **845** 93–119 (*Preprint* 1009.5228)
[10] Giuliani M, Gattringer C and Torek P 2016 *Nucl. Phys.* **B913** 627–642 (*Preprint* 1607.07340)
[11] Langfeld K and Lucini B 2014 *Phys. Rev.* **D90** 094502 (*Preprint* 1404.7187)
[12] Garron N and Langfeld K 2017 *Eur. Phys. J.* **C77** 470 (*Preprint* 1703.04649)
[13] Francesconi O, Holzmann M, Lucini B and Rago A 2020 *Phys. Rev. D* **101** 014504 (*Preprint* 1910.11026)
[14] Aarts G 2009 *JHEP* **05** 052 (*Preprint* 0902.4686)
[15] Francesconi O, Holzmann M, Lucini B, Rago A and Rantaharju J 2019 *PoS* **LATTICE2019** 200 (*Preprint* 1912.04190)

XXXII IUPAP Conference on Computational Physics IOP Publishing
Journal of Physics: Conference Series **2207** (2022) 012053 doi:10.1088/1742-6596/2207/1/012053

General purpose lattice QCD code set Bridge++ 2.0 for high performance computing

Yutaro Akahoshi[1], Sinya Aoki[1], Tatsumi Aoyama[2], Issaku Kanamori[3], Kazuyuki Kanaya[4], Hideo Matsufuru[2], Yusuke Namekawa[5], Hidekatsu Nemura[1], Yusuke Taniguchi[4]

[1] Yukawa Institute for Theoretical Physics, Kyoto University, Japan
[2] High Energy Accelerator Research Organization (KEK), Japan
[3] RIKEN Center for Computational Science (R-CCS), Japan
[4] University of Tsukuba, Japan
[5] Department of Physics, Kyoto University, Japan

E-mail: `kanamori-i@riken.jp`

Abstract.
Bridge++ is a general-purpose code set for a numerical simulation of lattice QCD aiming at a readable, extensible, and portable code while keeping practically high performance. The previous version of Bridge++ is implemented in double precision with a fixed data layout. To exploit the high arithmetic capability of new processor architecture, we extend the Bridge++ code so that optimized code is available as a new branch, *i.e.*, an alternative to the original code. This paper explains our strategy of implementation and displays application examples to the following architectures and systems: Intel AVX-512 on Xeon Phi Knights Landing, Arm A64FX-SVE on Fujitsu A64FX (Fugaku), NEC SX-Aurora TSUBASA, and GPU cluster with NVIDIA V100.

1. Introduction

Bridge++[1] is a code set for numerical simulations of lattice QCD[2], designed on the object-oriented programming and described in the C++ language. A goal of the project is to develop a readable, extensible, and portable code set with sufficiently high performance. When the development was launched in 2009, our major target platforms were parallel scalar systems represented by IBM Blue Gene/Q. Recent supercomputers, however, adopt a variety of architecture: multi-core parallel machines with wide SIMD (A64FX and Intel processors), and clusters with accelerator devices such as GPUs, PEZY-SC, and vector processors (NEC SX-Aurora). Soon after the first public release of Bridge++ in 2012 [2], we had started to investigate possible extensions of our code to exploit these new architectures while keeping the readability and portability, as well as to develop tuning techniques for them [3, 4, 5, 6, 7, 8]. Recently we have constructed a framework to incorporate the tuned codes as an alternative part to the previously developed Bridge++ code, and decided to release it as version 2.0.

In this paper, we describe the fundamental structure of this updated code set with several examples of application to recent architectures. In the next section, after a brief introduction of

[1] `https://bridge.kek.jp/Lattice-code/`
[2] Basics of lattice QCD are covered by many text textbooks, *e.g.*, [1].

the lattice QCD and its bottleneck, we describe the structure of Bridge++ ver.2.0 code set. In Sec. 3 we show the performance measured on available systems: the SIMD architectures (Intel AVX-512 and Arm A64FX), vector architecture (NEC SX-Aurora TSUBASA), and GPU devices with NVIDIA V100. The last section is our conclusion. Some details of the implementation have been reported in Ref. [8] together with that of a multi-grid solver.

2. Functionality and Code Structure

The lattice QCD is a field theory formulated on a four-dimensional Euclidean lattice. It consists of fermion (quark) fields and a gauge (gluon) field. The latter mediates the strong interaction among quarks and are represented by 'a link variable', $U_\mu(x) \in SU(3)$, where $x = (x_1, x_2, x_3, x_4)$ stands for a lattice site and $\mu = 1$–4 is the spacetime direction. In numerical simulations the lattice size is finite: $x_\mu = 1, 2, \ldots, L_\mu$. The fermion field is represented as a complex vector on lattice sites, which carries 3 components of 'color' and 4 components of 'spinor', thus in total 12, degrees of freedom on each site. We set the lattice spacing $a = 1$ throughout this paper.

The standard steps of lattice QCD simulations are as follows. One first generates gauge configurations $\{U_\mu(x)\}$ with a Monte Carlo method. Then physical observables are measured on $\{U_\mu(x)\}$, and their averages give the expectation values. Typical observables, such as hadronic matrix elements, are composed of fermion propagators, which are solutions of the linear equation for a fermion matrix $D[U]$. Since such a linear equation must be solved repeatedly also during the generation of $\{U_\mu(x)\}$, it is one of the main bottlenecks of lattice simulations. The fermion matrix, explicitly exemplified below, represents the interaction among quarks and gluons and is in general a large sparse matrix of a rank proportional to the lattice volume. The discretized fermion matrix has a variety since the requirement on $D[U]$ is to coincide with that of QCD only in the continuum limit (the lattice spacing $a \to 0$). In addition, efficient solver algorithms also have a variety depending on the system size, the condition number of the matrix, and so on.

Our code set Bridge++ is intended to cover a wide range of measurements, fermion matrices, and algorithms. It is parallelized with MPI and the time consuming parts are multi-threaded with OpenMP. The data structure initially adopted was in a fixed style with double precision. For further optimization to up-to-date architectures, we decided to replace the time consuming part with an alternative code leaving the other parts unchanged. The original code, called 'core library', provides building tools, reference results, and measurements that require less performance. On the other hand, the 'alternative code' offers the same functionality as that of the core library with higher flexibility to achieve better performance. It is possible to use multiple branches of the alternative code simultaneously if required.

The 'alternative code' is composed of the classes in two categories. One is those directly manipulate the data and thus are performance-sensitive, such as the fermion matrices. These classes are implemented and optimized in a way specific to each architecture. The other category contains the algorithms generically described by the C++ template. This structure allows incremental adoption to a new architecture. One only needs to implement the required operators and instantiate the desired algorithms.

Figure 1 shows a sample code which employs the extended Bridge++ with a branch for SIMD architecture. In line-3, `AField<double,SIMD>` indicates a class of field object in double precision for the 'SIMD' branch. The template class `Fprop_Standard_lex_alt_Mixedprec` determines a fermion propagator by a mixed precision solver algorithm. The sets of parameters for the fermion matrix and the linear equation solver are stored in the objects `params_fopr` and `params_solver`, respectively, including the types of fermion and algorithm. Once the object '`fprop`' is instantiated, the solution of the linear equation is determined and returned in the format of the core library, `Field_F`. One does not need to modify observables that use the solution '`sq[idx]`' to use the 'alternative' codes.

Journal of Physics: Conference Series **2207** (2022) 012053 doi:10.1088/1742-6596/2207/1/012053

```
// prepare a mixed precision solver for architecture SIMD
// details of the fermion op.: params_fopr; details of the solver: params_solver
Fprop_Standard_lex_alt_Mixedprec<AField<double,SIMD>, AField<float,SIMD>> fprop( params_fopr, params_solver );
fprop.set_config(U.get()); // set the gauge field
fprop.set_mode("D"); // mode is D, Ddag, DdagD,...

// source and solution are from core library
Source source("Local"); source.set_parameters(params_source);
std::vector<Field_F> sq(Nc * Nd); // Nc=3, Nd=4 : propagator (= 12 solution vectors)
Field_F b;

int nconv; double diff;
for (int id = 0; id < Nd; ++id) {
  for (int ic = 0; ic < Nc; ++ic) {
    int idx = ic + Nc * id;
    sq[idx].set(0.0);
    source.set(b, idx); // set the source
    fprop.invert(sq[idx], b, nconv, diff); // solve the linear equation
  }
}
```

Figure 1. A sample code to calculate a fermion propagator with a mixed precision solver. The blue-colored codes are from the 'alternative' code with the red-colored template parameter, while the orange-colored objects are from the 'core library'.

3. Implementation and Performance for each Architecture

We describe our code implementation and performance results for several architectures. While the code is available in both double precision and single precision, we mainly describe the latter since it plays a main role in multi-precision solver algorithms. We quote weak scaling behavior of the performance of matrix-vector multiplication without details of the tuning and execution setup.

A fermion matrix $D[U]$ acts on a fermion vector $\psi(x)$. As a typical example, we examine the $O(a)$-improved Wilson fermion matrix, also called clover fermion matrix,

$$D_{x,y} = [1 + F(x)]\delta_{x,y} - \kappa \sum_{\mu=1}^{4} \left[(1 - \gamma_\mu) U_\mu(x) \delta_{x+\hat{\mu},y} + (1 + \gamma_\mu) U_\mu^\dagger(x - \hat{\mu}) \delta_{x-\hat{\mu},y} \right], \tag{1}$$

where x, y are lattice sites, $\hat{\mu}$ the unit vector along μ-th axis, and the hopping parameter $\kappa = 1/(8 + 2m_0)$ related to the quark mass m_0. The link variable $U_\mu(x)$ is a 3×3 complex matrix acting on the color and γ_μ is a 4×4 matrix acting on the spinor degrees of freedom. $F(x)$ is a 12×12 Hermitian matrix made of the link variables, and helps to reduce the finite lattice spacing artifact. Thus D is a complex matrix of the rank $4 \cdot 3L_x L_y L_z L_t$. The boundary condition, such as a periodic boundary, is imposed in each direction.

3.1. SIMD architectures: Intel AVX-512 and Fujitsu A64FX

We start with the implementation for two SIMD architectures: Intel AVX-512 and Armv8.2-A with SVE (Scalable Vector Extension). Intel AVX-512 is the latest SIMD extension of x86 instruction set architecture with the 512-bit SIMD length. It is available on recent Xeon processors as well as Xeon Phi Knights Landing. The Armv8.2-A with SVE is adopted by the Fujitsu A64FX processor for the Fugaku supercomputer. While SVE enables variable SIMD length, currently 512-bit length is available on A64FX.

Although in both architectures the SIMD length corresponds to 16 single precision floating numbers, the efficient way of packing variables into the SIMD unit depends on the structure of arithmetic operation units. In the case of AVX-512, we pack 8 complex numbers which are consecutive in the x-direction as displayed in the left panel of Fig. 2. This requires that the

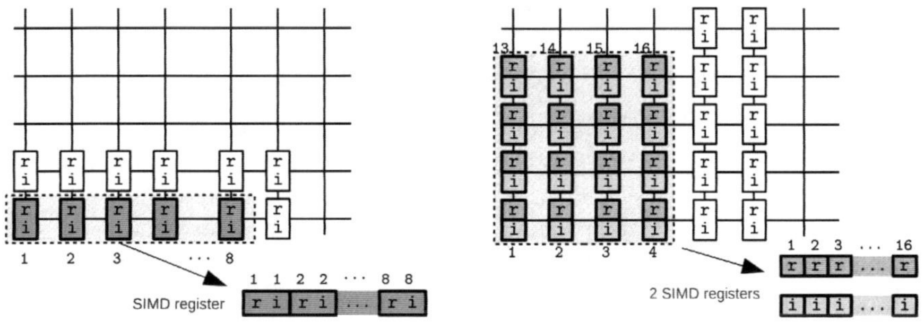

Figure 2. SIMD data layout we adopted for AVX-512 (left) and A64FX (right) architectures.

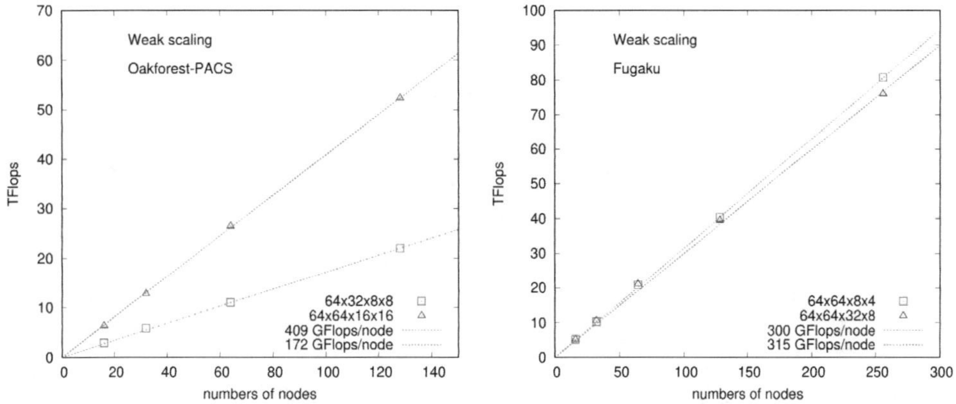

Figure 3. Weak scaling of the performance of the clover matrix multiplication on Oakforest-PACS (left) and Fugaku (right). The dotted lines are scaling from the largest numbers of node.

lattice size in x-direction must be a multiple of 8. The details of the tuning with the AVX-512 instruction set were presented in [7]. For Armv8.2-A-SVE, we adopt a different packing: as depicted in the right panel of Fig. 2, real and imaginary parts are separately stored. This packing shows better performance, found through the development of QCD Wide SIMD (QWS) library [9] as a product of the Post-K co-design project. With this layout, the field variables on 16 sites are executed in parallel. To keep flexibility in choosing the lattice size, we pack variables in the x-y plane into a SIMD vector, while a one-dimensional packing was adopted in QWS. Therefore we prepare two branches of code for the AVX-512 and A64FX architectures. Since the branch for A64FX also intends to call the QWS library, we need to convert the data layout and the physical convention that are different in Bridge++ and QWS.

We measure the performance of our code on the Oakforest-PACS system at JCAHPC, a cluster composed of the Intel Xeon Phi Knights Landing processors, and the supercomputer Fugaku at RIKEN. In Fig. 3, weak scaling behavior of the Clover matrix multiplication is displayed for these SIMD architectures. The results show good scaling on both systems, although Oakforest-PACS shows significant dependence on the lattice size per node as the smaller local volume is too small to hide the neighboring communication overlapped with the computation in bulk. The peak performance of each node is about 6 TFlops on both the systems so that the current sustained performance is around 5 %. This is reasonable performance for practical simulations, though there is still room for improvement.

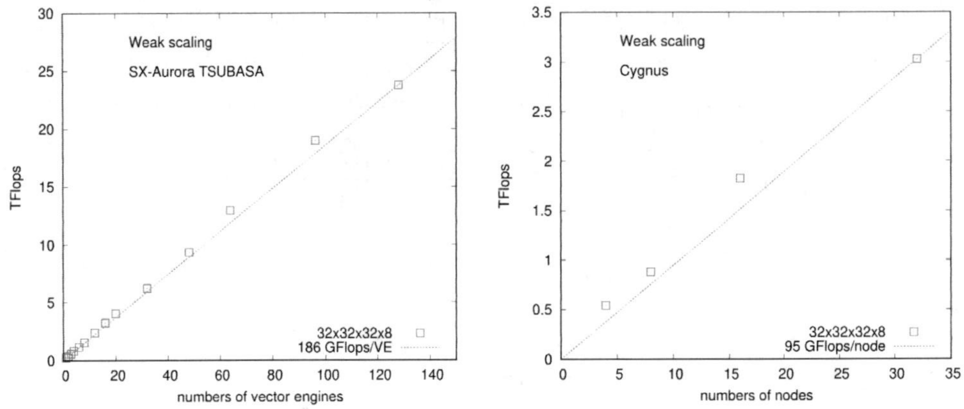

Figure 4. Weak scaling behavior of the performance of matrix-vector multiplication for the Wilson matrix on SX-Aurora TSUBASA (left) and the Clover matrix on the Cygnus system (right). The dotted lines are scaling from the largest numbers of nodes or vector engines.

3.2. Vector architecture: NEC SX-Aurora TSUBASA

The latest vector architecture of NEC is adopted in the NEC SX-Aurora TSUBASA system whose vector length is 256 in units of double precision floating-point number. The best performance is obtained for the loop size with multiples of this vector length, and thus we assign this vector index to the lattice sites. The data layout suitable to this setup is so-called Structure of Array (SoA) layout. We rearrange the data so that the site degrees of freedom are consecutively stored on the memory, including the padding to avoid the bank conflict.

We measure the performance on the NEC SX-Aurora TSUBASA at KEK. Each node of the system is composed of one Vector Host (Intel Xeon processor) and eight Vector Engines (VE). Each VE has 2.42 TFlops of the peak performance and 1.2 TB/s memory bandwidth, indicating the byte-per-flop of 0.5 as an advantage of this architecture. One can execute MPI jobs in units of core in VE (8 cores/VE).

In the left panel of Fig 4, we show the weak scaling behavior of the Wilson matrix multiplication (setting $F(x) = 0$) since the clover fermion matrix is in preparation. While sufficient performance is obtained on the single core, the performance decreases on multi-process for unspecified reasons. Results on multiple VE show good scaling behavior. The vector instruction ratio is 99.90% with average vector length 256.0.

3.3. GPU Cluster

On the clusters with accelerator devices, one needs to offload the data and tasks to the devices employing an offloading scheme. We implement such a code using OpenACC, which is a widely used directive-based framework. By calling an interface object, the data layout is rearranged to that respects so-called coalesced access, and the data are automatically transferred between the host and device. Thus one can generally compose the algorithms using these classes without worrying about the data transfer. The tasks to be offloaded are specified by OpenACC directives. We extract such kernel functions and collect them as a library so that one can replace them with more optimized functions with CUDA or OpenCL if necessary.

As an example of GPU cluster, we use the Cygnus system at the University of Tsukuba. Each node of Cygnus is composed of two Intel Xeon processors and four NVIDIA Tesla V100 GPUs. Each V100 GPU has 5120 CUDA cores which amount to 14 TFlops for FP32 arithmetics. While a GPU has high arithmetic performance, bottlenecks are the data transfer between the host and device (PCIe 3.0 with 16 lanes) and between the device global memory and device cores (900 GB/s for V100). Together with the Infiniband connection among nodes, one needs to choose the parameters considering the heterogeneous structure of the GPU cluster.

In the right panel of Fig. 4, we show the weak scaling of clover fermion matrix multiplication on the Cygnus system. The sustained performance is governed by the device memory bandwidth. Increasing the number of nodes, the overhead of inter-node communication becomes sizable.

4. Conclusions

We presented features of our forthcoming major update of Bridge++ that incorporates an optimized code for recent architectures as an alternative to the original code. We described our fundamental strategy of implementation and displayed several examples in practical application. These results demonstrate that our framework indeed works with sufficient performance for practical application. While we only showed sustained performance for the Wilson and Clover fermion matrices, other types of fermion and a variety of algorithms including multi-grid solvers and eigenvalue solvers are available. Now we are in the final stage toward the public release of the Bridge++ version 2.0.

Acknowledgments

The authors would like to thank the members in the lattice QCD working group in FS2020 project for post-K computer. Numerical simulations were performed on the Oakforest-PACS and Cygnus systems through Multidisciplinary Cooperative Research Program in CCS, University of Tsukuba (16a43, 17a41, ID14(FY2018), ID43(FY2019), ID79(FY2020)), supercomputer Fugaku at RIKEN Center for Computational Science, and SX-Aurora TSUBASA system at KEK through Particle, Nuclear, and Astro Physics Simulation Program (2019-T003, 2020-002, 2021-006). Some parts of the code development were performed on the supercomputer 'Flow' at Information Technology Center, Nagoya University, and Yukawa Institute Computer Facility. This work is supported by JSPS KAKENHI (JP20K03961, JP21K03553), the MEXT as 'Program for Promoting Researches on the Supercomputer Fugaku' (Simulation for basic science: from fundamental laws of particles to creation of nuclei) and 'Priority Issue 9 to be Tackled by Using the Post-K Computer' (Elucidation of the Fundamental Laws and Evolution of the Universe), and Joint Institute for Computational Fundamental Science (JICFuS).

References

[1] Knechtli F, Günther M and Peardon M 2017 *Lattice Quantum Chromodynamics: Practical Essentials* SpringerBriefs in Physics (Springer) ISBN 978-94-024-0997-0, 978-94-024-0999-4
[2] Ueda S, Aoki S, Aoyama T, Kanaya K, Matsufuru H, Motoki S, Namekawa Y, Nemura H, Taniguchi Y and Ukita N 2014 *J. Phys. Conf. Ser.* **523** 012046
[3] Motoki S *et al.* 2014 *Procedia Computer Science* **29** 1701–1710
[4] Matsufuru H *et al.* (Bridge++ Project) 2015 *Procedia Computer Science* **51** 1313–1322
[5] Aoyama T, Ishikawa K I, Kimura Y, Matsufuru H, Sato A, Suzuki T and Torii S 2016 *Procedia Computer Science* **80** 1418–1427
[6] Kanamori I and Matsufuru H 2017 *Proceedings of the Fifth International Symposium on Computing and Networking* **CANDAR'17** 66
[7] Kanamori I and Matsufuru H 2018 *Lecture Note on Computer Science* **10962** 456–471
[8] Kanamori I, Ishikawa K I and Matsufuru H 2021 *Lecture Note on Computer Science* **12953** 218–233
[9] Ishikawa K I, Kanamori I, Matsufuru H, Miyoshi I, Mukai Y, Nakamura Y, Nitadori K and Tsuji M 2021 (*Preprint* 2109.10687)

Tempered Lefschetz thimble method as a solution to the numerical sign problem

Masafumi Fukuma[1] **and Nobuyuki Matsumoto**[2]

[1] Department of Physics, Kyoto University, Koto 606-8502, Japan
[2] RIKEN/BNL Research center, Brookhaven National Laboratory, Upton, NY 11973, USA

E-mail: fukuma@gauge.scphys.kyoto-u.ac.jp, nobuyuki.matsumoto@riken.jp

Abstract. We review the tempered Lefschetz thimble method and its extension, which was recently proposed as a versatile solution to the numerical sign problem. We exemplify the effectiveness of the method by its successful application to a chiral random matrix model.

1. Introduction

The numerical sign problem has prevented the first-principles analysis of various important systems in physics, including (i) QCD at finite density, (ii) condensed matter systems (using Quantum Monte Carlo) such as strongly correlated electron systems and frustrated classical/quantum spin systems, (iii) real-time dynamics of quantum fields, and (iv) QCD with a finite θ angle. Among various approaches to the sign problem proposed so far, we concentrate on the Lefschetz thimble method [1, 2, 3, 4, 5, 6, 7, 8, 9], especially its tempered version [4] as well as the worldvolume extension [8].

In section 2, we explain the basics of the Lefschetz thimble method and point out that it suffers from the dilemma between the sign problem and the ergodicity problem. Section 3 describes our solution, the *tempered Lefschetz thimble method* (TLTM) [4], and its improved version, the *worldvolume tempered Lefschetz thimble method* (WV-TLTM) [8]. In section 4, we exemplify the effectiveness of the WV-TLTM by its successful application to the Stephanov model (a chiral random matrix model). Section 5 is devoted to conclusion and outlook.

2. Lefschetz thimble method

Our aim is to numerically estimate the expectation values of observables defined by the path integral

$$\langle \mathcal{O}(x) \rangle \equiv \int dx\, \rho(x)\, \mathcal{O}(x) \quad \Bigl(\rho(x) \equiv \frac{e^{-S(x)}}{\int dx\, e^{-S(x)}} \Bigr), \tag{1}$$

where $x = (x^i) \in \mathbb{R}^N$ is a dynamical variable, $S(x)$ the action, $dx \equiv \prod_{i=1}^N dx^i$ the measure of the path integral, and $\mathcal{O}(x)$ an observable of interest.

The basic idea of the Lefschetz thimble method is the complexification of the dynamical variable from $x = (x^i) \in \mathbb{R}^N$ to $z = (z^i) \in \mathbb{C}^N$. In the present article, we make an assumption (which usually holds for systems of interest) that $e^{-S(z)}$ and $e^{-S(z)}\,\mathcal{O}(z)$ are both entire functions in $\mathbb{C}^N$. Then, due to Cauchy's theorem, the integrals do not change under continuous deformations of integration surface (see the left panel of Fig. 1):

$$\langle \mathcal{O}(x) \rangle \equiv \frac{\int_{\Sigma_0} dx\, e^{-S(x)}\, \mathcal{O}(x)}{\int_{\Sigma_0} dx\, e^{-S(x)}} = \frac{\int_{\Sigma} dz\, e^{-S(z)}\, \mathcal{O}(z)}{\int_{\Sigma} dz\, e^{-S(z)}}. \tag{2}$$

Thus, even when the sign problem is very severe on $\Sigma_0 = \mathbb{R}^N$ due to the highly oscillatory phase

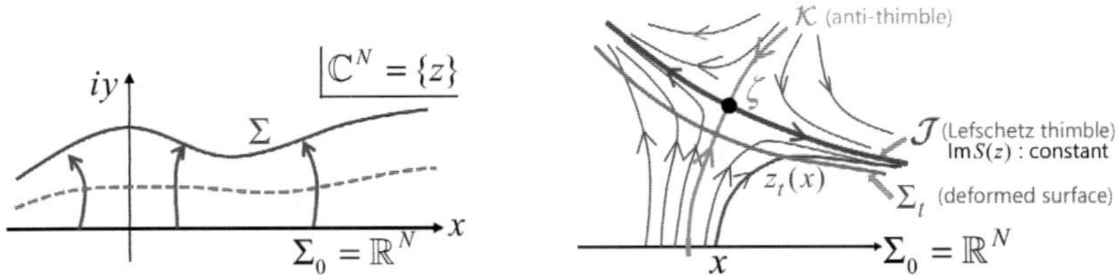

Figure 1. Deformation of integration surface.

factor $e^{-i\,\mathrm{Im}\,S(x)}$, the sign problem will be significantly reduced if the deformed integration surface Σ can be chosen such that $\mathrm{Im}\,S(z)$ is almost constant on Σ.

The prescription of the deformation is the following *anti-holomorphic gradient flow* (see the right panel of Fig. 1):

$$\dot{z}_t = \overline{\partial S(z_t)}, \quad z_{t=0} = x, \tag{3}$$

which defines a map from $x \in \Sigma_0 = \mathbb{R}^N$ to $z = z_t(x) \in \mathbb{C}^N$. We denote the deformed surface at flow time t by Σ_t ($\equiv \{z_t(x) \,|\, x \in \mathbb{R}^N\}$). From the inequality

$$[S(z_t)]^{\cdot} = \partial S(z_t) \cdot \dot{z}_t = |\partial S(z_t)|^2 \geq 0, \tag{4}$$

we see that

(a) $[\mathrm{Re}\,S(z_t)]^{\cdot} \geq 0$, i.e., $\mathrm{Re}\,S(z_t)$ always increases under the flow except at critical points (points where the gradient of the action vanishes),

(b) $[\mathrm{Im}\,S(z_t)]^{\cdot} = 0$, i.e., $\mathrm{Im}\,S(z_t)$ is constant under the flow.

Associated with a critical point ζ, we define the corresponding Lefschetz thimble $\mathcal{J}$ as a collection of orbits flowing out of ζ,

$$\mathcal{J} \equiv \{z \,|\, \lim_{t \to -\infty} z_t(z) = \zeta\}. \tag{5}$$

Due to the property (b), $\mathrm{Im}\,S(z)$ is constant on $\mathcal{J}$, $\mathrm{Im}\,S(z) = \mathrm{Im}\,S(\zeta)$ ($z \in \mathcal{J}$). Thus, when Σ_t approaches a single Lefschetz thimble in the large flow time limit, we expect that the integration on Σ_t becomes free from the sign problem if we take the flow time t to be sufficiently large.

Let us see what actually happens in the disappearance of the sign problem. We first recall that the expectation value is estimated by a ratio of reweighted averages with statistical errors:

$$\langle \mathcal{O}(x) \rangle = \frac{\langle e^{-i\,\mathrm{Im}\,S(x)}\, \mathcal{O}(x) \rangle_{\Sigma_0\,(\mathrm{rewt})}}{\langle e^{-i\,\mathrm{Im}\,S(x)} \rangle_{\Sigma_0\,(\mathrm{rewt})}} \approx \frac{e^{-O(N)} \pm O(1/\sqrt{N_{\mathrm{conf}}})}{e^{-O(N)} \pm O(1/\sqrt{N_{\mathrm{conf}}})} \quad (N_{\mathrm{conf}}\text{: sample size}) \tag{6}$$

Since the main parts are very small of order $e^{-O(N)}$, we need a sample of huge size $N_{\mathrm{conf}} \gtrsim e^{O(N)}$ in order to make the statistical errors smaller than the main parts. This is the sign problem.

On the other hand, if we make a reweighting at a flow time t, the main parts in the numerator and the denominator become of order $e^{-e^{-\lambda t}O(N)}$, and the expectation value is estimated as

$$\langle \mathcal{O}(x) \rangle = \frac{\langle e^{i\theta(z)}\,\mathcal{O}(z)\rangle_{\Sigma_t}}{\langle e^{i\theta(z)}\rangle_{\Sigma_t}} \approx \frac{e^{-e^{-\lambda t}O(N)} \pm O(1/\sqrt{N_{\rm conf}})}{e^{-e^{-\lambda t}O(N)} \pm O(1/\sqrt{N_{\rm conf}})}, \tag{7}$$

where $e^{i\theta(z)} \equiv e^{-i\,{\rm Im}\,S(z)}\,dz/|dz|$, and λ is a typical value of the matrix $\partial_i\partial_j S(\zeta)$. Therefore, if we take the flow time t to be sufficiently large so as to satisfy $e^{\lambda t} = O(N)$, the main parts become $O(1)$, and thus we no longer need a huge size of sample.

So far, so good. However, when multiple thimbles are relevant to estimation, there generically arises another problem at large flow times, the ergodicity problem. Figure 2 is a sketch of the deformation of integration surface for the case $e^{-S(x)} = e^{-\beta x^2/2}\,(x-i)^\beta$. In addition to two

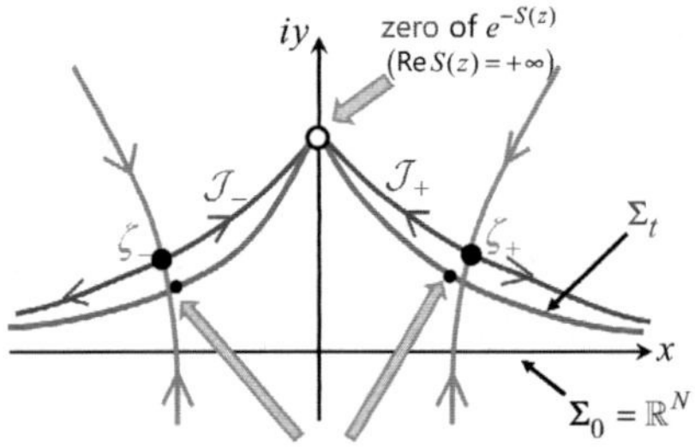

Figure 2. Ergodicity problem caused by a zero of $e^{-S(z)}$.

critical points at $\zeta_\pm = \pm\sqrt{3}/2 + i/2$ and the corresponding thimbles $\mathcal{J}_\pm$, there is a zero of $e^{-S(z)}$ at $z = i$, which behaves as an infinitely high potential barrier on Σ_t. Thus, in a MCMC simulation, the configuration space Σ_t is not well explored due to this potential barrier, and we need a huge computation time to realize global equilibrium. This is the ergodicity problem.

As a solution to this ergodicity problem, the *tempered Lefschetz thimble method* (TLTM) [4] was proposed to resolve both the sign problem (severe at small flow times) and the ergodicity problem (severe at large flow times) simultaneously by implementing the tempering algorithm [10, 11, 12, 13] to the LTM.

3. Tempered Lefschetz thimble method and its worldvolume extension

3.1. Tempered Lefschetz thimble method

The basic algorithm of the TLTM is as follows [4] (see the left panel of Fig. 3):

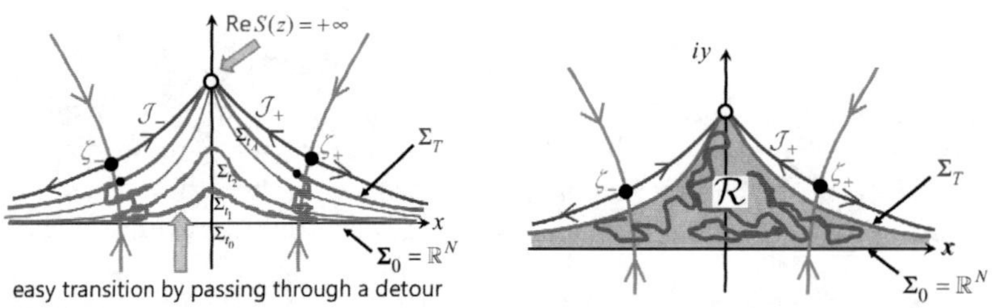

Figure 3. Configuration spaces of TLTM (left) and WV-TLTM (right).

(a) We set the target flow time T which is large enough to resolve the sign problem.

(b) We introduce replicas in between the original integration surface $\Sigma_0 = \mathbb{R}^N$ and the target integration surface Σ_T: $\{\Sigma_{t_0=0}, \Sigma_{t_1}, \Sigma_{t_2}, \ldots, \Sigma_{t_A=T}\}$.

(c) We construct a Markov chain which consists of (i) transitions on each surface Σ_{t_a} and (ii) exchanges of configurations between adjacent replicas Σ_{t_a} and $\Sigma_{t_{a+1}}$. The Markov chain is designed so that the equilibrium distribution on the extended configuration space $\mathbb{R}^N \times \mathcal{A} = \{(x, t_a) \,|\, x \in \mathbb{R}^N,\, a = 1, \ldots, A\}$ is given by $\rho(x, t_a) \propto e^{-S(z_{t_a}(x))} \,|\det J_{t_a}(x)|$, where $J_t(x) = \partial z_t(x)/\partial x$ is the Jacobian of the flow.

(d) After equilibration, we estimate observables with a subsample from the replica at $t_A = T$.

Note that estimates from replicas at large flow times (where the sign problem is reduced) must agree within statistical errors due to Cauchy's theorem. This observation enables us to enhance the precision of estimate by using the χ^2 fit with a constant as the fitting function [6].

3.2. Worldvolume Tempered Lefschetz Thimble Method (WV-TLTM)

Recall that in the original TLTM [4], we temper a system with the flow time introducing a discrete set of replicas $\{\Sigma_{t_0}, \Sigma_{t_1}, \ldots, \Sigma_{t_A}\}$. The advantage of this method over others is its versatility. In fact, the method can in principle be applied to any systems once the problem is defined in a path integral form over continuous variables. A drawback is its high numerical cost at large DOF. In fact, Lefschetz thimble methods generically require the computation of the Jacobian $J_t(x)$ in generating a configuration, whose cost is $O(N^3)$. This cost is further multiplied by the additional cost when the tempering is implemented, which is expected to be $O(N^{0-1})$.

To overcome this drawback, the worldvolume tempered Lefschetz thimble method (WV-TLTM) was invented in Ref. [8], where a HMC calculation is performed on a region (*worldvolume*) $\mathcal{R}$, that is a continuous accumulation of integration surfaces, $\mathcal{R} \equiv \bigcup_{0 \le t \le T} \Sigma_t$ (see the right panel of Fig. 3). This WV-TLTM keeps the aforementioned advantages intact, and significantly reduces the numerical cost at large DOF. In fact, in the WV-TLTM,

- we no longer need to introduce replicas,
- we no longer need to compute the Jacobian in generating a configuration,
- we can move configurations largely due to the use of the HMC algorithm.

The fundamental idea behind the WV-TLTM is that the t-independence (thanks to Cauchy's theorem) allows us to average the reweighted integrals over t with an arbitrary weight $e^{-W(t)}$:

$$\begin{aligned}\langle \mathcal{O}(x) \rangle &= \frac{\int_{\Sigma_0} dx\, e^{-S(x)}\, \mathcal{O}(x)}{\int_{\Sigma_0} dx\, e^{-S(x)}} = \frac{\int_{\Sigma_t} dz_t\, e^{-S(z_t)}\, \mathcal{O}(z_t)}{\int_{\Sigma_t} dz_t\, e^{-S(z_t)}} \\ &= \frac{\int_0^T dt\, e^{-W(t)} \int_{\Sigma_t} dz_t\, e^{-S(z_t)}\, \mathcal{O}(z_t)}{\int_0^T dt\, e^{-W(t)} \int_{\Sigma_t} dz_t\, e^{-S(z_t)}} = \frac{\int_{\mathcal{R}} dt\, dz_t\, e^{-W(t)-S(z_t)}\, \mathcal{O}(z_t)}{\int_{\mathcal{R}} dt\, dz_t\, e^{-W(t)-S(z_t)}}. \end{aligned} \quad (8)$$

The final expression does take a form of path integral over a region $\mathcal{R}$. Although $e^{-W(t)}$ can be an arbitrary function of t in principle, practically it is chosen such that the appearance ratios at flow times t are almost the same so as to ensure the region to be fully explored.

4. Applications

The TLTM has been successfully applied to various models, including

- $(0+1)$-dimensional massive Thirring model [4]
- two-dimensional Hubbard model [6, 7]
- a chiral random matrix model (Stephanov model) [8]

Journal of Physics: Conference Series **2207** (2022) 012054 doi:10.1088/1742-6596/2207/1/012054

- antiferromagnetic Ising model on a triangular lattice [M. Fukuma and N. Matsumoto, talk at JPS meeting 2020]

Below we discuss the application of WV-TLTM [8] to the Stephanov model.

The grand partition function of finite density QCD takes the form

$$Z_{\rm QCD} = \int [dA_\mu]\, e^{(1/2g^2)\int {\rm tr}\, F_{\mu\nu}^2}\, {\rm Det} \begin{pmatrix} m & \sigma_\mu(\partial_\mu + A_\mu) + \mu \\ \sigma_\mu^\dagger(\partial_\mu + A_\mu) + \mu & m \end{pmatrix}. \tag{9}$$

The Stephanov model [14] is obtained by replacing the field degrees of freedom with a complex matrix, and takes the following form at zero temperature:

$$Z_{\rm Steph} \equiv \int d^2W\, e^{-n\, {\rm tr}\, W^\dagger W}\, \det \begin{pmatrix} m & iW + \mu \\ iW^\dagger + \mu & m \end{pmatrix}. \tag{10}$$

Here, $W = (W_{ij} = X_{ij} + iY_{ij})$ is an $n \times n$ complex matrix, and thus the DOF is given by $N = 2n^2$. This corresponds to the DOF of the gauge field on a lattice, $4L^4(N_c^2 - 1)$ [L: linear size of four-dimensional periodic lattice, and N_c: color ($N_c = 3$ for QCD)]. This model is regarded as an important benchmark for algorithms towards solving the sign problem, because

- the model well describes the qualitative behavior of finite density QCD in the large n limit,
- the complex Langevin method fails for this model because of the incurable wrong convergence problem [15].

Figures 4 and 5 show the estimates of the chiral condensate and the baryon number density obtained with three methods: (i) the naive reweighting method, (ii) the complex Langevin method, and (iii) the WV-TLTM [8]. We see that the naive reweighting method fails due to

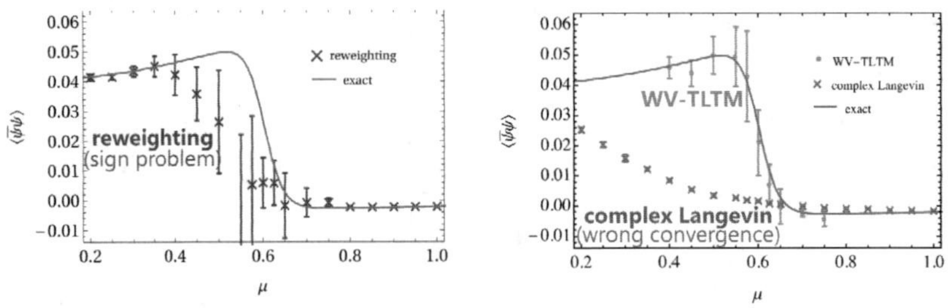

Figure 4. Chiral condensate $\langle\bar{\psi}\psi\rangle \equiv (1/2n)(\partial/\partial m) \ln Z_{\rm Steph}$ [8].

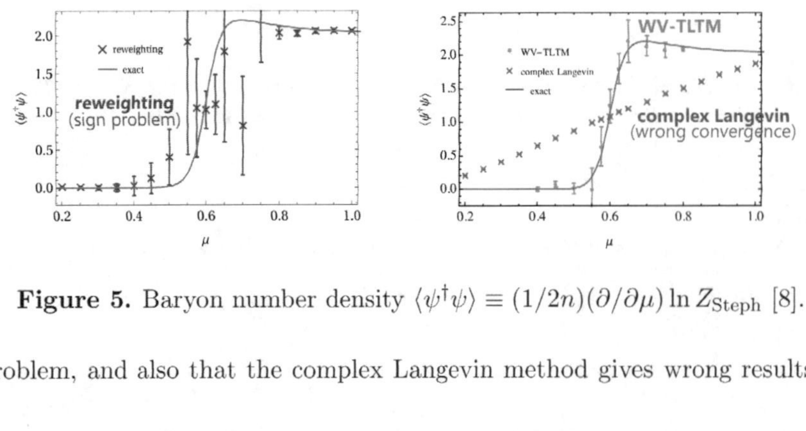

Figure 5. Baryon number density $\langle\psi^\dagger\psi\rangle \equiv (1/2n)(\partial/\partial\mu) \ln Z_{\rm Steph}$ [8].

the sign problem, and also that the complex Langevin method gives wrong results with small statistical errors. On the other hand, the WV-TLTM gives correct results agreeing with the exact values. In fact, this is the first (and only at the moment) successful example among the attempts using various methods to reproduce the correct results for the Stephanov model in parameter regions where the sign problem is severe.

5. Conclusion and Outlook

The sign problem has been an obstacle to the numerical calculations in various important systems in physics, and a versatile solution has long been awaited. In this article, we report that the (WV-)TLTM can be one of the most powerful candidate as a solution.

Our group including the present authors has already started the full-scale application of the WV-TLTM to systems with large DOF, porting the code so as to run on a supercomputer. We believe, however, that it is still important to keep developing the algorithm itself for further efficiency at large DOF. The most important in the future development will be the application to the real-time dynamics of quantum fields. If a Monte Carlo method is at hand for time-dependent quantum systems, the first-principles calculation will become possible for nonequilibrium systems, such as the very early Universe and heavy ion collision experiments.

Acknowledgments

M.F. thank the organizers of CCP2021. The authors thank Issaku Kanamori, Yoshio Kikukawa and Yusuke Namekawa for useful discussions. This work was partially supported by JSPS KAKENHI (Grand Numbers 20H01900, 18J22698) and by SPIRITS of Kyoto University (PI: M.F.). N.M. is supported by the Special Postdoctoral Researchers Program of RIKEN.

References

[1] Cristoforetti S, Di Renzo F and Scorzato L 2012 New approach to the sign problem in quantum field theories: High density QCD on a Lefschetz thimble *Phys. Rev.* D **86** 074506 (*Preprint* arXiv:1205.3996 [hep-lat])
[2] Fujii H, Honda D, Kato M, Kikukawa Y, Komatsu S and Sano T 2013 Hybrid Monte Carlo on Lefschetz thimbles - A study of the residual sign problem *JHEP* **1310** 147 (*Preprint* arXiv:1309.4371 [hep-lat])
[3] Alexandru A, Başar G, Bedaque P F, Ridgway G W and Warrington N C 2016 Sign problem and Monte Carlo calculations beyond Lefschetz thimbles *JHEP* **1605** 053 (*Preprint* arXiv:1512.08764 [hep-lat])
[4] Fukuma M and Umeda N 2017 Parallel tempering algorithm for integration over Lefschetz thimbles *PTEP* **2017** 073B01 (*Preprint* arXiv:1703.00861 [hep-lat])
[5] Alexandru A, Başar G, Bedaque P F and Warrington N C 2017 Tempered transitions between thimbles *Phys. Rev.* D **96** 034513 (*Preprint* arXiv:1703.02414 [hep-lat])
[6] Fukuma M, Matsumoto N and Umeda N 2019 Applying the tempered Lefschetz thimble method to the Hubbard model away from half filling *Phys. Rev.* D **100** 114510 (*Preprint* arXiv:1906.04243 [cond-mat.str-el])
[7] Fukuma M, Matsumoto N and Umeda N 2019 Implementation of the HMC algorithm on the tempered Lefschetz thimble method (*Preprint* arXiv:1912.13303 [hep-lat])
[8] Fukuma M and Matsumoto N 2021 Worldvolume approach to the tempered Lefschetz thimble method *PTEP* **2021** 023B08 (*Preprint* arXiv:2012.08468 [hep-lat])
[9] Fukuma M, Matsumoto N and Namekawa Y 2021 Statistical analysis method for the worldvolume hybrid Monte Carlo algorithm *to appear in PTEP* (*Preprint* arXiv:2107.06858 [hep-lat]).
[10] Marinari E and Parisi G 1992 Simulated tempering: A new Monte Carlo scheme *Europhys. Lett.* **19** 451 (*Preprint* hep-lat/9205018)
[11] Swendsen R H and Wang J S 1986 Replica Monte Carlo simulation of spin-glasses *Phys. Rev. Lett.* **57** 2607
[12] Geyer C J 1991 Markov chain Monte Carlo maximum likelihood *Computing science and statistics: Proceedings of the 23rd Symposium on the Interface* (New York: American Statistical Association) p 156
[13] Hukushima K and Nemoto K 1996 Exchange Monte Carlo method and application to spin glass simulations *J. Phys. Soc. Jpn.* **65** 1604
[14] Stephanov M A 1996 Random matrix model of QCD at finite density and the nature of the quenched limit *Phys. Rev. Lett.* **76** 4472 (*Preprint* hep-lat/9604003)
[15] Bloch J, Glesaaen J, Verbaarschot J J M and Zafeiropoulos S 2018 Complex Langevin Simulation of a Random Matrix Model at Nonzero Chemical Potential *JHEP* **03** 015 (*Preprint* arXiv:1712.07514 [hep-lat])

Journal of Physics: Conference Series **2207** (2022) 012055 doi:10.1088/1742-6596/2207/1/012055

Lattice QCD at nonzero temperature and density

G. Aarts[1,2], C. Allton[1], S. Hands[3], B. Jäger[4], S. Kim[5], M. P. Lombardo[6], A. A. Nikolaev[1], S. M. Ryan[7], J.-I. Skullerud[7,8]

[1] Department of Physics, Swansea University, SA2 8PP, Swansea, United Kingdom
[2] European Centre for Theoretical Studies in Nuclear Physics and Related Areas (ECT*) & Fondazione Bruno Kessler Strada delle Tabarelle 286, 38123 Villazzano (TN), Italy
3 Department of Mathematical Sciences, University of Liverpool, Liverpool L69 3BX, United Kingdom
[4] CP^3-Origins & Danish IAS, Department of Mathematics and Computer Science, University of Southern Denmark, 5230 Odense M, Denmark
[5] Department of Physics, Sejong University, Seoul 143-747, Korea
[6] INFN, Sezione di Firenze, 50019 Sesto Fiorentino (FI), Italy
[7] School of Mathematics and Hamilton Mathematics Institute, Trinity College, Dublin 2, Ireland
[8] Department of Theoretical Physics, National University of Ireland Maynooth, Maynooth, County Kildare, Ireland

E-mail: g.aarts@swansea.ac.uk

Abstract. We discuss some selected recent developments in the field of lattice QCD at nonzero temperature and density, describing in particular the transition from the hadronic gas to the quark-gluon plasma, as seen in simulations using Wilson fermions.

1. Introduction

The study of strongly-interacting matter under extreme conditions remains a topic of interest. Important questions, e.g. present on the list of STFC's Science Challenges [1], include:

(i) What are the basic constituents of matter and how do they interact?
(ii) How do quarks and gluons form hadrons?
(iii) What is the nature of nuclear matter?
(iv) Are there new phases of strongly interacting matter?

These questions have been studied over many years, both theoretically and experimentally, since, even though the underlying theory is well known – quantum chromodynamics (QCD) –, they are nonperturbative and difficult to answer.

Concerning the QCD phase diagram the thermal transition at the physical point is by now well understood [2–6]. The extension to small baryon density is accessible via an expansion in baryon chemical potential over temperature (μ_B/T) and/or via analytical continuation from imaginary chemical potential [7]. Properties at larger density are harder to access, due to the sign problem [8], but heavy-ion collisions [9] and compact stars [10] provide persistent experimental and observational motivation to pursue this direction. In this contribution, the focus is on the thermal transition, and in particular the quark mass (or pion mass) dependence as observed from simulations using $N_f = 2+1$ flavours of Wilson quarks. In addition the onset of parity doubling for baryons is studied, focussing again on pion mass dependence. After presenting some generalities, this contribution mostly follows Ref. [11]. For recent overviews of chiral properties with varying number of flavours, see e.g. Refs. [12, 13].

2. QCD thermal transition

The study of phase transitions in statistical mechanics often follows this standard route: identify an order parameter, discuss which symmetry is broken, and compute derivatives of the free energy (e.g. susceptibilities) to see whether and how they diverge in the infinite-volume limit. In QCD there are potentially various global symmetries to consider. For massless quarks it is chiral symmetry,

$$\psi \to e^{i\alpha\gamma_5}\psi, \qquad \bar{\psi} \to \bar{\psi}e^{i\alpha\gamma_5}, \tag{1}$$

with the chiral condensate $\langle\bar{\psi}\psi\rangle$ as order parameter. However, this symmetry is explicitly broken by the mass term $m\bar{\psi}\psi$ in the QCD lagrangian.

On the other hand, for infinitely heavy quarks centre symmetry holds, as a symmetry of the nonabelian SU(3) gauge theory. Here the gauge action and path integral measure are invariant under the centre of the SU(3) gauge group, $\mathbb{Z}_3$, when timelike links are multiplied by a phase factor. The order parameter in this case is the Polyakov loop, the gauge invariant trace of the product of timelike links. This symmetry is explicitly broken by the presence of not-infinitely-heavy quarks. We conclude that in nature both chiral and centre symmetry are broken explicitly and hence there is no (obvious) order parameter. The transition can then be smooth crossover, with no singularities in the free energy.

Indeed, for physical quark masses it is well established that the transition is a crossover [2]. Lattice QCD simulations have been carried out [3–6] directly at the physical point and the continuum limit is taken using at least four values of the lattice spacing, employing the relation

$$T = 1/(aN_\tau): \qquad a \to 0 \quad \Leftrightarrow \quad N_\tau \to \infty. \tag{2}$$

In practice the continuum limit is approached using $N_\tau = 6, 8, 10, 12, 16$.

The studies cited above are obtained using simulations with so-called staggered quarks. These are fast to simulate, have a remnant of chiral symmetry, such that no additive fermion mass renormalisation is needed, but require "rooting" to get correct number of flavours (a single staggered flavour realises four tastes, which have to be reduced to one). The correct fermion content is expected to emerge in the continuum limit.

One may ask what the situation is with Wilson-like formulations, if nothing else as an independent check on the staggered results. Conceptually, Wilson fermions do not need rooting, which eliminates lingering theoretical uncertainties. Since chiral symmetry is broken explicitly, a tuning of the bare quark mass is needed. Wilson fermions are four-component spinors, as in the continuum, and hence more expensive to simulate than one-component staggered quarks. Hence to date no results in the continuum limit at the physical point have been obtained. Nevertheless, an emerging consistency is observed, as we discuss here.

3. Wilson fermions on (an)isotropic lattices

Studies with $N_f = 2+1$ flavours of Wilson quark are available on both isotropic and anisotropic lattices. We denote the lattice size with $N_\tau \times N_s^3$ and the lattice spacings as a_τ and a_s. In the isotropic case, the spatial and temporal lattice spacing are identical, $a_\tau = a_s$. The requirement that the spatial extent should be larger than the temporal extent, $L \gg 1/T$ or $N_s \gg N_\tau$, leads to the need for large lattices, with, say, $N_s \geq 4N_\tau$. In the anisotropic case, $a_\tau \ll a_s$, it is possible to use $N_s \sim N_\tau$, while keeping $L \gg 1/T$, due to the coarseness of the spatial lattice. Since simulations at different lattice spacings require a nontrivial tuning of the bare parameters, no continuum limit has been taken (yet). However, using the fixed-scale approach, it is easy to simulate at many different temperatures, simply by changing N_τ using the relation $T = 1/(a_\tau N_\tau)$.

A handful of studies for $N_f = 2+1(+1)$ flavours of Wilson quarks exist. The fixed-scale approach is used on isotropic lattices by the Budapest-Wuppertal collaboration [14, 15], for pion masses $m_\pi = 545, 440, 285$ MeV, and taking a continuum extrapolation; by the WHOT collaboration [16–19] using gradient flow at a single lattice spacing; and using twisted mass fermions (with $N_f = 2+1+1$

	Gen2	Gen2L
a_τ [fm]	0.0350(2)	0.0330(2)
a_τ^{-1} [GeV]	5.63(4)	5.997(34)
$\xi = a_s/a_\tau$	3.444(6)	3.453(6)
a_s [fm]	0.1205(8)	0.1136(6)
N_s	24	32
m_π [MeV]	384(4)	236(2)
$m_\pi L$	5.63	4.36

Generation 2, $24^3 \times N_\tau$				
N_τ	T [MeV]	T/T_c	N_{cfg}	N_{stoch}
128*	44	0.24	305	100
48†	117	0.63	251	1200
40	141	0.76	502	800
36	156	0.84	501	400
32	176	0.95	1000	400
28	201	1.09	1001	400
24	235	1.27	1002	100
20	281	1.52	1000	100
16	352	1.90	1000	100

Generation 2L, $32^3 \times N_\tau$			
N_τ	T [MeV]	N_{cfg}	N_{stoch}
256*	23	750	–
128	47	1024	400
64	94	1041	1600
56	107	1042	1600
48	125	1123	1200
40	150	1102	1200
36	167	1119	800
32	187	1090	400
28	214	1031	400
24	250	1016	400
20	300	1030	100
16	375	1102	100
12	500	1267	–
8	750	1048	–

Figure 1. Lattice details (top left) for Generation 2 and 2L, and details of the finite-temperature ensembles for Generation 2 (bottom left) and 2L (right) [11].

flavours) [20–22], for multiple pion masses, including at the physical point at a single lattice spacing. All of these studies use isotropic lattices.

In this contribution we concentrate mainly on results obtained by the FASTSUM collaboration, employing a fixed-scale approach on anisotropic lattices [11]. Simulations are carried out using a Symanzik-improved gauge action and a Wilson tadpole-improved clover fermion action, with stout-smeared links. Details of the parameters and ensembles are given in Fig. 1. The tuning of the bare parameters and the ensembles at the lowest temperature, indicated with a *, have been constructed by the HadSpec collaboration [23–26]. We note that the renormalised anisotropy, $\xi = a_s/a_\tau \approx 3.45$, and that the temporal cutoff $a_\tau^{-1} \approx 6$ GeV. Ensemble information is given for 2 generations: while the strange quark mass m_s is at its physical value, the light quark mass is larger than in nature. The main difference between the 2 generations is the pion mass, with $m_\pi = 384(4)$ MeV (Generation 2) and $m_\pi = 236(2)$ MeV (Generation 2L, L = light). We emphasise that having many time slices available, also in the high-temperature phase, is useful for the analysis of temporal correlations functions, i.e. spectroscopy. The (pseudo)critical temperature in Gen 2 has been determined using the renormalised Polyakov loop and is estimated to be $T_{\text{pc}}^P = 185(4)$ MeV [27]. The transition temperature in Gen 2L will be discussed in this contribution. One aim is to analyse the dependence on the pion mass, i.e. the change going from Gen 2 to Gen 2L.

4. Physical observables

The dependence of the thermal transition on the light quark mass can be seen in observables involving the light degrees of freedom: (quark number) susceptibilities, the chiral condensate and the chiral susceptibility. The former are sensitive to the transition from hadronic to quark fluctuations, while

Journal of Physics: Conference Series **2207** (2022) 012055 doi:10.1088/1742-6596/2207/1/012055

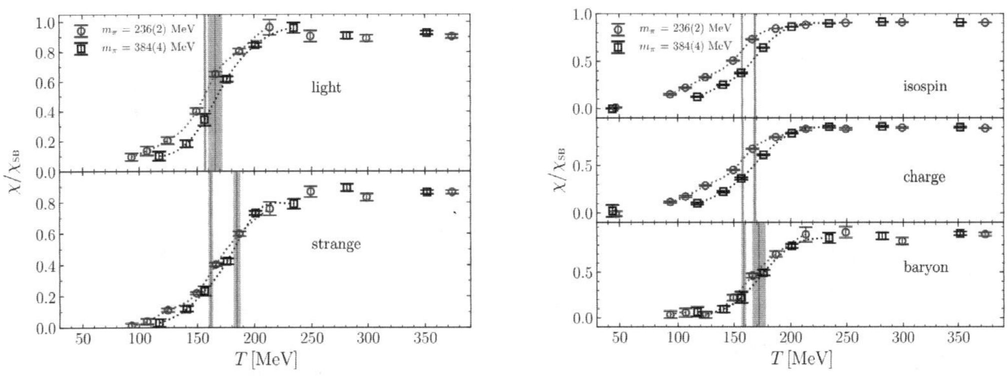

Figure 2. Susceptibilities of light and strange quark number (left) and of isospin, electric charge and baryon number (right), normalised with the susceptibilities $\chi_{\rm SB}$ for massless fermions on a lattice in absence of interactions, as a function of temperature. The vertical bands indicate the inflection point for Gen 2 ($m_\pi = 384(4)$ MeV) and Gen 2L ($m_\pi = 236(2)$ MeV) [11].

the latter are relevant for chiral symmetry restoration. In absence of a proper phase transition, details of the thermal transition depend on the observable and are hence characterised by observable-dependent pseudo-critical temperatures $T_{\rm pc}$. The expectation is that the $T_{\rm pc}$'s shift to lower values as m_π is reduced; eventually $T_{\rm pc}$'s for different observables should coincide when the transition becomes a proper (chiral) phase transition.

In Fig. 2 we show results for the fluctuations of light and strange quark number (left), and of isospin, electric charge and baryon number (right). These are normalised with the susceptibilities $\chi_{\rm SB}$ for massless fermions on a lattice in absence of interactions. Black symbols represent Gen 2, i.e. the heavier pion, and blue symbols Gen 2L, i.e. the lighter pion. The transition is identified by the inflection point of cubic spline fits and indicated with vertical bands; the thickness indicates statistical uncertainty. As the pion mass is reduced, a consistent shift of inflection points towards lower temperatures is observed. The actual values of $T_{\rm pc}$'s are collected in Table 1, to be discussed later.

The chiral condensate contains information on chiral symmetry breaking/restoration. The bare condensate is subject to additive and multiplicative renormalisation. Since we use a fixed-scale approach the renormalisation factors are identical at all temperatures. Nevertheless we consider the renormalised chiral condensate. Following the Budapest-Wuppertal collaboration [14] (which follows Ref. [28]), we consider

$$m_R\langle\bar{\psi}\psi\rangle_R(T) = \frac{\Delta^2_{\bar{\psi}\psi}(T)}{2N_f\Delta_{PP}(T)} + \dots \tag{3}$$

with the subtracted chiral condensate and pseudoscalar susceptibility

$$\Delta_{\bar{\psi}\psi}(T) = \langle\bar{\psi}_l\psi_l\rangle(T) - \langle\bar{\psi}_l\psi_l\rangle(T=0), \tag{4}$$

$$\Delta_{PP}(T) = \int d^4x\, \langle P(x)P(0)\rangle(T) - \int d^4x\, \langle P(x)P(0)\rangle(T=0). \tag{5}$$

Here it is important to note that the LHS of Eq. (3) is finite, while the RHS contains computable bare quantities. Finally we consider the dimensionless ratio,

$$\frac{m_R\langle\bar{\psi}\psi\rangle_R(T)}{m_\pi^2 m_\Omega^2}, \tag{6}$$

Journal of Physics: Conference Series **2207** (2022) 012055 doi:10.1088/1742-6596/2207/1/012055

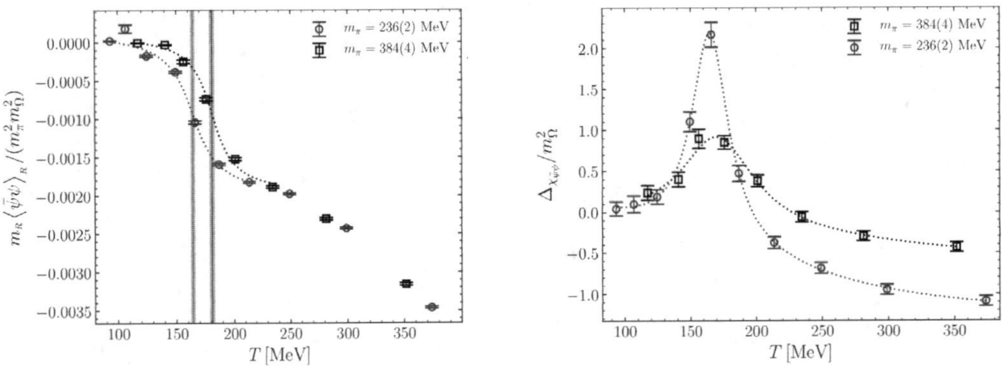

Figure 3. Dimensionless renormalised chiral condensate (left) and subtracted chiral susceptibility (right) as a function of temperature. The vertical bands on the left indicate the inflection points for Gen 2 and Gen 2L [11].

	$T_{\rm pc}$ [MeV]	
	Gen2L	Gen2
observable	$m_\pi = 236(2)$ MeV	$m_\pi = 384(4)$ MeV
$\chi_{\rm light}$	157(1)	166(6)
$\chi_{\rm strange}$	162(2)	184(3)
$\chi_{\rm I}$	157.2(4)	168.4(6)
$\chi_{\rm Q}$	157.5(6)	168.1(6)
$\chi_{\rm B}$	158(2)	172(5)
$\langle\bar\psi\psi\rangle_R$	164(2)	181(2)
$\chi_{\bar\psi\psi}$	165(2)(2)	170(3)(2)

Table 1. Estimates of the pseudocritical temperatures from a range of fermionic observables, for both values of the pion mass.

which is finite in chiral limit. Besides the condensate we also consider the subtracted susceptibility

$$\Delta_{\chi_{\bar\psi\psi}} = \chi_{\bar\psi\psi}(T) - \chi_{\bar\psi\psi}(T=0). \tag{7}$$

It is noted that there is a remaining multiplicative renormalisation, which is however T independent and hence only affects the vertical scale.

The results for the chiral condensate and the susceptibility are shown in Fig. 3. For the chiral condensate we observe a shift of the transition region to lower temperature. For the chiral susceptibility, we note in addition a more pronounced peak for the lighter pion. For both observables estimates of $T_{\rm pc}$'s can be extracted and they are collected in Table 1.

5. Pseudo-critical temperatures and comparison with other approaches

The estimates of the pseudocritical temperatures for the observables considered so far are collected in Table 1, for both values of the pion mass. They are also shown in Fig. 4 (most left pane). We observe that for the lighter pion the spread in temperature is reduced. This may be a sign of being closer to a proper phase transition. We also note that $\chi_{\bar\psi\psi}$ is somewhat of an outlier at the heavier pion, which may be due to the peak being quite broad in that case.

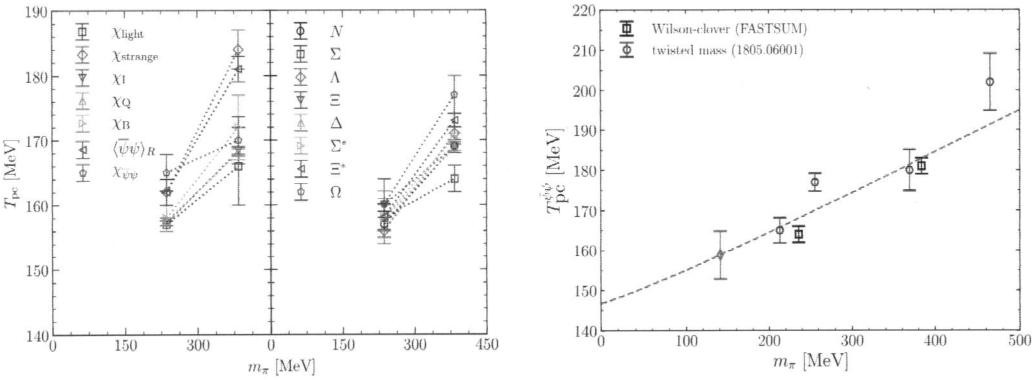

Figure 4. Left: Pseudocritical temperatures as a function of the pion mass, as obtained from susceptibilities and the chiral condensate (left pane) and from baryon parity doubling (right pane). Right: Pseudocritical temperatures as obtained from the inflection point of the renormalised chiral condensate, determined by twisted-mass fermions and this work. The fit given by the dashed line is discussed in the text; the diamond indicates the extrapolated value at the physical point [11].

One may wonder whether anything can be said about the physical point or even massless quarks. An extrapolation using only two pion masses is somewhat audacious, but it is possible to compare with other fermion formulations. Here we consider in particular twisted-mass fermions, with $N_f = 2+1+1$ [20], with a similar range of pion masses. The estimates for the pseudocritical temperature as obtained from the inflection point of the renormalised chiral condensate are shown in Fig. 4 (right). We may perform a combined extrapolation to physical point, using the Ansatz

$$T_{\rm pc}^{\bar{\psi}\psi}(m_\pi) = T_0 + \kappa m_\pi^{2/\Delta} \qquad \Delta = 1.833 \text{ fixed}, \tag{8}$$

(this value of Δ corresponds to the three-dimensional O(4) prediction). We find that $T_0 = 147(4)$ MeV and hence obtain at the physical point

$$T_{\rm pc}^{\bar{\psi}\psi} = 159(6) \text{ MeV} \qquad \text{(physical point)}. \tag{9}$$

This result is indicated with the diamond in Fig. 4 (right). We may compare this with the results obtained using simulations of staggered quarks directly at the physical point, $T_{\rm pc}^{\bar{\psi}\psi} = 155(3)(3)$ MeV – Budapest-Wuppertal [3] – and $T_{\rm pc} = 156.5(1.5)$ MeV – hotQCD [6] – although we hasten to state that in our case no continuum extrapolation has been carried out.

6. Baryons and parity doubling

Chiral symmetry breaking plays an important role in the hadronic spectrum at $T = 0$, for light mesons but also for baryons. In particular, the operators representing baryons come in pairs, related by parity. In nature, however, the corresponding states are not degenerate and there is no particle doubling in the QCD spectrum. As an example, consider the nucleon groundstate: the positive-parity state has mass $m_+ = m_N = 0.939$ GeV, while the negative-parity state is heavier, with $m_- = m_{N^*} = 1.535$ GeV. The absence of parity doubling is a direct consequence of chiral symmetry breaking.

One may wonder what happens at nonzero temperature when chiral symmetry is restored eventually. If the restoration takes place in the quark-gluon plasma phase, most likely the light mesons and baryons have disappeared from the spectrum. Nevertheless, a question to be asked is whether there is a precursor

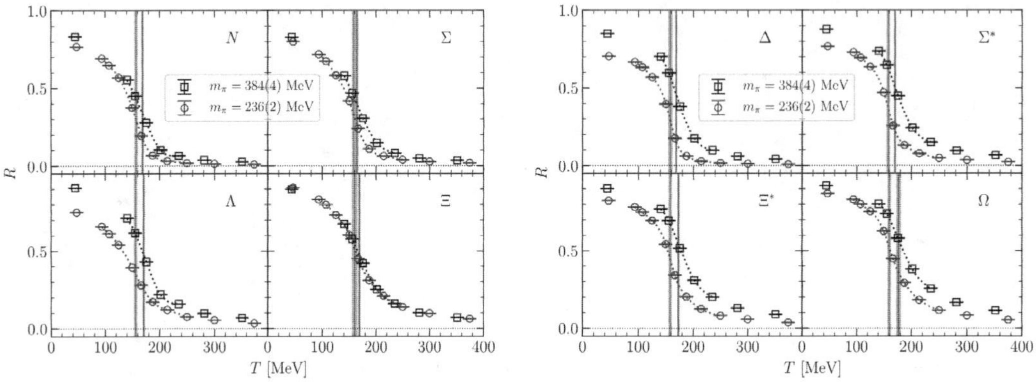

Figure 5. Parity-doubling R parameter as a function of temperature for octet (left) and decuplet (right) baryons. The vertical bands indicate the inflection points for Gen 2 and Gen 2L [11].

$T_{\rm inf}$[MeV]	N	Σ	Λ	Ξ
Gen2	169(1)	164(2)	171(1)	169(1)
Gen2L	157(2)	158(2)	156(2)	160(4)
$T_{\rm inf}$[MeV]	Δ	Σ^*	Ξ^*	Ω
Gen2	168.8(5)	170.3(7)	173(1)	177(3)
Gen2L	158(3)	158(2)	158(2)	160(2)

Table 2. Estimates of the transition temperatures obtained from the R parameter for the baryon channels.

to chiral symmetry restoration in the hadronic phase and hence an onset of parity doubling as well. To study this, we have investigated in-medium effects for the positive- and negative-parity baryonic groundstates in a series of papers [30–32]. We observed that there is an emerging degeneracy around the crossover temperature, with the negative-parity states becoming lighter as the temperature increases while the positive-parity states are seen to be nearly temperature independent.

Since it is hard to define a groundstate at nonzero temperature, and especially in the high-temperature phase, it is important to realise that parity doubling can already be seen at the level of baryonic correlation functions. Following Ref. [29], we may consider the ratio

$$R(\tau) = \frac{G_+(\tau) - G_-(\tau)}{G_+(\tau) + G_-(\tau)}, \tag{10}$$

where $G_\pm(\tau)$ is the positive/negative-parity baryon correlator, and construct a quasi-order parameter

$$R = \frac{\sum_n R(\tau_n)/\sigma^2(\tau_n)}{\sum_n 1/\sigma^2(\tau_n)}, \tag{11}$$

summed over the time slices (and weighted with the estimated error). In absence of parity doubling and when the positive-partner groundstate is much lighter than the negative-partner one, one finds that $R \sim 1$, while in presence of parity doubling $R = 0$. Details can be found in Refs. [30–32].

In Fig. 5 we show the parity-doubling R parameter as a function of temperature for octet and decuplet baryons. It is noted that R indeed changes from being $\mathcal{O}(1)$ at low temperature to being close to zero at high temperature. The presence of the strange quark, with a nonnegligible mass, is visible in the high-temperature phase, especially in the $\Xi^{(*)}$ and Ω channels, with strangeness 2 and 3. The effect of reducing the light quark masses, going from Gen 2 to Gen 2L, is equally visible for the N and Δ channels in the high-temperature phase.

As before we observe a shift of the transition region to lower temperature, indicated by the vertical bands representing the inflection points for Gen 2 and Gen 2L. The actual values for the temperature of the inflection points are given in Table 2 and presented in Fig. 4 (left figure, right pane). As before, we note a reduced spread of points, again possibly indicating the presence of a real phase transition at even smaller pion masses.

7. Summary

In the past 15 years or so a detailed understanding of the thermal transition in QCD has been obtained using numerical simulations, focussing in particular on thermodynamic observables. State-of-the-art studies use staggered quarks, directly at the physical point, and take the continuum limit. In this contribution the attention was instead on Wilson quarks, for which simulations are not yet simultaneously at the physical point and in the continuum limit. Nevertheless, a consistent picture is seen to emerge.

Chiral symmetry restoration has implications for the hadronic spectrum at nonzero temperature. Wilson fermions on anisotropic lattices are particularly well suited to study this. To demonstrate this, we showed results for parity doubling of light and strange baryons, providing a complimentary window into the thermal transition region. We stress that all results shown here have been obtained at a single lattice spacing, which indicates an important step for the future.

Acknowledgments

Results presented in this work were obtained with support from STFC via grants ST/L000369/1 and ST/P00055X/1, the Swansea Academy for Advanced Computing, SNF, ICHEC, the European Research Council (ERC) under the European Union's Horizon 2020 research and innovation programme under grant agreement No 813942. SK is supported by the National Research Foundation of Korea under grant NRF-2018R1A2A2A05018231 funded by the Korean government (MEST). AAN also acknowledges the support by RFBR grant 18-32-20172 mol_a_ved. We are grateful to DiRAC, HPC Wales, PRACE and Supercomputing Wales for the use of their computing resources. Results presented in this work were obtained using the PRACE Marconi-KNL resources hosted by CINECA, Italy and the DiRAC Extreme Scaling service and Blue Gene Q Shared Petaflop system at the University of Edinburgh operated by the Edinburgh Parallel Computing Centre. The DiRAC equipment is part of the UK's National e-Infrastructure and was funded by UK's BIS National e-infrastructure capital grant ST/K000411/1, STFC capital grants ST/H008845/1 and ST/R00238X/1, and STFC DiRAC Operations grants ST/K005804/1, ST/K005790/1 and ST/R001006/1.

References

[1] STFC Science Challenges, `stfc.ukri.org/research/science-challenges/`
[2] Y. Aoki, G. Endrodi, Z. Fodor, S. D. Katz and K. K. Szabo, Nature **443** (2006), 675-678 [arXiv:hep-lat/0611014 [hep-lat]].
[3] S. Borsanyi *et al.* [Wuppertal-Budapest], JHEP **09** (2010), 073 [arXiv:1005.3508 [hep-lat]].
[4] S. Borsanyi, G. Endrodi, Z. Fodor, A. Jakovac, S. D. Katz, S. Krieg, C. Ratti and K. K. Szabo, JHEP **11** (2010), 077 [arXiv:1007.2580 [hep-lat]].
[5] A. Bazavov *et al.* [HotQCD], Phys. Rev. D **90** (2014), 094503 [arXiv:1407.6387 [hep-lat]].
[6] A. Bazavov *et al.* [HotQCD], Phys. Lett. B **795** (2019), 15-21 [arXiv:1812.08235 [hep-lat]].
[7] P. de Forcrand, PoS **LAT2009** (2009), 010 [arXiv:1005.0539 [hep-lat]].
[8] G. Aarts, J. Phys. Conf. Ser. **706** (2016) no.2, 022004 [arXiv:1512.05145 [hep-lat]].
[9] C. Ratti, Rept. Prog. Phys. **81** (2018) no.8, 084301 [arXiv:1804.07810 [hep-lat]].
[10] E. Annala, T. Gorda, A. Kurkela and A. Vuorinen, Phys. Rev. Lett. **120** (2018) no.17, 172703 [arXiv:1711.02644 [astro-ph.HE]].
[11] G. Aarts, C. Allton, J. Glesaaen, S. Hands, B. Jäger, S. Kim, M. P. Lombardo, A. A. Nikolaev, S. M. Ryan and J. I. Skullerud, *et al.* [arXiv:2007.04188 [hep-lat]].
[12] O. Philipsen, Symmetry **13** (2021) no.11, 2079 [arXiv:2111.03590 [hep-lat]].

[13] A. Y. Kotov, M. P. Lombardo and A. Trunin, Symmetry **13** (2021) no.10, 1833 [arXiv:2111.00569 [hep-lat]].
[14] S. Borsanyi, S. Durr, Z. Fodor, C. Hoelbling, S. D. Katz, S. Krieg, D. Nogradi, K. K. Szabo, B. C. Toth and N. Trombitas, JHEP **08** (2012), 126 [arXiv:1205.0440 [hep-lat]].
[15] S. Borsanyi, S. Durr, Z. Fodor, C. Holbling, S. D. Katz, S. Krieg, D. Nogradi, K. K. Szabo, B. C. Toth and N. Trombitas, Phys. Rev. D **92** (2015) no.1, 014505 [arXiv:1504.03676 [hep-lat]].
[16] T. Umeda *et al.* [WHOT-QCD], Phys. Rev. D **85** (2012), 094508 [arXiv:1202.4719 [hep-lat]].
[17] Y. Taniguchi, S. Ejiri, R. Iwami, K. Kanaya, M. Kitazawa, H. Suzuki, T. Umeda and N. Wakabayashi, Phys. Rev. D **96** (2017) no.1, 014509 [erratum: Phys. Rev. D **99** (2019) no.5, 059904] [arXiv:1609.01417 [hep-lat]].
[18] K. Kanaya, A. Baba, A. Suzuki, S. Ejiri, M. Kitazawa, H. Suzuki, Y. Taniguchi and T. Umeda, PoS **LATTICE2019** (2019), 088 [arXiv:1910.13036 [hep-lat]].
[19] Y. Taniguchi *et al.* [WHOT-QCD], Phys. Rev. D **102** (2020) no.1, 014510 [erratum: Phys. Rev. D **102** (2020) no.5, 059903] [arXiv:2005.00251 [hep-lat]].
[20] F. Burger, E. M. Ilgenfritz, M. P. Lombardo and A. Trunin, Phys. Rev. D **98** (2018) no.9, 094501 [arXiv:1805.06001 [hep-lat]].
[21] A. Y. Kotov, M. P. Lombardo and A. M. Trunin, Eur. Phys. J. A **56** (2020) no.8, 203 [arXiv:2004.07122 [hep-lat]].
[22] A. Y. Kotov, M. P. Lombardo and A. Trunin, Phys. Lett. B **823** (2021), 136749 [arXiv:2105.09842 [hep-lat]].
[23] R. G. Edwards, B. Joo and H. W. Lin, Phys. Rev. D **78** (2008), 054501 [arXiv:0803.3960 [hep-lat]].
[24] H. W. Lin *et al.* [Hadron Spectrum], Phys. Rev. D **79** (2009), 034502 [arXiv:0810.3588 [hep-lat]].
[25] G. K. C. Cheung *et al.* [Hadron Spectrum], JHEP **12** (2016), 089 [arXiv:1610.01073 [hep-lat]].
[26] D. J. Wilson, R. A. Briceno, J. J. Dudek, R. G. Edwards and C. E. Thomas, Phys. Rev. Lett. **123** (2019) no.4, 042002 [arXiv:1904.03188 [hep-lat]].
[27] G. Aarts, C. Allton, A. Amato, P. Giudice, S. Hands and J. I. Skullerud, JHEP **02** (2015), 186 [arXiv:1412.6411 [hep-lat]].
[28] L. Giusti, F. Rapuano, M. Talevi and A. Vladikas, Nucl. Phys. B **538** (1999), 249-277 [arXiv:hep-lat/9807014 [hep-lat]].
[29] S. Datta, S. Gupta, M. Padmanath, J. Maiti and N. Mathur, JHEP **02** (2013), 145 [arXiv:1212.2927 [hep-lat]].
[30] G. Aarts, C. Allton, S. Hands, B. Jäger, C. Praki and J. I. Skullerud, Phys. Rev. D **92** (2015) no.1, 014503 [arXiv:1502.03603 [hep-lat]].
[31] G. Aarts, C. Allton, D. De Boni, S. Hands, B. Jäger, C. Praki and J. I. Skullerud, JHEP **06** (2017), 034 [arXiv:1703.09246 [hep-lat]].
[32] G. Aarts, C. Allton, D. De Boni and B. Jäger, Phys. Rev. D **99** (2019) no.7, 074503 [arXiv:1812.07393 [hep-lat]].

Journal of Physics: Conference Series **2207** (2022) 012056 doi:10.1088/1742-6596/2207/1/012056

Quantum field theories, Markov random fields and machine learning

Dimitrios Bachtis[1], **Gert Aarts**[2,3] **and Biagio Lucini**[1,4]

[1]Department of Mathematics, Swansea University, Bay Campus, SA1 8EN, Swansea, Wales, United Kingdom

[2]Department of Physics, Swansea University, Singleton Campus, SA2 8PP, Swansea, Wales, United Kingdom

[3]European Centre for Theoretical Studies in Nuclear Physics and Related Areas (ECT*) & Fondazione Bruno Kessler Strada delle Tabarelle 286, 38123 Villazzano (TN), Italy

[4]Swansea Academy of Advanced Computing, Swansea University, Bay Campus, SA1 8EN, Swansea, Wales, United Kingdom

E-mail: dimitrios.bachtis@swansea.ac.uk, g.aarts@swansea.ac.uk, b.lucini@swansea.ac.uk

Abstract. The transition to Euclidean space and the discretization of quantum field theories on spatial or space-time lattices opens up the opportunity to investigate probabilistic machine learning within quantum field theory. Here, we will discuss how discretized Euclidean field theories, such as the ϕ^4 lattice field theory on a square lattice, are mathematically equivalent to Markov fields, a notable class of probabilistic graphical models with applications in a variety of research areas, including machine learning. The results are established based on the Hammersley-Clifford theorem. We will then derive neural networks from quantum field theories and discuss applications pertinent to the minimization of the Kullback-Leibler divergence for the probability distribution of the ϕ^4 machine learning algorithms and other probability distributions.

1. Introduction

To construct a probability distribution in a high-dimensional space one can turn to the framework of probabilistic graphical models. Probabilistic graphical models comprise a set of random variables, positioned within a graph-based representation, that satisfy certain factorization as well as conditional dependence and independence properties. One notable case of probabilistic graphical models is a Markov field, in which the random variables are connected through undirected edges and satisfy a significant condition of locality, called the Markov property. Markov properties emerge as important mathematical conditions across distinct research fields, such as in machine learning [1] or in constructive quantum field theory [2].

In this contribution, we discuss the proof of Markov properties for discretized Euclidean field theories [3]. Specifically, we demonstrate, through the Hammersley-Clifford theorem, that the ϕ^4 scalar field theory on a square lattice satisfies the local Markov property, and is therefore mathematically equivalent to a Markov field. Based on this equivalence, we introduce algorithms which generalize a notable class of neural networks, specifically restricted Boltzmann machines. Finally, we present applications pertinent to the minimization of the Kullback-Leibler divergence

Journal of Physics: Conference Series **2207** (2022) 012056 doi:10.1088/1742-6596/2207/1/012056

for the probability distribution of the ϕ^4 machine learning algorithms and other probability distributions.

2. The ϕ^4 Markov field

We denote as Λ a finite set which is equivalently expressed as a graph $\mathcal{G} = (\Lambda, e)$, where the points of Λ correspond to the vertices of $\mathcal{G}$ and e denotes the edges of the graph. Two vertices $i, j \in \Lambda$ which are connected by an edge are neighbours. A clique is a set of neighbours, which is called maximal if no additional vertex can be included that is simultaneously a neighbour with all the vertices present in the clique, see Fig. 1. We now assign to each vertex i in the graph $\mathcal{G}$ a continuous-valued random variable, which is denoted as ϕ_i.

A Markov random field is defined as a set of random variables on a graph $\mathcal{G} = (\Lambda, e)$ whose associated probability distribution $p(\phi)$ satisfies the local Markov property:

$$p(\phi_i|(\phi_j)_{j\in\Lambda-i}) = p(\phi_i|(\phi_j)_{j\in\mathcal{N}_i}), \tag{1}$$

where $\mathcal{N}_i$ is the set of neighbours of a given point i. The local Markov property can be proven for a probability distribution that is encoded in a graph through the Hammersley-Clifford theorem:

Theorem 1 (Hammersley-Clifford) *A probability distribution p, satisfying the condition of positivity, is associated with the events generated by a Markov network, iff p can be factorized as a product of positive factors, or potential functions ψ_c, over the cliques of the associated graph structure $\mathcal{G}$:*

$$p(\phi) = \frac{1}{Z} \prod_{c\in C} \psi_c(\phi), \tag{2}$$

where $Z = \int_{\boldsymbol{\phi}} \prod_{c\in C} \psi_c(\boldsymbol{\phi}) d\boldsymbol{\phi}$ is a normalization constant, $c \in C$ is a maximal clique, and $\boldsymbol{\phi}$ denotes all configurations of the system.

The Euclidean action of the two-dimensional ϕ^4 scalar field theory is

$$S_E = -\kappa_L \sum_{\langle ij\rangle} \phi_i\phi_j + \frac{(\mu_L^2 + 4\kappa_L)}{2} \sum_i \phi_i^2 + \frac{\lambda_L}{4} \sum_i \phi_i^4, \tag{3}$$

where $\kappa_L, \mu_L^2, \lambda_L$ are dimensionless parameters. We redefine $w = \kappa_L$, $a = (\mu_L^2 + 4\kappa_L)/2$, $b = \lambda_L/4$ for simplicity, and consider them as inhomogeneous:

$$S(\phi;\theta) = -\sum_{\langle ij\rangle} w_{ij}\phi_i\phi_j + \sum_i a_i\phi_i^2 + \sum_i b_i\phi_i^4. \tag{4}$$

The ϕ^4 inhomogeneous action is described by the coupling constants $\theta = \{w_{ij}, a_i, b_i\}$, and the Boltzmann probability distribution:

$$p(\phi;\theta) = \frac{\exp\left[-S(\phi;\theta)\right]}{\int_{\boldsymbol{\phi}} \exp[-S(\boldsymbol{\phi},\theta)]d\boldsymbol{\phi}}. \tag{5}$$

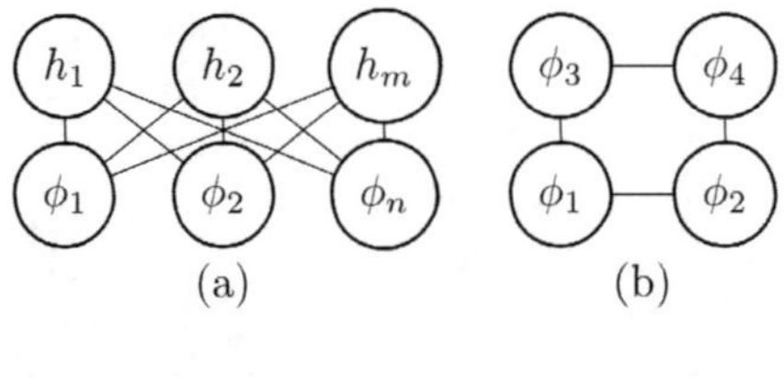

Figure 1. A bipartite graph (a) and a square lattice (b). Examples of maximal cliques are $\{\phi_1, h_1\}$ and $\{\phi_3, \phi_4\}$, respectively.

The lattice version of the ϕ^4 theory is, by definition, expressed as a graph. To verify that the ϕ^4 theory satisfies Markov properties we define the following potential function ψ_c that is able to factorize the probability distribution in terms of maximal cliques $c \in C$

$$\psi_c = \exp\Big[- w_{ij}\phi_i\phi_j + \frac{1}{4}(a_i\phi_i^2 + a_j\phi_j^2 + b_i\phi_i^4 + b_j\phi_j^4)\Big], \tag{6}$$

where i, j are nearest neighbours.

3. Machine learning with ϕ^4 Markov random fields

3.1. Learning without predefined data

To compare the probability distribution $p(\phi;\theta)$ of the Markov field with another probability distribution $q(\phi)$, we define the Kullback-Leibler divergence, a nonnegative quantity, as

$$KL(p||q) = \int_{-\infty}^{\infty} p(\phi;\theta) \ln \frac{p(\phi;\theta)}{q(\phi)} d\phi \geq 0. \tag{7}$$

We now consider a target Boltzmann probability distribution $q(\phi) = \exp[-\mathcal{A}]/Z_{\mathcal{A}}$ that describes an arbitrary statistical system and substitute the two probability distributions in the Kullback-Leibler divergence to obtain:

$$F_{\mathcal{A}} \leq \langle \mathcal{A} - S \rangle_{p(\phi;\theta)} + F \equiv \mathcal{F}, \tag{8}$$

where $\mathcal{F}$ is the variational free energy, $F_{\mathcal{A}} = -\ln Z_{\mathcal{A}}$, and $\langle O \rangle_{p(\phi;\theta)}$ denotes the expectation value of an observable O under the probability distribution $p(\phi;\theta)$. By minimizing this quantity the two probability distributions $p(\phi;\theta)$ and $q(\phi)$ will become equal and we can therefore use the distribution of the ϕ^4 theory to draw samples from the target probability distribution $q(\phi)$.

Based on a gradient-based approach, we are able to minimize Eq. 8 via

$$\frac{\partial \mathcal{F}}{\partial \theta_i} = \langle \mathcal{A} \rangle \Big\langle \frac{\partial S}{\partial \theta_i} \Big\rangle - \Big\langle \mathcal{A} \frac{\partial S}{\partial \theta_i} \Big\rangle + \Big\langle S \frac{\partial S}{\partial \theta_i} \Big\rangle - \langle S \rangle \Big\langle \frac{\partial S}{\partial \theta_i} \Big\rangle, \tag{9}$$

and the coupling constants θ are updated at each epoch t as:

$$\theta^{(t+1)} = \theta^{(t)} - \eta * \mathcal{L}, \tag{10}$$

where η is the learning rate and $\mathcal{L} = \partial \mathcal{F}/\partial \theta^{(t)}$.

We now consider a variation of the ϕ^4 theory which includes next-nearest neighbor interactions nnn, and with a complex action $\mathcal{A}$ defined as

$$\mathcal{A} = \sum_{k=1}^{5} g_k \mathcal{A}^{(k)} = g_1 \sum_{\langle ij \rangle_{nn}} \phi_i\phi_j + g_2 \sum_i \phi_i^2 + g_3 \sum_i \phi_i^4 + g_4 \sum_{\langle ij \rangle_{nnn}} \phi_i\phi_j + i g_5 \sum_i \phi_i^2, \tag{11}$$

where i denotes the imaginary unit. The coupling constants can have arbitrary values but for this example we consider $g_1 = g_4 = -1$, $g_2 = 1.52425$, $g_3 = 0.175$ and $g_5 = 0.15$, see Ref. [3].

We now utilize the ϕ^4 Markov field of action S, to approximate a probability distribution which is described by an action $\mathcal{A}_{\{4\}} = \sum_{k=1}^{4} g_k \mathcal{A}^{(k)}$. To investigate how accurate the equivalence between the two probability distributions is, we will implement reweighting based on the probability distribution of the ϕ^4 Markov field to calculate expectation values of the full complex action $\mathcal{A}$. The reweighting relation is given by

$$\langle O \rangle = \frac{\sum_{l=1}^{N} O_l \exp[S_l - g'_j \mathcal{A}_l^{(j)} - \sum_{k=1,k\neq j}^{5} g_k \mathcal{A}_l^{(k)}]}{\sum_{l=1}^{N} \exp[S_l - g'_j \mathcal{A}_l^{(j)} - \sum_{k=1,k\neq j}^{5} g_k \mathcal{A}_l^{(k)}]}, \tag{12}$$

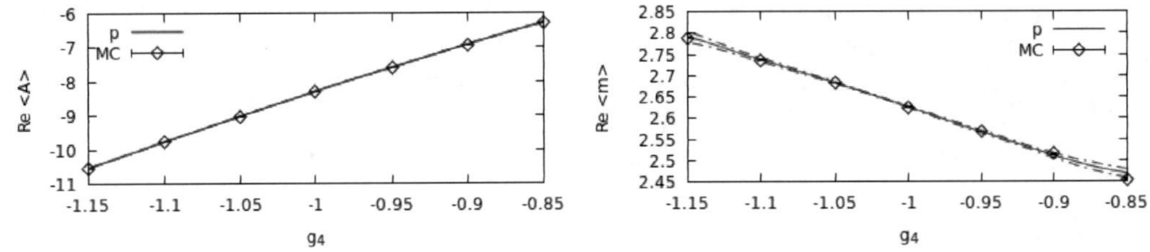

Figure 2. $\Re[\mathcal{A}]$ (left) and $\Re[m]$ (right) versus g_4. The statistical errors are comparable with the width of the line.

details of which can be found in Refs. [3, 4, 5, 6].

We consider $j = 4$ and calculate the expectation values of the action $\mathcal{A}$ and the magnetization m by extrapolating with reweighting in the range $g'_4 \in [-1.15, -0.85]$. The results, shown in Fig. 2, overlap within statistical uncertainty with independent calculations, therefore verifying that observables which would correspond to probability distribution of the full action $\mathcal{A}$ can be accurately calculated based on reweighting from the inhomogeneous action S.

3.2. *Learning with predefined data*

A different class of machine learning applications considers the case where the form of the target probability distribution is unknown but there exists a set of available data in which an empirical probability distribution is encoded. To explore this type of applications we consider the following expression of the Kullback-Leibler divergence:

$$KL(q||p) = \int_{-\infty}^{\infty} q(\phi) \ln \frac{q(\phi)}{p(\phi;\theta)} d\phi \geq 0. \tag{13}$$

By substituting and taking the derivative in terms of the variational parameters θ we obtain:

$$\frac{\partial \ln p(\phi;\theta)}{\partial \theta} = \left\langle \frac{\partial S}{\partial \theta} \right\rangle_{p(\phi;\theta)} - \frac{\partial S}{\partial \theta}, \tag{14}$$

and the update rule of the parameters θ at each epoch t is given based on Eq. 10, where $\mathcal{L} = -\partial \ln p(\phi;\theta^{(t)})/\partial\theta^{(t)}$.

As an example of a distribution to be learned by the machine learning algorithm, we consider the simple case of a Gaussian distribution with $\mu = -0.5$ and $\sigma = 0.05$. Since the lattice action is invariant under the Z_2 symmetry we expect that the symmetric values of the dataset are equiprobable in being reproduced. This invariance can be removed via the inclusion of a term $\sum_i r_i\phi_i$ which breaks the symmetry of the action explicitly. The results are shown in Fig. 3 (left) where the anticipated behaviour is observed.

Finally, we illustrate the approach in an image from the CIFAR-10 dataset. The thermalization of the trained Markov field is depicted in Fig. 3 (right), where the image emerges within the equilibrium probability distribution. Since the machine learning algorithm learns the correct values of coupling constants in the action that solve the considered problem, extensions towards learning the appropriate coupling constants which describe renormalized systems can potentially be explored [7].

3.3. *Machine learning with ϕ^4 neural networks*

To increase the expressivity of the machine learning algorithm, we will introduce a new set of latent or hidden variables h_j within the graph structure. In addition, we will restrict the

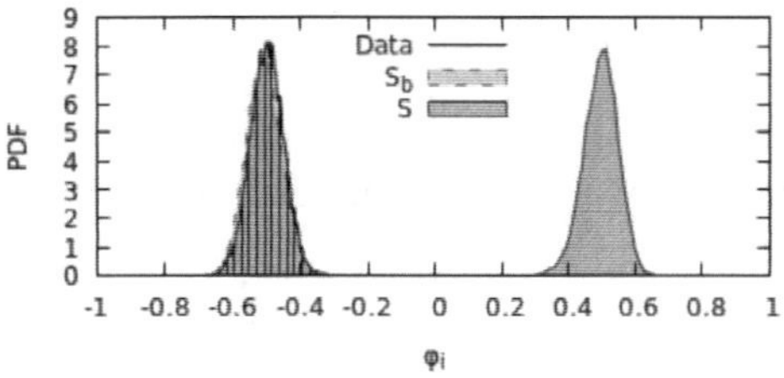

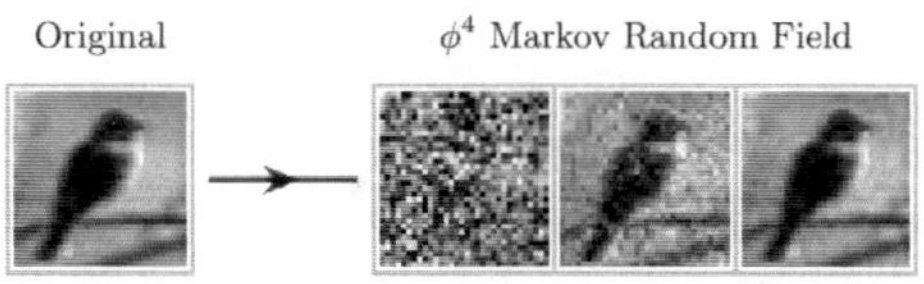

Figure 3. Probability density function (PDF) versus ϕ_i (left). The thermalization of the trained Markov field (right).

interactions to be exclusively between the visible ϕ_i and the hidden h_j variables, giving rise to the lattice action

$$S(\phi, h; \theta) = -\sum_{i,j} w_{ij}\phi_i h_j + \sum_i r_i \phi_i + \sum_i a_i \phi_i^2 + \sum_i b_i \phi_i^4 + \sum_j s_j h_j + \sum_j m_j h_j^2 + \sum_j n_j h_j^4. \quad (15)$$

The ϕ^4 neural network implemented above is a generalization of other neural network architectures: if $b_i = n_j = 0$ the ϕ^4 neural network reduces to a Gaussian-Gaussian restricted Boltzmann machine and if $b_i = n_j = m_j = 0$ and $h_j \in \{-1, 1\}$ then one obtains a Gaussian-Bernoulli restricted Boltzmann machine, see Refs [8, 9]. Restricted Boltzmann machines were inspired by Ising models and since the ϕ^4 theory can become equivalent to an Ising model [10], novel physical insights into how these machine learning algorithms work might emerge from the perspective of quantum field theory. We now train the ϕ^4 neural network on the Olivetti faces dataset. Some representative features, which resemble face structures, are shown in Fig. 4.

4. Conclusions

In this contribution we have shown that discretized Euclidean field theories are Markov fields, hence it is now possible to investigate machine learning directly within quantum field theory [3]. Based on the Hammersley-Clifford theorem, we demonstrated that the ϕ^4 theory, formulated on a square lattice, satisfies the local Markov property. In addition we have derived ϕ^4 neural networks which generalize restricted Boltzmann machines (see also Refs. [11, 12]). Finally, we have presented applications pertinent to physics and computer science, opening up the opportunity for nonperturbative investigations of machine learning within lattice field theory.

Acknowledgments

The authors received funding from the European Research Council (ERC) under the European Union's Horizon 2020 research and innovation programme under grant agreement No 813942. The work of GA and BL has been supported in part by the UKRI Science and Technology

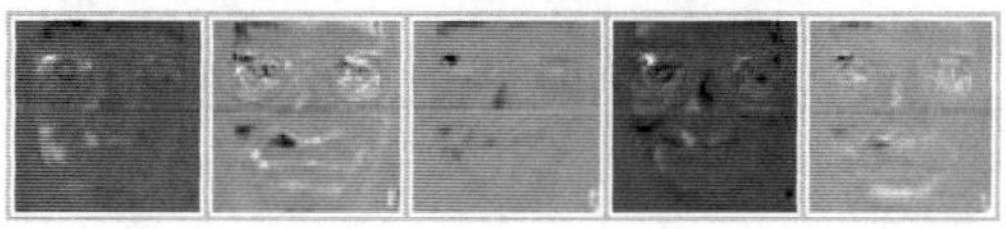

Figure 4. Extracted dependencies from the trained ϕ^4 neural network.

Facilities Council (STFC) Consolidated Grant ST/T000813/1. The work of BL is further supported in part by the Royal Society Wolfson Research Merit Award WM170010 and by the Leverhulme Foundation Research Fellowship RF-2020-461\9. Numerical simulations have been performed on the Swansea SUNBIRD system. This system is part of the Supercomputing Wales project, which is part-funded by the European Regional Development Fund (ERDF) via Welsh Government.

References

[1] D. Koller and N. Friedman, *Probabilistic Graphical Models: Principles and Techniques* (Cambridge, MA: The MIT Press).
[2] E. Nelson, *Construction of quantum fields from Markoff fields*, Journal of Functional Analysis **12**, 97 (1973).
[3] D. Bachtis, G. Aarts and B. Lucini, *Quantum field-theoretic machine learning*, Phys. Rev. D **103**, 074510 (2021).
[4] D. Bachtis, G. Aarts and B. Lucini, *Adding machine learning within Hamiltonians: Renormalization group transformations, symmetry breaking and restoration*, Phys. Rev. Research **3**, 013134 (2021).
[5] D. Bachtis, G. Aarts and B. Lucini, *Mapping distinct phase transitions to a neural network*, Phys. Rev. E **102**, 053306 (2020).
[6] D. Bachtis, G. Aarts and B. Lucini, *Extending machine learning classification capabilities with histogram reweighting*, Phys. Rev. E **102**, 033303 (2020).
[7] D. Bachtis, G. Aarts, F. Di Renzo and B. Lucini, *Inverse renormalization group in quantum field theory*, arXiv:2107.00466 [hep-lat] (2021).
[8] G. Hinton, *Training products of experts by minimizing contrastive divergence*, Neural Computation, **14**, 8 (2002).
[9] A. Fischer and C. Igel, *Training restricted Boltzmann machines: An introduction*, Pattern Recognition **47**, 25 (2014).
[10] A. Milchev, D. W. Heermann and K. Binder, *Finite-size scaling analysis of the ϕ^4 field theory on the square lattice*, J. Stat. Phys. **44**, 749 (1986).
[11] K. Hashimoto, S. Sugishita, A. Tanaka and A. Tomiya, *Deep learning and the AdS/CFT correspondence*, Phys. Rev. D **98**, 046019 (2018).
[12] K. Hashimoto, *AdS/CFT as a deep Boltzmann machine*, Phys. Rev. D **99**, 106017 (2019).

Machine learning the 2D percolation model

Djénabou Bayo[1,2], **Andreas Honecker**[2], **Rudolf A. Römer**[1]
[1]Department of Physics, University of Warwick, Coventry, CV4 7AL, United Kingdom
[2]Laboratoire de Physique Théorique et Modélisation, CNRS UMR 8089, CY Cergy Paris Université, Cergy-Pontoise, France

E-mail: djenabou.bayo@warwick.ac.uk

Abstract. We use deep-learning strategies to study the 2D percolation model on a square lattice. We employ standard image recognition tools with a multi-layered convolutional neural network. We test how well these strategies can characterise densities and correlation lengths of percolation states and whether the essential role of the percolating cluster is recognised.

1. Introduction

A popular application of machine learning (ML) is the identification and classification of phases of matter [1–4]. For example, many properties of the ferromagnetic-to-paramagnetic phase transition in the 2D Ising model were reconstructed in Ref. [2] while Anderson-type transitions and phase diagrams for models of topological matter have been considered in Ref. [5]. These promising results seem to pave the way for ML as a standard tool in condensed matter. ML approaches based on deep, i.e., multi-layered, neural nets fall into three broad classes namely, *supervised*, *un-supervised* and *reinforcement* learning approaches. Common to all approaches is the feed-forward/back-propagation cycle in which parameters of the neural nets are optimised such that suitably chosen loss functions become minimal. Various network architectures have been shown to be appropriate for tasks as varied as image recognition [6], deepfakes [7, 8] and board games [9]. Here, we shall employ supervised learning strategies in the 2D percolation model on a square lattice. It is well known that the percolation in 2D exhibits a phase transition as a function of occupation densities with a diverging two-point correlation length at a critical density p_c. Previous studies [4, 10] found supervised learning able to distinguish the two phases of the percolation model, while the utility of unsupervised learning seems less established [10, 11]. The aim of this paper is to develop a robust ML strategy to identify for finite lattices whether a given percolation lattice contains a spanning cluster or not. As such, we will (i) classify, with ML, percolation lattices according to densities and correlation lengths and (ii) use ML regression to predict densities and correlation lengths.

2. Model and Methods

The site percolation model on a 2D $L \times L$ lattice is defined as follows [12]: *sites* on the lattice are randomly occupied with probability p and left empty with probability $1 - p$. Each site has four *neighbours*, i.e., two along the horizontal and two along the vertical axes. A group of neighbouring occupied sites is called a *cluster*. For small p values, most clusters have size $< L$ while for $p \to 1$, the cluster size tends to L^2. At the *percolation threshold* p_c a spanning

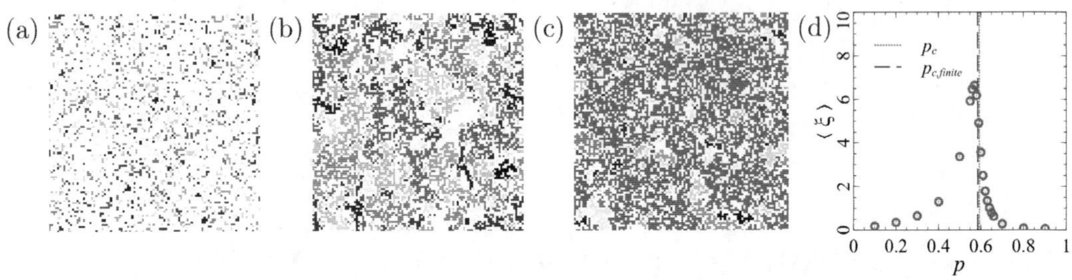

Figure 1. Percolation states on a 100×100 lattice with open boundary conditions for (a) the non-percolating phase ($p = 0.2 < p_c$), (b) close to the transition ($p = 0.5 \approx p_c$) and (c) in the spanning phase ($p = 0.6 > p_c$). Different clusters of occupied sites are denoted by grey squares and identical grey levels indicate a common cluster. There are 1235, 597 and 248 clusters in (a), (b) and (c), resp., and the largest cluster size is 12, 252 and 4422, resp., in each panel. (d) Average correlation length $\langle\xi\rangle(p)$ computed from 5000 percolation states for 20 p values and periodic boundaries. The vertical lines denote p_c in the thermodynamic limit [14] (green dotted) and the position $p = 0.5853$ for the largest $\langle\xi\rangle$ as a estimate of p_c for our finite-sized systems (black dashed). The standard error of the mean for each data point is within symbol size.

cluster emerges which connects opposite sides of the lattice. Much previous work [12, 13] has established that for $L \to \infty$, $p_c = 0.59274605079210(2)$ [14]. For finite L, one usually finds upon averaging a threshold value $p_c(L) < p_c$. Numerically, the Hoshen-Kopelman cluster identification algorithm [15] allows for a determination of $p_c(L)$ while also giving full information about the cluster geometries and shapes. We use it to compute configurations $\psi_n(\mathbf{r})$ of occupied and empty sites on 100×100 lattices with 20 densities $p = 0.1, 0.2, \ldots, 0.55, 0.56, \ldots, 0.66, 0.7, \ldots, 0.9$ and $N = 5000$ samples for each density. Here, n labels different configurations while $\mathbf{r}$ indicates positions of individual sites and $\psi_n(\mathbf{r}) \in \{0, 1\}$ denotes whether at $\mathbf{r}$ there is an empty (0) or occupied (1) site in the n state. Examples of such ψ_n are given in Fig. 1(a-c). Our results suggest that $p_c(100) \approx 0.59 \pm 0.03$ for our data, computed with open boundary conditions. The transition from non-spanning clusters at $p < p_c$ to a spanning cluster at $p > p_c$ is a continuous phase transition with the probability $P(p)$ that an arbitrary site belongs to the infinite cluster acting as the order parameter. Similarly, the connected correlation function $g(r)$, defined as the probability of a site at distance r from an occupied site to belong to the same cluster [16], diverges at p_c. Its associated correlation length $\xi = \sqrt{\sum_r r^2 g(r)/\sum_r g(r)}$ gives the average distance between two sites that belong to the same cluster. As $p \to p_c$, $\xi \propto | p - p_c |^{-4/3}$ for $L \to \infty$. An example of the behaviour of $\xi(r)$ in our data is shown in Fig. 1(d) where we can see that ξ indeed is maximal at $p_c(100)$ but does not diverge for the calculated p values.

Convolution neural nets (CNN) are a class of multi-layered (deep) neural nets in which spatial locality of data values is retained during the ML training. When coupled with a form of residual learning [17], the resulting residual networks (ResNets) have been shown to allow astonishing precision when classifying images, e.g., of animals [18] and handwritten characters [19], or when predicting numerical values, e.g., of market prices [20]. Here, we shall use a ResNet18 [17] network with 17 convolutional and 1 fully-connected layers, pretrained on the ImageNet dataset [21]. For our implementation we use the PyTorch suite of ML codes [22]. We train the ResNet18 on the ψ_n configurations, using a 90%/10% split into training and validation data. Additional ψ_n configurations are generated as necessary for test runs. We concentrate on two ML tasks. First, we classify percolation configurations ψ_n according to densities p, correlation lengths ξ and spanning non-spanning. In the second task, we aim to predict p and ξ values via ML regression. In both problems, the overall network architecture remains identical, we just adapt the last layer. For the classification the output layers have a number of neurons corresponding to the number of classes trained i.e, for the classification by density the $\mathcal{C} = 20$ values $p = 0.1, 0.2, \ldots, 0.55, 0.56, \ldots, 0.66, 0.7, 0.8, 0.9$, while for regression the output layer has only one neuron making the predictions. However, the loss functions are different. Let ω denote the parameters (weights) of the ResNet and let (ψ_n, χ_n) represent a given data sample with χ_n the classification/regression targets, i.e., classes p or ξ, and also χ'_n the predicted values, p' or ξ'. For classification of categorical data, the class names are denoted by a class index $a = 1, \ldots, \mathcal{C}$ and encoded as $\chi_{ai} = 1$ if $\chi_{ai} = i$, 0 otherwise. Then, for the (multi-class) classification problem, we choose the usual cross-entropy loss function, $l_c(\omega) = -\sum_{i=1}^{N\mathcal{C}} \sum_{a=1}^{\mathcal{C}} \chi_{ai} \log \chi'_{ai}(\omega) + (1 - \chi_{ai}) \log[1 - \chi_{ai}(\omega)]$ [23]. On the other hand, the loss function for the regression problem is given by the mean-squared error $l_r(w) = \frac{1}{n} \sum_{i=1}^{N\mathcal{C}} [\chi_i - \chi'_i(w)]^2$. When giving results for the values of the loss functions below, we always present those after averaging over at least 10 independent training and validation cycles. We also represent the quality of a prediction by confusion matrices [23]. These graphically represent the predicted class labels as a function of the true ones in matrix form, with an error-free prediction corresponding to a diagonal matrix. For comparison of the classification and regression calculations, we use in both cases a maximum number of $\epsilon_{\max} = 20$ epochs. Our ML calculations train with a batch size of 16 for classification and for regression on an NVIDIA Quadro RTX 5000 GPU card.

3. Results

Figure 2(a) shows the training and validation losses for classifying densities. An excellent training result is corroborated by a perfectly diagonal confusion matrix shown in Fig. 2(b). We note that this also holds when the spacing of densities is reduced as for the values $p = 0.55, 0.56, \ldots, 0.66$ around p_c. We now repeat the procedure of classification, but replace the density p labels with the computed average correlation lengths, $\langle \xi(p) \rangle$. These are $\langle \xi(0.1) \rangle, \langle \xi(0.2) \rangle, \ldots, \langle \xi(0.9) \rangle$ corresponding to 0.18 ± 0.07, 0.35 ± 0.07, 0.65 ± 0.05, 1.30 ± 0.05, 3.36 ± 0.07, 5.93 ± 0.06, 6.48 ± 0.04, 6.65 ± 0.07, 6.18 ± 0.05, 4.92 ± 0.04, 3.56 ± 0.06, 2.50 ± 0.03, 1.78 ± 0.05, 1.34 ± 0.07, 1.01 ± 0.07, 0.78 ± 0.05, 0.62 ± 0.04, 0.28 ± 0.04, 0.08 ± 0.07, 0.05 ± 0.05, respectively, in Fig. 1(d). In Figs. 2(c, d) we show $l_{\mathrm{c,train}}(\epsilon)$, $l_{\mathrm{c,val}}(\epsilon)$ and the confusion matrix, respectively. We again use the full set of 20 classes for the classification. As we can see in Fig. 2(c), the loss for the classification is as good as for the density classification shown in Fig. 2(a). Of course this should not be surprising since we effectively just replaced the p labels with ξ labels.

The percolation transition distinguishes non-spanning clusters at $p < p_c$ from spanning clusters at $p \geq p_c$. We therefore now also train the ResNet with all percolation data, but just using their classification in non-spanning and spanning percolation clusters as labels. In Figs. 2(e) and (f) we show $l_{\mathrm{c,train}}(\epsilon)$, $l_{\mathrm{c,val}}(\epsilon)$ and the confusion matrix, respectively. The confusion matrix shows good recognition for the majority of cases. Nevertheless, we also see that about 5% of the samples get misclassified, i.e., 298 of 4565 non-spanning clusters are classified as spanning while 279 of 5434 spanning clusters are wrongly identified as non-spanning. This suggests that the classification routine is struggling with correctly identifying the spanning cluster.

The classification approach is in principle restricted in its predictive accuracy for p by the number of available classes. While one can of course try to increase the number of classes, this will also lead to an increased chance in misclassifying percolation states from close by classes. The regression method outlined in section 2 allows to predict the p value for a given percolation state directly. We again start from a pre-trained ResNet in Fig. 3(a) we show the training and validation losses, $l_{\mathrm{r,train}}$ and $l_{\mathrm{r,val}}$, respectively. We train the ResNet for the evenly-spaced 9 densities $p = 0.1, 0.2, \ldots, 0.9$. We then use the data from the 12 densities $p = 0.55, 0.56, \ldots, 0.66$ as test of the regression model. We note that in this approach, the $p = 0.6$ data is used both in the training/validation cycle as well as in the test. In Fig. 3(b) we show the dependence of the true density p as a function of the average predicted density p'. We see that the predicted p'

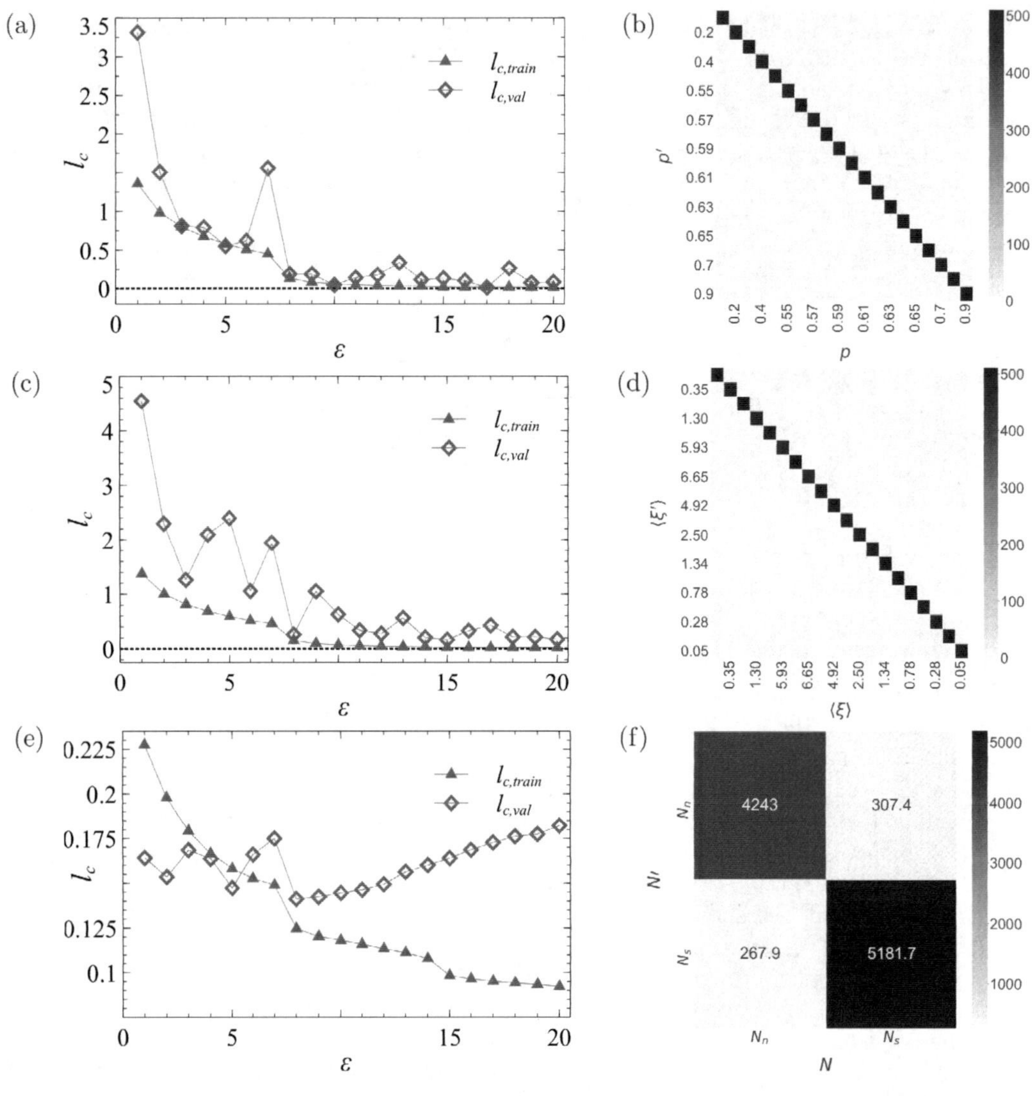

Figure 2. (a,c,e) Dependence of losses $l_{\mathrm{c,train}}$ and $l_{\mathrm{c,val}}$ on the number of epochs ϵ for classification according to (a) p's, (c) ξ's and (e) spanning/non-spanning. The diamonds (blue open) denote $l_{\mathrm{c,train}}$ while the triangles (red solid) show $l_{\mathrm{c,val}}$. The thin lines are guides to the eye only. The horizontal dotted lines in (a,c) indicate $l_{\mathrm{c}} = 0$. (b,d,f) Confusion matrices for the validation data used in (a,c,e) at $\epsilon = 20$. The true labels for p, ξ and N_s/N_n are indicated on the horizontal axes while the predicted labels are given on the vertical axes. The colour scale represents the number of samples in each matrix entry, for (f) these are given explicitly.

values closely follow the true p's. However, we also find that the predictions cluster somewhat around $p = 0.6$, a p value used in the training.

We can also perform a regression study using the correlation lengths ξ as regression target. However, we now have a distribution of computed ξ values for each p whereas for p-regression, each sample has exactly the correct p value. For a given p value, we therefore construct an average value $\langle\xi\rangle$. In order to capture this behaviour correctly, we train our ResNet on the individual ξ values for each sample at given p. Similarly, when computing the predicted ξ'

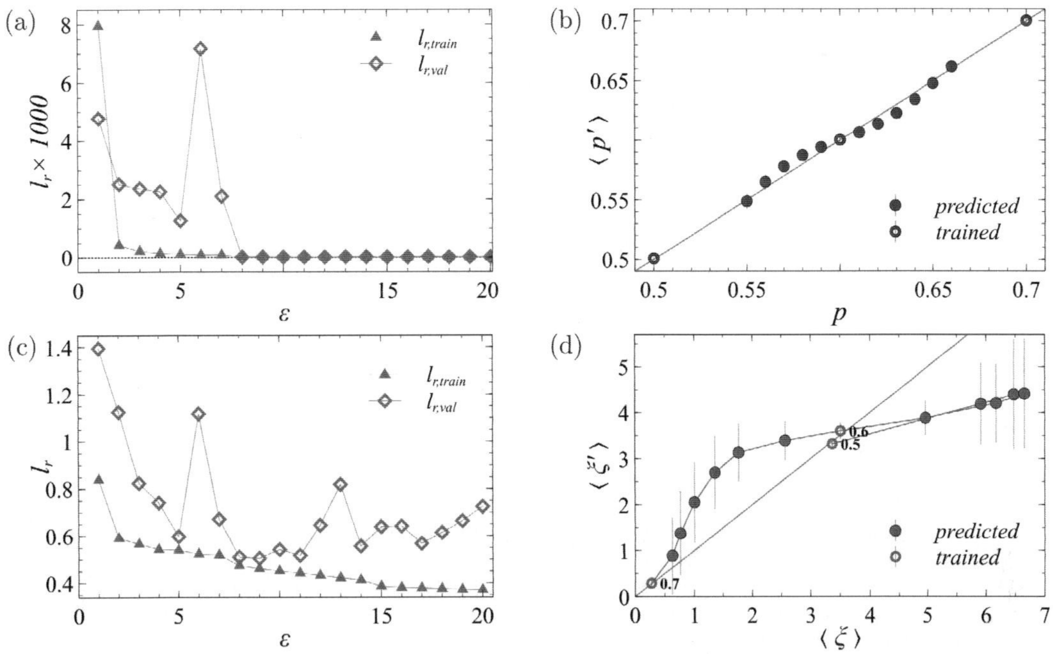

Figure 3. (a,c) Dependence of $l_{\rm r,train}$ and $l_{\rm r,val}$ on ϵ for the regression task with (a) p's and (c) ξ's. Similar to Fig. 2, triangles denote $l_{\rm r,train}$ and diamonds show $l_{\rm r,val}$. Lines are as before. (b, d) Dependence of the true (b) p on p' and (d) ξ on ξ'. The red lines denotes the perfect predictions $p = p'$ and $\langle\xi\rangle = \langle\xi\rangle'$. The three open circles highlight (c) $p = 0.5$, 0.6 and 0.7 and, respectively, (d) $\langle\xi\rangle = 3.36 \pm 0.07$, 3.56 ± 0.06 and 0.28 ± 0.04 included in the training sets while the solid circles are the results of the regression.

values, we average the individual predictions for given p to construct $\langle\xi'\rangle$. As before for p regression, we train using the ξ values of the percolation states for densities $p = 0.1, 0.2, \ldots, 0.9$ and predict ξ' from the 12 densities $p = 0.55, 0.56, \ldots, 0.66$. While all the previous training were performed on percolation states where empty sites were denoted by 0 and occupied site by 1, this dataset did not provide exploitable results. We therefore decided to train with percolation states containing the numbered cluster found through the Hoshen-Kopelman algorithm [15]. We obtain the dependence of the average predicted label $\langle\xi'\rangle$ on the average true label $\langle\xi\rangle$ as shown in Fig. 3(d). In Fig. 3(b), we see that there are important deviations from a perfect regression curve $\langle\xi'\rangle = \langle\xi\ \rangle$ with $\langle\xi\rangle$ corresponding to trained configuration $p = 0.5$, $p = 0.6$ and $p = 0.7$ on the line of perfect regression.

4. Conclusions

We have shown that CNNs, and in particular pretrained ResNets, are able to successfully classify percolation states according to densities p and correlation lengths ξ in the case of the 2D site percolation on a square lattice. When trying to classify spanning versus non-spanning percolation states, the results are less convincing. While the majority of samples are being correctly classified, there are about 5% of percolation states which are wrongly classified. This suggests that the ResNet is not able to identify the percolating/non-percolating nature of clusters correctly. For the regression analysis, when trained with evenly spaced densities $p = 0.1$, 0.2, ..., 0.9, we find good quantitative predictions, even for much closer spaced densities $p = 0.55$, 0.56, ..., 0.66. For the correlation length, the classification gave us perfect classes predictions, but the regression analysis showed that the network might not understand and therefore predict correctly correlation lengths.

In order to make sense of these results, one might want to ask whether the ML image recognition tools employed here simply count the number of occupied and unoccupied sites while disregarding the existing of a percolating cluster. Clearly, this would be consistent with the precision of the classification and regression results for p and ξ. However, then we would not expect, e.g., the deviations from $p = p'$ and $\xi = \xi'$ seen in Fig. 3(b,d). Similarly, when looking at which clusters are wrongly classified during the spanning/non-spanning classification, we find that these are mostly spanning for $p \lesssim p_c$ and non-spanning for $p \gtrsim p_c$. Clearly, more studies are needed to clarify these issues.

References

[1] Venderley J, Khemani V and Kim E A 2018 *Phys. Rev. Lett.* **120** 257204 URL https://doi.org/10.1103/PhysRevLett.120.257204
[2] Carrasquilla J and Melko R G 2017 *Nat. Phys.* **13** 431–434 URL https://doi.org/10.1038/nphys4035
[3] Ch'ng K, Carrasquilla J, Melko R G and Khatami E 2017 *Phys. Rev. X* **7** 031038 URL https://doi.org/10.1103/PhysRevX.7.031038
[4] Zhang W, Liu J and Wei T C 2019 *Phys. Rev. E* **99**(3) 032142 URL https://doi.org/10.1103/PhysRevE.99.032142
[5] Ohtsuki T and Mano T 2020 *J. Phys. Soc. Jpn.* **89** 022001 URL https://doi.org/10.7566/JPSJ.89.022001
[6] Fujiyoshi H, Hirakawa T and Yamashita T 2019 *IATSS Research* **43** 244–252 URL https://doi.org/10.1016/j.iatssr.2019.11.008
[7] Westerlund M 2019 *TIM Review* **9** 40–53 URL http://doi.org/10.22215/timreview/1282
[8] Karras T, Laine S, Aittala M, Hellsten J, Lehtinen J and Aila T 2020 Analyzing and improving the image quality of stylegan *2020 IEEE/CVF Conference on Computer Vision and Pattern Recognition (CVPR)* pp 8107–8116 URL http://doi.org/10.1109/CVPR42600.2020.00813
[9] Silver D, et al., 2017 *Nature* **550** 354–359 URL https://doi.org/10.1038/nature24270
[10] Cheng S, He F, Zhang H, Zhu K D and Shi Y 2021 Machine learning percolation model (*Preprint* 2101.08928)
[11] Yu W and Lyu P 2020 *Physica A* **559** 125065 URL https://doi.org/10.1016/j.physa.2020.125065
[12] Stauffer D and Aharony A 1992 *Introduction To Percolation Theory* (Taylor & Francis) ISBN 9781482272376 URL https://www.taylorfrancis.com/books/9781482272376
[13] Newman M E J and Ziff R M 2000 *Phys. Rev. Lett.* **85** 4104–4107 URL https://doi.org/10.1103/PhysRevLett.85.4104; *ibid*, 2001 *Phys. Rev. E* **64** 016706 URL https://doi.org/10.1103/PhysRevE.64.016706
[14] Jacobsen J 2015 *J. Phys. A Math. Theor.* **48** URL https://doi.org/10.1088/1751-8113/48/45/454003
[15] Hoshen J and Kopelman R 1976 *Phys. Rev. B* **14** 3438–3445 URL https://doi.org/10.1103/PhysRevB.14.3438
[16] Zhang J, Zhang B, Xu J, Zhang W and Deng Y 2021 Machine learning for percolation utilizing auxiliary Ising variables (*Preprint* 2110.06776)
[17] He K, Zhang X, Ren S and Sun J 2016 *2016 IEEE Conference on Computer Vision and Pattern Recognition (CVPR)* 770–778 URL https://doi.org/10.1109/CVPR.2016.90
[18] Tabak M A *et al.* 2019 *Methods Ecol. Evol.* **10** 585–590 URL https://doi.org/10.1111/2041-210X.13120
[19] Zhang R, Wang Q and Lu Y 2017 *2017 14th IAPR International Conference on Document Analysis and Recognition (ICDAR)* **05** 25–29 URL http://doi.org/10.1109/ICDAR.2017.324
[20] Zhao Y and Khushi M 2020 *2020 International Conference on Data Mining Workshops (ICDMW)* 385–391 URL http://doi.org/10.1109/ICDMW51313.2020.00060
[21] Deng J, Dong W, Socher R, Li L J, Li K and Fei-Fei L 2009 ImageNet: A Large-Scale Hierarchical Image Database *CVPR09*
[22] Paszke A *et al.* 2019 Pytorch: An imperative style, high-performance deep learning library *Advances in Neural Information Processing Systems 32* ed Wallach H, et al. (Curran Associates, Inc.) pp 8024–8035 URL http://papers.neurips.cc/paper/9015-pytorch-an-imperative-style-high-performance-deep-learning-library.pdf
[23] Mehta P, Bukov M, Wang C H, Day A G R, Richardson C, Fisher C K and Schwab D J 2019 *Phys. Rep.* **810** 1–124 URL https://doi.org/10.1016/j.physrep.2019.03.001

Journal of Physics: Conference Series **2207** (2022) 012058 doi:10.1088/1742-6596/2207/1/012058

Machine Learning the Square-Lattice Ising Model

Burak Çivitcioğlu[1], **Rudolf A. Römer**[2] **and Andreas Honecker**[1]

[1] Laboratoire de Physique Théorique et Modélisation, CNRS UMR 8089, CY Cergy Paris Université, Cergy-Pontoise, France

[2] Department of Physics, University of Warwick, Coventry, CV4 7AL, United Kingdom

E-mail: burak.civitcioglu@cyu.fr

Abstract. Recently, machine-learning methods have been shown to be successful in identifying and classifying different phases of the square-lattice Ising model. We study the performance and limits of classification and regression models. In particular, we investigate how accurately the correlation length, energy and magnetisation can be recovered from a given configuration. We find that a supervised learning study of a regression model yields good predictions for magnetisation and energy, and acceptable predictions for the correlation length.

1. Introduction

Condensed-matter physics studies the properties of matter with many constituents and interactions [1]. One central task in condensed-matter and statistical physics is to identify different phases of the system and locate the critical points separating these in the space of parameters. Machine learning (ML) is the study of algorithms that improve their performance by gaining experience from data. It has been used for image classification, object detection, self-driving cars, speech recognition and many other tasks that rely on data [2]. ML is commonly categorised into supervised, unsupervised and reinforcement learning. Recently, ML methods have been employed to tackle condensed-matter problems and have shown promising predictions in topics such as studying the phases of the Ising model [3–5], disordered quantum systems [6, 7], phase transitions in the Bose-Hubbard [8, 9], the Blume-Capel model [10], a highly degenerate biquadratic-exchange spin-one Ising variant, and the two-dimensional (2D) XY model [10], as well as material properties [11].

The square-lattice Ising model is an exactly solved model [12]. Notwithstanding its exact solution, it is commonly employed in the ML context [3–5, 13–18]. Motivations for such ML studies of the Ising model in 2D include the readily available comparison to the exact solution and that training data can be easily generated using standard Monte-Carlo simulations. Here, we continue along these lines and investigate the problem of predicting the temperature dependence of energy E, magnetisation m and correlation length ξ for the square-lattice Ising model by ML of configurations.

2. Ising Model, Methods, Data and the ML Model

The ferromagnetic Ising model is defined by the Hamiltonian

$$E = -\sum_{\langle i,j \rangle} s_i s_j, \tag{1}$$

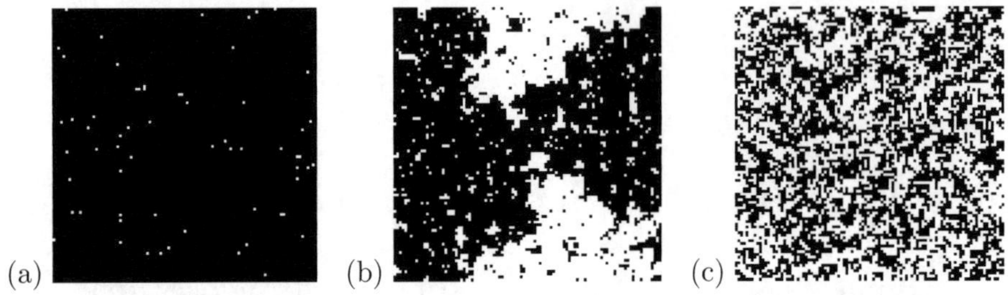

Figure 1. Configurations of the Ising model on a 100×100 square lattice with periodic boundary conditions at temperatures (a) $T = 1.5 < T_c$, (b) $2.3 \sim T_c$ and (c) $4 > T_c$. In each configuration, black and white pixels represent spin up and spin down states, respectively. Panel (a) shows a ferromagnetic configuration at low temperature, (b) a critical configuration close to T_c and (c) a paramagnetic configuration at high temperatures.

where s_i is the spin at site i that can point up $(+1)$ or down (-1) and $\langle i, j\rangle$ are nearest-neighbour pairs [19,20]. Here we focus on 100×100 square lattices subject to periodic boundary conditions. This choice of system size is mainly motivated by the ML context where the codes have been optimised for treating images whose size does not exceed a few hundred times a few hundred pixels. For a given configuration, we define the magnetisation m and the correlation length ξ, respectively, as

$$m = \frac{1}{N}\left|\sum_i s_i\right|, \quad \xi^2 = \frac{\sum_r r^2 g(r)}{\sum_r g(r)}, \tag{2}$$

where $g(r) = \sum_{i\neq j} s_i s_j - m^2$ denotes the connected two-point correlation function and $r = |i-j|$ is the distance between lattice sites i, j [21]. We note that the correlation length ξ quantifies the characteristic scale of an individual spin configuration.

We employ the Metropolis algorithm to generate many 100×100 configurations as data for the ML approach [19, 20, 22]. We start the simulations from an initial, hot temperature $T = 100$. This is subsequently slowly reduced to $T = 4$ in 1000 Monte-Carlo sweeps (as usual, a Monte-Carlo sweep is defined as one attempt on average to flip each spin in the system). We then collect five spin configurations at $T = 4$, again with 1000 MC sweeps between each measurement. Next, we reduce to $T = 3.95$ in 1000 further MC sweeps and then again collect another five, well-separated, spin configurations. This procedure then repeats until we reach $T = 1.5$. We then restart the data generation for another sample, with the overall ensemble consisting of 416 such samples. The training and cross-validation data hence consists of overall 2080 configurations for each of the 51 distinct temperatures $T = 1.5, 1.55, 1.6, \ldots, 4$. Ten percent of this data is reserved as cross-validation data for ML. The test set consists of 250 additional configurations for each T with the same parameters as the training and cross-validation data. Each sample consists of the spatial configuration and its m, E, ξ and $g(r)$ values. Examples of configurations are shown in Fig. 1 with (b) close to the critical temperature, exactly known in the thermodynamic limit to be $T_c = 2/\ln(1+\sqrt{2}) \approx 2.269$ [12], (a) in the ferromagnetic phase, $T < T_c$ and (b) in the paramagnetic phase, $T > T_c$.

For the ML classification and regression tasks with E, m and ξ, we use the same convolutional neural network (CNN) architecture with eight convolutional layers (with a 3×3 kernel, stride 1 in each layer and the number of feature maps/filters increasing as $16, 16, 32, 32, \ldots, 128$), and 2×2 pooling layers with $\mathrm{ReLU}(x) = \max(0, x)$ activation function, and 25% dropout between layers. This is capped by three dense layers (of size 512, 128, while the last dense layer has variable size K for classification and 1 for regression). For classification, K expresses the number of categories, and the softmax activation function $S : \mathbb{R}^K \to [0, 1]^K, S(\mathbf{r})_i = e^{r_i}/\sum_{j=1} e^{r_j}$, with

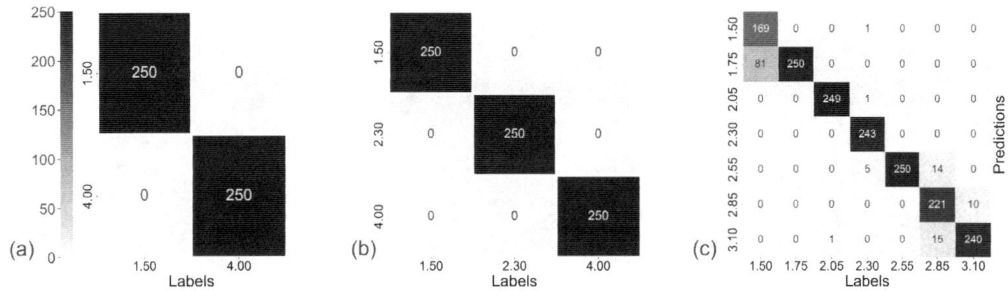

Figure 2. Confusion matrices for temperature classification. Horizontal axes denote the true values while the vertical axes indicate the predicted ones. The colour bar represents the number of samples in each entry. (a) Two classes: $T \in \{1.5, 4.0\}$, (b) three classes: $T = 1.5$, $2.3 \sim T_c$ and 4.0, (c) seven classes: $T \in \{1.50, 1.75, 2.05, 2.30, 2.55, 2.85, 3.10\}$.

$\mathbf{r} = (r_1, r_2, \ldots, r_K) \in \mathbb{R}^K$, $i = (1, 2, \ldots, K)$, is used in the output layer. The loss function is categorical cross entropy [2]. For regression, although we stay with the same CNN architecture, we adjust the output layer to a regression problem. Instead of the softmax activated output layer, a single dense node is used with no activation function. This single node returns the desired numerical estimate. As loss function we now work with the mean squared error (MSE). Overall, this results in six separately trained networks with separate weights, but shared architecture. We have used the TensorFlow implementation 2.4.1 for all machine learning tasks described here [23].

3. Results

Our results for the confusion matrices of the classification model for different numbers of categories are presented in Fig. 2. A "confusion matrix" represents the predicted label of the class as a function of the true one in matrix form, with an error-free prediction corresponding to a diagonal matrix. In Fig. 2(a), the categories of the phases, namely the ferromagnetic phase at $T = 1.5$ and the paramagnetic phase at $T = 4.0$, are correctly identified for all the test samples. Similarly, in Fig. 2(b), the categories are classified correctly for all the test samples, namely $T = 1.5$, $T = 2.3 \sim T_c$ and $T = 4.0$. When the number of categories is increased to seven, the confusion matrix develops off-diagonal terms, indicating "confusion", i.e. false positives and false negatives, mostly with neighbouring temperatures as shown in Fig. 2(c). Such misclassification in the $T > T_c$, paramagnetic phase is observed only among neighbouring temperatures, with a total of 39 misclassifications corresponding to 5.2% of all cases. In the $T < T_c$, ferromagnetic phase, there are 82 misclassifications corresponding to 10.9%, and only one of them is confused with a temperature that is not a neighbouring one, namely a $T = 2.05$ case is misclassified as $T = 3.1$. For $T = 2.3 \sim T_c$, only 7 out of 250 cases, i.e., 2.8% are misclassified, and only one of them ($T = 1.5$) is not a neighbouring temperature. We note that most of the confusion occurs at the highest and lowest T's. Indeed, the probability of a given configuration is controlled by its energy E. For a given T and on a finite system, the energy distribution has a certain width σ_E, and, as shown below in Fig. 4 for our case, these are consistent with the order of the confusion that we observe, at least between $T = 1.5 \leftrightarrow 1.75$ and $2.85 \leftrightarrow 3.1$, respectively. Thus, the classification according to 7 distinct temperatures reaches the physical limits of the 100×100 lattice.

We now turn to the regression models. Results for ξ, E and m are presented in Figs. 3, 4 and 5, respectively. Averages of the test data are denoted by black stars and their respective standard deviations σ_ξ, σ_E and σ_m, respectively, are indicated by the grey shaded region in

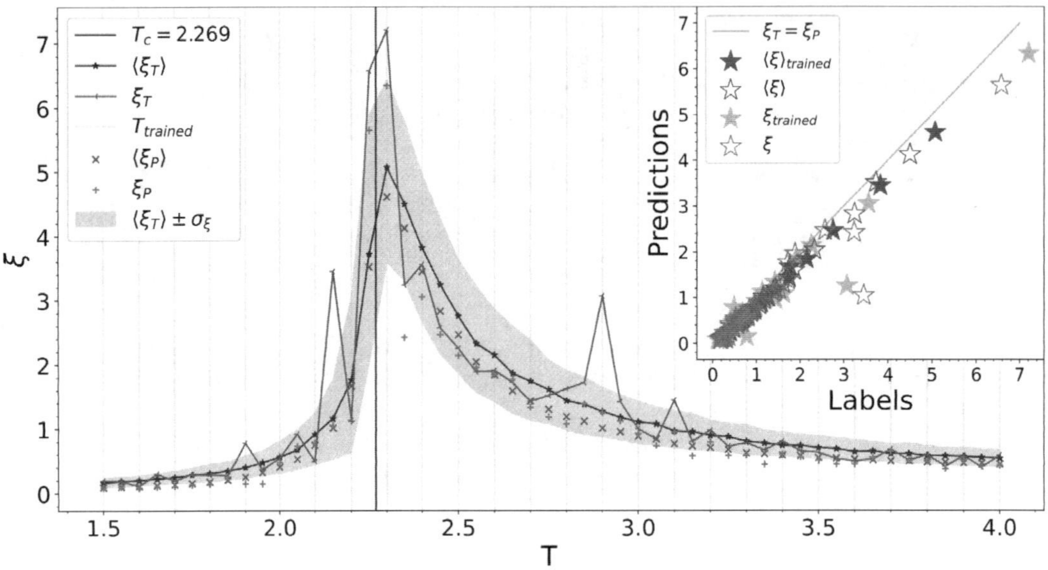

Figure 3. Correlation length ξ as a function of T. We show averages $\langle\xi_T\rangle$ ($\star$) obtained via Monte-Carlo simulations as well as predictions $\langle\xi_P\rangle$ (red $\times$) for the test data. The grey shaded region denotes the standard deviation of the training data. ξ_T (blue) and ξ_P (green $+$) are the Monte-Carlo results and ML predictions for a one specific sample, respectively. The thin black vertical line indicates the critical temperature T_c of an infinite system. Lines connecting data values are guides-to-the-eye only. The training was done only for the temperature values that are indicated by a grey vertical line in the background. The inset shows the relation of the true values and the predicted values with the diagonal giving the perfect result $\xi_T = \xi_P$.

the background. These standard deviations indicate the fluctuations for each quantity on the 100×100 lattice. The fluctuations are largest for ξ (Fig. 3), still significant for m around T_c (Fig. 5) and smallest but still non-negligible for E (Fig. 4). Values for one concrete configuration are shown in blue and as a function of T one observes indeed fluctuations on the scale of the standard deviations. We have used only every second temperature value for training; these are marked by the grey vertical lines in the background. Averages of the predictions are shown by red crosses. These lie within or close to the grey shaded region, showing that the network is indeed able to reconstruct ξ, E and m from a given configuration. This is also valid for the intermediate temperatures for which the network has not been trained, thus demonstrating the ability of the network to generalise. For more detail, we also show by green pluses the predicted values for the specific sample whose raw Monte-Carlo values were shown in blue. The fluctuations of these predictions generally follow those of the true values. Just for ξ (Fig. 3), the predictions do not fully reproduce the outliers of the raw data, i.e., ML appears to be trying to actually smooth out the sample-to-sample fluctuations. The insets of Figs. 3, 4 and 5 show the correlation between the predicted values and the true labels. Here, red symbols are for averages and green ones for the selected specific sample; closed symbols denote cases for which the CNN has been trained, open ones for temperature values where this is not the case. Again, one observes overall good performance of the network at measuring these values. Nevertheless, for ξ (inset of Fig. 3) one observes a systematic trend of the predicted values to lie below the true ones. For m, in particular intermediate values tend to be predicted slightly larger than their true value (inset of Fig. 5). We have observed other but comparable deviations in other trained CNNs and thus

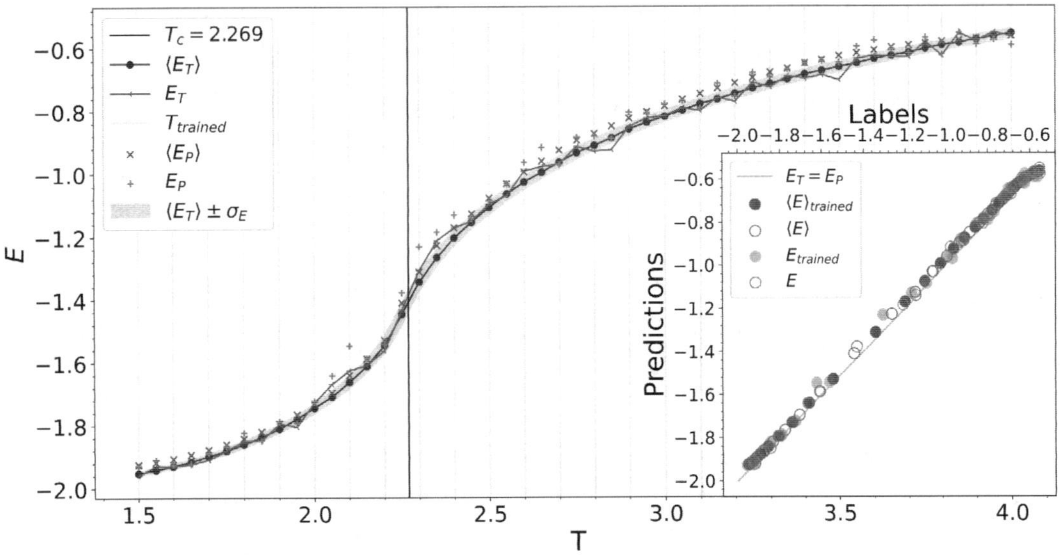

Figure 4. Energy per site E as a function of T. Averages $\langle E_T \rangle$, $\langle E_P \rangle$ and single-sample values E_T and E_P are indicated by symbols, as in Fig. 3. The inset shows the deviations from the perfect result $E_T = E_P$.

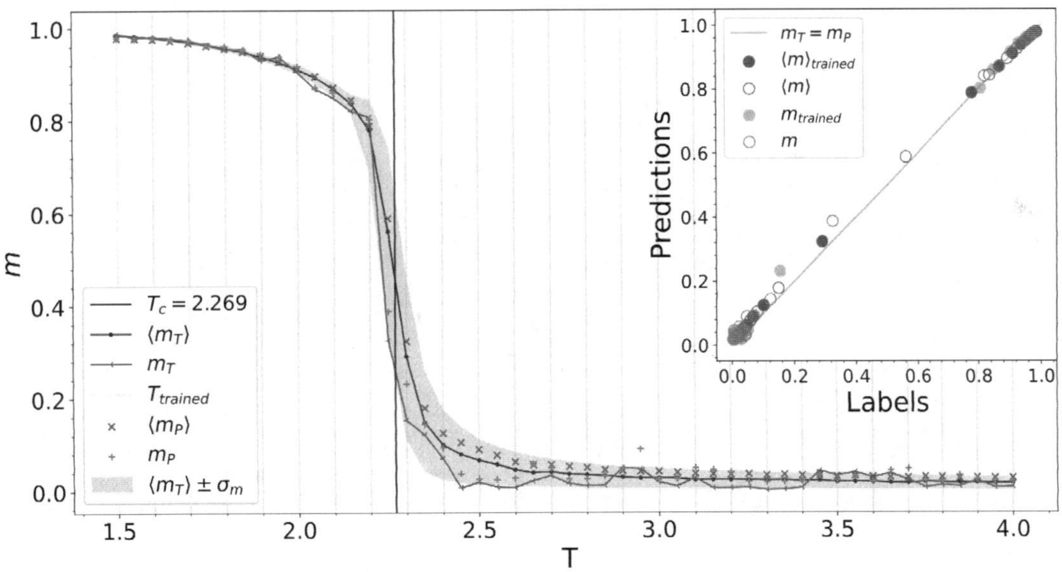

Figure 5. Magnetisation m as a function of T. Averages $\langle m_T \rangle$, $\langle m_P \rangle$ and single-sample values m_T and m_P are indicated by symbols, as in Fig. 3. The inset shows the deviations from a perfect result $m_T = m_P$.

believe these deviations from the diagonal to represent the limits of the accuracy of the neural network.

4. Discussion and Conclusions

We have shown that the specified CNN model is able to successfully classify configurations according to their temperatures until it reaches the physical limits of a 100×100 system for our choice of the 7 categories. We also found that a regression model is able to extract the correlation length ξ, energy E and magnetisation m with an accuracy that is consistent with the fluctuations of these quantities on the 100×100 lattice. The fact that this also works for intermediate temperatures, where the network has not been trained, demonstrates the ability of the CNN to learn more general features. One possible next step is to investigate transfer learning for the models that are trained here. Indeed, the case investigated here is the $J_1 = 1$, $J_2 = 0$ special case of the J_1-J_2 Ising model that has recently been investigated from the machine-learning perspective [17, 18], and it would be interesting to see how our neural networks trained on the case $J_2/J_1 = 0$ perform for other values of J_2/J_1.

References

[1] Domb C and Green M S 1972–1976 *Phase Transitions and Critical Phenomena* vols. 1, 2, 5B (Academic Press)
[2] Alpaydin E 2004 *Introduction to Machine Learning* 4th ed (The MIT Press) ISBN 9780262012119 URL https://mitpress.mit.edu/books/introduction-machine-learning
[3] Tanaka A and Tomiya A 2017 *J. Phys. Soc. Jpn.* **86** 063001 URL http://doi.org/10.7566/JPSJ.86.063001
[4] Walker N, Tam K M, Novak B and Jarrell M 2018 *Phys. Rev. E* **98** 053305 URL https://doi.org/10.1103/PhysRevE.98.053305
[5] Alexandrou C, Athenodorou A, Chrysostomou C and Paul S 2020 *Eur. Phys. J. B* **93** 226 URL https://doi.org/10.1140/epjb/e2020-100506-5
[6] Ohtsuki T and Ohtsuki T 2016 *J. Phys. Soc. Jpn.* **85** 123706 URL http://doi.org/10.7566/JPSJ.85.123706
[7] Ohtsuki T and Mano T 2020 *J. Phys. Soc. Jpn.* **89** 022001 URL http://doi.org/10.7566/JPSJ.89.022001
[8] Huembeli P, Dauphin A and Wittek P 2018 *Phys. Rev. B* **97** 134109 URL https://doi.org/10.1103/PhysRevB.97.134109
[9] Dong X Y, Pollmann F and Zhang X F 2019 *Phys. Rev. B* **99** 121104 URL https://doi.org/10.1103/PhysRevB.99.121104
[10] Hu W, Singh R R P and Scalettar R T 2017 *Phys. Rev. E* **95** 062122 URL https://doi.org/10.1103/PhysRevE.95.062122
[11] Pilania G, Wang C, Jiang X, Rajasekaran S and Ramprasad R 2013 *Sci. Rep.* **3** 2810 URL https://doi.org/10.1038/srep02810
[12] Onsager L 1944 *Phys. Rev.* **65** 117–149 URL https://doi.org/10.1103/PhysRev.65.117
[13] Morningstar A and Melko R G 2018 *J. Mach. Learn. Res.* **18** 1–17 URL http://jmlr.org/papers/v18/17-527.html
[14] Walker N, Tam K M and Jarrell M 2020 *Sci. Rep.* **10** 13047 URL https://doi.org/10.1038/s41598-020-69848-5
[15] D'Angelo F and Böttcher L 2020 *Phys. Rev. Research* **2** 023266 URL https://doi.org/10.1103/PhysRevResearch.2.023266
[16] Carrasquilla J and Melko R G 2017 *Nat. Phys.* **13** 431–434 URL http://doi.org/10.1038/nphys4035
[17] Corte I, Acevedo S, Arlego M and Lamas C 2021 *Comput. Mater. Sci.* **198** 110702 URL https://doi.org/10.1016/j.commatsci.2021.110702
[18] Acevedo S, Arlego M and Lamas C A 2021 *Phys. Rev. B* **103** 134422 URL https://doi.org/10.1103/PhysRevB.103.134422
[19] Berg, B. *Markov Chain Monte Carlo Simulations and Their Statistical Analysis with Web-Based Fortran Code.* (World Scientific Publishing Company,2004), URL https://doi.org/10.1142/5602
[20] Landau D P and Binder K 2014 *A Guide to Monte Carlo Simulations in Statistical Physics* 4th ed (Cambridge University Press) URL https://doi.org/10.1017/CBO9781139696463
[21] Montroll E W, Potts R B and Ward J C 1963 *J. Math. Phys.* **4** 308–322 URL https://doi.org/10.1063/1.1703955
[22] Metropolis N, Rosenbluth A W, Rosenbluth M N, Teller A H and Teller E 1953 *J. Chem. Phys.* **21** 1087–1092 URL https://doi.org/10.1063/1.1699114
[23] Abadi M *et al.* 2015 TensorFlow: Large-scale machine learning on heterogeneous systems software available from tensorflow.org URL https://www.tensorflow.org/

AUTHOR INDEX

Aarts, G. 324
Aarts, Gert 333
Adhikary, Sagarika 126
Agustin, Silva 186
Akahoshi, Yutaro 312
Akino, N 216
Akiyama, Takumi 73
Akritidis, Michail 19
Akutsu, Noriko 198
Allton, C. 324
Aoki, Sinya 312
Aoyama, Tatsumi 312
Arnett, W D 276
Arwansyah, M S 115
Bachtis, Dimitrios 333
Banno, H 61
Bayo, Djénabou 339
Bellini, Dom 89
Chattopadhyay, Sourav 25
Çivitcioglu, Burak 345
Constancio, M. Arizmendi 186
Corboz, Philippe 174
Cornelius, Sean P. 55
Farris, Alfred C. K. 7
Fitrasari, D 115
Francesconi, Olmo 306
Fujii, M. S. 288
Fukuma, Masafumi 318
Fytas, Nikolaos G 19, 43, 49
Georgy, C 276
Gessert, Denis 67
Gibson, N. L. 37
Goth, Florian 157
Hands, S. 324
Hirai, Y. 288
Hirashima, K. 288
Hirschi, R 276
Holzmann, Markus 306
Honda, Hisao 73
Honecker, Andreas 174, 339, 345
Hori, T 99
Hotta, Chisa 168
IIda, R 252
Ishida, Kunio 228
Ishii, Masahiro 210
Isobe, M 31, 61
Iwaki, Atsushi 168
Jäger, B. 324
Janke, Wolfhard 67, 145
Jayyinunnisya, H 120
Kageyama, Maya 73
Kanamori, Issaku 312
Kanaya, Kazuyuki 312
Kato, C 264
Kato, Masaru 222
Kawaguchi, K. 115, 120
Kim, S. 324
Kino, Yasushi 192
Kitaya, K 31
Kudo, Kazue 180
Kuo, Yu-Wei 105
Lancaster, David 306
Landau, David P. 1, 7
Lombardo, M. P. 324
Lucini, Biagio 306, 333
Mainou, Argyro 49
Majumder, Suman 145
Makino, J. 288
Malakis, Anastasios 43
Matsueda, Hiroaki 228
Matsufuru, Hideo 312
Matsukura, L 131
Matsumoto, Nobuyuki 318
Meakin, C 276
Mencagli, Mattia 294
Mila, Frédéric 174
Miura, S. 99
Miyashita, N 131, 135
Mizuhara, Yohitaka 222
Mocák, M 276
Mori, Masao 282
Moribayashi, Kengo 111
Moriwaki, K. 288
Mugita, D 61
Nagao, H 115, 120
Namekawa, Yusuke 312
Narumi, Takayuki 73
Nazarova, Natalia 294
Nemura, Hidekatsu 312
Niiyama, Motoaki 192
Nikolaev, A. A. 324
Noguchi, Hiroshi 95
Ohira, Toru 79
Ohnishi, N 264
Omar, G. Zabaleta 186
Onishi, Hiroaki 258

Osaki, Koichi 73
Otaki, Koki 282
Otsuka, Yuichi 163
Ottawa, M 135
Panayis, James 89
Paul, Subhajit 145
Phibanchon, Sarun 246
Purqon, A 115
Rago, Antonio 306
Rahmat, R. 83
Rajbhandari, E. 37
Rizzuti, F 276
Römer, Navodya S. 89
Römer, Rudolf A. 89, 339, 345
Ryan, S. M. 324
Sagara, Toshihiro 210
Saitoh, T. 288
Sakamoto, S 252
Sakamoto, Shoichi 210
Santra, S. B. 25, 126
Seki, Kazuhiro 163
Shiba, Hayato 139
Shinjo, Kazuya 151
Skullerud, J.-I. 324
Solihat, L S 120
Sorella, Sandro 163
Sota, Shigetoshi 151
Spera, Mario 294
Sugimori, K 120
Sun, Tao 234
Suprijadi 115
Surungan, T. 83
Suzuki, D 99
Suzuki, S 264
Takahashi, M 264
Takahashi, Masayuki 270
Tamemoto, Naoki 95
Tang, Ping-Han 105
Taniguchi, Yusuke 312
Tohyama, Takami 151
Toldin, F Parisen 13
Tomiya, M 252
Tomiya, Mitsuyoshi 210
Tsutsui, Kenji 151
Uemichi, Kenta 73
Vasilopoulos, Alexandros 43
Vatansever, Erol 43
Wallis, A. Katrine 89
Wang, Q 204
Weber, Lukas 174
Weigel, Martin 19, 43, 49, 67
Wentzcovitch, Renata M. 234
Wessel, Stefan 174
Wilson, Matthew S. 1
Woodside, C. R. 37
Wu, Ten-Ming 105
Yamamoto, Masaki 210
Yamashita, Takuma 192
Yasuda, Kazuhiro 192
Yoshihara, Sota 79
Yunita, S. 83
Yunoki, Seiji 163
Zempo, Y 204, 216
Zhang, Dong-Bo 234
Zhang, Huixin 55
Zhang, Zhen 234